Deutsche Forschungsgemeinschaft

High Intensity Combustors – Steady Isobaric Combustion

WILEY-VCH

Deutsche
Forschungsgemeinschaft

High Intensity Combustors – Steady Isobaric Combustion

Final Report of the Collaborative Research Centre 167
"Hochbelastete Brennräume – stationäre Gleichdruckverbrennung"

Edited by
Sigmar Wittig, Otmar Vöhringer and Soksik Kim

Collaborative Research Centres

Deutsche Forschungsgemeinschaft
Kennedyallee 40, D-53175 Bonn, Federal Republic of Germany
Postal address: D-53175 Bonn
Phone: ++49/228/885-1
Telefax: ++49/228/885-2777
E-Mail: postmaster@dfg.de
Internet: http://www.dfg.de

Library of Congress Card No.: applied for

A catalogue record for this book is available from the British Library.

Die Deutsche Bibliothek – CIP Cataloguing-in-Publication-Data
A catalogue record for this publication is available from Die Deutsche Bibliothek
ISBN 3-527-27731-5

Printed on acid-free paper.

Cover Design and Typography: Dieter Hüsken.
Composition: Hagedorn Kommunikation, Viernheim.
Printing: betz-druck gmbh, Darmstadt.
Bookbinding: Litges & Dopf Buchbinderei GmbH, Heppenheim.
Printed in the Federal Republic of Germany.

Contents

Preface

Due to the rising demand for improved environmental compatibility and economic efficiency in energy conversion of fossil fuels, extensive world wide research activities were initiated during the past decades. Significant progress in the development of high intensity combustors with steady isobaric combustion has been achieved in recent years, which has resulted in higher combustion efficiencies, lower fuel consumption, decreased pollutant formation as well as increased life times of the combustors. These developments affect not only heavy duty gas turbine and jet engine combustors, but also high efficiency steam generators, high temperature industrial kilns and furnaces in addition to industrial combustors and waste incinerators.

As a part of this development, comprehensive research activities were sponsored by the Deutsche Forschungsgemeinschaft for more than a decade (1984–1998) at the University of Karlsruhe (TH) within the interdisciplinary Collaborative Research Centre 167, "High Intensity Combustors – Steady Isobaric Combustion". A large number of research projects was successfully completed, directed towards the fundamental understanding of thermo- and fluid dynamics, chemical reaction kinetics, heat and mass transfer as well as the behavior of high temperature materials. The research report on the results of the Collaborative Research Centre 167 presents an overview summary of the newest developments and improvements of high performance combustors.

Primary research goals of the Collaborative Research Centre 167 were advances in the efficiency of the combustion processes, the utilization of a wider spectrum of fuels together with an improved environmental compatibility as well as an increased reliability of combustion chamber components due to a better durability of combustion chamber materials. The research efforts of the Collaborative Research Centre 167 with their far reaching results have received considerable recognition – internationally – and impact on newest design standards.

In this context, the following specific project areas play an important role. These include: preparation, atomization, mixing and reaction of fuel and air, reaction zone stabilization, pollutant formation and -emission, two phase flows, heat transfer, characteristics of high temperature materials under thermal and mechanical load, the life time and damage analysis of combustion chamber components. These topics are presented and discussed in detail in the present report.

Due to the extreme demands on the design of high intensity combustors under consideration of process control and material use, the projects become complex and cross-disciplinary between technical fields. Variations of the chemical-physical properties of the available fuels as well as increased high reaction densities determine the process controls and, furthermore, set the limits for low pollutant emissions with improved total efficiencies. They are combined with both, safety considerations and economical aspects, which require a better understanding of the basic phenomena and correlations. In this context, the development of new and advanced computer codes and modern diagnostics was of exceptional significance.

The topics covered by this report correspond to the particular needs of high intensity combustor systems and are characterized by the following six major areas:

- fuel preparation
- flow and combustion
- pollutant formation
- heat transfer and radiation
- high temperature materials
- thermal barrier coating systems

In order to gain a deeper scientific understanding, new and advanced equipment was designed and built, modern measurement techniques were developed, theoretical models formulated and numerical simulations developed in the first phase of the work. In the subsequent years the instrumentation was effectively applied. It should be mentioned that a large number of graduate students contributed to the program. Particularly the understanding of turbulent flames, the prevailing mode in combustion chambers of gas turbines and in industrial combustors is of dominating importance for the development of advanced combustion systems operating in premixed or staged arrangements. Therefore, the flow and the mixing processes in the combustion chamber which are strongly related to turbulence parameters, combined with the material heat transfer processes acted as an important link between the projects in the Collaborative Research Centre 167. As fundamental and methodical research aspects had priority, the investigations were focused on gas turbine relevant materials and material composites. Nevertheless, exact knowledge of the tolerable loading under operating conditions as well as the time history of the real loading sequences was essential for the development of improved combustion chambers. Real loading sequences are an extremely complex combination of low frequency, mainly thermally induced caused by engine start-up, load cycle, and shut-down procedures – of high frequency fatigue coupled with creep and relaxation processes and of oxidizing and hot gas corrosive material loads. Especially in this area an intensive exchange of experience with experts from industry was initiated.

Among others, the contributions of the Collaborative Research Centre 167 in the areas of fuel preparation and combustion gained international attention. New measurement techniques for particle sizing as well as the exact determination of thin wavy liquid films of single and multi component fuels can be mentioned as typical examples. Experimental studies and the development of numerical models for the evaporation of fuel droplet clouds under realistic conditions identify important problems in liquid fuel preparation. Novel techniques for advanced low pollution combustion concepts like the utilization of film vaporization on porous ceramic surfaces were studied. Furthermore, comprehensive investigations of soot formation, growth and burnout were performed in order to close a gap in the description of the processes in real flames. Another main focus of research was the study on NO_x formation in technically relevant flames. Experimental and theoretical research efforts on turbulence exchange, its interaction with exothermal reactions and its influence on the pollutant formation in swirl stabilized flames achieved highly successful results and international attention as well.

The combustion intensity and the heat release in combustors is closely linked with their design in general and specifically with the heat load at the bounding walls. Therefore, the prediction of the complex three dimensional flow inside combustion chambers and the related heat transfer problems was one of the priority areas of the Collaborative Research Centre. As an example, newly developed numerical simulation techniques were applied to three dimensional flows in combustion chambers with complex geometry using parallel computers. Furthermore, basic understanding of the complicated flow behavior had to be gained for typical combustion chamber flow situations, wall-parallel and wall-normal cooling air jets as well as for the interaction between cooling films and mixing air jets. This also is true for the flow through the entrance diffusers of the combustion chambers taking into account the variation of the diffuser geometry and the shape of the entrance section. The ongoing increase of pressure and temperature levels in modern gas turbines made radiative heat transfer a focus of attention. The lack of an appropriate basic knowledge made it necessary to initiate extensive studies on multidimensional radiative heat transfer. In the course of several projects, new experimental, analytical and numerical techniques were successfully developed, which allow for the first time an exact prediction of the local radiative heat load for combustion chamber walls.

As indicated earlier, the design of high intensity combustors is determined by the high temperature deformation and failure characteristics of the materials used under the extreme conditions prevailing inside combustion chambers. Accordingly, the research effort concerning the combustor materials was concentrated along the following main topics:

- deformation and failure behavior of typical materials for combustion chambers under realistic operating conditions

- development, testing and evaluation of the properties of ceramic barrier coatings as well as their influence on the deformation and failure behavior of the substrates
- modeling of the deformation and failure behavior of uncoated and coated components of combustion chambers with suitable material laws

The analyses of materials and composites for combustor applications were performed mainly under static, isothermal cyclic and thermal-mechanical cyclic loading. In addition mechanical properties and the deformation behavior were also characterized micro-structurally and micro-analytically. Following the test conditions, the mechanical behavior was modeled considering static and cyclic loading as well.

The second main topic includes the development of new ceramic thermal barrier coatings as well as the careful scientific study of industrially produced thermal barrier coating systems. Extensive experience with coated specimen demanded that not only the coatings itself but the influence of the thermal barrier coating systems on the deformation and failure behavior of the substrate-materials had to be studied. The key issue of the third topic was the inelastic analyses and damage prediction to evaluate the failure of thermo-cyclically loaded materials and components of combustion chambers. This topic was of crucial importance for the whole project as the results obtained from the other research areas and, in particular, the heat transfer and radiation data were utilized. Taking into account the results of the first two main topics the parameters of the material laws in the predictions could be adapted to realistic material and loading conditions leading to reliable simulations and predictions of the behavior of highly loaded combustion chambers. Finally, the quantitative microstructure analysis enabled an important contribution to the physical and structure-mechanically based modeling of the deformation behavior of combustion chamber materials.

The present book appeals to scientists, engineers as well as technicians working on modern combustion systems. It offers specific knowledge on high intensity combustors and provides those who are interested in the area of fuel preparation and combustion, flow and heat transfer, and particularly in high temperature materials with a deep understanding of fundamental physics.

The participants in the Collaborative Research Centre 167, contributing to this final report, are members of eight Institutes of the Departments of Mechanical Engineering, Chemical Engineering and Chemistry at the University of Karlsruhe, and of the Faculty of Materials Science and Technology at the University of Mining and Technology (TU Bergakademie) Freiberg/Sachsen (1991 to 1995). They are:

- Institut für Thermische Strömungsmaschinen
  (Institute for Thermal Turbomachinery)
- Engler-Bunte-Institut, Bereich III – Feuerungstechnik
  (Engler-Bunte-Institute, Department III: Combustion Technology)

- Institut für Chemische Technik
  (Institute for Chemical Technology)
- Institut für Meß- und Regeltechnik
  (Institute for Measurement and Automatic Control)
- Institut für Keramik im Maschinenbau
  (Institute for Ceramics in Mechanical Engineering)
- Institut für Werkstoffkunde I
  (Institute for Materials Science and Engineering I)
- Institut für Zuverlässigkeit und Schadenskunde im Maschinenbau
  (Institute for Reliability and Failure Analysis in Mechanical Engineering)
- Institut für Metallkunde der TU Bergakademie Freiberg
  (Institute for Physical Metallurgy of the University of Mining and Technology)

The members of the Collaborative Research Centre 167 express their sincere appreciation to the Deutsche Forschungsgemeinschaft for the generous support over five funding periods. In particular, the advice and assistance of Prof. Mayinger, Prof. Wagner, Prof. Hennecke, and Prof. Renz as chairmen of the evaluation committees and the support of Dr. Lachenmeier, Dr. Lange, Dr. Nießen, Dr. May, and Dr. Rohe from the Headquarters in Bonn were essential for the success of the research work. Special thanks are due to the members of the technical and office staff of the Institutes for their efficient and ingenious contributions.

Karlsruhe, May 2000 S. Wittig, O. Vöhringer, S. Kim

1 Fuel Preparation

Advances in Fuel Preparation

Sigmar Wittig* and Georg Maier

The ambitious aim in combustor development of the combined effort to attack the basic problems of thermo- and fluid dynamics, reaction kinetics as well as heat and mass transfer together with those of high temperature material behavior is reflected in the structure of the Collaborative Research Centre 167 "High Intensity Combustors – Steady Isobaric Combustion". In this general concept, the project group A: "Fuel Preparation, Combustion and Formation of Emission" was of major importance as the physical and chemical processes during fuel preparation and combustion determine the thermal loading of combustor liner and turbine components.

The majority of technical combustion devices has the common characteristic that the fuel is supplied as a liquid with varying physical and chemical properties. In standard applications the formation of a combustible mixture of fuel vapor and air decisively depends on fuel atomization into tiny droplets, spray dispersion and evaporation. An alternative way of liquid fuel preparation is based on fuel evaporation and mixture formation from shear driven liquid wall films. An example of a combination of film and spray evaporation is realized in prefilming airblast atomizers.

In order to create improved design criteria and tools for highly loaded combustors, it was necessary to extend the knowledge of the governing processes and their interactions by means of detailed theoretical and experimental analysis. The numerous experimental, analytical, and theoretical results of these analyses were used in various modeling approaches and finally combined in a powerful computational scheme for complex two phase flows in combustors.

Basic studies of various atomization technologies were performed in the first three project phases to establish a comprehensive data base. The investigations were focused on the atomization process by means of airblast atomizers as well as pressure swirl atomizers. In both cases it was necessary to obtain detailed information on the liquid sheet leaving the injector as a pre-

* Institut für Thermische Strömungsmaschinen, Universität Karlsruhe, Kaiserstr. 12, 76128 Karlsruhe, Germany

requisite of an improved description of the atomization process. Thus, the internal flow of pressure swirl atomizers was analyzed experimentally and theoretically in subtask A3. Starting from simple 2d flow computations, the final project phase comprised numerical simulations considering the turbulence in the fluid as well as the free surface of the hollow spray cone and the back flow of the nozzle flow. In the subtasks A4, A5, and B6 prefilming airblast atomizers have been studied in an approach comparable to subtask A3. Particularly the fluid dynamic processes occurring in airblast atomizers were analyzed carefully. For this purpose the development of a measurement system to determine the thickness and structure of liquid films based on the extinction of infrared light was an essential step to gain comprehensive experimental data for shear driven liquid wall films. Parallel to the experimental investigations numerical codes were developed and evaluated. The initial project phase was characterized by coupled computation of liquid wall film and air flow. Experimentally supported state of the art predictions of shear driven liquid wall films describe the realistic behavior of the multiple layer structure of a wavy liquid film. The models developed were successfully implemented in the general CFD code. As mentioned earlier, the crucial point for this progress was the development of a complex measurement system which allows simultaneous determination of thickness, surface structure, and velocity profile of thin wavy wall films.

The next logical step in the attempt to predict the atomization and spray propagation process was to investigate the physical parameters which determine the atomization of liquid sheets in detail. The results of the comprehensive study in the Collaborative Research Centre 167 allowed the simulation of essential steps of the fuel preparation process utilizing combined numerical prediction of liquid film flow and liquid film atomization. The last step to a complete prediction of fuel preparation was the simulation of spray propagation i. e. dispersion and evaporation of fuel droplets within a complex turbulent flow field. Its successful completion is of major importance for the correct prediction of combustion processes as it represents the link between the two phase flow and the chemical reaction of an air-fuel mixture. Particularly the development of improved models for a better prediction of the dispersion and the evaporation of fuel droplets was a major intention within the subtasks A3 and A5. The objective of the early project phases was to develop a basic understanding of the physical processes and to derive first numerical models often assuming simplified respectively academic boundary conditions. However, in the last two project phases more realistic operating conditions could be realized to improve and verify the models. Outstanding examples for the success of the Collaborative Research Centre 167 can be found in the subtasks A5 and A12 which dealt with the improvement of models for the evaporation of multi component fluids at supercritical pressure conditions.

In addition to the purposeful and successful effort in the area of numerical simulation, the use of advanced laser based measurement techniques was an important precondition for the accurate experimental results. These non-intrusive techniques are particularly suitable to characterize droplet, film and gas flows even at exceedingly difficult conditions like reacting two phase flows. Nevertheless, before applying these techniques substantial fundamental work had to be carried out to identify potential sources of errors and to find out how to avoid these errors by suitable measures. Thus the accuracy of measurement necessary for validation experiments could be ensured. The fundamental work on the Phase Doppler method in the subtasks A5 and A13 and the subsequent successful employment of this technique is one exceptional example for the significant progress in the characterization of reacting two phase flows obtained within the Collaborative Research Centre 167. The large number of publications in internationally recognized journals and the transfer of the results to practical applications in industry illustrates the impact of the work.

1.1 Atomization and Spray Propagation in Gas Turbine Combustors

Georg Maier, Robert Meier, Michael Willmann, Reinhold Kneer, Johann Himmelsbach, Hans-Jörg Bauer, and Sigmar Wittig*

Abstract

Liquid atomization is the most decisive step in fuel preparation for gas turbine combustors, because the droplet sizes and the achieved degree of mixing between liquid fuel and combustion air has an important influence on flame stability and pollutant emissions. The fuel preparation process itself can be divided into the specific parts: flow and evaporation of liquid wall films, fuel atomization, droplet dispersion and evaporation. The development of physical models and numerical codes describing these processes determines decisively the scientific work at the Institute for Thermal Turbo Machinery (ITS). The development of measurement techniques and the exemplary investigation of both, single phenomena and complex two phase flows, is an essential part of this work. Thus, the work presented here compares two different combustor concepts and their specific techniques of atomization and mixture formation. The main focus is on spray visualization and measurement of droplet velocities and size distributions and the utilized measurement techniques. In order to get practice orientated results the measurements were carried out at elevated pressure and temperature within the high pressure high temperature test facility (HDT) of the ITS. The atomizers under investigation were developed and analyzed within the Collaborative Research Centre 167 "High Intensity Combustors – Steady Isobaric Combustion" at the University of Karlsruhe, Germany. In addition some of the given results are due to projects which base on the work or the results from the Collaborative Research Centre 167.

* Institut für Thermische Strömungsmaschinen, Universität Karlsruhe, Kaiserstr. 12, 76128 Karlsruhe, Germany

Nomenclature

ALR	[–]	air liquid ratio
c	[m/s]	velocity
D_{32}	[µm]	Sauter mean diameter
f	[1/s]	frequency
l	[m]	length
p	[N/m^2]	pressure
Re	[–]	Reynolds number
S	[–]	swirl number
Str	[–]	Strouhal number
u	[m/s]	velocity
ϕ	[–]	equivalence ratio
λ	[–]	$1/\phi$
μ	[–]	flow split

Subscripts

prim	primary
sec	secondary
air	air flow
ax	axial flow direction
m	mean
n	nominal
p	pulse

1.1.1 Introduction

The increased awareness of global environmental issues has led to concentrated efforts to limit gas turbine emissions since the 1970's. Main emphasis of the recent developments is the reduction of NO_x since combustion chamber pressures and temperatures are continuously increasing in order to improve engine performance and efficiency which results in greater NO_x production. A number of strategies for new combustion concepts, such as staged combustion and lean premixed prevaporized combustion have been considered (Fig. 1.1.1). All of these new concepts are based on the fact that thermal NO_x production is reduced when operating away from stoichiometric conditions. Both, RQL-[1] and LPP[2] concepts have been investigated within the Collaborative Research Centre 167.

The success of the LPP- as well as the RQL concept strongly depends on a suitable homogeneity of the air fuel mixture in the reaction zone. At the ITS mainly the atomization and the two phase flow in gas turbine combustors have been investigated both theoretically and experimentally. Airblast atomization [1, 2] the disintegration of liquid fuel films [3], spray propagation [4–6] and droplet vaporization [7] have been the main tasks.

Measurement systems which were used to analyze the spray propagation in gas turbine combustion chambers have been improved and applied. Chemical analysis of the exhaust gases under fuel rich and fuel lean conditions have been conducted. New numerical tools for the simulation of two

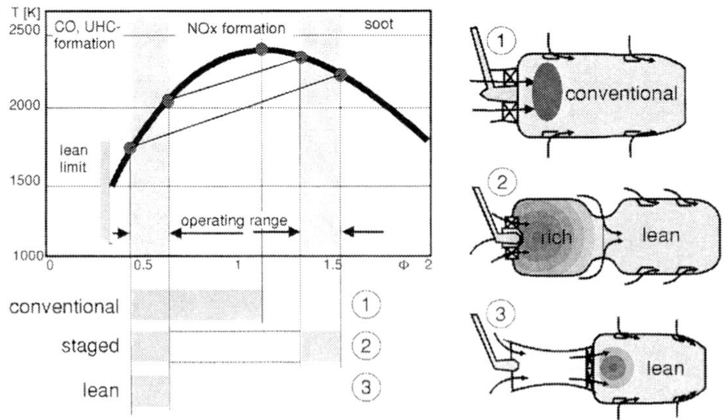

Figure 1.1.1: Schemes of the combustor concepts.

1 **R**ich Burn **Q**uick Quench **L**ean Burnout

2 **L**ean **P**remixed **P**revaporized

phase flows and shear driven liquid wall films have been developed and validated using experimental acquired test cases.

In this contribution a lean premixed prevaporized (LPP) and a lean internal mixing (LIM) combustion concept are compared. Both concepts are investigated using the high temperature, high pressure test rig (HDT) of the ITS. The visualization and measurement techniques required for the investigation of these combustion concepts and the results of this research work are described in detail.

1.1.2 Experimental Setup

As mentioned above, the combustion concepts are investigated in the HDT test facility. This test facility offers near engine conditions with maximum temperatures of 1120 K and maximum pressure of 10 bar and is therefore suitable to investigate new combustion concepts at realistic boundary conditions. The air is preheated by a heat exchanger which is fired by four natural gas burners with a total thermal power of 2 MW. The ambient pressure in the test rig is automatically controlled and the mass flows of preheated and cooling air are adjusted by electromechanical valves.

For the investigation of the two combustion concepts qualitative and quantitative measurement techniques are used. A cross section of the casing in which both combustion concepts are tested is shown in Fig. 1.1.2. It offers full optical access to the fuel preparation zone and to the combustion chamber and is used for spray and flame visualizations as well as for LDA and PDPA measurements. A first characterization is conducted by the visualization of the spray and the flame luminescence. For the spray visualization an improved laser light sheet (LLS)-technique is applied [8, 9].

To visualize fuel droplets in flames it is necessary to illuminate the droplets by high intensive light pulses of a copper vapor laser (CVL) and to capture the scattered light by means of a CCD-camera which is synchronized

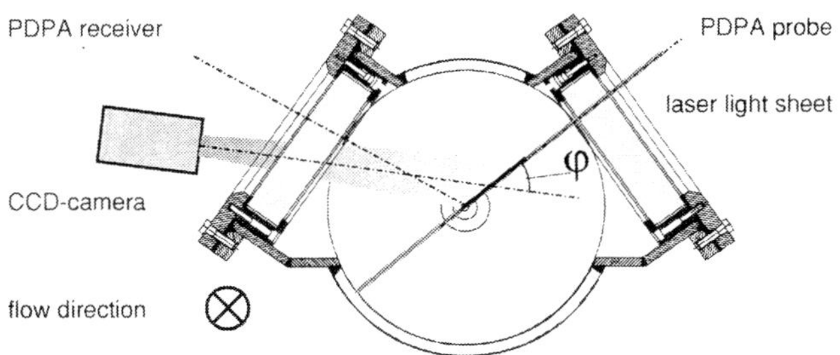

Figure 1.1.2: Cross section of the combustor casing.

with the laser pulses. To separate the flame luminescence from the scattered light narrow bandpass filters with maximum transmission at the wavelength of the laser were used. Using this technique single pictures of the spray are captured. Frequency distributions of a multitude of pictures are calculated by means of a picture processing software in order to achieve an integral information about penetration angle and depth of the spray cone [10].

For a more detailed investigation quantitative measurement systems like Laser Doppler Anemometry (LDA) and species analysis of the exhaust gases have to be applied. For the investigation of the two phase flow a Phase Doppler Particle Analyzer (PDPA) is used. Due to refractive index variations of evaporating fuel droplets PDPA measurements especially in the LIM combustion chamber may be objectionable. In project A5 of the Collaborative Research Centre 167 theoretical work has been undertaken to find suitable PDPA configurations for this task [5, 6, 11].

1.1.3 Combustion Concepts

Figure 1.1.3 shows schemes of the two investigated combustion concepts. In the LPP concept the liquid fuel is atomized by a simplex air assist atomizer and premixed and prevaporized in the premix duct. An axial swirler is used to stabilize the flame at the end of the premix duct.

As the premix duct is not cooled, the presence of a flame inside this duct would cause severe damage. Therefore the flow velocity of the air inside the premix duct has to exceed the flame velocity to prevent the flame from burning into the duct. In addition the residence time of the air-fuel mixture has to be shorter than the ignition delay time to avoid autoignition. In order to avoid carbon formations and inhomogeneous air-fuel mixture droplet impingement on the wall of the premix duct must be avoided for all operating conditions.

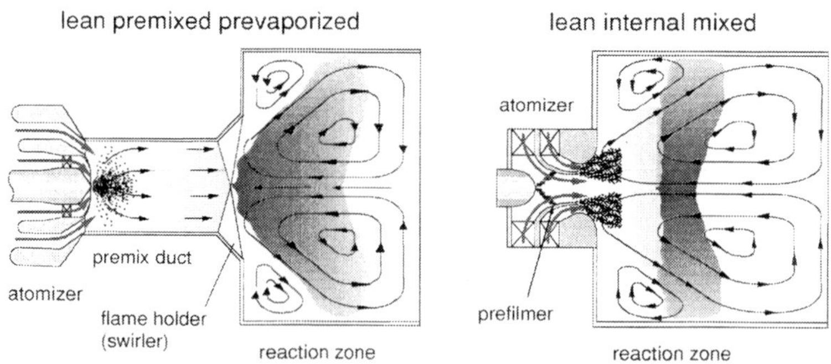

Figure 1.1.3: Schemes of LPP and LIM concept.

Hence, the range of conditions for a secure operation of a premix duct is limited.

In the LIM concept there is a spatial separation of the mixing area and the combustion zone which is induced only by aerodynamic effects. For the atomization of the liquid fuel an airblast atomizer is used which consists of a primary inner air channel and a coannular secondary air channel. These two radial swirlers which are separately supplied with combustion air differ in swirl number. Therefore the overall swirl number which results from the superposition of both flows depends on the selected flow split. A reduced overall swirl number enlarges the distance between the prefilmer lip and the reaction zone which results in a better mixing and vaporization.

1.1.4 Atomization Systems

From the numerous existing atomizer designs, it is possible to extract two major classes for the application in gas turbine combustors: Airblast atomizers and pressure swirl atomizers, both having their inherent advantages and detriments. Airblast atomization, for example is roughly unattached by the fuel flow rate, but decisively governed by the density and velocity of the atomization air [12, 2]. Pressure atomizers, on the other hand, are characterized by superior atomization at high fuel pressures within a limited range of fuel flow [13]. Similar to the atomizer concepts of Lefebvre [14] and Rizk [15], the pressure swirl atomizer in the present study is surrounded by a converging air flow. The basic idea of this atomizer is an intensification of the aerodynamic interaction between the liquid sheet and the surrounding air flow. The left hand scheme in Fig. 1.1.4 shows the air assisted pressure swirl atomizer. Analogous to the airblast type, the primary air is partitioned into a bypass flow without swirl, and the atomization air itself. To reduce the pressure loss $\Delta p/p$ down to 3 % at 60 m/s air velocity, which is typical for applications in gas turbines, the flow acceleration is realized very close to the pressure swirl atomizer. Due to the small amount of air passing the inner channel, no global air recirculation can be observed within the premix duct. Thus, for this flow channel swirlers with different directions and angles relative to the axis may be used. A series of investigations was carried out in order to separate the influence of liquid properties, velocity levels, and geometry factors which are all influencing the performance of this atomizer concept [16, 6, 17], and some of the results are presented here.

The right hand scheme in Fig. 1.1.4 shows the conventional prefilming airblast atomizer which is used for atmospheric comparison tests. The performance of this atomizer type is dominated by air friction forces and consequently depends on a high air velocity level at the atomization lip. In general, the atomization process can be improved by a swirled flow which increases

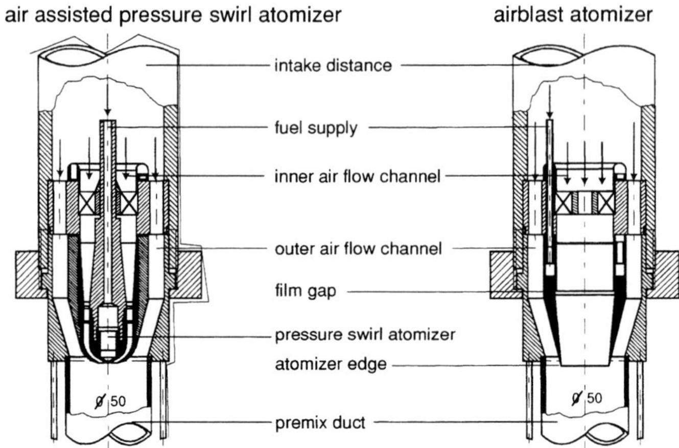

Figure 1.1.4: Schemes of the atomization concepts.

the interaction between liquid film and air flow. For LPP applications the swirl number is strongly limited by the appearance of an axial recirculating flow pattern and the formation of a liquid wall film within the premix duct. Both phenomena may initiate autoignition or flashback. First of all, a detailed visualization of the atomization mechanisms of both concepts was carried out by means of flash light shadowgraphy, see Figs. 1.1.5 and 1.1.6.

The first line represents the atomization process of the airblast atomizer at 20 m/s mean axial air velocity. The visualization confirms that the ob-

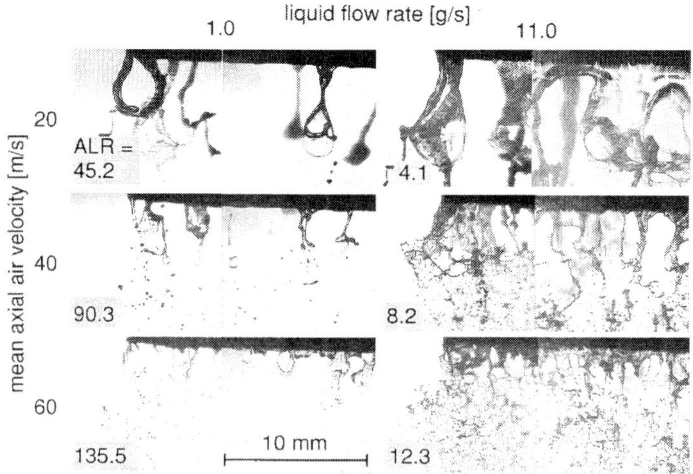

Figure 1.1.5: Motion and disintegration of fluid ligaments at the prefilming airblast atomizers. (Pictures show the region between the left side and the centerline of the atomizer lip, see Fig. 1.1.4).

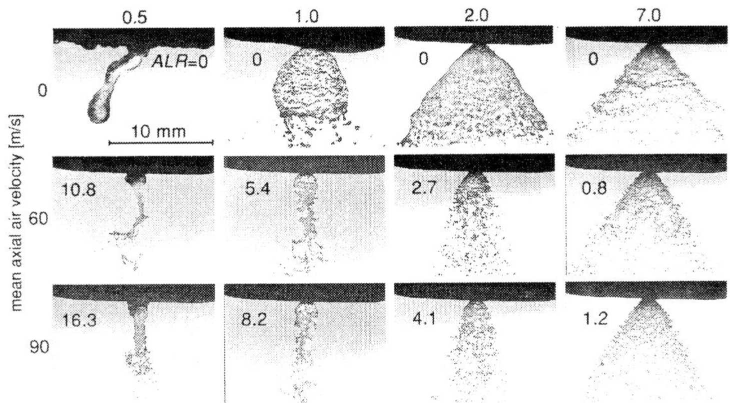

Figure 1.1.6: Liquid disintegration at the air assisted pressure swirl atomizer.

served mechanisms were insensitive towards liquid flow rate. In the first line for example a bag type atomization mechanism can be observed both, at 1 g/s and 11 g/s liquid flow rate. For higher air velocity levels, the disintegration distance is considerable shortened. On the other hand, there is still a radial motion in the ligament which initiates the radial dispersion of the spray, even if there is an axial air flow without a swirl component.

Corresponding to this flow visualizations integral droplet size measurements were carried out by means of a laser diffraction device in order to get a data base to compare the concepts. The left hand diagram in Fig. 1.1.7 shows the dependency of the Sauter mean diameter on liquid flow rate for different air velocity levels. As shown in the visualization and known from the literature the droplets sizes are primarily determined by the velocity level. In spite of this, the liquid flow rate has only minor influence on droplet sizes.

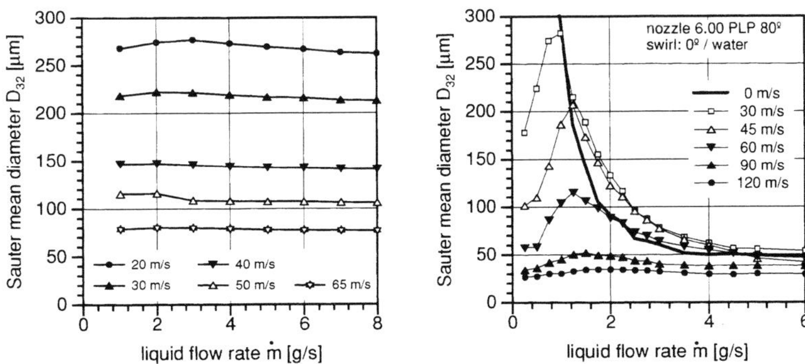

Figure 1.1.7: Performance of airblast atomizer (left) and air assisted pressure swirl atomizer (right): dependency of Sauter mean diameter on liquid flow rate for different air velocities.

The different operation regimes of the air assisted pressure swirl atomizer are more complex as shown in Fig. 1.1.6. At low liquid flow rates and in stagnant air, the liquid dribbles from the orifice. With increasing pressure, a hollow cone is formed and contracted later by surface tension forces. If the cone is completely closed it is called the "onion" stage. At this condition liquid disintegration results in fairly large droplets. At higher liquid mass flow rates a large tulip shaped film is generated (third picture). A further increase in atomization pressure leads to the formation of unstable waves with an exponentially increasing amplitude. The growth rate of the wave amplitude is governed by sheet properties, thickness and velocity, and air properties and velocity, Fraser et al. [18]. As a result, an increase in water mass flow directly results in faster atomization and reduced distance between the orifice of the nozzle and the position of atomization. This can be seen in the last two pictures of the first line. At higher air velocities (lower pictures) the liquid disintegration occurs closer to the orifice. For minimum liquid flow, the atomization process substantially depends on the air velocity level (first column) and is similar to airblast coaxial atomization, Eroglu and Chigier [19].

At 60 m/s axial air velocity and 0.5 g/s water mass flow, the liquid jet is transformed into a thin sheet which curls up forming shallow hemispheres or ladles. The air flow then stretches the liquid membrane, which subsequently bursts into ligaments and drops. This phenomenon is similar to the bag mechanism of single droplet atomization investigated by Samenfink et. al. [20]. At increased air velocities above (90 m/s) a cloud of ligaments and droplets is situated directly at the orifice of the atomizer. At these conditions and higher air velocities prompt atomization occurs.

Figure 1.1.7 presents droplet size measurements corresponding to the above flow visualizations. An increase in mean air velocity improves the atomization quality, especially at water flow rates between 0.5 and 3.0 g/s. An increase in liquid flow rate, on the other hand, initially increases the Sauter mean diameter up to a maximum followed by a continuous reduction of the value. With increasing air velocity the maximum of a curve is moved to higher water flow rates, accompanied by a simultaneous reduction of the maximum value. Due to highest air velocities, the maximum declines and the curves rise decently with increasing water flow rate. Even if the atomization process of the air assisted pressure swirl atomizer still depends on the liquid flow rate droplet sizes are essentially smaller in comparison to an usual pressure swirl atomizer. Anyway, the comparison with the airblast atomizer shows disadvantages in the range between 1 and 3 g/s water mass flow and low air velocities. Within this range, the angle of the liquid cone is small and consequently the liquid moves in the wake flow of the nozzle. The present peak values of the SMD curves should disappear if a spill-return nozzle with a liquid cone always fully developed is used. However, the present arrangement enlarges the dynamic range of a pressure swirl atomizer enormously and guaranties smallest droplet sizes when the air velocity is high which is typical of LPP applications.

1.1.5 Spray Propagation/Mixture Formation

1.1.5.1 LPP Concept

In the following paragraph a comparison of spray propagation and mixture formation is given for a LPP and a LIM combustor. The degree of mixture formation is judged by droplet dispersion which is determined by the mentioned LLS technique and the PDPA device. The measurements were carried out under practice orientated conditions at elevated pressures and temperatures.

The premix duct of the LPP combustor is a cylinder of 44.6 mm diameter with a length of 124 mm. Depending on the requirements of the measurement technique employed, this may be made of segments of quartz glass or metal. The metal segments are designed for use with PDPA measurement and are provided with observation ports in various axial locations. This allows the use of laser visualization and measurement techniques for the determination of droplet size and velocity as well as the droplet dispersion in the test section.

The heart of the combustion chamber is the air assisted pressure swirl atomizer ("simplex air assist") as shown in Fig. 1.1.8. The bypass air is necessary in this concept to provide the high total air supply required for a lean combustion and to avoid liquid wall film. Flame stabilization is realized by means of an axial swirler at the end of the premix duct.

LLS visualization results are given in Fig. 1.1.9. The small pictures show the near nozzle region, the right hand pictures show frequency distributions of the droplets in the middle plane of the entire premix duct. The pictures show the influence of increasing flow velocity with the operating conditions remaining constant. For the 60 m/s case two distinct regions close to the nozzle are visible, an outer region with relatively large droplets and an inner region with much smaller droplets such that individual droplets were smaller than the resolution of the camera system. The path of the droplets

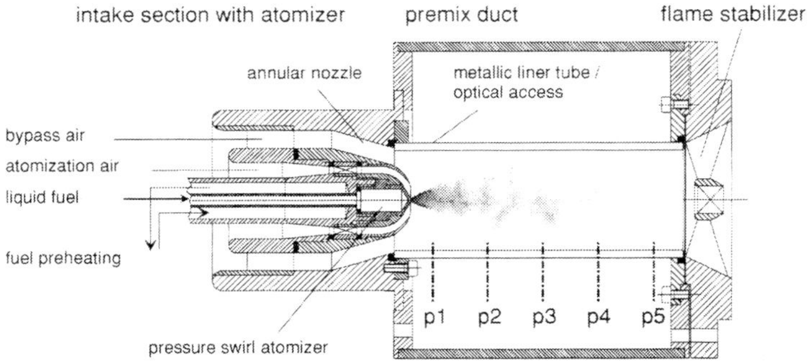

Figure 1.1.8: Atomizer, premix duct and flame holder of the LPP combustor.

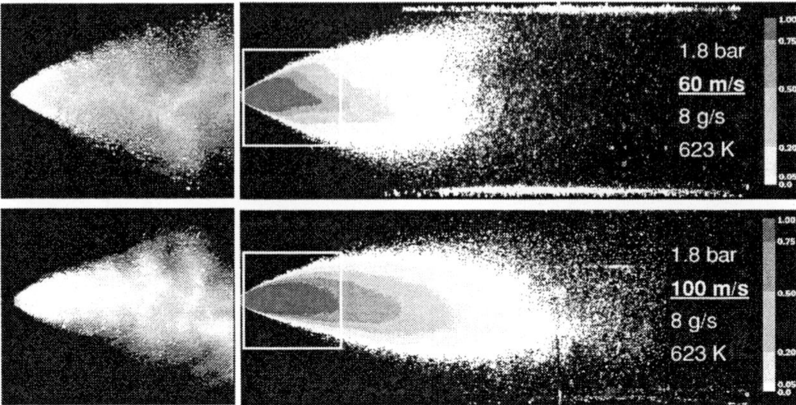

Figure 1.1.9: Frequency distribution of the fuel spray at different air velocity levels.

in the outer region describes the conic angle of the hollow spray jet without external flow, while droplets of the inner region are deflected in an axial direction by the atomizing air flow. No instabilities could be detected close to the nozzle, however, periodic flow instabilities downstream of the jet were observed. As these instabilities are also present without combustion, it may be concluded that they are caused by the turbulent mixing of the spray and the atomization air with the bypass air.

The spatial separation of the areas of higher fuel concentration combined with the mean flow velocity (see Fig. 1.1.8) allows the calculation of a typical frequency for this instabilities. It is known that for level flows with subcritical Reynolds numbers ($Re < 2 \cdot 10^5$) the Strouhal number, a function of the frequency f, the characteristic flow velocity c_m, and the typical length l, has a constant value of 0.2. The visualization of the jet dispersion in the premix duct allows for the following equation based upon the frequency of the fuel pockets and the mean axial flow velocity. This equation is based on a series of pictures captured at different operation conditions.

$$l = \mathrm{Str}\ c_m / f = 2.2 \ldots 2.6 \text{ mm} \tag{1}$$

This value is a good approximation of the diameter of the compact fluid lamella directly behind the atomizer and allows the conclusion to be drawn that the frequency of the visible instability in the mixture is directly proportional to the flow velocity.

In the area near the jet, an increase of the flow velocity from 60 to 100 m/s causes a reduction of the conic angle. As the radial momentum of the fuel droplets remains constant due to the constant fuel mass flow rate, this clearly shows an increase of the momentum of atomization air. The right pictures in Fig. 1.1.9 are the result of the superposition of one hundred single spray pictures. The light grey pixels represent areas with a droplet frequency of less than 5 %. The darkest grey pixels show that in more than 95 % of the compared single pictures scattered light was detected on the pixels. If there is no scattered light the pixels are black. However, this contour plots allow to evaluate the influence of air velocity, pressure or temperature on the dispersion of the spray.

Close to the atomizer most of the fuel mass flow is illuminated by the laser light sheet. With increasing radial dispersion downstream the nozzle the amount of fuel within the light sheet decreases. This leads to the erroneous impression that the fuel is entirely evaporated at the end of the premix duct. It is therefore not possible to base a decision about the mixture homogeneity at the entrance to the reaction zone solely on the basis of visualization.

1.1.5.2 LIM Concept

The LIM concept uses a prefilming airblast atomizer for the fuel preparation (Fig. 1.1.10). The fuel is spread onto the prefilming tube by a pressure atomizer.

At the atomization edge the film disintegrates due to the interaction of the film with the inner and outer swirling air flow. The fuel film is preheated by recirculating hot gases which leads to smaller droplets due to the reduced surface tension of the film. The thickness of the fuel film on the prefilming tube has only little influence on the droplet size distribution produced by the atomizer. Hence, this atomizer offers good atomization over a wide range of fuel mass flow. The swirl number of the primary swirler is significantly lower then the secondary one. On the test rig the two swirlers are separately supplied by combustion air which allows to vary the ratio of primary and secondary air (flow split μ).

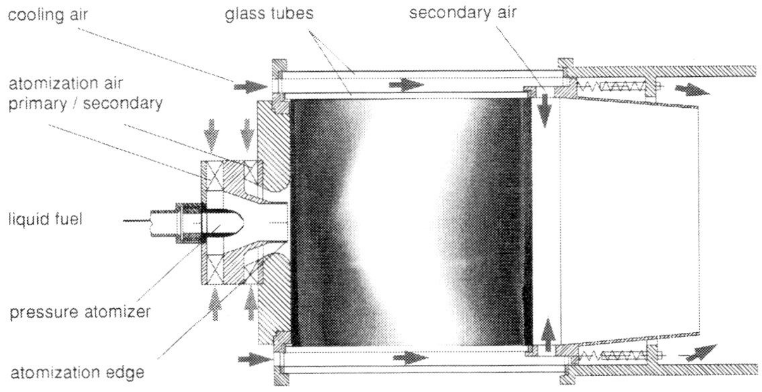

Figure 1.1.10: Cross section of the LIM combustor without casing.

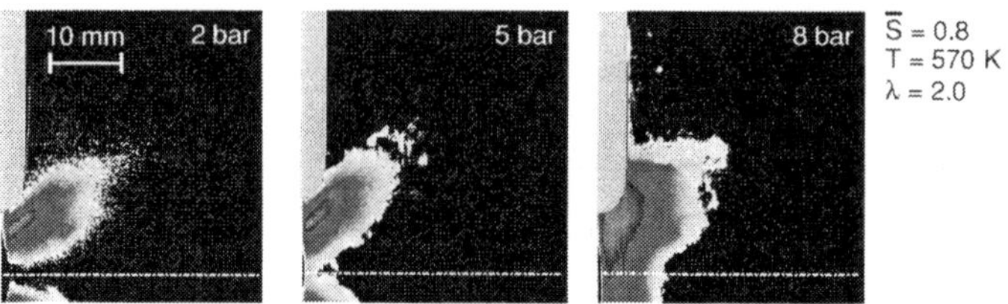

Figure 1.1.11: Frequency distribution of the spray close to the atomizer lip at different ambient pressures.

The nominal flow split is $\mu_n = 0.6$. The overall swirl number $\overline{S}$ in the combustion chamber can be approximated by Eq. (2).

$$\overline{S} = \left|\frac{\mu}{\mu+1} \cdot S_{\text{prim}}\right| + \left|\frac{1}{\mu+1} \cdot S_{\text{sek}}\right| \tag{2}$$

Hence a variation of the flow split μ leads to a change of the overall swirl number $\overline{S}$ without changes of the hardware. $\overline{S}$ determines the gas flow field and thereby the stabilization form of the flame.

The influence of ambient pressure on the spray propagation has been investigated, keeping ALR and the gas velocities at the atomizer exit constant (Fig. 1.1.11). Consequently an increase in ambient pressure leads to an increase in fuel mass flow and thereby in thermal power. As the heat loss through the glass tubes does not increase as strong as thermal power the flame temperature and thereby the flame velocity increases. Due to this higher turbulent flame velocity and because of an intensified droplet evaporation the flame stabilizes closer to the atomizer. This causes the spray penetration angle to increase with increasing ambient pressure. Additionally

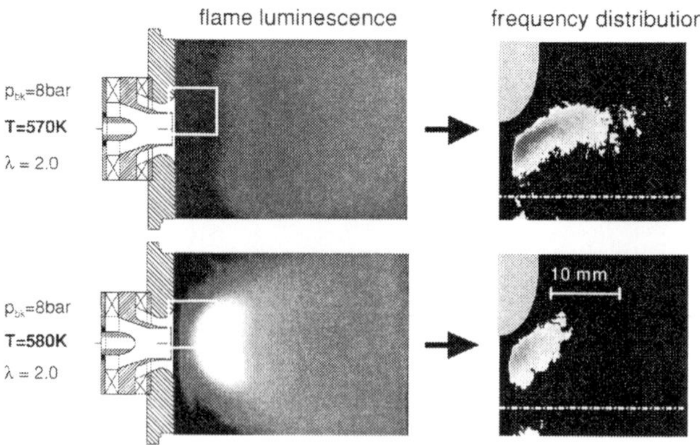

Figure 1.1.12: Frequency distributions and flame luminescence for an attached and a lifted flame (p = 8 bar, λ = 2.0).

a higher amount of droplets from the simplex atomizer does not reach the prefilmer as a result of the higher aerodynamic drag forces. These droplets are visible at the exit of the atomizer on the combustor axis.

To examine the influence of the flame position on the fuel penetration the reacting two phase flow of a lifted and an attached flame have been investigated. As test conditions an ambient pressure of 8 bar and a combustor inlet temperature of 570 K and 580 K respectively have been chosen. At these test conditions increasing the inlet temperature from 570 K to 580 K causes the flame to change from a lifted to an attached stabilization form (Fig. 1.1.12). The mass flow of air and fuel has been kept constant.

The frequency distribution of the spray shows a higher penetration depth of the droplets in the lifted flame due to the lower ambient flame temperature. The attached flame causes a higher entrance angle of the spray because of the high gradients in density close to the nozzle.

With increasing combustor inlet temperature in general the frequency distributions show less penetration depth combined with smaller droplets and a higher amount of fuel evaporation inside the atomizer. Due to the more rapid evaporation and mixing of the spray and because of the higher flame velocity the flame stabilizes closer to the atomizer exit.

1.1.6 Performance Characterization with PDPA Results

1.1.6.1 LPP Concept

The quantitative investigation of the spatial mixture homogeneity was carried out using the PDPA technique. Comparison of the droplet sizes at various planes in the premix duct show how the mixture homogenizes. The data for plane 1 in Fig. 1.1.13 clearly shows the hollow cone character of the atomizer as seen in the visualization with the smallest Sauter mean diameter of 35 μm measured on the axis and the largest of roughly 70 μm measured at a radius of 8 mm. In flow direction it is possible to see the equalizing of the radial droplet size distribution as well as the radial dispersion of the spray. Calculations of droplet vaporization which were carried out by Schmehl et al. [21] show that due to the short residence time of the droplets in the premix duct, only smallest droplets (<10 μm) can vaporize in this region. The strong reduction in the maximum droplet size is therefore not due to vaporization of the largest droplets, rather it is result of secondary atomization and a more equal radial distribution of the liquid phase. The enormous influence of the flow velocity on the spray dispersion was already shown in Fig. 1.1.9, in which it may be seen that an increase in the flow velocity results in a more narrow shape of the spray. A similar result can

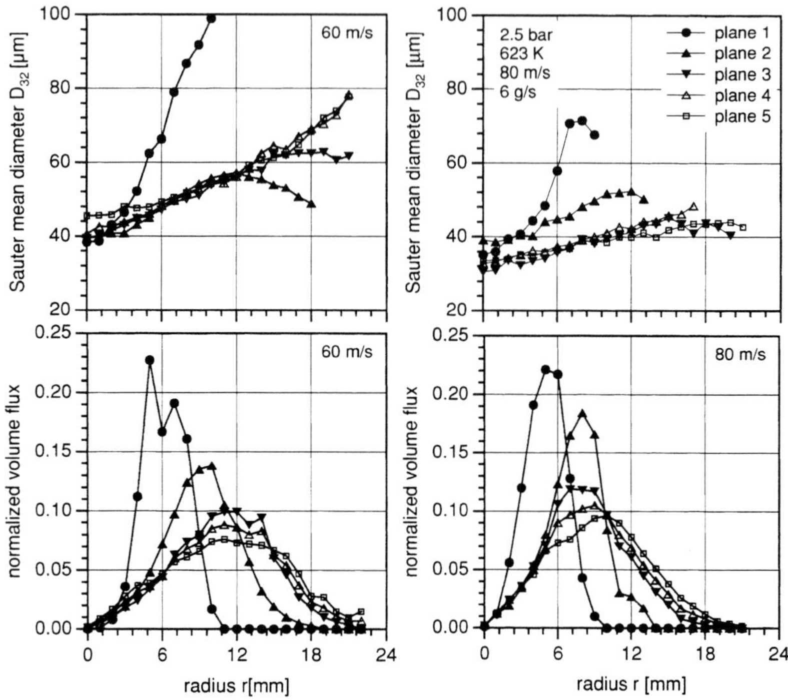

Figure 1.1.13: Sauter mean diameter and normalized volume flux within different planes at 60 and 80 m/s mean axial air velocity.

be achieved by means of PDPA measurements within different planes of the premix duct.

Figure 1.1.13 also shows the spatial distribution of the normalized volume flux within the different planes of the premix duct. The volume flux of each measurement point was weighted with the area of an angular ring which encloses the measurement point. The latitude of the ring is fixed to 1 mm. Unfortunately, the determination of the volume flux by means of the PDPA technique is still objectionable. Therefore, the global liquid flow rates were normalized to 100 % within each axial plane. Nevertheless, an increase of mean air velocity from 60 m/s to 80 m/s chiefly focuses the spray towards the centerline of the premix duct. However, the given results demonstrate the performance of the premix duct which has the primary function to attain a homogenous and lean fuel/air mixture.

1.1.6.2 LIM Concept

In Fig. 1.1.14 the Sauter mean diameter and the normalized volume flux within different planes in the LIM combustor at $\lambda = 1.0$ and 2.0 are shown. The measurements have been conducted at atmospheric pressure and with the combustion air preheated to 640 K. The volume flux is normalized in the same way as described for the LPP combustor.

Downstream of the atomization edge smallest droplet size distributions with a SMD of 15 μm were found. Smallest droplets directly follow the gas flow which leads to smaller Sauter mean diameters in this region. Larger droplets having a higher momentum are less deviated by the gas flow and therefore reach the outer parts of the combustor and the zone close to the combustor axis. Furthermore, recirculating hot gases on the combustor axis lead to an increased evaporation of small droplets and thereby again cause the mean droplet diameter to increase. A comparison of the atomization at stoichiometric conditions ($\lambda = 1.0$) and at fuel lean conditions ($\lambda = 2.0$) shows that the fuel load on the prefilmer has negligible influence on the atomization result. At stoichiometric conditions even slightly smaller mean diameters are measured. The volume flux graphs show the hollow cone shape of the fuel spray and a broadening of the spray towards the reaction zone. While

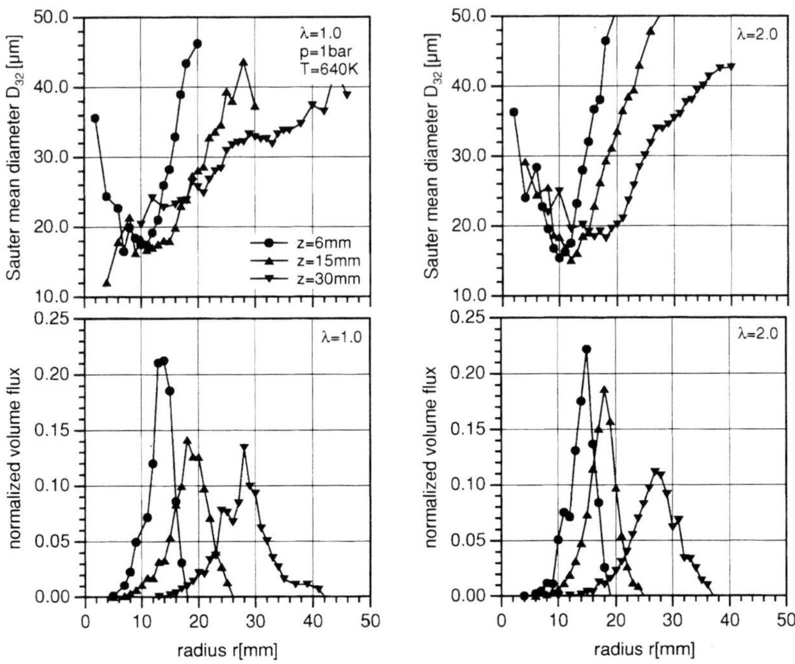

Figure 1.1.14: Sauter mean diameter and normalized volume flux within different planes in the LIM combustor at $\lambda = 1.0$ and 2.0.

at z = 6 mm the major part of the spray is build by droplets of SMD = 15–22 μm at z = 30 mm the peak of volume flux can be at droplet diameters of SMD = 30–40 μm. This again indicates the rapid vaporization of the small droplets.

1.1.7 Summary and Outlook

In this contribution two low NO_x lean burning combustion concepts have been presented. A lean premixed prevaporized (LPP) combustion concept and a lean internal mixing (LIM) concept and their specific techniques of atomization and mixture formation have been compared. The experimental setup for atmospheric tests and the high pressure high temperature test facility (HDT) for tests under real engine conditions have been described in detail. Utilizing measurement techniques which partly have been developed within the Collaborative Research Centre 167 the processes of atomization and mixture formation have been investigated by means of visualization, laser diffraction and phase Doppler measurements.

The LPP concept uses an air assist atomizer whereas the LIM concept utilizes an airblast atomizer. The differences in operation and in the dynamic range of these atomizers have been investigated on an atmospheric test rig using water as test fluid. The visualizations as well as the laser diffraction measurements of the air assist atomizer showed a strong influence of the liquid flow rate on the atomization process for low air velocities. The comparison with the airblast atomizer shows clear disadvantages of the air assist atomizer for liquid flow rates from 1 to 3 g/s water mass flow due to an incomplete developed liquid cone of the simplex atomizer. Nevertheless, for higher air velocities which are typical for LPP applications good atomization is realized over a wide range of liquid flow rate. In order to enlarge the dynamic range to low air velocity levels a spill-return nozzle producing an always fully developed liquid lamella of conical shape should be used.

The mixture formation of both concepts has been examined by laser light sheet technique and by phase Doppler measurements. In the rear measurements planes of the premix duct very small droplets of SMD = 30–40 μm have been found. Volume flux measurements of the different planes showed close to the exit rather small gradients indicating a good homogeneity of mixture. In the LIM concept for increasing ambient pressure a lack in mixture formation has been detected due to an increasing spray penetration angle.

Acknowledgements

The Collaborative Research Centre 167 was supported by grant from the Deutsche Forschungsgemeinschaft (DFG) which is gratefully acknowledged.

References

1. Aigner, M., Wittig, S. (1987): Swirl and Counterswirl Effects in Prefilming Airblast Atomizers, ASME-Paper 87-GT-204.
2. Sattelmayer, T., Wittig, S. (1989): Performance Characteristics of Prefilming Atomizers in Comparison with other Airblast Nozzles, Encyclopedia of Fluid-Mechanics, Vol. 8, Aerodynamics and Compressible Flows, Gulf Publ. Company.
3. Himmelsbach, J. (1992): Zweiphasenströmung mit schubspannungsgetriebenen welligen Flüssigkeitsfilmen in turbulenter Heißluftströmung – Meßtechnische Erfassung und numerische Beschreibung, Dissertation Universität Karlsruhe, Institut für Thermische Strömungsmaschinen.
4. Kneer, R., Willmann, M., Schneider, M., Hirleman, E. D., Koch, R.: Theoretical Studies on the Influence of Refractive Index Gradients within Multicomponent Droplets on Size Measurements by Phase-Doppler Anemometry. In: Proceedings of the 6th International Conference on Liquid Atomization and Spray Systems, Rouen, France, 451–458.
5. Maier, G., Bauer, H. J., Wittig, S. (1995): In Situ-Charakterisierung des Sprühstrahls einer druckbeaufschlagten Gasturbinenbrennkammer – Einfluß von Druck und Luftmassenstrom, 11. Tecflam Seminar Heidelberg.
6. Maier, G., Willmann, M., Wittig, S. (1997): Development and Optimization of Advanced Atomizers for Application in Premix Ducts, ASME-Paper 97-GT-56.
7. Stengele, J., Bauer, H.-J., Wittig, S. (1996): Numerical Study of Bicomponent Droplet Vaporization in a High Pressure Environment, ASME-Paper, 96-GT-442.
8. Meier, R., Maier, G., Wittig, S. (1998): Sichtbarmachung der Sprühstrahlausbreitung von Airblast-Zerstäubern unter brennkammertypischen Bedingungen, Workshop Spray '98, GH Essen.
9. Willmann, M., Meier, R., Wittig, S. (1997): Visualisation of the instationary atomization and spray propagation inside a reacting model gasturbine combustor, ILASS, 9.–11. July 1997, Florence, Italy.
10. Rottenkolber, G., Dullenkopf, K., Wittig, S. (1998): Visualization and PIV-Measurements inside the Intake Port of an IC Engine, 9th International Symposium on the Application of Laser Techniques to Fluid Mechanics, 13.–16. July 1998, Lisbon, Portugal.
11. Willmann, M., Eigenmann, L., Wittig, S., Hirleman, D. (1994): Experimental Investigations on the Effect of Trajectory Dependent Scattering on Phase Doppler Particle Sizing with a Standard Instrument, Proc. of the 7th Int. Symp. on the Application of Laser Techniques to Fluid Mechanics, 20.–23. July 1994, Lisbon, Portugal.
12. Aigner, M., Wittig, S. (1985): Performance and Optimization of an Airblast Nozzle: Drop Size Distribution and Volumetric Air Flow: Proc. Int. Conf. on Liquid Atomization and Spray Systems, London, IIC/3, 1–8.

13. Lefebvre, A. H. (1989): Atomization and Sprays, Taylor and Francis, 234–238.
14. Lefebvre, A. H. (1992): Twin Fluid Atomization: Factors Influencing Mean Drop Size, *Atomization and Sprays*, **2**, 101–119.
15. Rizk, N. K., Chin, J. S., Razdan, M. K. (1995): Modeling of Gas Turbine Fuel Nozzle Spray, ASME paper 95-GT-225.
16. Maier, G., Willmann, M., Wittig, S. (1997): Performance of Air-Assisted Pressure Swirl Atomizers, ILASS, 13th International Conference of Liquid Atomization and Spray Systems, Florence, Italy, July 1997.
17. Maier, G., Wittig, S. (1998): Effects of Liquid Properties on the Operating Performance of Air-Assisted Pressure Swirl Atomizers, ILASS, 14th International Conference of Liquid Atomization and Spray Systems, Manchester, England, July 1998.
18. Fraser, R. P., Eisenklam, P., Dombrowski, N., Hasson, D. (1962): Drop Formation from Rapidly Moving Liquid Sheets, *A. I. Ch. E. Journal*, **8**, No. 5, 672–680.
19. Eroglu, H., Chigier, N. (1991): Initial Drop Size and Velocity Distributions for Airblast Coaxial Atomizers, *J. of Fluids Engineering*, 453–459.
20. Samenfink, W., Hallmann, M., Elsäßer, A., Wittig, S. (1994): Secondary Break-up of Liquid Droplets: Experimental Investigation for a Numerical Description, ILASS 1994, Rouen, France, Paper No.: I.21.
21. Schmehl, R., Klose, G., Maier, G., Wittig, S. (1998): Efficient Numerical Calculation of Evaporating Sprays in Combustion Chamber Flows, 92nd Symp. on Gas Turbine Engine Combustion, Emissions and Alternative Fuels, Lisbon, Portugal, 1998.

1.2 Calculation of Two Phase Flows in Combustors

Roland Schmehl, Göran Klose, Georg Maier, and Sigmar Wittig*

Abstract

Representing two different conceptual approaches, either Eulerian continuum models or Lagrangian particle models are commonly applied for the numerical description of dispersed two phase flows. Taking advantage of the positive features inherent to each model, a combination approach is presented in this study for the efficient computation of liquid fuel sprays in combustor flows. In the preconditioning stage, Eulerian transport equations for gas phase and droplet phase are solved simultaneously in a block-iterative scheme based on a coarse discretization of spray boundary conditions at the nozzle. Due to the close coupling of both phases, the time expense of this approximate flow field computation is not much higher as for single phase flows. In the refinement stage, Lagrangian droplet tracking is applied with a detailed discretization of initial conditions. To account for complete interaction between gas phase and droplets, gas flow solution and droplet tracking are concatenated by an iterative procedure. In this stage, the numerical description of the spray is enhanced by additional modeling of droplet breakup. Results of numerical simulations are compared with measurements of the two phase flow in a premix duct of a LPP research combustor.

* Institut für Thermische Strömungsmaschinen, Universität Karlsruhe, Kaiserstr. 12, 76128 Karlsruhe, Germany

Nomenclature

c_p	specific heat capacity
D	droplet diameter
$D_{0.5}$	mass median diameter
D_{32}	Sauter mean diameter
$D_{0.632}$	characteristic diameter
f	body force
h	enthalpy
$\dot{H}$	enthalpy flux
H	energy transfer rate
I	momentum transfer rate
k	turbulent kinetic energy
$\dot{m}$	mass flux
M	mass transfer rate
On	Ohnesorge number
P	Pressure
Pr	Prandtl number
$\dot{Q}$	conductive heat flux
Re	Reynolds number
S	source term
Sc	Schmidt number
T	temperature
Tu	degree of turbulence
U	velocity component
We	Weber number
Y	mass fraction

Greek Symbols

α	heat transfer coefficient
α_k	liquid volume fraction
β	off axis angle
ε	dissipation rate of k
Γ	diffusion coefficient
μ	dynamic viscosity
ν	kinematic viscosity
ρ	density
τ	shear stress

Subscripts

0	initial state
g, d	gas, droplet
int	interface
t	turbulent
vap	vapor

1.2.1 Introduction

Improving modern gas turbine efficiencies by increasing pressure and temperature levels of the combustion process, essentially requires sophisticated combustion concepts in order to meet todays strict limitations on pollutant emissions. Fundamental to these low emissions concepts is a characteristic strategy to inject and mix the liquid fuel with the compressed air flow, avoiding local stoichiometric combustion conditions as far as possible. Two promising approaches in this context are the concepts of Lean-Premix-Prevaporize (LPP) and Rich-Quench-Lean (RQL) combustion. In order to develop advanced combustor designs with the required flow characteristics, a better understanding of the two phase flow physics is necessary. Two phase flow effects typical for premix ducts of LPP combustors or prefilming airblast atomizers are summarized in Fig. 1.2.1.

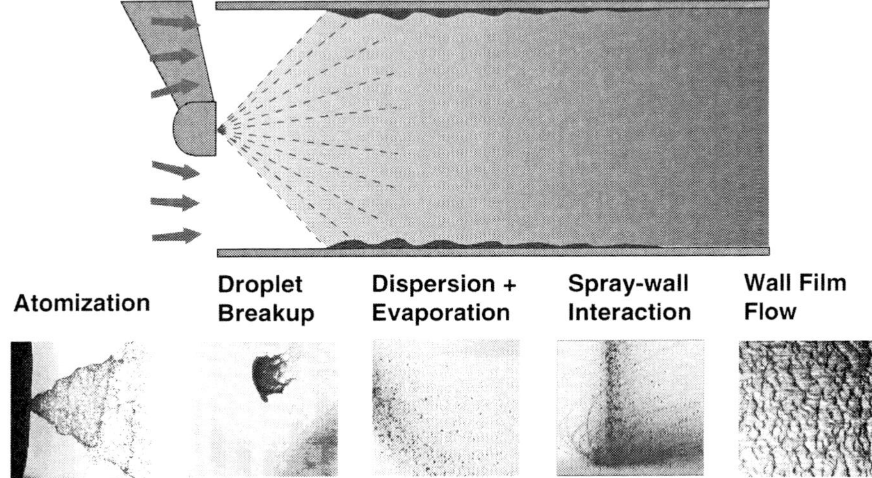

Figure 1.2.1: Two phase flow effects in a LPP premix duct.

Due to the enormous increase in computing performance, Computational Fluid Dynamics (CFD) offers a promising potential for efficient combustor design and optimization. In particular when compared to experimental studies at elevated pressures, CFD analysis may be employed to reduce turnaround times and costs of combustor design significantly. On the other hand, complex flow phenomena such as turbulence, atomization or chemical reaction still represent some of the most challenging topics for CFD tools.

Basically, two different conceptual approaches may be employed for the numerical description of dispersed two phase flows [3]. In analogy to single phase gas flow, the Eulerian approach is based on a continuum model of the spray, resulting in transport equations describing the propagation and evaporation of this droplet phase [23, 6]. In the Lagrangian approach, the spray is modeled by superposition of trajectories calculated for large numbers of representative droplets. Each of the two basic approaches is characterized by specific advantages and restrictions.

In the Eulerian method, the transport equations of the droplet phase are appended to the gas phase transport equations, resulting in a compact description of the interacting two phase flow system. The essential advantage is a simultaneous solution of the interacting flow fields of gas phase and spray by a single numerical method. Applying a standard block-iterative solver for systems of linearized equations, the information exchange between phases is realized on the level of the non-linear iterations. As a consequence, computation times are generally small compared to the Lagrangian approach. However, each droplet initial condition to be simulated requires the solution of an individual set of 6 transport equations. This conceptual feature is a severe limitation for the discretization of complex sprays with wide ranges of initial droplet size, injection angles and velocities. Furthermore, Eulerian methods are generally not suited for the modeling of complex two phase phenomena like secondary atomization or spray-wall interaction. A detailed description of the Eulerian model used in this study is given in [19].

In the Lagrangian method, a large number of droplet trajectories has to be tracked to achieve a continuous distribution of the liquid phase. Due to the stochastic simulation of turbulent spray dispersion by random sampling of gas velocity fluctuations, identical droplet initial conditions lead to different trajectories. As a consequence, the liquid phase flow field is a statistical quantity. To limit the maximum field deviations, the number of simulated droplet trajectories has to be increased to values up to 10^4-10^6 for typical combustor flows. Effects of the spray back on the gas flow such as aerodynamic dragging or evaporation cooling are recorded in droplet source terms, describing the local interfacial transfer of mass, momentum and energy. Including these source terms in the gas flow transport equations, complete phase interaction is taken into account by an iterative concatenation of gas flow computation and Lagrangian droplet tracking. Besides increased computation times, this iterative procedure entails an artificial decoupling of gas flow and spray. In particular for flows with intense phase interaction, this may cause severe convergence problems, requiring strong relaxation of source term fields. Nevertheless, Lagrangian methods are commonly preferred for practical CFD analysis due to significant advantages regarding complex spray discretization and modeling of flow phenomena such as secondary droplet breakup or droplet-wall interaction.

In this study, a new hybrid approach is presented combining Eulerian and Lagrangian methods to take advantage of the capabilities inherent to both methods. In the first stage of this procedure, the Eulerian method is used for an efficient computation of an approximate two phase flow field. A coarse discretization of spray boundary conditions at the nozzle limits the size of the system of transport equations to a practical dimension. Good convergence rates are achieved due to the close coupling between gas flow and spray. In the refinement stage, iterative cycles of single phase gas flow computation and subsequent droplet tracking are employed to improve the quality of the preconditioned two phase flow field. Taking advantage of the stochastical nature of the tracking approach, a fine discretization of polydisperse sprays is achieved by random sampling of droplet initial conditions at the nozzle. The numerical description of the spray is enhanced by optional modeling of secondary droplet breakup and spray-wall interaction. Since modeling of spray-wall interaction has been described in detail in [20], only secondary atomization of droplets is considered in this study.

To demonstrate the performance and accuracy of the numerical methods discussed in this paper, the evaporating spray in the premix duct of a LPP research combustor is simulated and assessed by measured droplet data. The experimental investigation of this premix section has been a focus of various studies [12, 13, 10] and represents a valuable source of experimental data.

1.2.2 Lagrangian Approach

1.2.2.1 Spray Dispersion

The Lagrangian simulation of dispersed two phase flow is based on the tracking of statistically significant droplet parcels in the gas flow. Each parcel is represented by one droplet and is determined by discretization of the continuous spectra of droplet initial conditions in the near field of the atomizer. The tracking is based on the integration of the droplets equation of motion combined with an empirical correlation for the aerodynamic drag coefficient C_D.

$$\frac{\mathrm{d}\vec{u}_\mathrm{d}}{\mathrm{d}t} = -\frac{3\rho_\mathrm{g}}{4\rho_\mathrm{d}}\frac{C_\mathrm{D}}{D}\left|\vec{u}_\mathrm{d} - \vec{u}_\mathrm{g}\right|(\vec{u}_\mathrm{d} - \vec{u}_\mathrm{g}) \tag{1}$$

$$C_D = 0.36 + 5.48\,\mathrm{Re}_d^{-0.573} + \frac{24}{\mathrm{Re}_d} \tag{2}$$

In order to simulate the effect of turbulent spray dispersion, the turbulence structure of the gas flow field is modeled by a random process along the droplet trajectories [5, 14]. In this concept, the local turbulence structure is characterized by the length scale l_e and dissipation time scale t_e of eddies representing the coherent flow structures

$$l_e = C_\mu^{\frac{1}{2}} \frac{k^{\frac{3}{2}}}{\varepsilon} \quad , \quad t_e = \frac{l_e}{\sigma} \tag{3}$$

In addition to the life time scale t_e, a crossing time scale t_c is calculated from

$$\left\| \int_{t_0}^{t_c} (\vec{u}_g - \vec{u}_d)\mathrm{d}t \right\| = l_e \tag{4}$$

taking into account the droplet dynamics. Each time the smaller one of both time scales is elapsed, the droplet enters a new eddy. Consequently, the random process generates a new velocity fluctuation $\vec{u}_g'$ from a Gaussian distribution which is determined by

$$\mu = 0 \quad , \quad \sigma = \sqrt{\frac{2}{3}k} \tag{5}$$

This velocity fluctuation remains constant for the period of droplet-eddy interaction and is added to the local value of the gas flow velocity.

1.2.2.2 Spray Evaporation

In this study, droplet evaporation is simulated by means of the Uniform Temperature model [4, 22, 2]. This computationally effective droplet model is based on the assumption of a homogeneous internal temperature distribution in the droplet and phase equilibrium conditions at the surface. The analytical derivation of this model does not consider contributions to heat and mass transport by forced convection by the gas flow around the droplet. Since diffusive time scales in the surrounding gas phase are much smaller than in the droplet fluid, a quasi stationary description of the gas phase is applied. Using reference values for variable fluid properties (1/3-rule), an integration of the radially symmetric differential equations yields analytical expressions for the transport fluxes $\dot{m}_{\mathrm{vap}}$, $\dot{Q}_{\mathrm{cond,s}}$ and $\dot{H}_{\mathrm{vap,s}}$. At this point, convective transport is taken into account by two empirical factors [1] resulting in the corrected fluxes $\dot{m}^*_{\mathrm{vap}}$, $\dot{Q}^*_{\mathrm{cond,s}}$ and $\dot{H}^*_{\mathrm{vap,s}}$ [18].

$$\dot{m}^*_{\mathrm{vap}} = cfm\ \dot{m}_{\mathrm{vap}} \tag{6}$$

$$\dot{Q}^*_{\mathrm{cond,s}} = \pi D^2 \alpha^* (T_\mathrm{d} - T_\mathrm{g}) \tag{7}$$

$$\dot{H}^*_{\mathrm{vap,s}} = \dot{m}^*_{\mathrm{vap}}\ c_{\mathrm{p,vap,ref}}\ (T_\mathrm{d} - T_\mathrm{g}) \tag{8}$$

Vapor mass flux and heat transfer coefficient are calculated as follows:

$$\dot{m}_{\mathrm{vap}} = 2\pi D\ \rho_{\mathrm{g,ref}} \Gamma_{\mathrm{im,ref}} \ln \frac{1 - Y_{\mathrm{vap,g}}}{1 - Y_{\mathrm{vap,s}}} \tag{9}$$

$$\alpha^* = cfh\ \frac{\dfrac{\dot{m}_{\mathrm{vap}}\ c_{\mathrm{p,vap,ref}}}{\pi D^2}}{\exp\left(\dfrac{\dot{m}_{\mathrm{vap}}\ c_{\mathrm{p,vap,ref}}}{2\pi D \lambda_{\mathrm{g,ref}}}\right) - 1} \tag{10}$$

$$cfm = 1 + 0.276\ \mathrm{Re}^{\frac{1}{2}}\ \mathrm{Sc}^{\frac{1}{3}} \tag{11}$$

$$cfh = 1 + 0.276\ \mathrm{Re}^{\frac{1}{2}}\ \mathrm{Pr}^{\frac{1}{3}} \tag{12}$$

The balance equations of the droplet reduce to ordinary differential equations,

$$\frac{\mathrm{d}}{\mathrm{d}t} m_\mathrm{d} = -\ \dot{m}^*_{\mathrm{vap}} \tag{13}$$

$$\frac{\mathrm{d}}{\mathrm{d}t} (m_\mathrm{d}\ h_\mathrm{d}) = -\ \dot{Q}^*_{\mathrm{cond,s}} - \dot{H}^*_{\mathrm{vap,s}} \tag{14}$$

which can be appended to the differential equations describing the droplet motion, Eq. (1).

1.2.2.3 Secondary Droplet Breakup

At low relative velocities, the spherical shape of the droplets is preserved by the dominating effects of surface tension and viscous forces in the liquid. With increasing velocities, the destabilizing aerodynamic forces on the droplet surface are intensifying, resulting in deformation, oscillations and disintegration of the droplets.

1.2.2.3.1 Classification of Breakup Mechanisms

A common practice to classify secondary droplet atomization processes is based on two characteristic groups of parameters,

$$\mathrm{We} = \frac{\rho_\mathrm{g}\ u^2_{\mathrm{rel}}\ D}{\sigma_\mathrm{d}}, \quad \mathrm{On} = \frac{\mu_\mathrm{d}}{\sqrt{\rho_\mathrm{d}\ D\ \sigma_\mathrm{d}}} \tag{15}$$

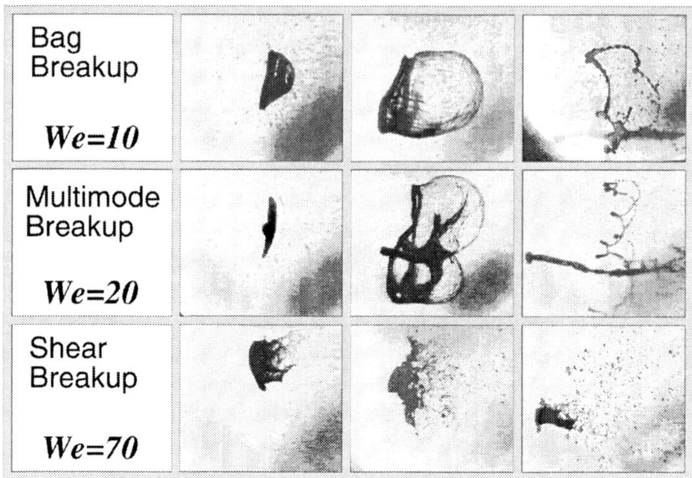

Figure 1.2.2: Breakup mechanisms of water droplets [17].

The Weber number is a measure of the strength of aerodynamic forces relative to surface tension forces, whereas the Ohnesorge number assesses the damping effect of viscous friction in the droplet against surface tension. In the Weber number range from We = 1 up to a critical value We = We_c, non-destructive droplet deformation and oscillation is observed. As illustrated in Fig. 1.2.2 three different mechanisms govern the breakup of droplets for increased Weber numbers typical for flows in gas turbine combustors. From these visualizations it is obvious that a common feature of all three mechanisms is an initial deformation of the droplet into a disc shape. After this deformation period, various complex flow phenomena lead to the final droplet breakup depending on the intensity of the aerodynamic forces.

Exceeding the critical Weber number, the first mechanism observed is bag breakup. This process is characterized by the formation of a thin hollow bag of droplet fluid stretching from a toroidal rim. The thin film of this bag is eventually bursting into a cloud of tiny droplets, followed by a disintegration of the toroidal rim into significantly larger fragments. With increasing aerodynamic forces, a transition to more complex bag structures is observed. In this multimode or stamen breakup regime the aerodynamic flow interaction is forming an additional fluid filament in the center of the bag structures which is aligned with the relative flow velocity. For even higher Weber numbers, shear breakup is observed. This mechanism is fundamentally different to the preceding mechanisms and is characterized by a rapidly disintegrating film of fluid continuously stripped off the rim of the disc shaped droplet by shear forces.

Figure 1.2.3 summarizes the results of various experimental studies [9, 15] in a breakup regime map, indicating the relevant mechanism corresponding to a specific combination of Ohnesorge and Weber number. For On $>$ 0.1

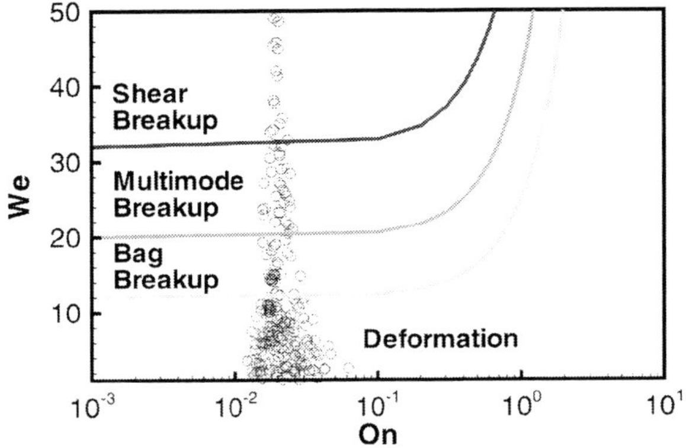

Figure 1.2.3: Breakup regime map (○ Present flow simulation).

a significant influence of viscosity is observed. The transitions between the different mechanisms are given by the following functions

- Critical Weber number (Transition to bag breakup):

$$\mathrm{We_c} = 12\,(1 + 1.077\,\mathrm{On}^{1.6}) \tag{16}$$

- Transition to multimode breakup:

$$\mathrm{We} = 20\,(1 + 1.200\,\mathrm{On}^{1.5}) \tag{17}$$

- Transition to shear breakup:

$$\mathrm{We} = 40\,(1 + 1.500\,\mathrm{On}^{1.4}) \tag{18}$$

1.2.2.3.2 Deformation and Breakup Times

Basically, the breakup process can be subdivided in two stages: Initial deformation and further deformation with disintegration. It is convenient to express the relevant times of the breakup process in terms of the characteristic time of shear breakup.

$$t^* = \frac{D_0}{u_{\mathrm{rel}}}\sqrt{\frac{\rho_\mathrm{d}}{\rho_\mathrm{g}}} \tag{19}$$

As stated in [8], the time of initial droplet deformation t_{def} has a constant value independent of the specific breakup mechanism.

$$\frac{t_{\mathrm{def}}}{t^*} = 1.6 \tag{20}$$

However, the breakup time t_b measured from begin of deformation until final destruction strongly depends on the specific mechanism. A fit to a large number of experimental data is given in [15].

$$\frac{t_b}{t^*} = \begin{cases} 6 & (\mathrm{We}-12)^{-0.25} & 12 < \mathrm{We} < 18 \\ 2.45 & (\mathrm{We}-12)^{0.25} & 18 < \mathrm{We} < 45 \\ 14.1 & (\mathrm{We}-12)^{-0.25} & 45 < \mathrm{We} < 351 \\ 0.766 & (\mathrm{We}-12)^{0.25} & 351 < \mathrm{We} < 2670 \\ 5.5 & & 2670 < \mathrm{We} \end{cases} \tag{21}$$

For On > 1, liquid viscosity is the dominating parameter of the breakup process resulting in the following correlation.

$$\frac{t_b}{t^*} = 4.5(1 + 1.2\,\mathrm{On}^{0.74}) \tag{22}$$

1.2.2.3.3 Droplet Drag

Deformation of the droplet prior to breakup leads to a significant increase of aerodynamic drag. Due to the resulting acceleration, the droplet generally experiences substantial displacement from initial deformation until final breakup. With respect to the deformation period, several authors [8] report a linear increase of the droplet size from D_0 up to a maximum value given by (On < 0.1, We < 100).

$$\frac{D_{max}}{D_0} = 1 + 0.19\sqrt{\mathrm{We}} \tag{23}$$

The effect of higher Ohnesorge numbers is taken into account by using a corrected Weber number in the above and following equations.

$$\mathrm{We}_{corr} = \frac{\mathrm{We}}{1 + 1.077\,\mathrm{On}^{1.6}} \tag{24}$$

As suggested in [8], a linear transition of the drag coefficient from the sphere shape value to the disc shape value is used in the present study to model the aerodynamic properties of the flattening droplet.

In the following period of disintegration, droplet drag depends on the specific breakup mechanism. As illustrated in Fig. 1.2.3, the bag and filament structures observed prior to breakup are very complex. According to [9, 17] the toroidal rim evolving in bag breakup is expanding to seven times the initial droplet diameter, whereas in multimode breakup a maximum diameter of six times the initial diameter is reached (see Fig. 1.2.6). At this time however, a major part of the droplet cross section consists of a thin fluid film accelerated in direction of the relative velocity thus reducing the aerodynamic drag. To bypass the difficulties of describing these opposing effects, the drag of the disintegrating droplet is calculated from the constant disc state reached at the end of the deformation period. In shear breakup, the size of the disc shaped droplet is continuously decreasing to its final value at t_b.

1.2.2.3.4 Secondary Droplet Sizes

Based on an extended experimental study covering the complete range of breakup mechanisms, a single correlation for the Sauter mean diameter D_{32} has been derived in [8] for all three mechanisms (On < 0.1, We < 1000).

$$\frac{D_{32}}{D_0} = 6.2 \ \mathrm{On}^{0.5} \ \mathrm{We}^{-0.25} \tag{25}$$

The exponents in this correlation have been determined by the authors from an approximate analysis of the droplet internal flow during shear breakup, leaving only a constant factor as a parameter for the fitting to experimental data. Using the Weber number given by Eq. (24) to account for viscosity effects, additional fitting of the exponents leads to an improved correlation.

$$\frac{D_{32}}{D_0} = 1.5 \ \mathrm{On}^{0.2} \ \mathrm{We}_{\mathrm{corr}}^{-0.25} \tag{26}$$

This correlation which is used for the flow simulations in the present study and the experimental data is shown in Fig. 1.2.4.

Although the above correlation is valid for the complete range of breakup conditions, the distribution function of the droplet diameter is substantially different for the various mechanisms of secondary breakup.

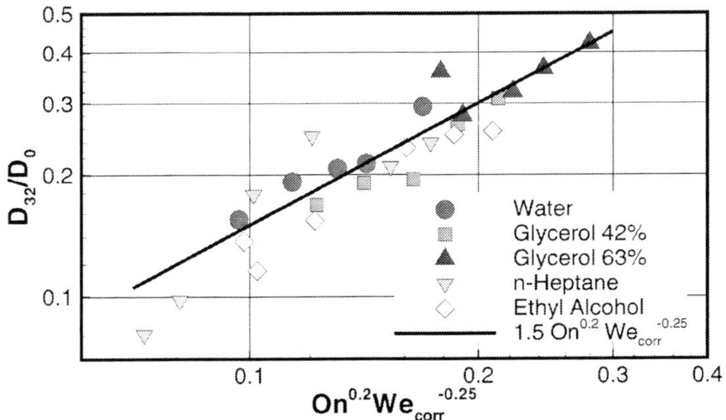

Figure 1.2.4: Improved correlation for D_{32} (Data from [8]).

Bag and Multimode Breakup

Considering bag or multimode breakup, the volume distribution of the droplet fragments is approximated by a root normal distribution [21], given by the following density function

$$f(D) = \frac{x}{2\sqrt{2\pi}\,\sigma\,D}\,\exp\left\{-\frac{1}{2}\left[\frac{x-\mu}{\sigma}\right]^2\right\} \tag{27}$$

and the parameters

$$x = \sqrt{\frac{D}{D_{0.5}}}, \qquad \mu = 1, \qquad \sigma = 0.238 \tag{28}$$

For a volume distribution with this distribution function, the mass median diameter $D_{0.5}$ is related to the Sauter mean diameter D_{32} by

$$\frac{D_{0.5}}{D_{32}} = 1.2 \tag{29}$$

Shear Breakup

According to [8], the volume distribution resulting from shear breakup is characterized by a bimodal density function with a maximum at small diameters and a second maximum at large diameters. From the experimental data illustrated in Fig. 1.2.5, it is concluded that the fine fraction of the droplet fragments corresponds to approximately 80 % of the total cumulated volume.

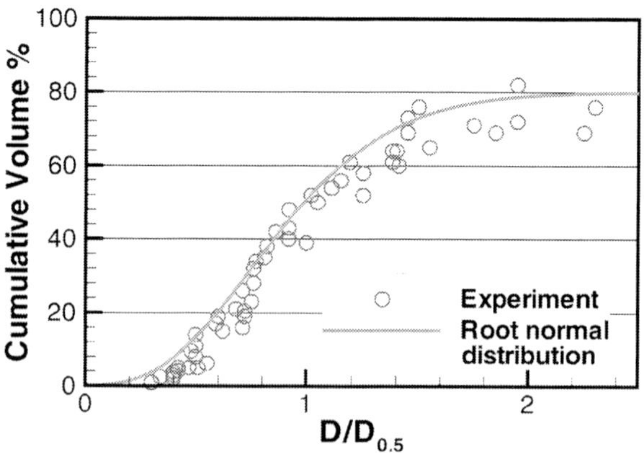

Figure 1.2.5: Cumulative droplet volume (Data from [8]).

These droplets result from film fragments stripped off the disc-shaped droplet by shear forces. The 20 % in the large diameter range not specified in Fig. 1.2.5 represent the contribution of the core droplet fragment left by the stripping process. The diameter D_c of this core droplet is estimated from the critical Weber number, evaluating Eq. (16) at flow conditions at the instant of breakup. As illustrated by the curve in Fig. 1.2.5, the volume distribution of the fine fraction of the droplet spectrum after shear breakup can be approximated by a root normal distribution based on a reduced Sauter mean diameter.

$$D_{32,\mathrm{red}} = \frac{4\, D_{32}\, D_c}{5\, D_c - D_{32}} \tag{30}$$

In this equation, the Sauter mean diameter of the complete droplet spectrum is evaluated from Eq. (26).

1.2.2.3.5 Secondary Droplet Velocities

Due to the dominating influence of aerodynamic forces on small droplets, tiny breakup products are immediately dragged with the gas flow. So, an accurate modeling of initial velocities is not necessary in general. These considerations apply in particular to the tiny droplet fragments generated by bursting of film bags or by shear induced film stripping. The motion of large droplets in turn is dominated by inertia forces. As a consequence, the modeling of initial velocities of large droplet fragments has a significant influence on the dispersion behavior of the spray.

As a first approximation, the fragments generated by droplet breakup inherit the velocity of the parent droplet due to momentum conservation. In bag or multimode breakup, a transverse velocity component of droplet fragments is observed induced by the transverse spreading motion of droplet fluid during the expansion of the toroidal rim. This transverse velocity component is responsible for increased dispersion of sprays with secondary atomization. For an approximate estimation, the growth velocity of the rim is determined from time-resolved visualizations of breakup processes. Fig. 1.2.6 indicates that the ring has an extension of about seven times the original droplet diameter in bag breakup against six times in multimode breakup. These observations agree with the values reported in [9]. Based on these results, the transverse velocity component is estimated as

$$v_t = \frac{D_{r,\max} - D_0}{2(t_b - t_{\mathrm{def}})} \tag{31}$$

With respect to multimode breakup, a fraction of the droplet fluid is concentrating on the axis of the disintegrating droplet (see Fig. 1.2.2, We = 20). Due to the alignment of this prolate filament with the flow, the fragments of this

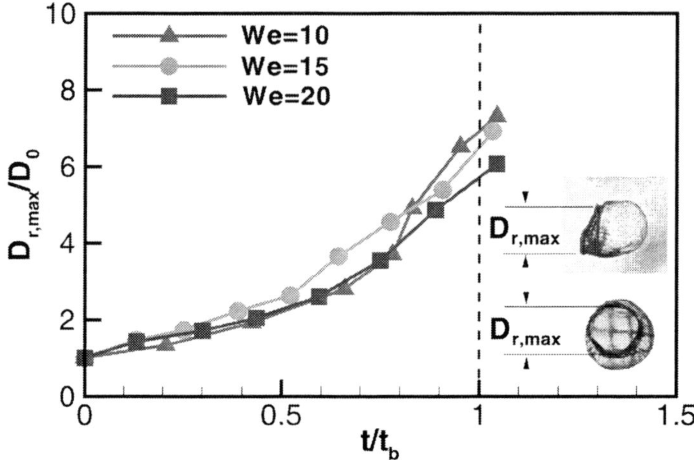

Figure 1.2.6: Growth of the toroidal rim (Water droplet, [17]).

structure have no transverse velocity component. The volume fraction of the filament is estimated from a Weber number based interpolation between the limiting values of 0 % for bag breakup and 20 % (core droplet) for shear breakup.

1.2.2.3.6 Stochastical Simulation of Droplet Breakup

The period of droplet disintegration is specified by the characteristic times t_{def}, and t_b of the breakup process. In shear breakup mode,the secondary droplets are continuously generated in the time from t_{def} to t_b, whereas in bag or multimode breakup mode significant generation of fragments is observed during the second half of this time period [17]. Instead of focusing on a realistic simulation of each individual breakup event, the computational implementation makes use of the Lagrangian trajectory superposition approach involving large numbers of droplet parcels.

During droplet deformation, the cross sectional area of the droplet and the drag coefficient are continuously increased up to their maximum values at t_{def}. A certain time later, the parent droplet trajectory is terminated and a fixed number of child droplets is generated by random sampling of initial conditions. Each secondary trajectory is assigned an equal fraction of the volume flux. To limit the number of secondary droplets to be tracked, only three child droplets have been modeled per breakup event in the present flow simulation in which 10 000 parent droplets are injected per Lagrangian step. In analogy to the stochastical modeling of fuel atomization and gas flow turbulence effects, the superposition of large numbers of trajectories leads to continuous and thus realistic representation of secondary atomization.

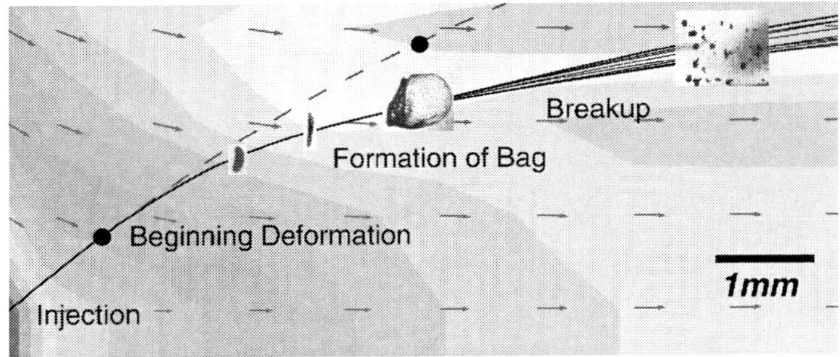

Figure 1.2.7: Breakup of a 60 µm droplet.

Modeling a single shear breakup event, the time of droplet disintegration is determined as a random number with uniform distribution between t_{def} and t_{b}. To model bag and multimode breakup events, the time of disintegration is sampled in the second half of this interval. The initial size of the child droplets is determined as a random number with a root normal distribution or from stability criteria (core droplet in shear breakup). Droplets generated by rim fragmentation are provided with an additional transverse velocity component. In bag breakup mode, virtually all larger fragments possess a transverse velocity component whereas in multimode breakup a volume fraction of up to 20 % of the largest fragments is starting without dispersive transverse momentum. The limiting value of 20 % represents the transition to shear breakup and corresponds to the volume fraction of the core droplet. In this breakup regime all secondary droplets inherit the velocity of the parent droplet since no significant spreading motion is observed.

Fig. 1.2.7 illustrates the computation of a deforming and disintegrating droplet in the near field of the nozzle. Compared to the trajectory of a rigid spherical droplet (dashed line), a significant deflection of the deforming droplet due to increased aerodynamic forces is observed. This single droplet computation clearly demonstrates that advanced modeling of secondary breakup has to take into account the time scales of the breakup process since the droplets experience considerable displacements before their final disintegration.

1.2.2.4 Iterative Solution Procedure

Tracking of a statistically significant number of droplet parcels and superposition of their trajectories yields a flow field approximation of the dispersed liquid phase. However, a simple Lagrangian two step calculation consisting of a gas flow computation and subsequent droplet tracking does not take account of spray effects on the gas flow such as aerodynamic dragging or evaporation cooling. In particular for evaporating sprays in combustor flows, these effects have a significant influence on the overall two phase flow field. To establish mutual information exchange between both phases, Lagrangian two step cycles are concatenated in an iterative procedure with droplet source terms being updated during each tracking step. Representing local transfer rates of mass, momentum and energy from spray to gas flow, droplet source terms are included in the gas flow computation of the following iteration cycle. This iterative approach is illustrated in the lower part of Fig. 1.2.8.

The separated computation of gas and liquid phase flow fields and the iterative exchange of interfacial transfer data entails an artificial decoupling of both phases. In particular for two phase flows with intense phase interaction, this effect leads to a critical overestimation of droplet source terms in the first iteration cycles. To achieve convergence of the iterative procedure, a relaxation of the droplet source term fields is employed. Recursive

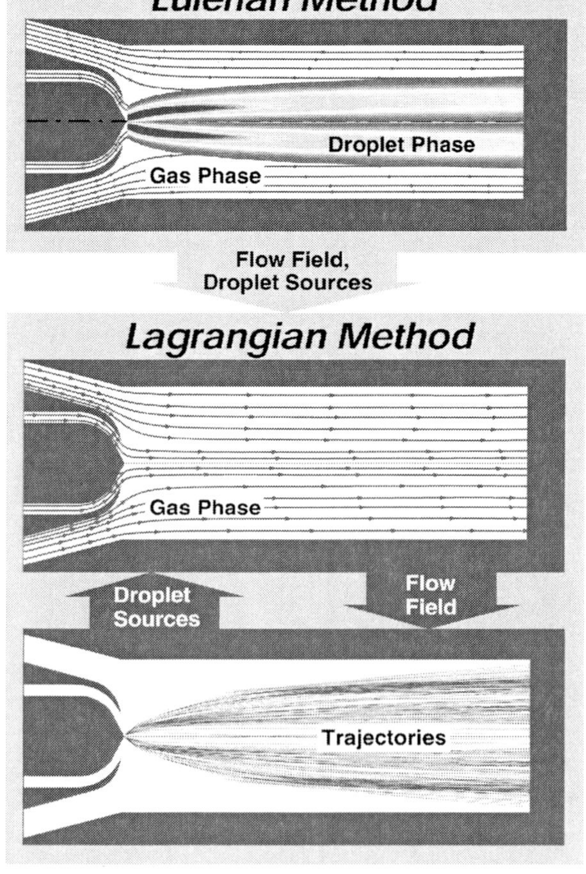

Figure 1.2.8: Structure of the hybrid procedure.

damping of the source terms on the level of the iteration cycles is realized by the following equation.

$$\overline{S}^{i+1}_{\mathrm{d},\phi} = \alpha_\phi S^{i}_{\mathrm{d},\phi} + (1 - \alpha_\phi)\overline{S}^{i}_{\mathrm{d},\phi} \qquad (32)$$

According to Eq. (32), only a fraction of the source terms recorded during the previous droplet tracking, $S^{i}_{\mathrm{d},\phi}$, is contributing to the source term $\overline{S}^{i+1}_{\mathrm{d},\phi}$ included in the current gas gas flow solution. The second contribution is calculated from the source term included in the previous gas flow computation. For gas flows which are substantially influenced by the fuel spray, strong relaxation of droplet source terms may be necessary to enforce convergence of the iterative procedure. In these cases an increased number of two stage iterations may be necessary for complete consideration of phase interaction [20]. In the present flow simulation, only weak relaxation ($\alpha_\phi > 0.5$) is required to achieve a convergent solution for the two phase flow field within 10 two stage iterations.

1.2.3 The Hybrid Procedure

As indicated in the preceding sections, the continuum description of the Eulerian method has the advantage of close coupling of gas and liquid phase in a single numerical scheme. Thus, computation times are rather short as long as the number of droplet classes used for the discretization of the spray is small. Consequently, Eulerian methods are particularly suited for an approximate but efficient computation of polydisperse two phase flows in combustors. Lagrangian methods in turn achieve a high resolution discretization of complex spray structures by tracking large numbers of droplet parcels of various initial sizes and velocities. However, the price to be payed for a realistic spray representation is high. Due to the artificial decoupling of the two phase flow computation by separate solution schemes for each phase and iterative realization of phase interaction, total computation times are rather large [20]. For flow cases where strong relaxation of droplet source terms is required, time expenses can grow to practically unmanagable extents.

For such two phase flows, a reduction of computational effort is achieved by preconditioning the two phase flow field by means of an Eulerian method based on a coarse discretization of the spray. This is in fact the basic idea of the Hybrid procedure: A two stage combination of both methods in order to reduce total computation times by Eulerian preconditioning yet maintaining the detailed modeling of spray physics in a Lagrangian refinement stage. This approach is schematically illustrated in Fig. 1.2.8. Following an approximate computation of the two phase flow, the flow field and the droplet source terms are passed to the refinement stage. Here, Lagrangian iteration cycles are based on a fine discretization of droplet injection conditions and an advanced modeling of secondary droplet breakup. Since the two phase flow field calculated by the Eulerian method already accounts for spray effects on the gas flow, the droplet source terms recorded during subsequent tracking steps are rather close to the final flow result. Consequently, the number of iterations is significantly reduced compared to a standard Lagrangian simulation.

1.2.4 Simulation of a LPP Premix Duct Flow

The performance and accuracy of the numerical methods presented is demonstrated by a simulation of the two phase flow in the premix duct of a LPP combustor. The experimental investigation of this combustor has been part of an extended research project on low emission combustion concepts [12, 13, 10]. A detailed description of the test rig and the measurement techniques is presented in a parallel study [16]. The combustor section of interest for the present flow simulation is illustrated in Fig. 1.2.9. Compressed air is supplied to the cylindrical duct (l = 124 mm, d_i = 44.6 mm) by two coaxial annular ducts. The fuel is injected into the gas flow by a pressure swirl atomizer aligned with the duct axis. The nozzle diameter is about 1 mm. In order to perform PDPA measurements, optical access to the flow is given by circumferencial slits in the duct liner at various axial positions. For spray visualizations, the metallic liner is substituted by a quartz glass cylinder. Premix and reaction zones are separated by an arrangement of swirler vanes acting as a flame stabilizer. The complete configuration illustrated in Fig. 1.2.9 is enclosed in a pressure casing with water cooled window ports. The outer

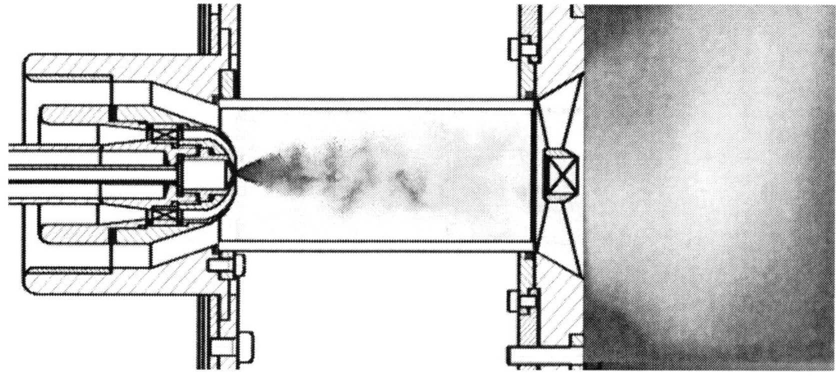

Figure 1.2.9: Premix zone of the LPP research combustor.

Table 1.2.1: Parameters at the inlet of the premix duct ($z = 0$ mm).

Gas flow (air)		Fuel (diesel)	
$\dot{m}_g$	213 g/s	$\dot{m}_{fuel}$	6 g/s
T_g	753 K	T_{fuel}	350 K
p_g	4 bar	v_{sheet}	30 m/s
Tu_g	15 %	β_{sheet}	40°

coaxial annular air flow (bypass air) is shielding the duct liner from droplet impact and film formation. The highly accelerated inner flow (atomization air) is focused directly onto the conical fuel sheet generated by the pressure swirl atomizer. The fundamental idea of this atomization concept is a further reduction of droplet sizes by high velocity aerodynamic interaction between spray and gas flow.

The operating point of the premix duct investigated in this study is specified by the flow parameters summarized in Tab. 1.2.1. The computational domain of the flow simulation is illustrated in Fig. 1.2.10 including (from left to right) intake section, coaxial annular ducts, premix zone, swirler vanes, reaction zone, dilution holes and burnout zone. The axial coordinate is measured from the atomizer nozzle. To model the evaporation behavior of the diesel spray, tetradecane is used as a single component diesel substitute in the present flow simulation. A detailed description of the calculated non-reacting single phase gas flow in the intake and premix zone is given in [16].

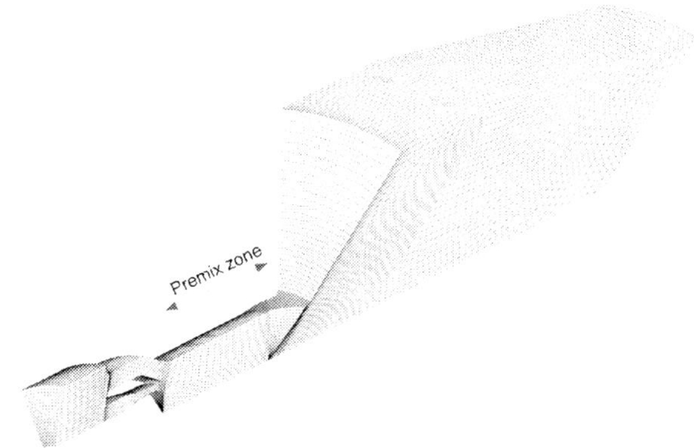

Figure 1.2.10: Computational domain (30°-segment).

1.2.4.1 Discretization of the Spray

In order to derive droplet phase boundary conditions for the Eulerian method and droplet initial conditions for the Lagrangian method, the spray is visualized in the near field of the atomizer ($0 < z < 10$ mm). A side view on the three dimensional spray cone is given by the flashlight shadowgraph in Fig. 1.2.11(a). The picture was taken under atmospheric, cold conditions without bypass air flow, using water as a fuel substitute. It is evident that the outer region of the spray is dominated by larger fuel fragments. To get an impression of the spray structure inside the cone, a laser light sheet photograph is shown in Fig. 1.2.11(b), which was taken under real operation conditions of the combustor. In contrast to the coarse structure of the outer spray region, this cross view reveals a very fine droplet distribution in the spray cone.

The spray visualizations indicate two basic processes that govern the atomization of the liquid fuel [12]. Due to its high swirl, the fuel is leaving the nozzle as a conical sheet. According to Fig. 1.2.11(a), this sheet is completely disintegrating along a distance of 1 to 2 mm. In this immediate near zone of the nozzle, prompt atomization is the governing process. This mechanism is controlled mainly by the internal sheet dynamics [11]. Further disintegration of sheet fragments is induced by aerodynamic interaction with the high velocity gas flow which penetrates the spray. The resulting small secondary fragments are dragged by the gas flow into the core flow as indicated in Fig. 1.2.11(b).

With respect to the numerical simulation, it is obvious that a representation of the complex two phase flow in the nozzle near field by non-interacting droplets is a rather crude approximation of the physical reality. But despite of this simplified modeling of fuel atomization, the simulation of spray dispersion and evaporation in the premix duct agrees well with the experimental data. In the following sections, two strategies are presented to derive droplet

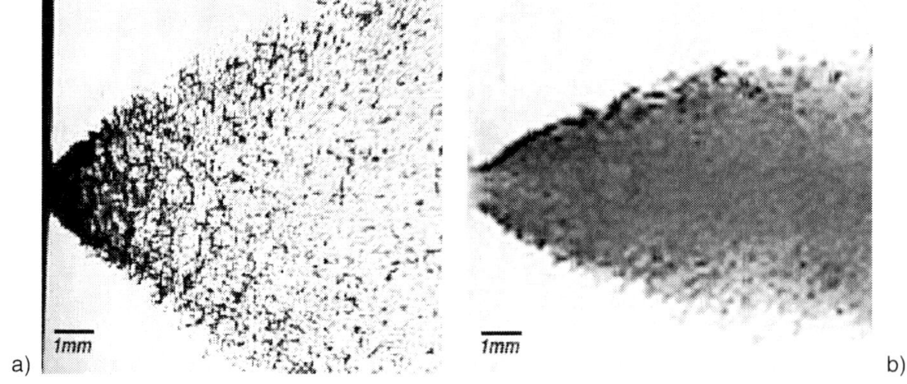

Figure 1.2.11: (a) Flashlight shadowgraphy, (b) laser light sheet.

injection conditions at the nozzle. The Eulerian calculation is essentially based on droplet data measured at a downstream position of the atomizer. It is important, that secondary atomization has ceased completely and droplets are spherical at this position. Basically, the procedure starts from an assumed discretization of the injection conditions of the droplet phase. In subsequent optimization iterations, the computed droplet data is fitted to the corresponding measured data. In contrast to this approach, the Lagrangian calculation is based on a rather crude approximation of the fuel sheet disintegration in the nozzle near field. Here, a far more physical description of the spray in the secondary atomization region is achieved by modeling droplet deformation and breakup. A similar strategy is described in [7] where pressure swirl fuel injection into a diesel engine combustion chamber is discussed.

1.2.4.1.1 Eulerian Method, Droplet Phase Boundary Conditions

Since secondary droplet breakup is observed in a spray region up to 50 mm downstream of the atomizer, measured droplet data at $z = 55$ mm is used for an optimization of droplet phase boundary conditions. The droplet size distribution of the spray is discretized by means of three diameter classes. The volume flux fraction (or normalized volume flux) of each class is described by a Rosin-Rammler distribution evaluated at the representative class diameter D_i.

$$\frac{\dot{V}_i}{\dot{V}} = 1 - \exp\left[-\left(\frac{D_i}{D_{0.632}}\right)^n\right], \; D_{0.632} = \frac{D_{0.5}}{0.693^{\frac{1}{n}}} \tag{33}$$

Values of $D_{0.5} = 46$ µm for the mass median diameter and $n = 4.8$ for the spreading parameter are determined as a good fit to the reference droplet data at $z = 55$ mm. The mean injection velocity of the droplet phase of 30 m/s is derived from the mean velocity of the liquid sheet. The remaining parameters with significant influence on the spray structure are the injection angles of the droplet phases. Introducing three angle classes per diameter class leads to a final discretization of the spray by means of nine droplet phases with different boundary conditions at the nozzle. Table 1.2.2 summarizes the correlations between droplet phase diameter, normalized volume flux and injection angle employed for the Eulerian two phase flow simulation in the present study.

Table 1.2.2: Droplet phase boundary conditions.

D_i [µm]	$\dot{V}_i/\dot{V}$ [l]	β_i [°]
0–41.2	4/15	34
	1/30	7
	1/30	21
41.2–50.7	4/15	30
	1/30	10
	1/30	20
50.7–∞	4/15	27
	1/30	13.5
	1/30	20.5

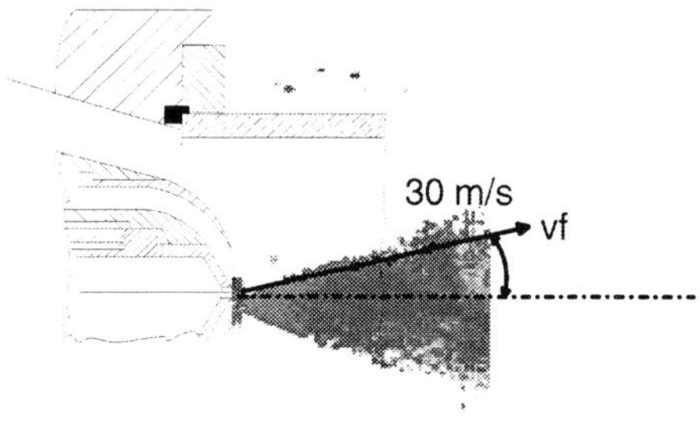

1.2.4.1.2 Lagrangian Method, Droplet Initial Conditions

In the Eulerian calculation, the boundary conditions of the droplet phase at the nozzle have to account implicitly for secondary atomization in an extended downstream flow region. It is evident that small secondary droplets originating from breakup of larger sheet fragments in outer flow regions may not be reproduced by a spray representation as described in the previous section. The approximation by rigid spherical particles of different size injected into the contracting, high velocity gas flow leads to a separation of droplet sizes. As a consequence, tiny and small droplets are captured in the axis region of the core flow unless they are injected with unphysically large radial velocity components.

In the Lagrangian calculation, secondary breakup of droplets is taken into account during trajectory integration. Thus, only prompt atomization of the conical sheet has to be considered for the formulation of droplet initial conditions. The basic idea is to inject most of the fuel in form of large droplets with sizes similar to the characteristic sheet thickness of about 100 to 200 µm. Due to this coarse primary spray structure and the high relative velocities in the nozzle near zone, the critical conditions of droplet breakup are significantly exceeded as indicated by the data points mapped in Fig. 1.2.3. Modeling of delayed droplet deformation, drag increase and breakup results in a fine secondary spray contribution to the core flow region in the premix duct. Very good agreement to the experimental droplet data is achieved by using the spray discretization summarized in Tab. 1.2.3. Droplet velocity and injection angle are sampled as random numbers with Gaussian distributions. From Fig. 1.2.11(b) it is obvious that the disintegration of the conical sheet is responsible for a fine primary contribution to the spray in the core flow. To model this effect approximately, the variance $\sigma_{\beta i}$ of the injection angle β is correlated with droplet size resulting in values of 30° for the smallest droplets (D from 0 to 20 µm) up to 5° for the largest droplets (D from 180 µm to 200 µm).

Table 1.2.3: Droplet initial conditions.

Droplet diameter:
- ten size classes equally spaced from 0 to 200 μm
- Rosin-Rammler distribution of $\dot{V}_i/\dot{V}$
- $D_{0.5} = 88.5$ μm, $n = 3$

Droplet velocities:
- Sampled, Gaussian distribution
- $\overline{v} = 30$ m/s, $\sigma_v = 5$ m/s

Injection angle:
- Sampled, Gaussian distribution
- $\overline{\beta} = 40°$, $\sigma_{\beta i} = 30°, \ldots, 5°$
- $\beta_{max} = 45°$ (clipping value)

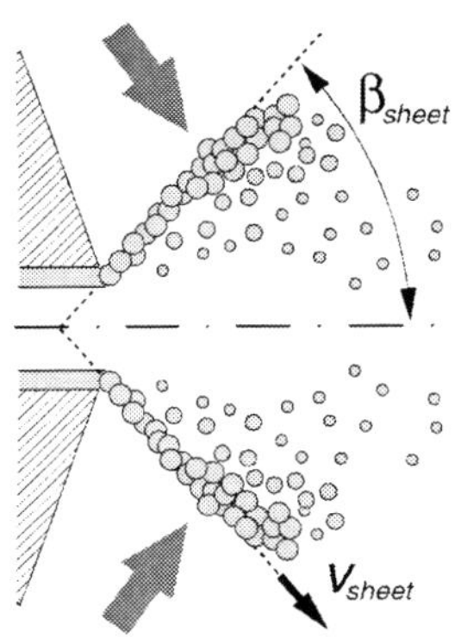

1.2.4.2 Results

To illustrate the calculated two phase flow in the premix duct, contour plots of the Eulerian flow simulation are discussed first. Comparing the axial gas velocities of the single phase and the two phase flow calculations from Fig. 1.2.12 indicates that spray-induced deceleration of the gas flow is limited to the core flow of the duct. In particular in the nozzle near zone, the axial gas velocity is decreased by up to 40 m/s due to the aerodynamic acceleration of the droplet phase.

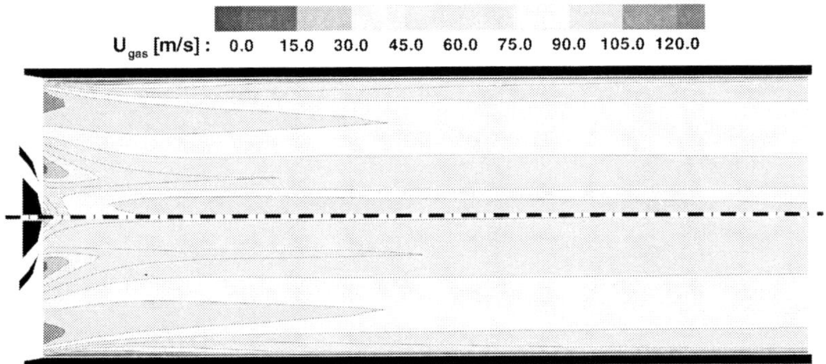

Figure 1.2.12: Axial gas velocity: Single phase calculation (top half) and two phase calculation (bottom half).

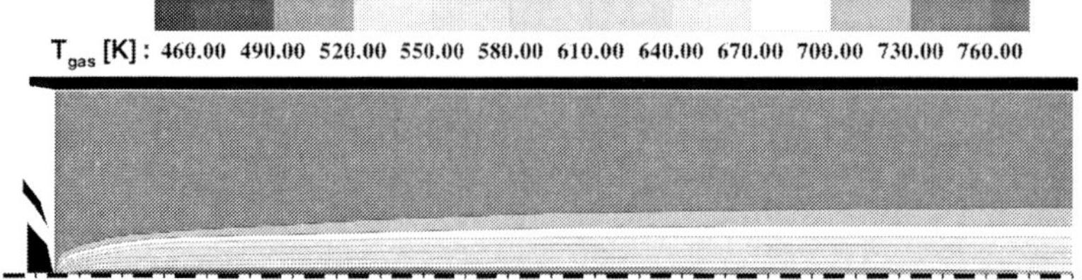

Figure 1.2.13: Calculated gas temperature.

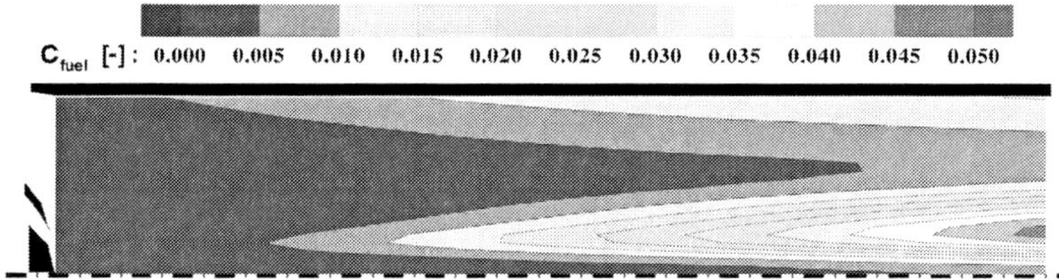

Figure 1.2.14: Calculated vapor concentration.

The influence of fuel evaporation is indicated by the gas flow temperature and fuel vapor concentration in Figs. 1.2.13 and 1.2.14. Although the gas phase experiences a substantial temperature drop across the whole core flow region, significant concentrations of fuel vapor are not calculated in the first half of the duct. This delay in vapor generation is a consequence of low evaporation rates in the transient heating phase of the droplets. This conclusion is confirmed by an analysis of Lagrangian single droplet computations, which indicate that heatup and propagation time scales of typical droplets are of the same order. In total, 28 % of the injected fuel is evaporated in the Eulerian simulation, in contrast to a value of 42 % in the Lagrangian simulation. The difference is caused by the secondary atomization modeling in the tracking algorithm, resulting in considerable numbers of small, rapidly evaporating droplet fragments. At this point it should be noted that the total fraction of evaporated fuel substantially depends on the D_{32}-correlation used for the secondary breakup modeling. Summarized over all calculated breakup events, Eq. (25), which is actually not used, leads to a fragment mass median diameter of $D_{0.5} = 53$ µm, whereas Eq. (26) results in a value of $D_{0.5} = 38$ µm.

To compare the calculated axial volume flux density $\alpha_{\mathrm{d}}\ U_{\mathrm{d}}$ with PDPA measurements, it is weighted by the annular area and normalized by the total axial volume flux

$$\frac{\dot{V}_{\mathrm{r}}}{\dot{V}} = \frac{2\pi \int\limits_{r-0.5\ \mathrm{mm}}^{r+0.5\ \mathrm{mm}} \alpha_d\, U_d\, r\, \mathrm{d}r}{2\pi \int\limits_{0}^{24\ \mathrm{mm}} \alpha_d\, U_d\, r\, \mathrm{d}r} \tag{34}$$

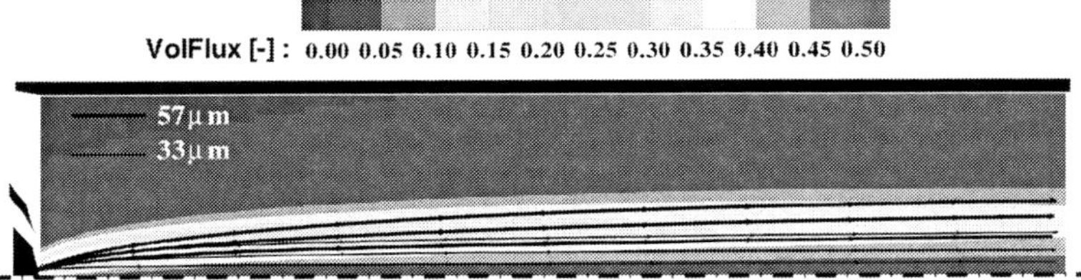

Figure 1.2.15: Normalized volume flux and mean trajectories.

This normalized volume flux is illustrated in Fig. 1.2.15 and represents the fraction of the total liquid volume flux which passes an annular fraction of the duct cross section. The contour plot includes mean trajectories determined by integration of the mean axial velocity of the droplet phase. The trajectories are evaluated for the two limiting size classes and clearly demonstrate the influence of initial droplet momentum on the propagation of the droplet phase.

The second point of discussion is concerning the comparison of experimental droplet data with Eulerian and Lagrangian two phase flow simulations. In this context, the Hybrid procedure is used as a computational tool to accelerate the Lagrangian calculation. The overall reduction of computation time achieved is about 30 % of the time required for a standard Lagrangian flow simulation without Eulerian preconditioning. Radial profiles of number averaged two phase flow variables are presented at axial positions $z = 20$, 55 and 90 mm. With respect to the mean axial velocities of the droplets shown in Figs. 1.2.16 and 1.2.17, both numerical methods predict a maximum in the core flow ($r < 6$ mm) which is not observed in experiment. In particular at $z = 55$ and 90 mm the axial droplet velocities in the core

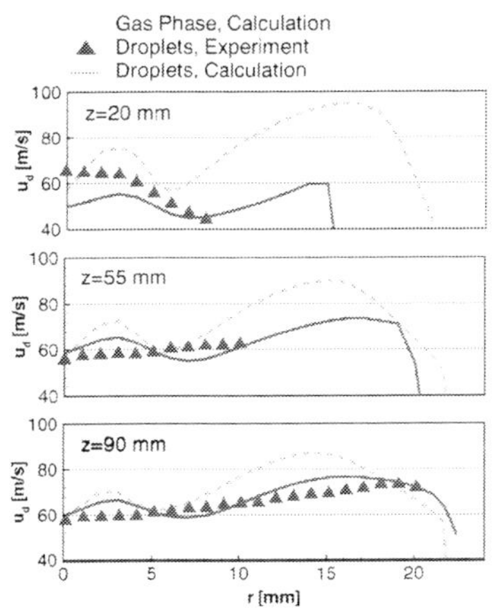

Figure 1.2.16: Mean axial velocities (Euler).

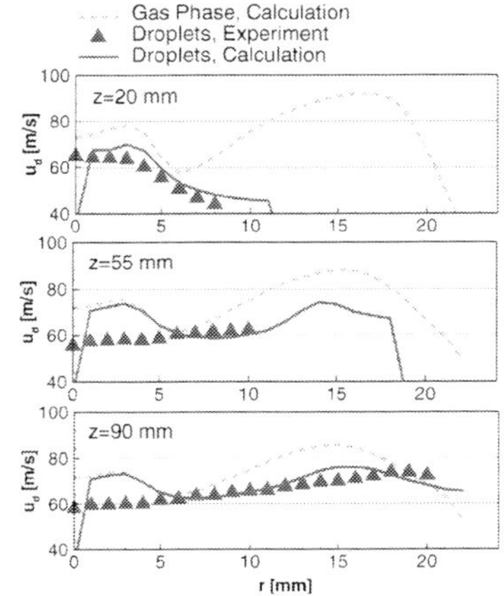

Figure 1.2.17: Mean axial velocities (Lagrange).

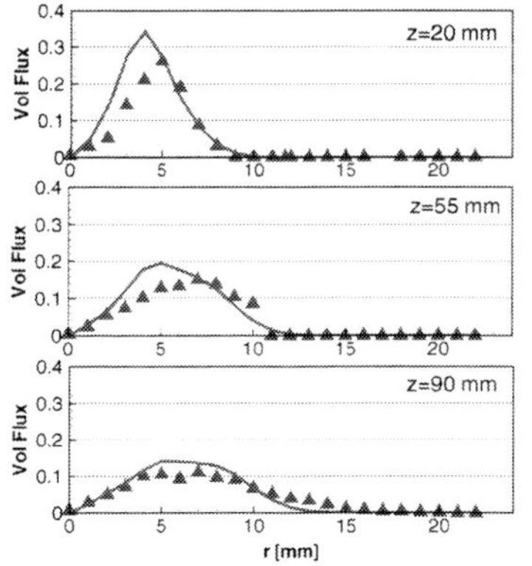

Figure 1.2.18: Normalized volume flux (Euler).

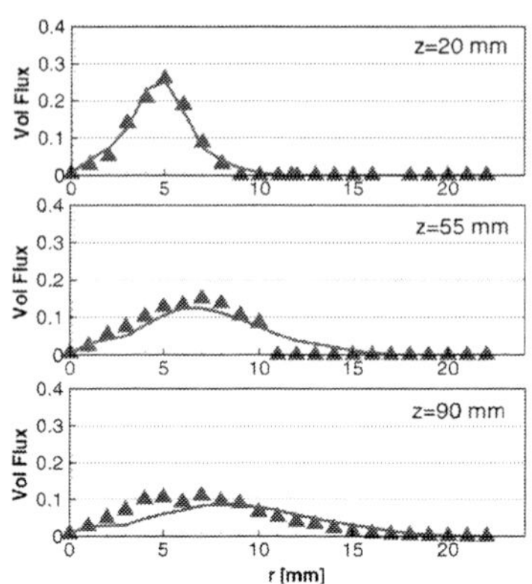

Figure 1.2.19: Normalized volume flux (Lagrange).

flow region are overestimated by the Lagrangian calculation. Basically, this effect is a result of an insufficient spray-induced flow deceleration. This conclusion is supported by the underestimated liquid volume flux in the core flow region at the corresponding axial positions as shown in Fig. 1.2.19. With respect to the Eulerian result shown in Fig. 1.2.18, it is evident that the unavailability of secondary breakup models requires small injection angles of the droplet phase to meet the radial volume flux profile in the second half of the duct flow. Consequently, substantial deviations are observed in the upstream flow region where secondary atomization occurs. As illustrated by Figs. 1.2.20 and 1.2.21, Eulerian and Lagrangian flow simulations predict rather similar radial profiles of the Sauter mean diameter. Although the calculated values are deviating from experimental data by up to 20 μm, both flow simulations reproduce the trend in the evolution of the droplet size spectrum.

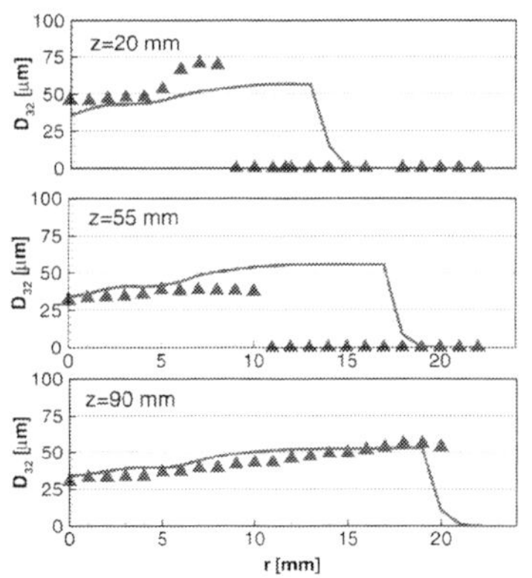

Figure 1.2.20: Sauter mean diameter (Euler).

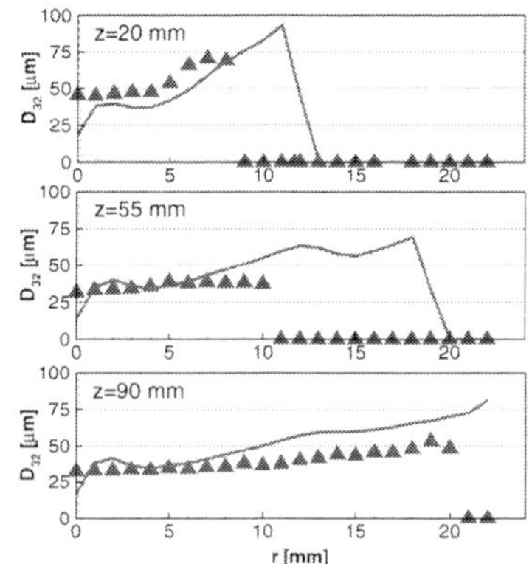

Figure 1.2.21: Sauter mean diameter (Lagrange).

1.2.5 Conclusions

Evaporating fuel sprays in combustor flows are characterized by high rates of mass, momentum and enthalpy transfer between spray and gas flow. Fuel atomization, spray dispersion and evaporation are often complicated by additional physical effects such as droplet-wall interaction, shear driven evaporating wall films or secondary breakup of droplets.

The primary objective in this study is the design of a computational tool for an efficient numerical simulation of combustor two phase flows including advanced modeling of secondary atomization physics. Compared to state of the art Lagrangian spray simulations, a significant reduction of computation time is achieved by the presented Hybrid procedure. Basically, this computational strategy is a two stage combination of an Eulerian and a Lagrangian method. The Eulerian method is used as an efficient preconditioner for the interacting two phase flow field, in order to reduce the number of subsequent Lagrangian gas flow computation – droplet tracking iterations. In this refinement stage, advanced modeling of secondary breakup of droplets including bag, multimode and shear mechanisms is used to improve the physical description of the spray.

To assess the accuracy of the two fundamental numerical approaches, Eulerian and Lagrangian flow simulations are compared with droplet data measured in the two phase flow of a LPP combustor premix duct. The atomization concept employed in this premix duct achieves a substantial improvement of fuel atomization by aerodynamic breakup of droplets in an extended flow region downstream of the nozzle. With respect to the Eulerian flow simulation, extrapolation of measured droplet data to the point of injection with implicit consideration of secondary atomization effects results in a rather crude approximation of the spray structure. Since secondary breakup of droplets is modeled in detail by the tracking algorithm, the numerical description of the spray structure in the Lagrangian flow simulation is significantly improved.

It is evident that the computational acceleration achieved by the Hybrid procedure significantly depends on the structure of the two phase flow. With respect to the rather uncritical premix duct flow presented in this study, it is certainly questionable whether the moderate time savings are worth the additional complexity of the CFD program. However, technical practice offers a sufficient number of critical two phase flow applications, where strong relaxation of droplet source terms requires an excessive number of Lagrangian coupling iterations [18]. In these flow cases, the Hybrid procedure has an increased potential for substantial speedup of flow simulations.

Acknowledgements

This work has been founded by the Department of Education, Science, Research and Technology of Germany under contract No. 50-807642. The support is gratefully acknowledged.

The authors would like to thank Mr. Michael Willmann, BMW Rolls-Royce, Germany, for his support concerning the development of droplet breakup models.

References

1. B. Abramzon and W.A. Sirignano (1989): Droplet Vaporisation Models for Spray Combustion Calculations. *Int. J. Heat and Mass Transfer*, **32**, 1605–1618.
2. S.K. Aggarwal and F. Peng (1995): A Review of Droplet Dynamics and Vaporization Modeling for Engineering Calculations. *ASME-Journal of Engineering for Gas Turbines and Power*, **117**, 453–461.
3. C.T. Crowe (1982): Review – Numerical Models for Dilute Gas-Particle Flows. *J. of Fluids Engineering*, **104**, 297–303.
4. G.M. Faeth (1983): Evaporation and Combustion of Sprays. *Progress in energy and combustion science, Pergamon Press*, **9**, 1–76.
5. A.D. Gosman, E. Ioannides (1983): Aspects of Computer Simulation of Liquid-Fueled Combustors. *J. of Energy*, **7(6)**, 482–490.
6. M. Hallmann, M. Scheurlen, and S. Wittig (1995): Computation of Turbulent Evaporating Sprays: Eulerian Versus Lagrangian Approach. *Transactions of the ASME*, **117**, 112–119.
7. Z. Han, S. Parrish, P.V. Farrell, and R.D. Reitz (1997): Modeling Atomization Processes of Pressure-Swirl Hollow-Cone Fuel Sprays. *Atomization and Sprays*, **7**, 663–684.
8. L.-P. Hsiang and G.M. Faeth (1992): Near-Limit Drop Deformation and Secondary Breakup. *Int. J. Multiphase Flow*, **18(5)**, 635–652.
9. S.A. Krzecskowski (1980): Measurements of Liquid Droplet Disintegration Mechanisms. *Int. J. Multiphase Flow*, **6** 227–239.
10. W. Layher, G. Maier, and S. Wittig (1998): Schadstoffemissionen und Betriebsverhalten einer LPP-Brennkammer. In: *DGLR Jahrestagung in Bremen, Germany.*
11. A.H. Lefebvre (1998): Atomization and Sprays. Hemisphere Publications, New York.
12. G. Maier, M. Willmann, and S. Wittig (1997): Development and Optimization of Advanced Atomizers for Application in Premix Ducts. In *97-GT-56*. ASME.
13. G. Maier and S. Wittig (1998): Effects of Liquid Properties on the Operating Performance of Air-Assisted Pressure Swirl Atomizers. In: *ILASS Europe*, 193–199.
14. D. Milojević (1990): Lagrangian Stochastic-Deterministic (LSD) Predictions of Particle Dispersion in Turbulence. *Particle and Particle Systems Characterization*, **7**, 181–190.

15. M. Pilch and C. A. Erdman (1987): Use of Breakup Time Data and Velocity History Data to Predict the Maximum Size of Stable Fragments for Acceleration-Induced Breakup of a Liquid Drop. *Int. J. Multiphase Flow,* **13(6)**, 741–757.
16. K. Prommersberger, G. Maier, and S. Wittig (1998): Validation and Application of a Droplet Evaporation Model for Real Aviation Fuel. *Presented at the 92nd Symp. on Gas Turbine Engine Combustion, Emissions and Alternative Fuels, Lisbon, Portugal.*
17. W. Samenfink (1995): Sekundärzerfall von Tropfen. In: *Atomization and Sprays, Short Course 1995.* Institut für Thermische Strömungsmaschinen, Universität Karlsruhe.
18. R. Schmehl (1998): Theory and Application of Single Component Evaporation Models. Technical report, Institut für Thermische Strömungsmaschinen, Universität Karlsruhe, http://itsnova.mach.uni-karlsruhe.de/~schmehl.
19. R. Schmehl, G. Klose, G. Maier, and S. Wittig (1998): Efficient numerical calculation of evaporating sprays in combustion chamber flows. In: *Presented at the 92nd Symp. on Gas Turbine Engine Combustion, Emissions and Alternative Fuels, Lisbon, Portugal.*
20. R. Schmehl, H. Rosskamp, M. Willmann, and S. Wittig (1998): CFD Analysis of Spray Propagation and Evaporation Including Wall Film Formation and Spray/Film Interactions. In: *ILASS Europe*, 546–555.
21. H. Simmons (1977): The Correlation of Drop-Size Distributions in Fuel Nozzle Sprays; Part I: The Drop-Size/Volume-Fraction Distribution. *ASME-Journal of Engineering for Power*, **99**, 309–314.
22. W. A. Sirignano (1984): Fuel Droplet Vaporization and Spray Combustion Theory. *Progress in energy and combustion science, Pergamon Press*, **9**, 291–322.
23. S. Wittig, M. Hallmann, M. Scheurlen, and S. Schmehl (1993): A New Eulerian Model for Turbulent Evaporating Sprays in Recirculating Flows. *AGARD-CP-536.*

1.3 Investigations of Droplet Evaporation at Elevated Pressures

Klaus Prommersberger, Jörg Stengele, Klaus Dullenkopf*, Johann Himmelsbach, and Sigmar Wittig

Abstract

A new experimental setup is introduced where the evaporation of free falling droplets is investigated. Monodisperse droplets are generated in the upper part of the test rig and fall due to gravity through the stagnant high pressure gas inside the pressure chamber. Due to the relative velocity between droplet and gas, convective effects have to be considered in this study which are taken into account by experimental correlations. The droplet diameter and the droplet velocity are measured simultaneously by means of video technique and a stroboscope lamp. Detailed measurements with various model fuels are presented, concerning as well single component fuels as bicomponent mixtures. The measurements were carried out for different pressures ($p = 20, 30, 40$ bar), gas temperatures ($T = 550, 650$ K) and droplet initial diameters ($d_0 = 680, 780$ µm) which results in Reynolds numbers comparable to two phase flow conditions in gas turbine combustors (Re ~ 100–1000).

The experimental results are compared with numerical calculations based on the most common droplet evaporation models. For this purpose a numerical code has been developed, which is capable to calculate droplet evaporation under supercritical gas phase conditions. The best correspondence with the experiments was found for the Conduction (Diffusion) Limit Model (Law and Sirignano [1]), which assumes a diffusive heat and mass transport within the droplet. The assumption of an infinite rapid mixing process inside the droplet was found unrealistic (Uniform Temperature Model, Faeth [3]). The convective transport of heat and mass at the droplet surface was calculated according to the film theory of Abramzon and Sirignano [4] in combination with the correlations of Ranz and Marshall [5].

* Institut für Thermische Strömungsmaschinen, Universität Karlsruhe, Kaiserstr. 12, 76128 Karlsruhe, Germany

Nomenclature

B_m mass transfer number
B_T heat transfer number
c_p specific heat at constant pressure
c_w drag coefficient
d_d droplet diameter
D_g binary gaseous diffusivity
$D_{A,B}$ binary liquid diffusivity
F correction factor for relative change of film thickness
g gravity constant
L enthalpy of evaporation
Le Lewis number
$\dot{m}$ mass flow
n number of fuel components
Nu Nusselt number
p pressure
Pr Prandtl number
$\dot{Q}$ heat flux
r radial coordinate
r_d droplet radius
Re Reynolds number
t time
Sc Schmidt number
Sh Sherwood number
T temperature
u velocity
x distance
Y mass fraction

Greek Symbols

ε evaporation rate
λ thermal conductivity
ρ density
ω dimensionless radial coordinate

Subscripts

crit critical
d droplet
f fuel vapor
g gas phase
i fuel component
ref reference
rel relative
s droplet surface
vap vaporization
∞ infinity
0 initial state
0 without evaporating mass flow
* evaporating mass flow included

1.3.1 Introduction

The motion of droplets and the simultaneous heat and mass transfer inside a fuel spray has been found to influence the fuel preparation process in gas turbine combustors intensively [6–8]. The objective of the air fuel mixing process is to generate a homogenous mixture of fuel vapor and air which provides a stable combustion with low pollutant emissions for the complete operating range of the engine. An important condition for the successful design of a low emission combustor is therefore to understand the processes of droplet dispersion, evaporation and combustion of the air fuel mixture. For the optimization of modern combustors the evaporation characteristic of fuel has to be known for all load levels of the engine. The evaporation characteristics of different fuel types vary in large extent, which has to be considered for the design of the combustion chamber.

Since the pressure level inside the combustion chambers of modern gas turbine engines has continuously increased during the past years reaching or even exceeding the critical pressure of the fuel used, the present study is intended to describe the evaporation of droplets in a high pressure environment by numerical and experimental methods.

Typical fuels consist of many different hydrocarbons and it is shown e. g. in [5, 6] or [2], that the evaporation process of the fuel mixture cannot be described correctly by one-component fuels. This is due to the great variety of the thermophysical and chemical properties and the different volatility of the fuel components. Especially during cold starting conditions of the engine, when the temperature of the ambient gas is low, the highly volatile components of the fuel play an important role for ignition and burning processes of the engine.

Numerous theoretical studies have been carried out describing droplet evaporation in a high pressure gas [9–13]. They take into account the real gas behaviour, the variation of thermophysical properties, the non-ideality of the latent heat of evaporation and the non-ideal phase equilibrium including the solubility of the ambient gas inside the droplet. These points are found to be extremely important for the correct prediction of droplet evaporation in high pressure environments. Since the droplet temperature rises with increasing pressures, the unsteady heating of the droplet significantly affects the droplet evaporation process under high pressure conditions and steady state evaporation cannot be obtained. If the pressure and the temperature of the gaseous environment are sufficiently high, the droplet may reach the critical point during its evaporation time [14]. These conditions are reached e. g. in liquid oxygen combustion systems. Delplanque and Sirignano [15] apply new evaporation models for this situation, since the surface tension and the heat of evaporation become zero. However, Hsieh et al. [10] state in a theoretical study that for pentane droplets subcritical models can be used up to 65 bar.

Prior to this study only few experimental studies have been published describing the droplet evaporation process in high pressure environments [16–18]. In order to exclude gravity and convective effects on the evaporation process some experiments on high pressure droplet vaporization were carried out in microgravity environments [19, 20]. These experiments have been conducted with droplets attached to a thin fiber causing an impairment of the droplet evaporation process due to the suspension unit. Shih and Megaridis [21] have shown that these kinds of experiments will lead to an overestimation of the liquid evaporation rates and to an underprediction of the droplet lifetimes.

The present study introduces a new experimental setup to investigate the evaporation of free falling droplets in a stagnant high pressure gas. There is no need for suspension of the droplet and the evaporation process can be observed without any disturbing influence. Since the distance between the evaporating droplets is large enough, there is no interaction between the droplets.

The present study is intended to describe the evaporation of one- and two-component droplets in a high pressure environment by experimental methods and numerical simulation. Two-component droplets are chosen as a simplification to describe the influence of the liquid mixture and the different volatilities of the components on the evaporation process.

Based on the experimental results theoretical droplet evaporation models can be established and validated. Two different droplet models have been investigated. The first model assumes a diffusive heat and mass transport within the droplet (Conduction and Diffusion Limit Model). The second model assumes an infinite rapid mixing process inside the droplet (Uniform Temperature and Rapid Mixing Model).

1.3.2 Droplet Evaporation Models

The present model describes the evaporation process of an isolated evaporating droplet in a convective flow field. The flow around the droplet is assumed to be laminar. Variable thermophysical properties are used inside the droplet and the surrounding gas to account for the variation of temperature and concentration [22]. The correlations necessary to determine the physical properties are taken from Stengele et al. [23–25]. The following simplifications are assumed in the model: The droplet evaporates in an inert environment. The interface between liquid and gas phase is assumed to be in thermodynamic equilibrium. Radiative heat transfer is neglected [26]. Fick's law is used to calculate mass diffusion. Dufour and Soret effects are not considered. The evaporation process is assumed to be spherically symmetric, the asymmetric convective effects on the evaporation process are included implicitly in the correlations of Ranz and Marshall [5].

1.3.2.1 Gas Phase Equations

In the present model the gas phase is assumed to be quasi-steady. Delplanque [15] proved that this assumption is valid for typical high pressure spray application, since the characteristic time scale of the flow around the droplet is much shorter than the droplet lifetime. Applying the integral solution of the governing equations (Hubbard et al. [27]) and the extended film theory of Abramzon and Sirignano [4] the evaporating mass flow of the droplet and the heat transfer from the hot environment towards the droplet are calculated as described below.

The convective effects on the heat and mass transfer of the evaporating droplet are determined by Eqs. (1) and (2).

$$\mathrm{Nu}^* = 2 + \frac{\mathrm{Nu}_0 - 2}{F(B_\mathrm{T})} \tag{1}$$

$$\mathrm{Sh}^* = 2 + \frac{\mathrm{Sh}_0 - 2}{F(B_\mathrm{m})} \tag{2}$$

The radially outward directed flow from the evaporating droplet is described by correction factors for heat transfer $F(B_\mathrm{T})$ and mass transfer $F(B_\mathrm{m})$.

$$F(B_\mathrm{m}) = (1 + B_\mathrm{m})^{0.7} \frac{\ln(1 + B_\mathrm{m})}{B_\mathrm{m}} \tag{3a}$$

$$F(B_\mathrm{T}) = (1 + B_\mathrm{T})^{0.7} \frac{\ln(1 + B_\mathrm{T})}{B_\mathrm{T}} \tag{3b}$$

To calculate Nu_0 and Sh_0 the correlations of Ranz and Marshall [5] are used in Eqs. (1) and (2).

$$Nu_0 = 2 + 0.6\sqrt{Re_d}Pr^{1/3} \tag{4}$$

$$Sh_0 = 2 + 0.6\sqrt{Re_d}Sc^{1/3} \tag{5}$$

The evaporating mass flow of the droplet is obtained by integrating the quasi-steady energy and mass equations.

$$\dot{m}_{vap} = 2\pi r_d \frac{\lambda_{g,ref}}{c_{p_{g,ref}}} Nu^* \ln(1 + B_T) \tag{6}$$

$$\dot{m}_{vap} = 2\pi r_d \rho_{g,ref} D_{g,ref} Sh^* \ln(1 + B_m) \tag{7}$$

where B_T and B_m are the Spalding heat and mass transfer coefficients.

$$B_T = \frac{\dot{m}_{vap} c_{p_{g,ref}} \left(T_{g,\infty} - T_{g,s}\right)}{\dot{Q}_{tot}} \tag{8}$$

The mass flow rate of the fuel vapor components and the mass transfer number is determined by Eqs. (9) and (10).

$$1 + B_m = \frac{\sum_{i=1}^{n} Y_{i,f,\infty} - 1}{\sum_{i=1}^{n} Y_{i,f,s} - 1} = \frac{Y_{i,f,\infty} - \varepsilon_i}{Y_{i,f,s} - \varepsilon_i} \tag{9}$$

$$\varepsilon_i = \frac{\dot{m}_{vap,i}}{\dot{m}_{vap}}, \qquad \sum_{i}^{n} \varepsilon_i = 1 \tag{10}$$

In order to get a more realistic modeling, the specific heat capacity of the pure vapor is replaced by the specific heat capacity of the vapor/gas mixture. The thermophysical properties necessary are obtained by using the "1/2-Rule" of Renksizbulut and Yuen [28]. The only exception is the gas density of the Reynolds number, which is calculated at free stream conditions [29]. Combining Eqs. (6) and (7) the relation (11) is obtained.

$$B_T = (1 + B_m)^{\frac{1}{Le_g}\frac{Sh^*}{Nu^*}} - 1 \tag{11}$$

The calculation procedure of the droplet evaporation process starts from Eq. (7) where the evaporating mass flow of the liquid fuel is calculated as a starting value for the iterative calculation of B_T in Eq. (11). Then, the total heat transfer from the surrounding gas towards the droplet can be calculated using Eq. (8).

1.3.2.2 Droplet Motion

The general equation of droplet motion can be obtained by balancing the forces. For reasons of simplicity, the motion of the droplet and the gas flow is assumed to be one-dimensional. The gravity forces and the buoyancy forces are also included, since they affect the evaporation process of the droplet in the experimental setup used in this study.

$$\frac{du_d}{dt} = -\frac{3\rho_g c_w}{4\rho_d d_d} u_{rel}\left(u_d - u_g\right) + \left(1 - \frac{\rho_g}{\rho_d}\right) g \tag{12}$$

The drag coefficient is calculated by the correlation of Renksizbulut and Haywood [30] where the radially outward flow of the evaporating fuel is included. As previously mentioned, the Reynolds number is calculated again using the free stream density [31].

$$c_w(1 + B_m)^{0.2} = 0.36 + 5.48 \cdot Re_d^{-0.573} + \frac{24}{Re_d} \tag{13}$$

After solving Eq. (12) the one-dimensional displacement of the evaporating droplet can be determined from simple kinematics.

$$\frac{dx_d}{dt} = u_d \tag{14}$$

1.3.2.3 Description of the Heat and Mass Transport Model of the Liquid Phase

1.3.2.3.1 Conduction Limit Model and Diffusion Limit Model

Both models of the droplet interior consider the diffusive transport inside the droplet. The Diffusion Limit Model is in general an extension of the Conduction Limit Model for bicomponent mixtures. Since convection within the droplet is neglected, the liquid phase equations reduce to the conservation of species and energy. For one-component droplets (Conduction Limit Model) only the conservation of energy has to be considered. The differential equations of the conservation of mass and species have to be solved numerically.

$$\rho_d c_{p,d} \frac{\partial T_d}{\partial t} = \frac{1}{r^2} \frac{\partial}{\partial r}\left(r_d^2 \lambda_d \frac{\partial T_d}{\partial r}\right) \tag{15}$$

$$\frac{\partial (Y_{i,d}\rho_d)}{\partial t} = \frac{1}{r^2} \frac{\partial}{\partial r}\left(\rho_d r^2 D_{A,B} \frac{\partial Y_{i,d}}{\partial r}\right) \tag{16}$$

The time depending droplet diameter can be derived from a mass balance around the droplet.

$$\frac{\mathrm{d}r_\mathrm{d}}{\mathrm{d}t} = -\frac{1}{\rho_{\mathrm{d,s}} r_\mathrm{d}^2}\left(\frac{\dot{m}_\mathrm{vap}}{4\pi} + \int\limits_0^{r_\mathrm{d}} \frac{\partial \rho_\mathrm{d}}{\partial t} r^2 \mathrm{d}r\right) \tag{17}$$

Since the droplet diameter changes during the evaporation process, the radial coordinate in the gas phase and in the liquid phase is non-dimensionalized by the time depending droplet diameter. With this transformation the droplet surface is always located at $\omega = 1$ resulting in simpler solution procedures of Eqs. (15) and (16).

$$\omega = \frac{r}{r_\mathrm{d}(t)} \tag{18}$$

In order to numerically determine the temperature and fuel distribution inside the droplet the following initial and boundary conditions are applied. The initial temperature and concentration within the droplet is assumed to be uniform. The boundary conditions result from symmetry at the droplet center and the conservation of energy and mass at the droplet surface. They can be expressed in terms of Neumann conditions.

$$\left.\frac{\partial T_\mathrm{d}}{\partial r}\right|_{r=0} = 0 \tag{19}$$

$$\left.\frac{\partial Y_\mathrm{i,d}}{\partial r}\right|_{0} = 0 \tag{20}$$

$$\rho_\mathrm{d} D_\mathrm{AB} \left.\frac{\partial Y_\mathrm{i,d}}{\partial r}\right|_{s} = \frac{1}{4\pi r_\mathrm{d}^2}\left(\dot{m}_\mathrm{vap} Y_\mathrm{i,d,s} - \dot{m}_\mathrm{vap,i}\right) \tag{21}$$

$$4\pi r_\mathrm{d}^2 \lambda_\mathrm{d} \left.\frac{\partial T_\mathrm{d}}{\partial r}\right|_{r_\mathrm{d}} = \dot{Q}_\mathrm{tot} - \dot{m}_\mathrm{vap} L \tag{22}$$

where $\dot{Q}_\mathrm{tot}$ is the total heat flux from Eq. (9) and $\dot{m}_\mathrm{vap} L$ represents the latent heat of evaporation of the fluid.

1.3.2.3.2 Uniform Temperature and Rapid Mixing Model

The Rapid Mixing Model and the Uniform Temperature Model assume a flat temperature profile in the droplet interior. The Uniform Temperature Model considers a single droplet component, the Rapid Mixing Model considers a mixture of two or more droplet components. The assumption of an infinite diffusivity inside the droplet results in a permanently equalized species concentration profile for the Rapid Mixing Model. With the simplification of an uniform although time varying temperature and concentration profile the droplet temperature can be calculated directly from a balance of the transferred energy:

$$m_d c_{p,d} \frac{\partial T_d}{\partial t} = \dot{Q}_{tot} - \dot{m}_{vap} L \tag{23}$$

The time depending droplet diameter can be calculated from:

$$\frac{dr_d}{dt} = -\frac{1}{\rho_d}\left(\frac{\dot{m}_{vap}}{4\pi r_d^2} + \frac{r_d}{3}\frac{\partial \rho_d}{\partial t}\right) \tag{24}$$

1.3.2.4 Phase Equilibrium at High Pressures

All droplet evaporation models assume that the gas/liquid interface is in thermodynamic equilibrium. In contrast to low pressure applications, where Raoult's law is valid, the determination of the phase equilibrium in high pressure environments has to consider real gas effects and the solubility of the ambient gas inside the droplet. Additionally, in high pressure applications the latent heat of evaporation is a function of pressure and temperature as well. These effects are described by Stengele et al. [23, 24] in detail. In the present study it is assumed that the gas dissolved within the droplet is confined in the outermost layer of the droplet. It turned out that the gas diffusing towards the droplet center does not affect the evaporation process of the droplet.

1.3.3 Experimental Setup

For the experimental verification of the droplet evaporation model a new test section was built where single droplets evaporate under well defined conditions [32]. The maximum pressure inside the test chamber was 40 bar, the maximum temperature 700 K. Nitrogen was chosen as the test gas in order to exclude any combustion processes.

Figure 1.3.1 shows the cross-sectional view of the cylindrical test section. The droplets are falling vertically downwards on the centerline of the tubular chamber and evaporate. The heating tube was operated by two electrical resistance heaters with 2200 W each. One element heats the upper part of the evaporation zone, the other element the lower part. In order to minimize

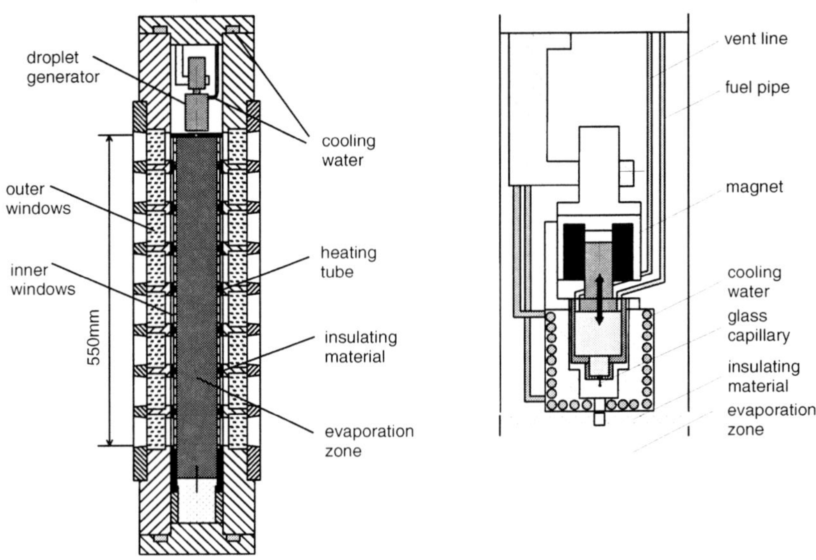

Figure 1.3.1: Pressure chamber (left) and droplet generator.

heat losses and temperature gradients within the evaporation zone, insulating material was installed between the heaters and the pressure chamber wall. To reduce heat losses and provide a homogeneous temperature field in the windows area additional inner windows were inserted in the heating tube. In order to avoid interactions of the droplets with the wall and to establish a sufficient evaporation time the total length of the evaporation zone was more than 500 mm and the diameter was 65 mm. The evaporation of the droplets could be observed through eight sets of quartz glass windows. The size history and the velocity of the droplets were measured simultaneously by a CCD-camera combined with a stroboscope lamp. The diameter of the droplet could easily be determined by the fixed enlargement factor of the optical setup. The biggest lens aperture was chosen to minimize errors by the depth of focus. The velocity of the droplet was calculated from the known flash frequency of the stroboscope lamp and the distance between the two shadow images of the same droplet in one picture.

Since the droplets enter the evaporation zone with identical conditions established by the droplet generator, the droplet chain could be studied at different axial positions representing different evaporation times.

The smallest droplet size detectable was approximately 300 μm. At this point more than 90 % of the initial droplet mass was already evaporated. Exact measurements with smaller droplets turned out to be difficult since minor inhomogenities in the gas phase affect the movement of the droplet and the evaporation distance significantly. The resolution of the measuring systems was ±20 µm for the droplet size and ±0.05 m/s for the droplet velocity. Consequently, the standard deviation of the measurements varied between 2 % and 8 %.

As presented in Fig. 1.3.1, the droplet generator was located in the upper part of the pressure chamber. The droplets move down vertically due to gravity through the high temperature stagnant nitrogen gas. On the right side of Fig. 1.3.1 a schematic cross-section of the droplet generator is shown. The working principle is based on a constant fuel mass flow through a thin glass capillary tube where a droplet is generated at the tip. The diameter of the droplet increases until the weight of the droplet exceeds the cohesion forces, the droplet leaves the capillary tube and falls into the evaporation zone. With this static technique only droplet diameters larger than 1 mm could be generated, since the diameter of the capillary could not be decreased.

In order to further reduce the initial droplet diameter a technique to force the separation of droplets from the capillary was applied. It was realized by a movable droplet generator lifted by a electromagnet and hitting a stroke device when released again. The droplet falls off the glass capillary tube and enters the evaporation zone. With this technique droplets in a diameter range between 600 µm and 900 µm could be generated. If the mass flow through the capillary tube was kept constant and the frequency of the up and down movement of the droplet generator was adjusted properly, monodisperse droplets were produced.

Another crucial parameter to obtain constant mass flows was the constant differential pressure between the fuel tank and the pressure chamber. Depending on the frequency and the mass flow, the distance between the droplets can be adjusted arbitrarily within certain limits. In the present study, the droplet distance was more than 100 times the initial droplet diameter. Therefore, any interaction between the evaporating droplets was excluded. Inside the cavity surrounding the glass capillary, a thermocouple sensor was installed close to the capillary tip to determine the initial temperature of the droplet. The knowledge about the initial temperature is important for the comparison of numerical and experimental results. Under high temperature conditions the droplet generator was cooled by water in order to keep the droplet initial temperature constant.

In Fig. 1.3.2 a schematic overview of the whole test section is shown including temperature and pressure measuring locations. In addition, valves, piping, heaters (H 1 and H 2) and control units are presented. In order to compensate pressure fluctuations a gas reservoir was connected to the fuel tank. Before the test section was filled with nitrogen gas it was evacuated to avoid any residual oxygen inside the evaporation zone.

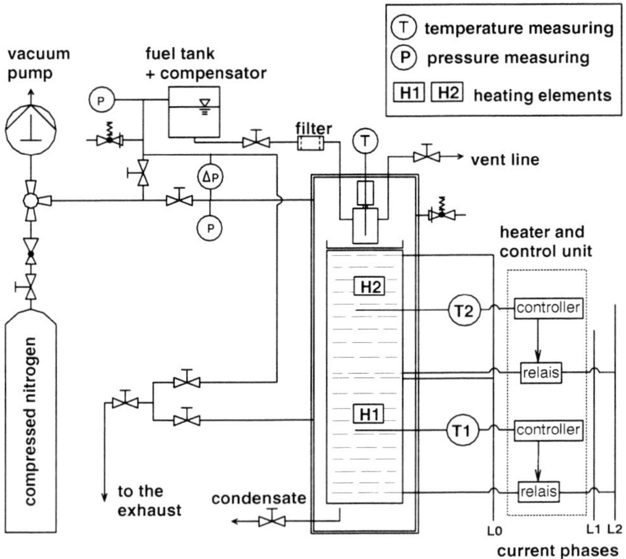

Figure 1.3.2: Test rig.

1.3.4 Results

As mentioned, the high pressure test chamber provides the possibility to conduct measurements of evaporating single- and multi-component droplets in order to validate the theoretical results of the droplet evaporation models applied in the theoretical part of this study, especially in a high pressure atmosphere. To our knowledge no experiments considering two- or multi-component droplet evaporation in high pressure atmosphere have been published before (compare Stengele et al. [25]).

To match the evaporation behavior of typical fuels *n*-alkanes were used, because their thermophysical properties are exactly known and provide a realistic basis for the numerical study. The experiments were conducted for three different pressure levels (p = 20 bar, 30 bar and 40 bar), two gas temperatures (T_∞ = 550 K and 650 K). For p = 40 bar and T_∞ = 650 K, the critical pressure and the critical temperature of the droplet components were exceeded ($T_{crit,pentane}$ = 470 K, $p_{crit,pentane}$ = 33.7 bar and $T_{crit,nonane}$ = 595 K, $p_{crit,nonane}$ = 23.1 bar). However, in all experiments no supercritical conditions of the droplets, i. e. a fading of the phase boundary could be observed. The calculated droplet temperatures were also well below the critical temperature of the fuel mixture. In the following figures, the diameter and the velocity of the droplet are plotted over the evaporation distance. The symbols indicate the measured values, and the solid lines represent the calculated values based on the Conduction Limit Model. The initial droplet velocity results from the droplet acceleration between the glass capillary and the entrance of the evaporation zone. It is for all conditions approximately the same, $u_{d,0}$ = 0.5 m/s. The initial droplet temperature rises slightly with higher pressures and higher gas temperatures. This was caused by intensified heat transfer from the evaporation zone towards the droplet generator and could not be compensated totally by the water cooling.

1.3.4.1 Single Component Droplets

The experiments were conducted with heptane droplets for three typical pressure levels (p = 20, 30, 40 bar), two gas temperatures (T_∞ = 550 K and 650 K) and two initial droplet diameters ($d_{d,0}$ = 680 μm and 780 μm). At p = 40 bar the critical pressure of heptane was exceeded by a factor of 1.5 and at T_∞ = 650 K the critical temperature was exceeded by a factor of 1.2.

1.3.4.1.1 Variation of Pressure

A comparison between experimental and theoretical results for different pressures and at a gas temperature of T_∞ = 550 K is shown in Fig. 1.3.3. Increasing the pressure the evaporation distance and the velocity of the droplet decreases. This results from the increased aerodynamic force at higher pressures due to higher gas density.

For all pressure levels similar velocity and diameter distributions were observed. Due to the large droplet diameters at the beginning of the evaporation process the droplet velocity increases strongly reaching a velocity maximum. Its magnitude depends on the gas pressure, the higher the gas pressure the lower the velocity maximum. Passing this point the droplet velocity decreases, since the aerodynamic drag exceeds the force of gravity due to smaller droplet diameters. Towards the end of the evaporation process velocity decreases rapidly. In this region the influence of gravity become negligible. For all pressure levels investigated, an excellent agreement between the measured and calculated droplet diameter and velocity distributions could be achieved.

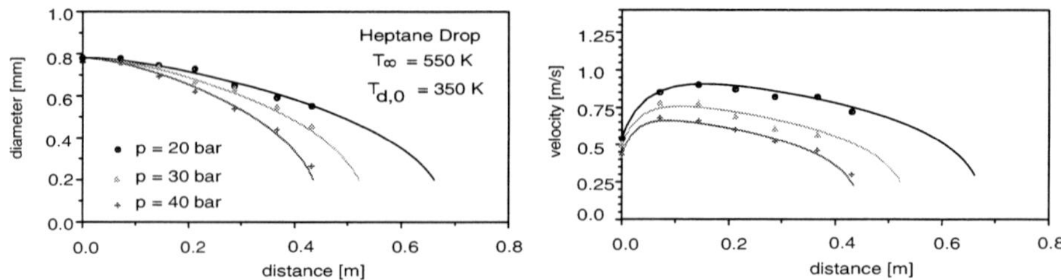

Figure 1.3.3: Evaporation of one-component droplets at different pressure levels.

1.3.4.1.2 Variation of Gas Temperature

Excellent agreement between calculated and measured results for different gas temperatures could also be demonstrated in Fig. 1.3.4. Elevating the gas temperature, the evaporation distance of the droplet shortens. During the first part of the evaporation process the velocity increases with higher gas temperatures resulting from smaller aerodynamic drag. However, the following reduction in droplet velocities is much steeper for higher temperatures due to the faster decrease of the droplet diameter.

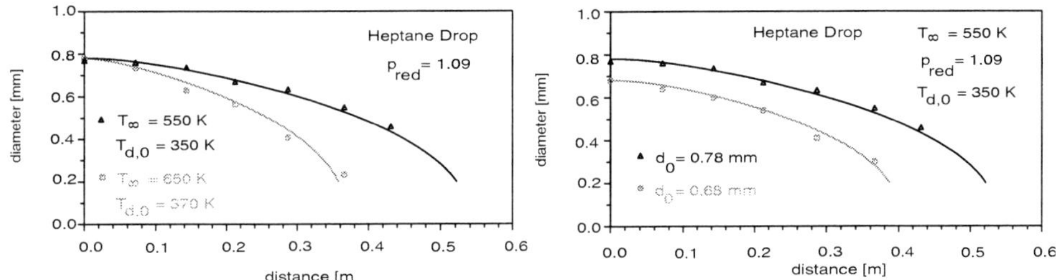

Figure 1.3.4: Evaporation of one-component droplets at different temperature levels.

1.3.4.1.3 Comparison of Uniform Temperature and Conduction Limit Model

The calculations presented in the previous figures had been done using the Conduction Limit Model. In Fig. 1.3.5 the calculated results of the Uniform Temperature and Conduction Limit Model are compared for p = 30 bar and T = 650 K. The initial droplet diameter was 840 μm and the initial droplet temperature was 370 K. The behavior of both models can be clearly seen in the diameter-distance plot in Fig. 1.3.5.

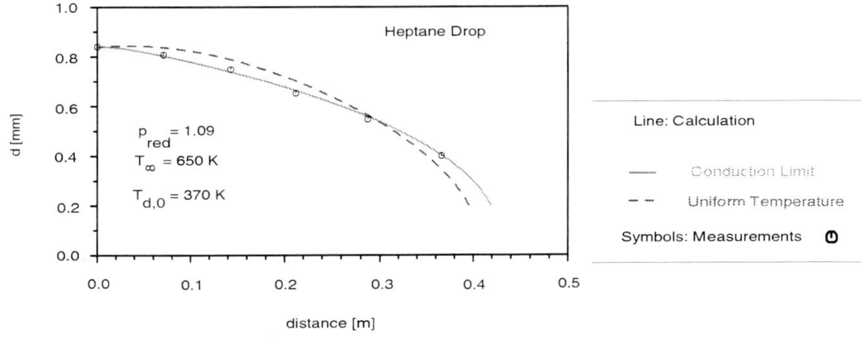

Figure 1.3.5: Comparison of Uniform Temperature and Conduction Limit Model.

It is also noticeable that the measurements are in close agreement with the results of the Conduction Limit Model. In contrast the Uniform Temperature Model predicts too large droplet diameters at the beginning of the evaporation process followed by an overprediction of the evaporation rate.

1.3.4.2 Two-Component Droplets

To gain a more realistic insight to the behavior of typical fuels droplets experiments using two-component droplets were carried out with binary mixtures of *n*-pentane and *n*-nonane. Two different initial droplet mixtures ($Y_{d,0,pentane} = 0.3$, $Y_{d,0,nonane} = 0.7$ and $Y_{d,0,pentane} = 0.7$, $Y_{d,0,nonane} = 0.3$) were investigated. For comparison experiments with one-component droplets consisting solely of *n*-pentane or *n*-nonane were also conducted. The gas pressure of the nitrogen environment was again varied from $p = 20$ bar to 40 bar and the gas temperature T_∞ was 550 K and 650 K. The initial droplet diameters varied between $d_{d,0} = 640$ µm and 820 µm.

For $p = 40$ bar and $T_\infty = 650$ K the critical pressure and the critical temperature of the droplet components were exceeded ($T_{crit,pentane} = 470$ K, $p_{crit,pentane} = 33.7$ bar and $T_{crit,nonane} = 595$ K, $p_{crit,nonane} = 23.1$ bar). However, the calculated droplet temperature was well below the critical temperature of the fuel mixture.

1.3.4.2.1 Variation of Gas Pressure

A comparison of the experimental and theoretical results of the Diffusion Limit Model is shown in Fig. 1.3.6. In this case the initial droplet composition was $Y_{d,0,pentane} = 0.3$ and $Y_{d,0,nonane} = 0.7$ and the gas temperature was $T_\infty =$ 550 K. The gas pressure varied between 20 bar and 40 bar, the initial droplet

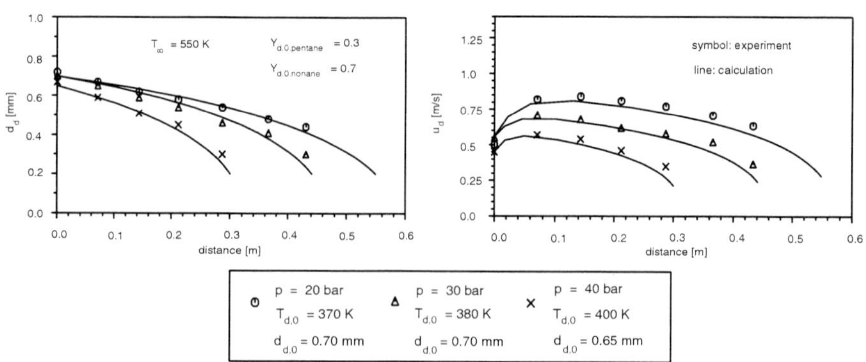

Figure 1.3.6: Evaporation of multi-component droplets at different pressure levels.

temperature between $T_{d,0} = 370$ K and $T_{d,0} = 400$ K, and the initial droplet diameter between $d_{d,0} = 650$ µm and 700 µm. As for the single component experiments the agreement of the measured and calculated droplet diameter and velocity distributions is excellent at all pressure levels investigated.

1.3.4.2.2 Variation of Gas Temperature

The calculated and measured results also coincide very well for different gas temperatures (see Fig. 1.3.7). In this case the initial droplet composition was $Y_{d,0,pentane} = 0.7$ and $Y_{d,0,nonane} = 0.3$ and the gas pressure was $p = 30$ bar. The gas temperature was $T_\infty = 550$ K and $T_\infty = 650$ K, the initial droplet temperature $T_{d,0} = 370$ K and $T_{d,0} = 380$ K, and the initial droplet diameter $d_{d,0} = 800$ µm and 820 µm.

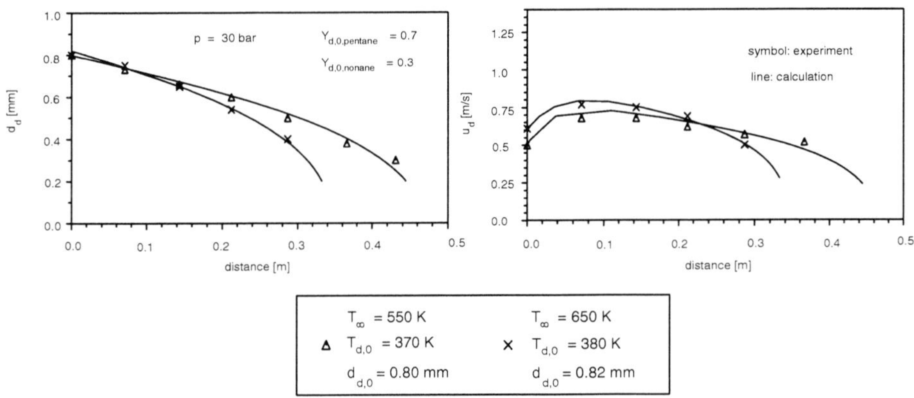

Figure 1.3.7: Evaporation of multi-component droplets at different temperature levels.

1.3.4.2.3 Variation of Fuel Volatility

In Fig. 1.3.8 the influence of the fuel volatility on the evaporation process of the droplet as determined from the numerical and experimental results for one- and two-component droplets are shown. Different initial mixtures of two-component droplets consisting of $Y_{d,0,pentane} = 0.7$ and $Y_{d,0,nonane} = 0.3$ or $Y_{d,0,pentane} = 0.3$ and $Y_{d,0,nonane} = 0.7$ and the limiting cases of one-component droplets consisting of solely pentane or nonane are investigated.

From Fig. 1.3.8 it becomes obvious, that increasing the volatility of the fuel intensifies the evaporation and as a consequence shortens the evaporation distance of the droplet. In this case the gas temperature was 550 K, the gas pressure 30 bar, the initial droplet temperature 380 K. The initial

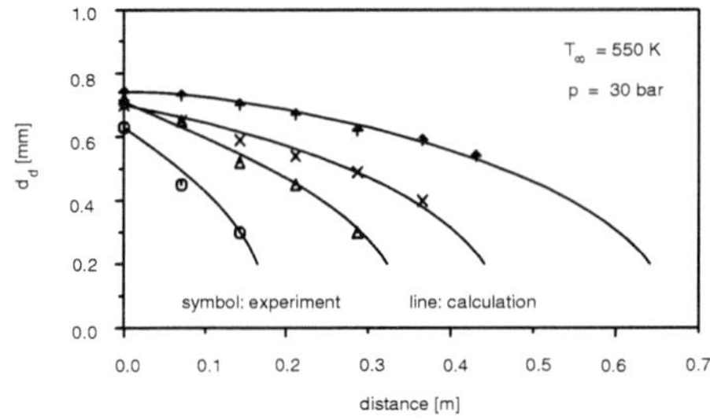

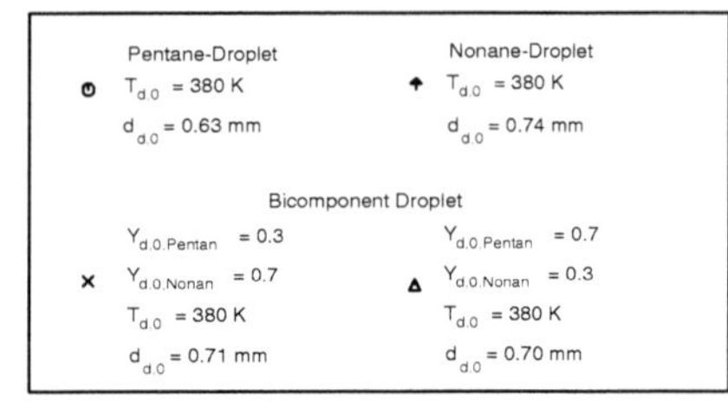

Figure 1.3.8: Effect of fuel volatility on evaporation.

droplet diameter varied between $d_{d,0} = 630$ µm and 740 µm. Again, experiment and theory agree very well (the velocity distribution is not shown explicitly for this case).

1.3.5 Conclusion

Within the subtask "Investigations of Droplet Evaporation at Elevated Pressures" a comprehensive experimental and analytical effort was conducted to develop and validate models to predict the evaporation of fuels droplets at elevated pressure. In the context of this study a new experimental setup was designed and used where the evaporation of free falling, non-interacting droplets is investigated without any disturbing influence. After extensive tests a droplet generator was built allowing the production of monodisperse droplets in the appropriate size range of 600 µm and 900 µm initial diameter. Since the distance between the droplets is more than 100 droplet diameters, interaction of the different droplets does not occur during the evaporation process.

Detailed measurements were conducted with one- and two-component droplets at different gas pressures (p = 20, 30 and 40 bar) and gas temperatures of 550 K and 650 K, respectively. Additionally the initial droplet diameter was varied between 640 µm and 820 µm.

Experiments with various test liquids had been performed. Heptane for the one-component droplet tests and binary mixtures of n-pentane and n-nonane for two-component experiments were used.

The experimental results were compared with an evaporation model based on the Conduction Limit Model and the Diffusion Limit Model extended for high gas pressures. The best agreement of measured and calculated values was obtained, when the theory of Abramzon and Sirignano [4] was combined with the correlation of Ranz and Marshall [5], and the specific heat capacity of the pure vapor was replaced by the specific heat capacity of the vapor/gas mixture. In general the agreement was found to be excellent providing a basis for application of results of this study to the prediction of fuel droplet evaporation under realistic conditions.

References

1. K. Law, W. A. Sirignano (1977): Unsteady droplet combustion with droplet heating – II: conduction limit, *Combust. Flame* **28**, 175–186.
2. B. Landis, A. F. Mills (1974): Effect of diffusional resistance on the evaporation of binary droplets, Proc. 5th Int. Heat Transfer Conf., Tokyo, Paper B7.9.
3. G. M. Faeth (1977): Current status of droplet and liquid combustion, *Prog. Energy Combust. Sci.* **3**, 191–224.
4. B. Abramzon, W. A. Sirignano (1989): Droplet vaporization model for spray combustion calculations, *Int. J. Heat Mass Transfer* **32**, 1605–1618.
5. E. Ranz, W. R. Marshall (1952): Evaporation from drops: part I + II, *Chem. Eng. Progress* **48**, 141–146, 173–180.
6. S. Wittig, W. Klausmann, B. Noll, J. Himmelsbach (1988): Evaporation of fuel droplets in turbulent combustor flow, ASME-88-GT-107.
7. M. Hallmann, M. Scheurlen, S. Wittig (1993): Computation of turbulent evaporating sprays: Eulerian versus Lagrangian approach, ASME-93-GT-333.
8. M. Kurreck, M. Willmann, S. Wittig (1996): Prediction of the three-dimensional reacting two-phase flow within a jet-stabilized combustion, ASME-96-GT-468.
9. A. Manrique, G. L. Borman (1969): Calculations of steady state droplet vaporization at high ambient pressures, *Int. J. Heat Mass Transfer* **12**, 1081–1095.
10. C. Hsieh, J. S. Shuen, V. Yang (1991): Droplet vaporization in high pressure environments I: near critical conditions, *Combust. Sci. Tech.* **76**, 111–132
11. W. Curtis, P. V. Farrell (1992): A numerical study of high-pressure droplet vaporization, *Combust. Flame* **90**, 85–102.
12. H. Jia, G. Gogos (1993): High pressure droplet vaporization; effects of liquid-phase gas solubility, *Int. J. Heat Mass Transfer* **36**, 4419–4431.
13. J. Stengele, H.-J. Bauer, S. Wittig (1996): Numerical study of bicomponent droplet vaporization in a high pressure environment, ASME-96-GT-442.
14. R. Wieber (1963): Calculated temperature histories of vaporizing droplet to the critical point, *AIAA Journal* **1**, 2764–2770.
15. P. Delplanque (1993): Liquid-oxygen droplet vaporization and combustion: analysis of transcritical behaviour and application to liquid-rocket combustion instability, Ph. D. Thesis, University of California, Irvine.
16. L. Matlosz, S. Leipziger, T. P. Torda (1972): Investigation of liquid drop evaporation in a high temperature and high pressure environment, *Int. J. Heat Mass Transfer* **15**, 831–852.
17. T. Kadota, H. Hiroyasu (1976): Evaporation of a single droplet at elevated pressures and temperatures, *Bull. JSME* **19**, 1515–1521.
18. P. Olthoff (1994): Modellierung des Tropfenverdunstungsprozesses bei überkritischem Umgebungsdruck, *DLR-Forschungsbericht*, DLR-FB 93-54.
19. J. Sato, M. Tsue, M. Niwa, M. Kono (1990): Effects of natural convection on high pressure droplet combustion, *Combust. Flame* **82**, 142–151.
20. P. Hartfield, P. V. Farrell (1993): Droplet vaporization in a high-pressure gas, *Transactions of ASME* **115**, 699–706.

21. T. Shih, C. M. Megaridis (1995): Suspended droplet evaporation modeling in a laminar convective environment, *Combust. Flame* **102**, 256–270.
22. R. Kneer, M. Schneider, B. Noll, S. Wittig (1993): Effects of variable liquid properties on multicomponent droplet vaporization, *Transactions of ASME* **115**, 467–472.
23. J. Stengele, M. Willmann, S. Wittig (1997): Experimental and theoretical study of droplet vaporization in a high pressure environment, ASME-97-GT-151.
24. J. Stengele, M. Willmann, S. Wittig (1997): Tropfenverdunstung in Hochdruckatmosphäre, *Chemie Ingenieur Technik* **69**, VCH, 962–966.
25. J. Stengele, K. Prommersberger, M. Willmann, S. Wittig (1999): Experimental and theoretical study of one- and two-component droplet vaporization in a high pressure environment, *Int. J. Heat Mass Transfer*, **42**, 2683–2694.
26. L. C. Lage, R. H. Rangel (1993): Single droplet vaporization including thermal radiation absorption, *J. of Thermophysics and Heat Transfer* **7**, 502–509.
27. L. Hubbard, V. E. Denny, A. F. Mills (1975): Droplet evaporation: effects of transient and variable properties, *Int. J. Heat Mass Transfer* **18**, 1003–1008.
28. M. Renksizbulut, M. C. Yuen (1983): Experimental study of droplet evaporation in a high temperature air stream, *J. Heat Transfer* **105**, 384–388.
29. C. Yuen, L. W. Chen (1977): Heat transfer measurements of evaporating liquid droplets, *Int. J. Heat Mass Transfer* **21**, 537–542.
30. M. Renksizbulut, R. J. Haywood (1988): Transient droplet evaporation with variable properties and internal circulation at intermediate Reynolds numbers, *Int. J. Multiphase Flow* **14**, 189–202.
31. C. Yuen, L. W. Chen (1976): On drag of evaporating liquid droplets, *Comb. Sci. and Technol.* **14**, 147–154.
32. J. Stengele, K. Dullenkopf, S. Wittig (1995): Brennstoffverdunstung bei überkritischen Drücken, Forschungsbericht, SFB 167, "Hochbelastete Brennräume – Stationäre Gleichdruckverbrennung", Teilprojekt A 12.

1.4 Shear-Driven Liquid Wall Films in Combustor Flows: Recent Advances in Experiment and Numerical Simulation

Heiko Rosskamp, Alfred Elsäßer, Joachim Ebner, Georg Maier, Berthold Noll, and Soksik Kim*

Abstract

Thin shear-driven liquid wall films are present in many combustion devices. They play a major role in the fuel preparation of modern gas turbine engine combustors and can even significantly affect mixture formation and pollutant emissions.

This paper highlights some important applications of prefilming combustor technology and reviews the research work done within the Collaborative Research Centre 167 at the University of Karlsruhe, Germany. It also comprises the results of related research projects that were performed at the Institut für Thermische Strömungsmaschinen based upon the work done within the Collaborative Research Centre 167.

Shear-driven liquid wall films have been investigated in detail aiming at a profound understanding of the global and internal flow behaviour as well as the interaction with the gas phase boundary layer. The experimental work provides the basis for an advanced numerical modelling of the combustor two-phase flow entailing the treatment of liquid wall films. The films focused on are very thin (typically less than 250 µm in mean height), are volatile Newtonian fluids very similar to water, alcohols or hydrocarbons and are driven by the shear stress exerted from a co-flowing gas flow.

This paper summarizes the relevant information in the fields of measurement techniques, experimental results as well as numerical modelling. Furthermore, it provides researchers and development engineers with the tools needed for describing combustor related film flows.

* Institut für Thermische Strömungsmaschinen, Universität Karlsruhe, Kaiserstr. 12, 76128 Karlsruhe, Germany

Nomenclature

Symbols

a, b	wave geometry parameter	[–]
b, B	film width	[m]
c_p	specific heat	[J/(kg K)]
d	diameter	[m]
h_F	film thickness	[m]
$\dot{H}$, $\dot{h}$	specific enthalpy flux	[J/(s m^2)]
I	laser light intensity	[W/m^2]
k'	absorption coefficient	[1/m]
$\dot{m}$	mass flow rate	[kg/(s m^2)]
$\dot{m}_{eff}$	effective vaporization rate	[kg/(s m^2)]
n	refraction index	[–]
$\dot{q}$	specific heat flux	[W/m^2]
p	pressure	[Pa]
T	temperature	[K]
u	velocity	[m/s]
$\dot{V}_F$	volume flux	[m^3/s]
x,y,z	cartesian coordinates	[m]
α	heat transfer coefficient	[W/m^{2}K]
α	wave back length	[m]
β	rolling wave length	[m]
υ	angle to receiving optics	[°]
λ	wave length	[m]
λ_f	thermal conductivity	[W/mK]
ν	kinematic viscosity	[m^2/s]
ν_f	eddy viscosity	[m^2/s]
ρ	density	[kg/ m^3]
τ	shear stress	[N/m^2]

Subscripts

+	non-dimensionalized variable
cl	continuous layer
cond	conductive
conv	convective
drop	droplet
eff	effective
f, F	film

G	gas, glass
M	measured
O	original
s	surface
t	turbulent
vap	vapour
w, W	wall
wb	wave back

1.4.1 Introduction

Shear-driven liquid fuel films on the walls inside a gas turbine combustor play an important role in the fuel preparation process. As the liquid wall films highly interact with the driving air flow, they may alter the flow field and affect the mixture formation to a great extent. The liquid wall films considered in this paper occur as a major parameter for the fuel preparation in prefilming LPP combustors (Lean Premixed Prevaporized) or as a side effect in LPP combustors utilizing spray evaporation when droplets impinge on the walls. The most common application today where liquid wall films are present are prefilming airblast atomizers for low emission combustors.

Figure 1.4.1(a) shows a sketch of a prefilming airblast atomizer which is applied in many modern aircraft engines. Here, the fuel is supplied by means of a hollow cone pressure atomizer. The spray which impinges on the atomizer lip forms a liquid film. This film is accelerated and driven downstream by the swirling air flow. At the atomizer edge, the film disintegrates and is dispersed into a very fine spray over a wide range of operating conditions. Therefore, this atomizer type is state of the art for advanced clean combustion

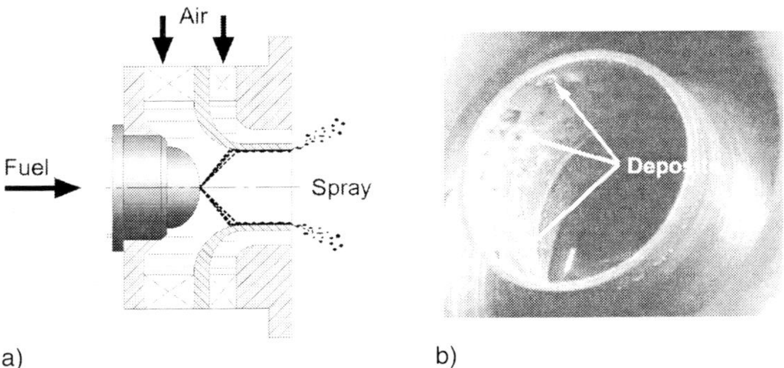

Figure 1.4.1: (a) Sketch of a prefilming airblast atomizer, (b) coke deposits on the prefilming lip of an airblast atomizer.

chambers. Nevertheless, the homogeneous mixture of evaporated fuel and air in the combustion chamber, which is a major necessity for a clean combustion, depends strongly on the droplet size distribution of this device. This distribution is affected by the propagation, the evaporation and the fluid properties of the film at the end of the lip. Thus, to predict the amount of evaporated fuel, the velocity and the temperature of the liquid film by numerical simulations, a detailed knowledge of the dynamics and the heat transfer of thin shear-driven films is needed.

A key problem of the film vaporization technique for LPP combustors as well as for prefilming airblast fuel nozzles is the formation of coke deposits on the prefilming surface if the fuel is heated above a critical temperature [1]. Figure 1.4.1(b) shows such coke formations on the prefilming lip of a generic airblast atomizer. They have to be avoided for several reasons. When the coke deposits become too thick the film propagation is strongly disturbed, leading to an inhomogeneous fuel distribution. Moreover, carbon fragments which break off and hit the combustor liner or the turbine inlet guide vanes may cause mechanical damage of these components. Therefore, coke formation is a major reason why the technique of film vaporization has found only limited application in combustor flows even though it became well developed in the past.

However, in the late 1980's, fuel film evaporation has been suggested for heavily loaded industrial gas turbines. Figure 1.4.2 gives a sketch of such a fuel injection device. It features film formation on the outer wall and on the centre tube. The gas turbines equipped with this fuel injector emitted less than 20 ppm_V NO_X [2].

To give a general idea of film vaporizer performance, Fig. 1.4.3 presents the characteristics of different film vaporization combustors. As the designs are very different the data in Fig. 1.4.3 has to be compared with care. The research premix duct of Kuck [3], for example, features a very strong swirl and a strongly heated prefilming plate which leads to effective vaporization rates even beyond $\dot{m}_{\mathrm{eff}} = 1.0$ kg/(sm^2). However, a clear distinction can be made between designs that utilize swirling air flow (light grey) and those

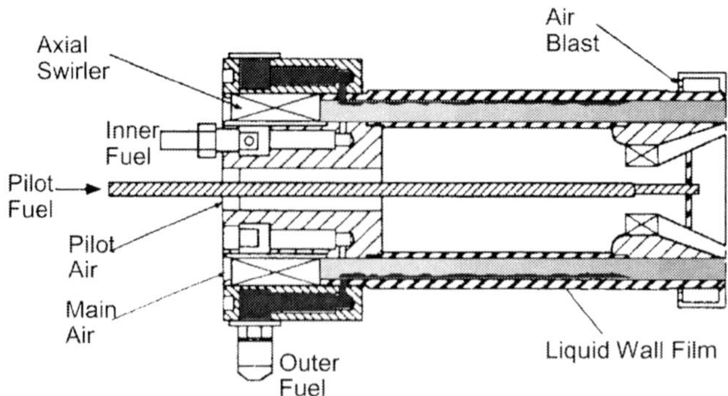

Figure 1.4.2: Prefilming fuel vaporizer for an industrial LPP combustion chamber [2].

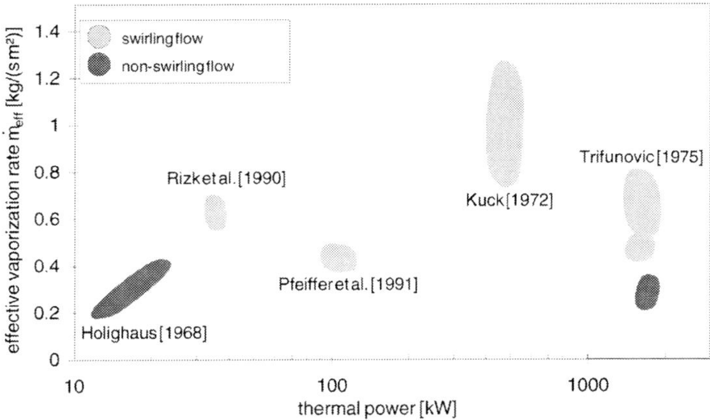

Figure 1.4.3: Performance of film vaporizing combustor assemblies [4–7].

who does not (dark grey). Even though the boundary conditions for film evaporation are very different a "typical" range of vaporizer performance which is approximately $0.2 < \dot{m}_{\text{eff}} \leq 1.0$ kg/(sm^2) may be identified. The performance of a vaporizer section in terms of vaporized liquid fuel per unit size or volume, depends on the thermodynamic parameters as well as on the geometrical details.

Since both types of wall films in airblast atomizers (Fig. 1.4.1) and in prefilming fuel vaporizers (Fig. 1.4.2) are very similar, they may be described by the same physical basics and numerical means. This paper focuses especially on this kind of wall films. Thereby, special emphasis is laid on the experimental work in this field and on the realistic numerical modelling of these flows.

1.4.2 Measurement Techniques

A detailed understanding of the complex film transport mechanisms requires adequate measurement systems to obtain accurate information about film thickness and film velocity under changing flow conditions. In the past many different measurement techniques were used. The film thickness was determined with mechanical probes or by capacitive or conductivity sensors. However, these techniques are not suited for very thin and wavy liquid films. This is mainly due to the problems with calibrating the sensors under difficult operating conditions. Also the dynamic range and the resolution are very limited. Measurements of mean film velocities were carried out by tracking the transport of particles on the film surface or by transient film behaviour observations after short interruptions in the liquid supply. In this case, especially the wetting effects can not be quantified, so the experimental results are difficult to appraise. To avoid such insufficiencies for films, a new film thickness measurement system based upon laser light absorption has been developed at the Institut für Thermische Strömungsmaschinen for analyzing the film transport mechanism. In addition, a modified LDV system served for investigating the internal film flow.

1.4.2.1 Film Thickness Measurement System

Due to the lack of commercially available film thickness measurement systems, a novel device was developed by Sattelmayer [8] within the Collaborative Research Centre 167. The technique is based on the absorption of light by the liquid layer on the wall, as shown in Fig. 1.4.4(a). By means of the Lambert-Beer law, the layer thickness can be calculated from the intensity ratio of the initial light intensity I_0 and the measured intensity I_M with absorption in the film:

$$h_F = -\frac{1}{k'} \ln \left(\frac{I_M}{I_0} \right) \tag{1}$$

To reach a sufficient intensity reduction, in previous work the film was dyed and the actual absorption coefficient k' of the liquid was measured with a calibration cell at the film entry. The addition of dye restricts the usability to non-evaporating films because when evaporation occurs, the absorbing colour remains in the non-evaporated liquid increasing the light absorption during the measurement. Therefore, an important further development of this system for evaporating films was made by Himmelsbach [9] at the Institut

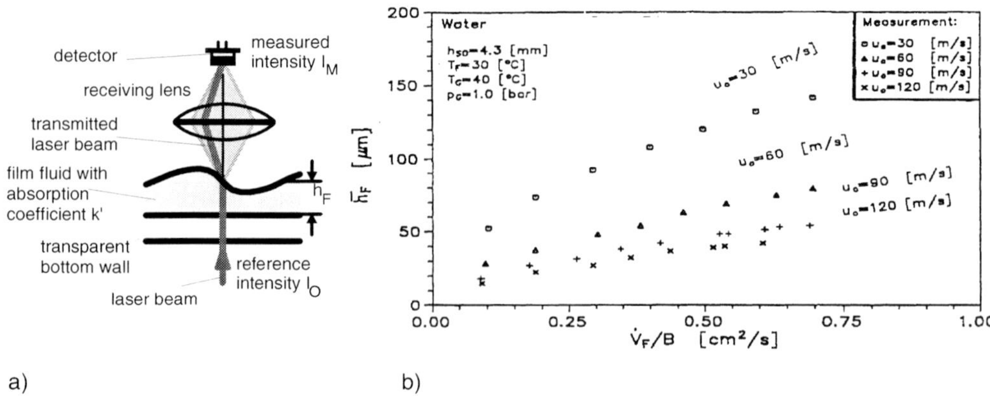

a) b)

Figure 1.4.4: (a) Light absorption in the liquid film, (b) mean film thickness versus liquid flow rate [9].

für Thermische Strömungsmaschinen. Instead of dying the liquid, he used light of a near infrared wavelength $\lambda = 1530$ nm. For this wavelength the absorption in water and alcohols is very strong. With this approach, the measurement effect relies exclusively on the absorption coefficient k' of the liquid which is a constant fluid parameter. Thus, thickness measurements in thin shear-driven films can be carried out with a high level of accuracy.

Typical results of optical film thickness measurements [9] are presented in Fig. 1.4.4(b). It shows the characteristic degressive increase of film thickness with the liquid flow rate. This effect results from the increasing roughness at the film surface as well as from the rising displacement of the air flow in the small gap. This again leads to higher shear stress at the liquid/gas interface and causes film acceleration. With the higher liquid transport velocity the increase of the film thickness is consequently diminished and results in the observed degressive behaviour.

1.4.2.2 Velocity Profile Measurements in Liquid Films

Since no adequate technique was available to measure the velocity profile in thin, wavy liquid films a novel LDV system had to be developed at the Institut für Thermische Strömungsmaschinen (University of Karlsruhe), especially adapted to the required conditions. In the following the design of the transmitting as well as the receiving optics of the laser system to fulfil the essential requirements will be introduced.

The newly established set-up consists of a transmitting optics looking vertically up through a transparent bottom wall of the horizontal test section as illustrated in Fig. 1.4.5(a). This provides a good optical access to the film

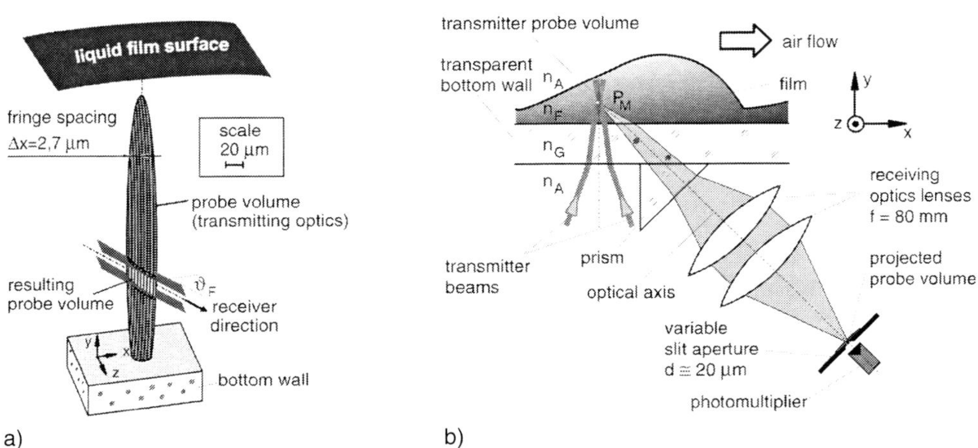

Figure 1.4.5: (a) Effective probe volume for velocity measurements in wavy liquid films, (b) reduction of the resulting probe volume length.

flow. The probe volume can be located exactly within the thin liquid layer without any disturbances due to the wavy film surface. As basic system a commercial two-component LDV system with a 2D-fiber probe is used. To achieve a spatial resolution suitable for the shear-driven films under investigation, a beam expander and a short focal length front lens have been adapted to the transmitter. They reduce the probe volume size to a length of $\Delta y_F \approx 250$ µm and a diameter of $\Delta x_F \approx \Delta z_F \approx 25$ µm at the $1/e^2$ light intensity point. However, the size Δy_F is still larger than the desired spatial resolution of about 25–30 µm in film height direction. Therefore, the probe volume length has to be reduced further by the receiver design.

The probe volume would be too large when using the standard backscatter configuration and, furthermore, the reflections at the film surface would cause serious problems considering the signal to noise ratio in a wide range of surface angles. To solve these problems, the detector was shifted to an off-axis angle with respect to the optical axis of the transmitter (Fig. 1.4.5(b)), thus cutting a narrow section of the probe volume as shown in Fig. 1.4.5(a). To observe only the defined section of the probe volume in y-axis direction a slit aperture is positioned in front of the colour splitter in the receiving optics. Additionally, the off-axis angle helps to solve the reflection problem at the air/liquid interface. An angle of 45° to the optical axis was chosen, because previous investigations had shown that reflections in this angle range occur relatively seldom.

The signals have been detected by a two-component photomultiplier set-up in combination with counter processors and a newly developed software for data acquisition and analysis. By means of a prism which is fixed perpendicularly to the optical axis of the receiver, a minimal aberration for the projection is achieved. It is an absolute necessity to protect the photomultipliers from intensive light directly reflected at the liquid/air interface to avoid overload and erroneous Doppler signals. This is obtained by switching off the laser light for reflecting conditions by means of a Bragg cell deflecting the emitted LDV laser beams out of the critical area. The reflection conditions are determined by an additional continuous laser beam of different wavelength (pilot beam) [10]. It is superposed to the LDV beams in the transmitter optics and is therefore reflected in the same manner. To separate the different colours in the receiving optics a wavelength selective mirror is used. Scattered light of the LDV beams wavelength can pass onto the photomultipliers and light of the pilot beam wavelength is reflected to a detector giving the signal for the switch-off device. When the light reflected at the wave surface leaves the pilot detector the measurement can be continued by switching the main beams back on.

Now, for the first time, a measurement technique is available for the detailed analysis of film hydrodynamics under a wide range of flow conditions. With this tool, comprehensive investigations can be realized, giving a deep insight into the transport mechanisms of shear-driven liquid wall films.

The left diagram of Fig. 1.4.6 shows measured time mean velocity profiles in main stream direction. A comparison with film thickness data indi-

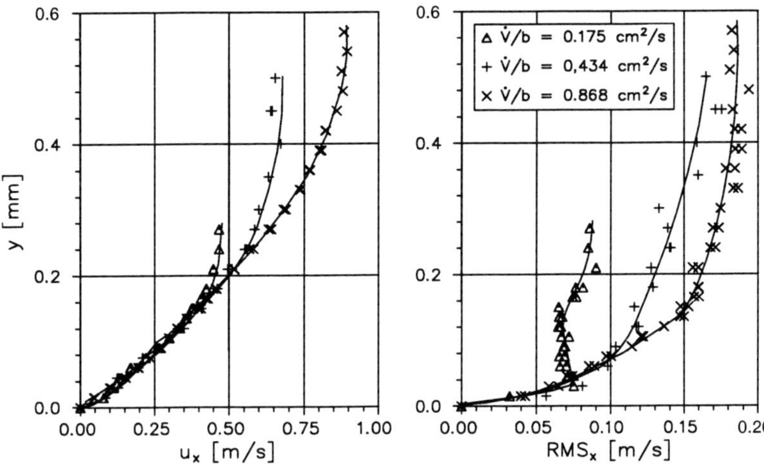

Figure 1.4.6: Time mean velocities and velocity fluctuations for water films ($u_G = 30$ m/s, $\nu = 1$ mm^2/s) [11].

cates, that the profile measurements with the novel LDV set-up can be carried out even up to the maximum wave tips due to the photomultiplier protection mentioned above. A characteristic feature of the investigated films with constant dynamic viscosity and under the influence of constant air velocity ($\nu = 1$ mm^2/s, $u_G = 30$ m/s) is, that the velocity profiles are nearly congruent up to a distinct height (here about $y = 0.180$ mm). Above this value the curves diverge significantly for the different volume fluxes. In contrast to that, the RMS values (Root Mean Square) show a different behaviour. The strong divergence of the RMS curves starts at a height of $y = 0.040$ mm. From film thickness data of accompanying measurements, this height can be identified as the transition regime from a thin continuous sublayer to the wavy film flow moving on its top.

The wavy film is usually treated as a liquid layer of constant thickness and driven by a constant shear force at its surface. Therefore, also the mean velocity profile inside the film has to be known for numerical computation. The profile information is normally determined by the assumption of the turbulent viscosity ν_t in the boundary layer close to the wall [12–14]. Velocity, wall distance and fluid properties have been non-dimensionalized using the shear velocity $u_\tau = (\tau/\rho)^{1/2}$. By means of u_τ the velocity and the distance to the wall are given as $u^+ = u/u_\tau$ and $y^+ = yu_\tau/\nu$. For a comparison of experimental and analytical values the knowledge of the shear stress τ at the phase boundary is absolutely necessary. The values of τ used for the transformation to non-dimensional film velocities are determined from air flow profiles measured in the test duct following the suggestion from Ellis and Gay [15]. By means of the shear stress data τ the measured mean film velocity profiles may be plotted in terms of generalized coordinates, as shown in Fig. 1.4.7.

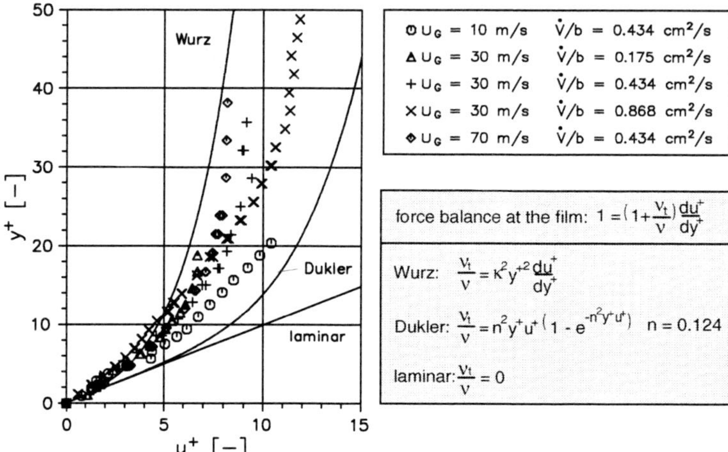

Figure 1.4.7: Generalized velocity distribution within shear-driven liquid films.

The diagram shows that the film flow regime spans from the laminar sublayer to the outer region of turbulent flow for large volume fluxes ($0 \leq y^+ \leq 50$). Additionally, various analytical profiles are applied in Fig. 1.4.7: The fully turbulent approach established by Wurz [16], a semi-empirical approach by Dukler [12], and the laminar theory for the ratio of eddy viscosity and molecular viscosity ν_t/ν are given. This comparison shows that the measured profiles lay between the distributions based on the approaches of Wurz and Dukler for shear-driven wall films. Furthermore, films exposed to

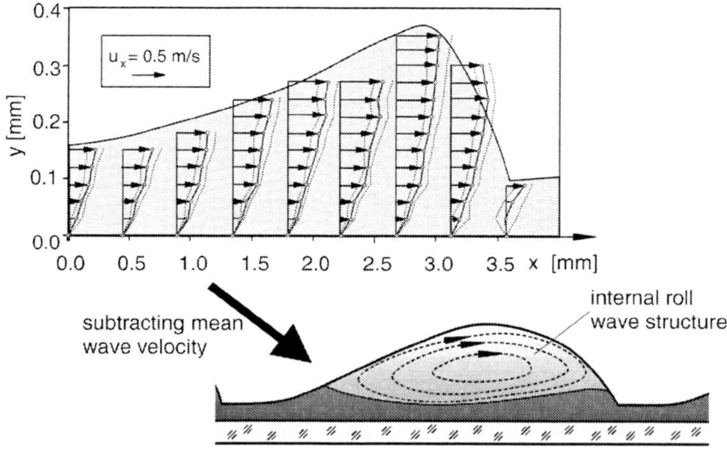

Figure 1.4.8: Internal wave velocity distribution from simultaneous measurements of film thickness and velocity.

higher shear stresses at the surface, tend to a more turbulent profile shape due to the change of internal liquid transport mechanisms. This has to be studied in detail by simultaneous measurements of film thickness and film velocity leading to a closer insight on the phenomena governing the film transport. An example of this analysis is shown in Fig. 1.4.8.

The presentation of the measured velocities within the liquid film shows a distinctive deformation of the profiles above a height $y \approx 0.100$ mm, particularly in the area of the wave crest [17]. Subtracting the mean velocities of the wave gives the wave internal flow structure as can be seen in the lower part of Fig. 1.4.8. Thus, for the first time the view of the film flow as a tumbling fluid bulk travelling on a thin sublayer is confirmed by an experimental investigation. The relevancy of the roll wave characteristics for the theoretical modelling of the film flow, especially with respect to the heat transfer, will be discussed in the following.

1.4.3 Computation of Wall Film Flows

1.4.3.1 Improved Flow Models for Liquid Wall Films

Within the Collaborative Research Centre 167 comprehensive flow simulation tools for shear-driven wall films have been developed. Brandauer et al. [18] used a flat plate boundary layer approach for the calculation of the film flow in the project B6. This computer code was intended for predicting the thermal boundary conditions for the coke formation from heated hydrocarbon films on diabatic prefilmers. The physical modelling of the wall film flow features a laminar velocity profile and a linear temperature profile in the fluid along with a one-dimensional heat transfer balance through the prefilming wall. The gas phase has been described by flat plate correlations.

The basic methodology for modelling the film flow had been chosen taking pattern from the work performed within project A4 [19, 20]. From the beginning, the work in project A4 aimed at providing the experimental and numerical basics for a comprehensive and fully coupled two-phase flow simulation. Himmelsbach et al. [21] proposed a numerical method for the internal airblast atomizer two-phase flow for the first time. The gas phase was calculated in body-fitted coordinates with a finite-volume approach. This computer code accounted for the film roughness effect, film heat-up and film evaporation. Figure 1.4.9 presents a sketch of the model film highlighting the important integral film characteristics that were considered in the calculation.

Himmelsbach [9] could prove that shear-driven liquid wall films can be efficiently calculated for a wide range of flow conditions when presupposing

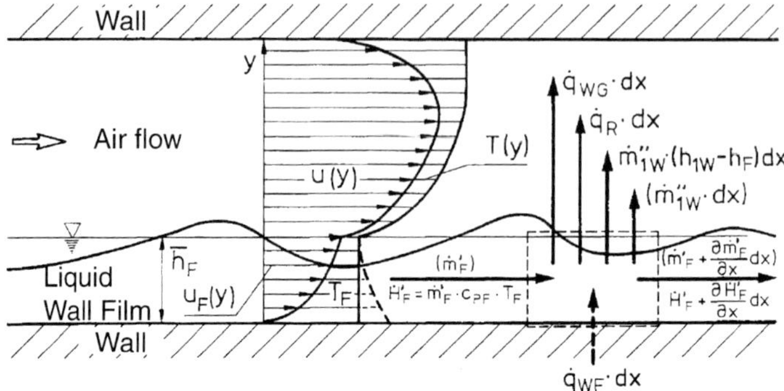

Figure 1.4.9: Model film for predicting the coupled two-phase flow of gas and wall film.

a linear velocity profile within the film. This is shown by an example in Fig. 1.4.10, where measured and computed film thickness values are plotted for a water film.

As mentioned above, the previous models calculated the heat flux through the liquid film by the arbitrary assumption of a linear temperature profile within a flat model film layer of constant thickness. This assumption has been made with view to the presupposed laminar flow behaviour of the wall film. In continuation of the program of the Collaborative Research Centre 167 the new experimental investigations, described in the previous section, revealed that the internal film flow is highly turbulent. It is also proved that the mass transport within the film is associated mainly with rolling waves travelling on a thin sublayer as indicated in Fig. 1.4.8. The pronounced internal vortex structure within the waves and its resulting high internal fluid exchange demand for a new heat transfer model that accounts for this complex flow structure in a realistic manner.

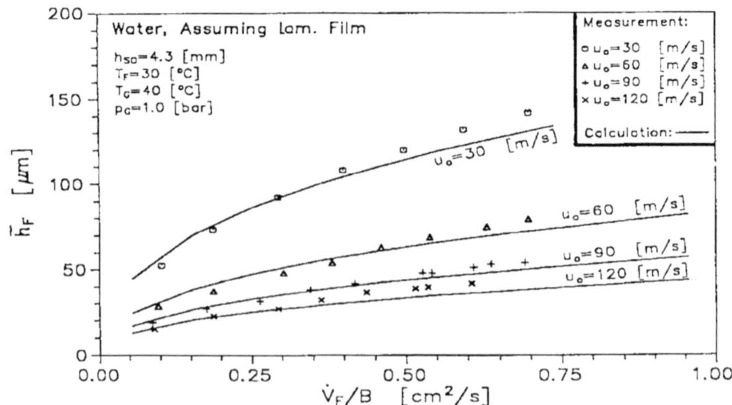

Figure 1.4.10: Comparison of measured and predicted film thickness with a laminar film.

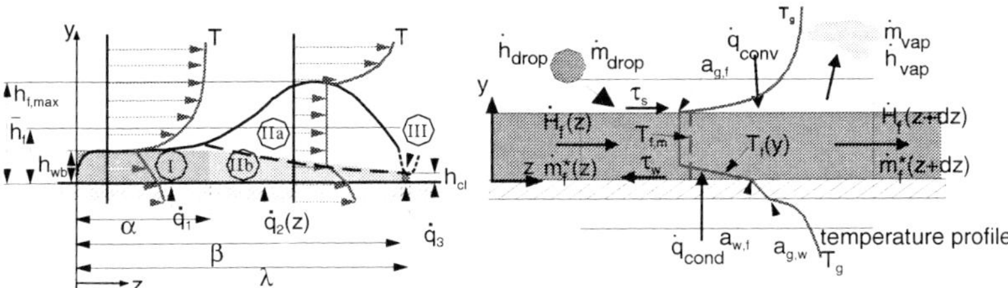

Figure 1.4.11: New film model for heat transfer computation.

An enhanced model for the heat transfer from the wall to the film which satisfies the major need of a more realistic representation of the film internal temperature distribution and its influence on the wall heat transfer was presented by Rosskamp et al. [22]. The new heat transfer approach applies to a model film of average thickness with two given temperatures, the film surface temperature and the wall temperature. Therefore, the idea is to define a virtual near wall fluid layer with a linear temperature profile representing the real average temperature gradient close to the wall. This gradient is determined to yield the same resulting heat flux for the flat model film and the real wavy film flow. To get the time averaged wall heat flux for a wavy film, the wave surface structure is subdivided into three regimes I–III as shown on the left side of Fig. 1.4.11.

All waves consist of a "tumbling" fluid bulk (IIa) travelling on a sublayer (IIb), of an impingement area right in front of the wave (III) and of a prolonged wave back (I). Depending on the fluid properties and on the shear force, the single regimes are more or less pronounced. Within the roll wave itself rapid mixing and therefore a constant temperature is assumed. The areas beneath the wave vortex, within the prolonged wave back and within the impingement zone are described by a linear temperature profile. The resulting heat flux is calculated from a superposition according to characteristic geometrical wave parameters occurring at the actual flow conditions. As the heat flux with this modelling can be determined from an arithmetic average over all waves, the geometry parameters themselves may be directly averaged from detailed measurements of film wave shapes. To obtain the parameter values with high accuracy, the optical film thickness measurement system already described in Section 1.4.4 is used (cp. Fig. 1.4.4).

The model parameters are determined from single wave analysis. As may be seen in Fig. 1.4.12, the main waves are of a distinctive geometrical similarity. The set of measured waves under the specified flow conditions demonstrates, that a representative wave form can be found. A comparison of this averaged wave shape with the over-all data of the film thickness measurement reveals that about 70 % of the wave forms can be represented sufficiently by a representative model wave for the actual operation conditions. A statistical analysis indicated that the waves basically scale with their wave length and crest height. For details of the experimental methodology and the

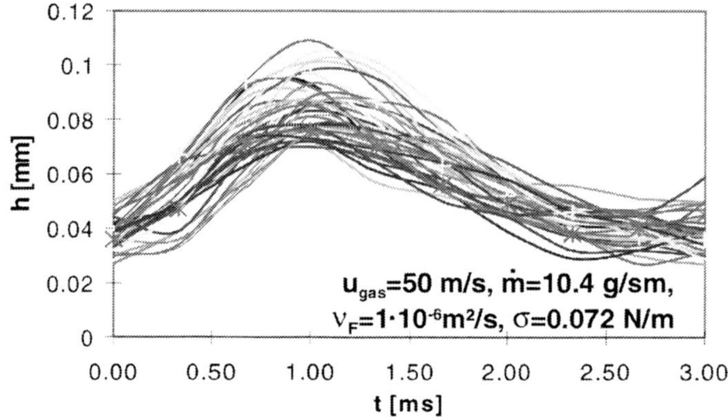

Figure 1.4.12: Typical wave pattern; wave fronts are to the left, wave backs to the right of the crest.

data processing, the reader is referred to [17]. The wavy film structure may be represented by the following non-dimensionalized parameters, according to the variables introduced in Fig. 1.4.11.

$$a := \frac{\alpha}{\lambda} = 0.182 \qquad b := \frac{\beta}{\lambda} = 0.870$$

$$h_1 := \frac{h_{\mathrm{wb}}}{h_{\max}} = 0.370 \qquad h_2 := \frac{h_{\mathrm{cl}}}{h_{\max}} = 0.120$$

$$\Phi := \frac{h_{\max}}{\bar{h}_{\mathrm{f}}} = 1.358$$

The values for the parameters result from several thousands individual measurements of wave shape comprising liquids and flow conditions typically for fuel preparation systems ($\nu \leq 1\ \mathrm{mm^2/s}$).

The heat flux to the wall film can now be calculated by averaging over the wave regimes (Fig. 1.4.11). In the middle section beneath the rolling wave crest, the region with the linear temperature profile is represented by a linear ramp between h_{wb} and h_{cl} (IIb). Now, depending on the geometry parameters discussed above, the resulting average wall heat flux may be calculated according to Eq. (2).

$$\dot{q}_{\mathrm{wavy}} = \overline{\dot{q}_1 + \dot{q}_2 + \dot{q}_3} = \frac{\lambda_{\mathrm{f}}}{\Phi \bar{h}_{\mathrm{f}}} \Delta T \left[\frac{a}{h_1} + \frac{b-a}{h_1 - h_2} \ln\left(\frac{h_1}{h_2}\right) + \frac{1-b}{h_2} \right] \tag{2}$$

In this expression, ΔT is the temperature difference between the wall and the film surface, λ_{f} the thermal conductivity of the film and $\bar{h}_{\mathrm{f}}$ the time averaged film thickness. The temperature levels are assumed to be constant over one wave period. Following the discussion about the different regimes of rapid mixing and conduction limitzones within the waves, the assumed internal temperature profile within the flat model film is composed of a linear and a block profile part. The block profile temperature equals the surface temperature, while the temperature gradient at the wall follows from Eq. (2). This new model for wall film flow computation is presented on the right side of Fig. 1.4.11.

It is evident, that the near wall temperature gradient of the turbulent model film is steeper than that one of a laminar flowing film. The heat transfer enhancement due to the high wave-internal mixing may be derived from the following contemplation.

$$\dot{q}_{\text{wavy}} = \frac{[\ldots]}{\Phi}\frac{\lambda_{\text{f}}}{\bar{h}_{\text{f}}}\Delta T \qquad \alpha_{\text{wavy}} = \frac{[\ldots]}{\Phi}\frac{\lambda_{\text{f}}}{\bar{h}_{\text{f}}}$$

$$\dot{q}_{\text{lam}} = \frac{\lambda_{\text{f}}}{\bar{h}_{\text{f}}}\Delta T \qquad \alpha_{\text{lam}} = \frac{\lambda_{\text{f}}}{\bar{h}_{\text{f}}}$$

$$\Rightarrow \frac{\alpha_{\text{wavy}}}{\alpha_{\text{lam}}} = \frac{[\ldots]}{\Phi} = 3.44$$

By applying the model constants as discussed before, the wall heat transfer coefficient of wavy films appears to be 3.44 times higher than that one of a flat laminar flowing film. In other words: The wavy flow pattern of shear-driven films yields a significant increase of the heat transfer which may be traced back to the high vertical exchange mechanism in the roll waves.

1.4.3.2 Coupled Wall Film and Gas Phase CFD-Code

Based upon the extensive numerical work within projects A4 and A13 of the Collaborative Research Centre 167, the new general computer code for combustor flows with wall films consists of a 3D finite-volume code for the gas phase (EPOS) and a 2D finite-difference code for the shear-driven wall film (PROFILM). Figure 1.4.13 illustrates the structure and the main features of the CFD-code. EPOS and PROFILM exchange sources for momentum, heat

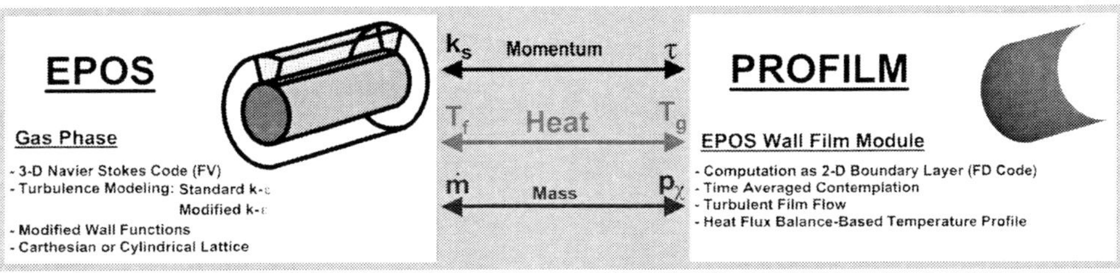

Figure 1.4.13: Structure of the two-phase flow CFD-code at the Institut für Thermische Strömungsmaschinen.

and mass transfer to allow for a fully coupled flow field solution. For a detailed description of the code the reader is referred to [23].

The gas phase code features special wall functions to account for the interface properties over shear-driven liquid wall films. In swirling flows, the near wall turbulence characteristics significantly change and affect wall film propagation and evaporation. Therefore, the calculational method for the interface shear stress in swirling flows is implemented following Kind et al. [24]. The swirl corrected k,ε-model that has been developed within the project A10 serves for a correct calculation of the gas flow field in swirling flow applications [25]. This swirl-corrected gas phase code in combination with the newly developed film flow model now gives the opportunity to analyse complex combustor flows that are dominated by two-phase flow effects.

The following example illustrates the capabilities of flow simulations using the new two-phase models developed. The test case is chosen taking pattern from a model combustor under investigation at the Institut für Thermische Strömungsmaschinen [26]. A prefilming airblast atomizer releases the liquid fuel into a cylindrical flame tube. The design of the atomizer is basically similar to the design shown in Fig. 1.4.1. The geometry of the atomizer including the swirlers is critical to both, the flow field in the combustor and the film heat-up on the prefilming lip. As the spray quality depends strongly on the fuel state (i. e. the film temperature) at the atomizer edge, the film heat-up process is of major importance for the spray formation. The new film model allows to calculate the fuel state at the atomizer lip and thus, to estimate the atomizer performance [27]. Figure 1.4.14 shows the vapour concentration of the fuel in the area of the atomisation zone. The contour plot reveals that the evaporation of the film can not be neglected in this device,

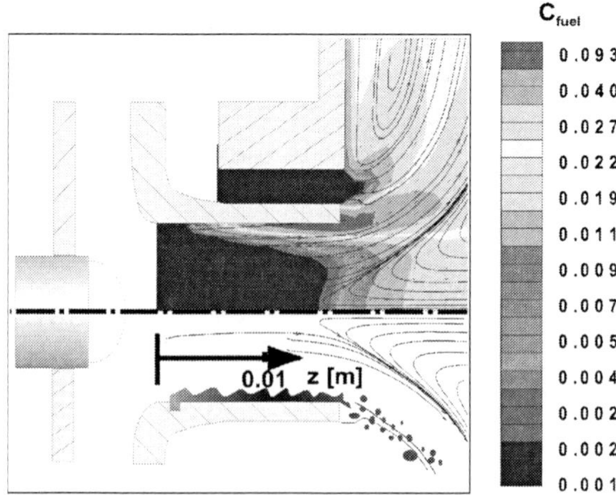

Figure 1.4.14: Internal nozzle flow, the wall film starts at $z = 2$ mm, the contour plot illustrates the fuel vapour concentration.

since nearly stoichiometric conditions can be found in the region behind the lip. These results are of high importance to understand the near-atomizer flow and serve as an important boundary condition for a subsequent spray simulation.

The newly developed computer code is a valuable help for optimising these atomizer geometries as it delivers both, the gas flow field data including pressure drop and fuel concentration as well as the film temperature and vaporization. Based upon comparison with the comprehensive experimental data sets discussed before it may be stated that the main propagation features of the film flow as well as the heat and mass transfer can now be described and predicted with a high level of accuracy.

1.4.4 Summary and Outlook

The present paper summarizes the experimental and theoretical work on shear-driven liquid wall films performed within the projects A4, A5 and A13 of the Collaborative Research Centre 167. Many pioneering results concerning the fluid dynamic aspect of the wall film flows as well as its numerical modelling have been presented. New and very advanced measurement techniques have been developed to attack fundamental questions that are connected with the global and local flow structure of the film. The unique laser-optical device proved superior qualification for investigations of film thickness and interface structure. In related work a modified LDV set-up was built up to investigate for the first time the velocity profile in thin shear-driven liquid wall films. Thus, by the completion of the Collaborative Research Centre 167, a comprehensive understanding of the physics of the wall film flow exists and important basics for the numerical treatment of these flow had been acquired. Within the last years, the results of the Collaborative Research Centre 167 also contributed to the success of many related projects dealing with the flow simulation of combustor two-phase flows.

The connection of the results with the other research projects has set the current state of the art in the two-phase flow computation of turbulent combustor flows as well as in the experimental investigation of thin liquid wall films.

Acknowledgements

The Collaborative Research Centre 167 was supported by grant from the Deutsche Forschungsgemeinschaft (DFG) which is gratefully acknowledged.

References

1. Brandauer, M., Schulz, A., Wittig, S.(1995): Mechanisms of Coke Formation in Gas Turbine Combustion Chambers, ASME Paper 95-GT-049.
2. Cowell, L. H. and Smith, K. O. (1992): Development of a Liquid-Fueled, Lean-Premixed Gas Turbine Combustor, ASME Paper 92-GT-112, 1–10.
3. Kuck, H. A. (1972): Experimentelle und theoretische Untersuchung der Vorgänge in einer zylindrischen Modellbrennkammer bei Wandfilmauftragung des Kraftstoffs, Ph. D. Thesis, University RWTH Aachen.
4. Holighaus, R. (1968): Untersuchungen zur filmförmigen Kraftstoffverdampfung in Bennkammern, Ph. D. Thesis, University of Darmstadt.
5. Trifunovic, R. (1975): Brennstofffilmverdampfung in Brennkammern von Verbrennungskraftmaschinen, Ph. D. Thesis, University of Munich.
6. Rizk, N. K., Mongia, A. C. (1990): Lean Low NO_X Combustion Concept Evaluation, 23rd (International) Symposion on Combustion, The Combustion Institute, 1063–1070.
7. Pfeiffer, A., Schulz, A., Wittig, S. (1991): Design and Development of the ITS Ceramic Research Combustor, Conference Proceedings of the Yokohama International Gas Turbine Congress, Oct. 27th to Nov. 1st 1991, III-9–III-14.
8. Sattelmayer, Th., Sill, K. H., Wittig, S. (1987): Optisches Meßgerät zur Bestimmung der Eigenschaften welliger Flüssigkeitsfilme, *Technisches Messen (tm)* **54**, No. 4/1987.
9. Himmelsbach, J. (1992): Zweiphasenströmungen mit schubspannungsgetriebenen welligen Flüssigkeitsfilmen in turbulenter Heißgasströmung – Meßtechnische Erfassung und numerische Beschreibung, Ph. D. Thesis, Institut für Thermische Strömungsmaschinen, University of Karlsruhe.
10. Wittig. S., Elsäßer, A., Samenfink, W., Ebner, J., and Dullenkopf, K. (1996): Velocity Profiles in Shear-driven Liquid Films: LDV-Measurements, Eighth International Symposium on Applications of Laser Techniques to Fluid Mechanics, July 8.–11, 1996, Lisbon, Portugal.
11. Elsäßer, A., Samenfink, W., Ebner, J., Dullenkopf, K., and Wittig, S. (1997): Dynamics of Shear-driven Liquid Films, Proceedings of the 7th International Conference on Laser Anemometry – Advances and Applications, Karlsruhe, Germany, September 8–11, 1997.
12. Dukler, A. E. (1960): Fluid Mechanics and Heat Transfer in Vertical Falling-Film Systems, *Chem. Eng. Progress Symposium Series* **56**, 1–10.
13. Mudawar, I. A. and Houpt, R. A. (1993): Mass and Momentum Transport in Smooth Falling Liquid Films Laminarized at Relatively High Reynolds Numbers, *Int. J. Heat Mass Transfer* **36**, No. 14, 3437–3448.
14. Glahn, A. and Wittig, S. (1995): Two-Phase Air/Oil Flow in Aero Engine Bearing Chambers – Characterization of Oil Film Flows, ASME-Paper 95-GT-114.
15. Ellis, S. R., Gay, B. (1959): The Parallel Flow of Two Fluid Streams: Interfacial Shear and Fluid-Fluid Interaction, *Trans. Inst. Chem. Eng.* **37**, 206–215.
16. Wurz, D. (1971): Experimentelle Untersuchung des Strömungsverhaltens dünner Wasserfilme und deren Rückwirkung auf einen gleichgerichteten Luftstrom mäßiger bis hoher Unterschallgeschwindigkeit, Ph. D. Thesis, Institut für Thermische Strömungsmaschinen, University of Karlsruhe.
17. Elsäßer, A. (1998): Gemischbildung in Verbrennungskraftmaschinen: Grundlagen der Strömung schubspannungsgetriebener Wandfilme, Ph. D. Thesis, Institut für Thermische Strömungsmaschinen, University of Karlsruhe.

18. Brandauer, M., Schulz, A., Wittig, S. (1995): Filmbildung und -verdampfung auf keramischen Oberflächen, Research Report for Project B6, Forschungsbericht 1993–1995, Sonderforschungsbereich 167, "Hochbelastete Brennräume – Stationäre Gleichdruckverbrennung", 411–426.
19. Kim, S., Noll, B., Himmelsbach, J. (1989): Brennstoffaufbereitung durch Zerstäubung und Verdunstung schubspannungsgetriebener Flüssigkeitsfilme bei erhöhten Drücken und Temperaturen, Research Report for Project A4, Forschungsbericht 1987–1989, Sonderforschungsbereich 167, "Hochbelastete Brennräume – Stationäre Gleichdruckverbrennung", 75–106.
20. Gerendás, M., Himmelsbach, J., Kim, S. (1995): Experimentelle und numerische Untersuchungen der Zweiphasenströmung bei luftunterstützten Zerstäubern unter besonderer Berücksichtigung von Mehrkomponenten-Effekten, Research Report for Project A4, Forschungsbericht 1993–1995, Sonderforschungsbereich 167, "Hochbelastete Brennräume – Stationäre Gleichdruckverbrennung", 21–50.
21. Himmelsbach, J., Noll, B., Kim, S. (1992): Experimentelle und numerische Untersuchung zur Vorhersage der düseninternen Zweiphasenströmung in luftunterstützten Zerstäubern bei brennkammertypischen Bedingungen, Research Report for Project A4, Forschungsbericht 1990–1992, Sonderforschungsbereich 167, "Hochbelastete Brennräume – Stationäre Gleichdruckverbrennung", 81–106.
22. Rosskamp, H., Elsäßer, A., Samenfink, W., Meisl, J., Willmann, M., Wittig, S. (1998): An Enhanced Model for Predicting the Heat Transfer to Wavy Shear-driven Liquid Wall Films, 3rd International Conference on Multiphase Flow, June 8–12, 1998, Lyon.
23. Rosskamp, H., Willmann, M., Wittig, S. (1997): Computation of Two-Phase Flows in Low NOx Combustor Premix Ducts Utilizing Fuel Film Evaporation, ASME Paper 97-GT-226.
24. Kind, R.J., Yowakim, F.M., Sjolander, S.A. (1989): The Law of the Wall for Swirling Flow in Annular Ducts, *J. of Fluids Engineering* **111**, 160–164.
25. Ganz, B., Schmittel, P., Koch, R., Wittig, S. (1998): Validation of Numerical Methods at a Confined Turbulent Natural Gas Diffusion Flame Considering Radiative Transfer, 42nd ASME Gas Turbine & Aeroengine Congress, June 2–5, 1998, Stockholm.
26. Willmann, M., Meier, R., Wittig, S. (1997): Visualization of the Instationary Atomization and Spray Propagation inside a Reacting Model Gasturbine Combustor, ILASS-Europe, 9–11 July 1997, Florence, Italy.
27. Rosskamp, H., Willmann, M., Meisl, J., Meier, R., Maier, G., Wittig, S. (1998): Effect of the Shear-driven Liquid Wall Film on the Performance of Prefilming Airblast Atomizers, ASME-Paper 98-GT-500.

1.5 Pressure-Swirl and Twin-Fluid Atomization with Regard to Industrial Liquid Fuel Combustion

Andreas Kufferath, Martin Löffler-Mang, Andreas Horvay, and Wolfgang Leuckel*

1.5.1 Introduction

The process of atomization of liquids, i. e. the process by which an initially continuous bulk of liquid is finally disintegrated into a large number of small droplets, is a key operation process applied in many areas of industry, e. g. the combustion of liquid fuels, the treatment of flue gases or spray drying processes. For the above mentioned applications the immediate goal of the atomization process is to increase the specific surface area of the liquid to improve heat and mass transfer. In order to optimize the process and to minimize pollutant formation, detailed knowledge of spray generation, spray dynamics and spray interaction with the turbulent flow field are required. In spite of all the work already done, there are still many unsolved problems in the field of atomization, and there still exists uncertainty as regards some of the fundamental processes of liquid jet or film disintegration [1]. Therefore, the investigation of fundamental two-phase flow structures inside the nozzles and of their correlation with detailed measurements of drop sizes and velocity patterns in the generated spray downstream of the nozzle remains of primary importance.

The selection of the best suited atomizer for any desired application is of prime importance for the optimization of the whole process. According to Fig. 1.5.1, the general parameters of inlet conditions and required spray conditions enter into the proper selection of the atomizer.

Main inlet conditions are the mass flow rate, usually in a range between 1 and 2000 kg/h for one individual atomizer, the control range required for the process (turndown ratio), the liquid properties like density, viscosity, surface tension, and the possibility or requirement of using additional energy for the atomization process. Also a possible content of solid particles, like e. g. in lime suspension used for the desulphurization of flue gases, is an excluding

* Engler-Bunte-Institut, Bereich III – Verbrennungstechnik, Universität Karlsruhe, Kaiserstr. 12, 76128 Karlsruhe, Germany

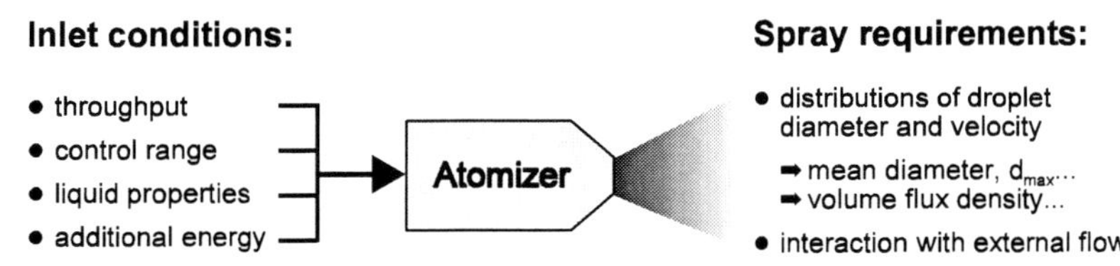

Figure 1.5.1: Selection criteria for atomizers.

criterion for some atomizer types. Depending on the particular problem, the desired spray characteristics differ from each other. Spray characteristics of primary relevance are the required spatial distributions of the droplet diameter, the droplet velocity and the volume-flux density in the spray. Especially for combustion processes, the knowledge of the upper end of the drop size distribution can be extremely important with respect to the completeness of burnout obtainable. In addition, the interaction of the spray with the surrounding flow field, for example the interaction with the complex turbulent rotating flow field in a furnace, is another criterion for the selection of the atomizer.

The classification of atomizers is currently based on the type of energy used for the atomization process. The conversion of the pressure drop of the feeding liquid into kinetic energy is most commonly used for atomization. This operation principle is the basis for all simple jet atomizers, including pressure-swirl nozzles and spill controlled pressure-swirl nozzles. These types are mostly used in technical systems due to their simplicity, their comparatively low capital investments and their low total energy consumption. For the second large atomizer group, the twin-fluid atomizers, the energy required for disintegration of the liquid is provided by a secondary compressible fluid, usually compressed air or steam. As typical examples of this "air-assist" atomizer we mention the internal and external mixing type, the Y-jet atomizer, the mixing-chamber type, and the airblast atomizer. Other systems make use of the mechanical energy of a rotating disc, or the energy of a ultrasonic sound field. For details of other atomizer types and their fields of applications see [2]. There is a large variety of different atomization systems with different atomizing mechanisms, but we will deal only with systems investigated in the frame of the Collaborative Research Centre 167 (Fig. 1.5.2).

Due to their relatively simple design and their good atomization characteristics by consuming moderate pressure drop, pressure-swirl atomizers are most widely applied. Their use is, however, restricted to applications with nearly constant flow rates, because their spray characteristics depend strongly on the mass throughput rate. Common applications are oil fired domestic and small industrial burners, where the throughput is normally in a range of 1 up to several hundred kg/h. The relatively small outlet orifice and the high outlet velocities exclude the acceptance of even small solid particles in the liquid, which would generate heavy erosion.

Atomizer	Advantages/disadvantages	Applications
Pressure-swirl $\dot{V}_{F\,liq}$	• control range $\approx$ 1 : 0.5 • low energy consumption • spray characteristic: – spray angle = f(load) – droplet diameter = f(load) – hollow cone spray	• domestic burners at nearly const. load • surface coating • agricultural spraying
Pressure-swirl, spill return $\dot{V}_{F\,liq}$ $\dot{V}_{SR\,liq}$	• control range > 1 : 0.5 • low energy consumption • additional control required • spray characteristic: – spray angle = f(load, RV) – droplet diameter = f(load, RV) – hollow cone spray	• industrial burners with variable load • gas turbine combustion
Air-assist, internal-mixing $\dot{V}_{liq}$ $\dot{V}_{air}$	• control range $\leq$ 1 : 0.05 • additional energy required • high jet momentum • all liquids (emulsions, CWS...) • spray characteristic: – droplet diameter = f(load, V_{air}) – full cone spray, const. angle	• industrial burners with variable load • liq. waste incineration • slurry combustion • gas treatment (washing, scrubbing...)

Figure 1.5.2: Investigated atomizers, their advantages/disadvantages and application areas.

To overcome the restriction of small control ranges, pressure-swirl atomizers with spill return can be used. With the adjustment of the spill return flow nearly constant droplet diameter distributions can be obtained within a control range of 0.3 : 1 for the total output fluid flow. Due to the need of an additional control mechanism, these atomizers are used only in industrial burners with a total throughput of one individual atomizer up to 500 kg/h or even 5000 kg/h for large oil fired steam boilers.

If the requirements as to spray quality and control range are high, an internal mixing air-assist atomizer must be used. In those twin-fluid atomizers, the energy required for the disintegration of the liquid is provided by a secondary fluid, usually compressed air or steam. The significant advantages of air-assisted internal-mixing atomizers are quite homogeneous sprays and low sensitivity as to physical properties of the liquid. Furthermore, by adjusting the atomizing gas flow rate to the liquid flow rate it is possible to maintain the droplet size distribution approximately constant over a load range of 0.05 : 1. This atomizer type has applications in all areas where its advantages prevail compared to the requirement of additional atomizing energy. Examples are the incineration of liquid wastes, the combustion of heavy fuel oil, process gas treatment, thermal quenching, separation of particles or gaseous species in all areas of chemical industry. The total design flow rates of those atomizers are in a range between 1 and 2000 kg/h or even higher.

Results reported here only provide a survey of the work which has been carried out within the project of the Collaborative Research Centre 167 on pressure-swirl nozzles with and without spill return, and on internal mixing twin-fluid atomizers for industrial flames (Fig. 1.5.1). In the first part, results concerning the investigations of the pressure-swirl nozzle will be presented. The second part will deal with pressure swirl nozzles with spill return. Finally, results regarding the internal-mixing air-assist nozzle will be discussed.

1.5.2 Results and Discussion for Pressure-Swirl Atomization

1.5.2.1 Pressure-Swirl Atomization without Spill Return

According to Fig. 1.5.3 the pressure-swirl atomization process can be characterized as follows. The liquid to be atomized enters the swirl chamber of the nozzle through several tangential channels. The energy necessary to generate a free film surface is the pressure-induced kinetic energy of the liquid. As a result of the radial pressure gradient caused by the swirling motion (centrifugal forces) within the nozzle, an air-filled hollow core develops about the nozzle chamber axis. Hence, the liquid issues from the nozzle exit as a rather thin concentric annular film, and proceeds as an expanding hollow cone downstream of the nozzle exit, which finally disintegrates into an atomized spray as a result of inertia and viscous forces acting upon the film flow.

The investigations made [3] can be correlated with three zones of the process (Fig. 1.5.3). Measurements in region I, the internal flow in the nozzle chamber, focuses on the flow field of the liquid and the formation of the hollow cone as depending on the inlet conditions and the geometric parameters of the nozzle. Region II comprises the development of the liquid film at and downstream the nozzle orifice, being directly controlled by the internal

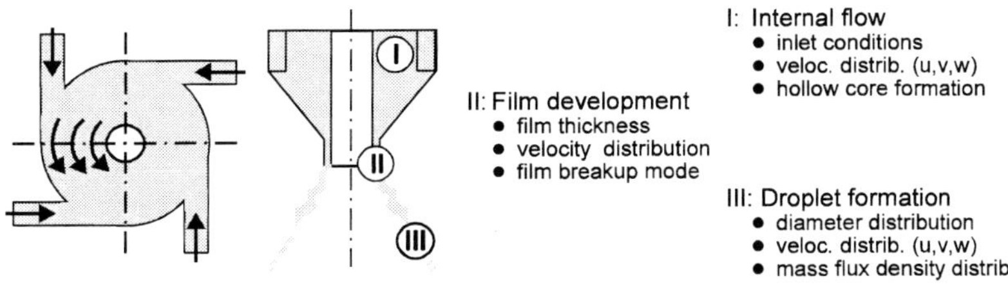

Figure 1.5.3: Scheme of the pressure-swirl atomizer.

nozzle flow. Of major interest in this zone are the film thickness and the breakup mode of the film. Region III refers to the droplet formation after disintegration of the liquid film and the development of the two-phase flow further downstream of the nozzle.

The flow field in the nozzle was investigated by means of Laser Doppler Anemometry (LDA) applied to the flow inside an optically accessible nozzle. Small air bubbles injected into the liquid were used as a tracer for the LDA measurements. The determination of the film thickness at the nozzle outlet and also the disintegration mechanism of the film were analyzed by high speed photography. The resulting integral droplet diameter distributions downstream of the nozzle were measured by means of a Malvern particle sizer, based on Fraunhofer diffraction of a parallel beam of monochromatic light by small particles.

Region I, Internal Flow

The formation of the hollow cone inside the nozzle can be seen from Fig. 1.5.4, left side. For constant initial tangential momentum, the development of the cone remains nearly unaffected by variation of the liquid Reynolds number. In the lower range of initial tangential momentum, a not fully developed cone was observed. Figure 1.5.4, right side, exemplifies the corresponding flow field, i. e. the profiles of the axial (u) and tangential (w) velocity components at different axial and radial positions in the nozzle. Both velocity components were expressed as ratios of the mean tangential inlet velocity $\overline{w_0}$. From the axial velocity component profiles, an increase of velocity in the downstream direction, especially near the hollow cone, may be observed. Due to the effect of angular momentum and centrifugal forces, a steep increase of axial velocity towards the hollow cone boundary occurs in the outlet port. The tangential velocity profiles show a monotone decrease towards the chamber periphery in the upper zones, whereas in the nozzle outlet a nearly flat profile was obtained. These general observations proved valid for both, small and high initial tangential momentum.

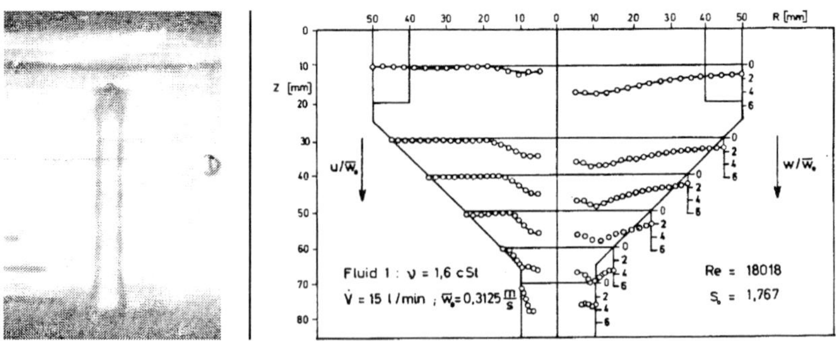

Figure 1.5.4: Formation of a hollow core inside the nozzle (left), and radial profiles of tangential and axial velocity for internal pressure-swirl nozzle flow (right).

Region II, Liquid Film Development

Development and disintegration of the film strongly depend on the volume flow rate. Therefore, the available control range for optimal atomizer performance is only about 0.7 : 1. In Fig. 1.5.5 high speed photos of four characteristic stages of film development, depending upon the liquid volume flow rate, are shown. For the smallest volume flux of 15 l/h, the liquid issues from the nozzle as a thin distorted jet. At 20 l/h a cone forms at the orifice, but is contracted by surface tension forces into a closed bubble. With further increase up to 30 l/h the bubble opens into a hollow tulip shape terminating in a ragged edge where the liquid disintegrates into big droplets. At the real operation liquid flow rate, the surface straightens to form a conical sheet. As the sheet expands its thickness diminishes rapidly, gets unstable and disintegrates into ligaments and, finally, droplets in the shape of a well defined hollow-cone spray. The cone angle of the film is being controlled by the axial and tangential velocity components at the nozzle outlet and, therefore, this angle is defined by the nozzle inlet conditions and geometry.

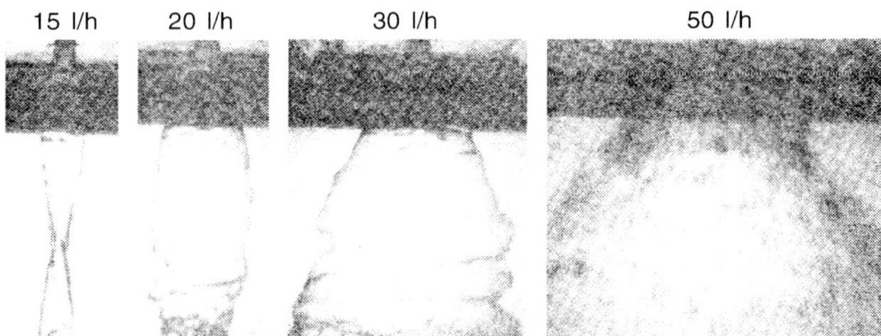

Figure 1.5.5: Different stages of film disintegration.

Region III, Film Disintegration and Droplet Formation

As mentioned above, the useful operation range of pressure-swirl nozzles is very small, since droplet diameter distributions are strongly dependent on nozzle pressure or throughput, respectively. In Fig. 1.5.6 the volume based density function of the droplet diameter is plotted as a function of nozzle pressure. It can be seen that with increasing pressure drop, the size distributions become narrower and the peak value is simultaneously shifted to smaller droplet diameters. The reason is an increase in initial tangential momentum with increasing nozzle pressure drop and, therefore, a decreasing film thickness at the nozzle orifice. This influence depends on the Weber number, calculated for the conditions at the nozzle orifice, but turns out to be only a weak function of the Reynolds number (i. e. effect of fluid viscosity is of minor importance only).

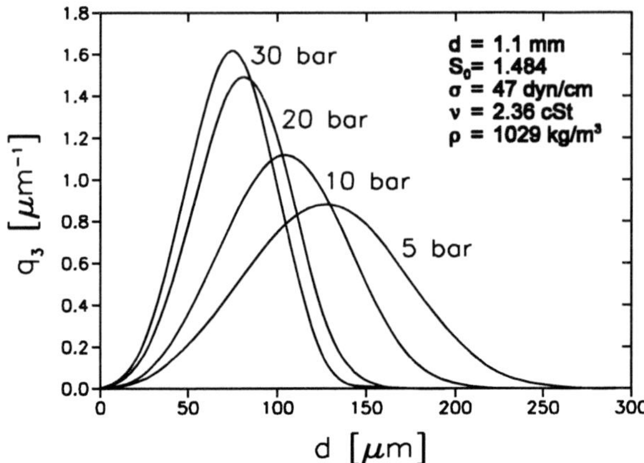

Figure 1.5.6: Droplet size distributions for different liquid flow pressure drop.

1.5.2.2 Pressure-Swirl Atomizer with Spill Return

The general atomizing mechanism for pressure-swirl atomizers with spill return is equivalent to the mechanism of nozzles without spill return. Also the nozzle design is similar, except that for spill controlled nozzles the rear wall of the swirl chamber, instead of being solid, has an opening through which part of the liquid stream can be "spilled" back from the atomizer to the liquid reservoir (Fig. 1.5.7, left side). The operation is such that the liquid is always supplied to the nozzle at its maximum flow rate ($\dot{V}_{\mathrm{F\,liq}}$). If maximum spray throughput is desired, the valve of the spill-flow is closed so that the spill feed ratio RV is zero ($RV = \dot{V}_{\mathrm{SR\,liq}}/\dot{V}_{\mathrm{F\,liq}}$). Opening the valve at the back allows part of the liquid to spill out the nozzle with a flow rate $\dot{V}_{\mathrm{SR\,liq}}$. In this way it is possible to operate this atomizer even at low output flow

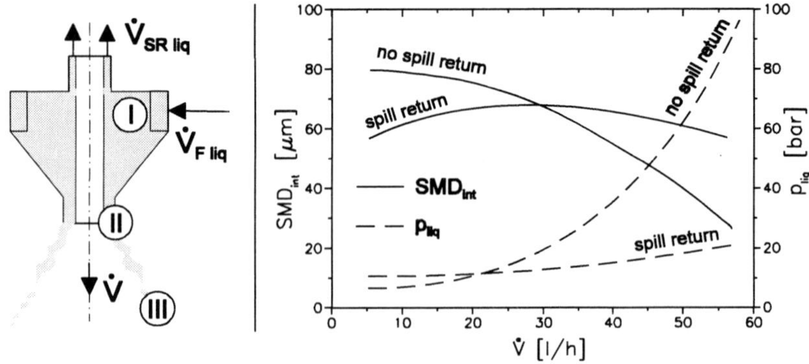

Figure 1.5.7: Scheme of a pressure-swirl atomizer with spill return (left), comparison of general behavior of pressure-swirl atomizer with and without spill return (right).

rates ($\dot{V}$) with adequate swirl to provide a thin film at the nozzle orifice and, therefore, good atomization. Hence, the essential advantage of spill-controlled nozzles compared to those without spill return is the much wider control range and the fairly constant spray quality over a wide throughput range, normally in a range of 0.3 : 1 (Fig. 1.5.7, right side).

Investigations of internal nozzle flow, film development and droplet formation were carried out with the same measuring techniques and procedures as used for the pressure nozzle without spill return. For measuring spatial distributions of velocity and droplet diameter downstream of the nozzle, a Phase Doppler Particle Analyzer has been used in addition [4].

Region I, Internal Nozzle Flow

Figure 1.5.8 exemplifies the flow fields for two different spill ratios RV at a constant nozzle feed rate of 35 l/h. The behavior of the hollow cone, marked at the left side of each graph, as affected by increasing spill ratio is obvious. At a ratio of $RV = 0$ the hollow cone radial width increases progressively towards the nozzle outlet, whilst for $RV = 0.86$ the hollow cone assumes an almost cylindrical shape of large diameter. In all cases, the tangential velocity profile resembles a Rankine vortex with solid body vortex characteristic in the vicinity of the nozzle axis and the tangential velocity decreasing towards the chamber periphery. As to the axial component profiles, a clear influence of spill return can be observed. Whilst at low spill ratios two radial zones of forward flow appear which merge with each other near to the orifice, for high values of RV ($RV > 0.75$) a pronounced backflow along the hollow cone establishes for nearly the whole nozzle chamber length. Near to the nozzle outlet the values of tangential and axial velocities are then rather small [5], due to high wall friction. Hence, there is a control range limitation of about 30 % exit flow rate also with spill return.

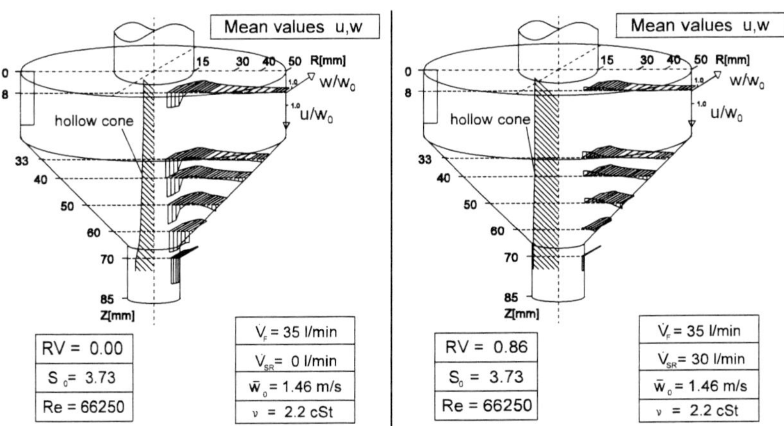

Figure 1.5.8: Radial profiles of tangential and axial velocity for internal pressure-swirl nozzle flow, $RV = 0$ (left), $RV = 0.86$ (right).

Region II, Liquid Film Development

Two main regimes are relevant for film disintegration and resulting droplet formation. For small Reynolds numbers the conical sheet gets periodically unstable, and wave formation can be observed. For higher Reynolds numbers aerodynamic forces act upon the sheet and lead to much faster disintegration which starts at small distance after the nozzle exit. That means that atomization by wave formation forms large droplets, whereas the latter mechanism more favorably forms smaller droplets. The disintegration mechanism is qualitatively unaffected by film thickness at the nozzle exit orifice. On the other hand, film thickness is responsible for resulting droplet diameters. According to Fig. 1.5.9 film thickness at the nozzle outlet depends on the spill return ratio and the feed ratio $\dot{V}_{F\,liq}$. For the highest spill-return ratios film thickness was found almost independent of the feed ratio, a film thickness of about only 100 μm was observed. Maximum film thickness, coupled with maximum droplet diameters downstream of the nozzle, occurred at a spill return ratio of about $RV = 0.4$. These results are nearly unaffected by nozzle geometry [6].

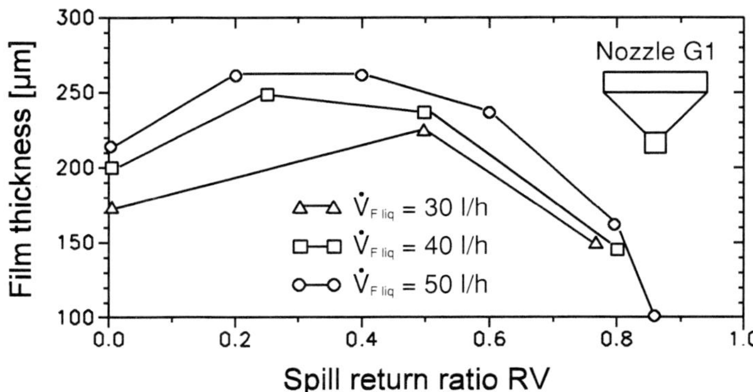

Figure 1.5.9: Influence of spill return ratio RV on film thickness at the nozzle outlet orifice.

Region III, Film Disintegration and Droplet Formation

For spill return ratios of about $RV = 0.4$ minimum angular momentum occurs at the nozzle exit, leading to maximum film thickness (see Fig. 1.5.9). This may be assumed leading to coarser atomization in this range of spill return ratio, which indeed is confirmed by increasing spatial diameter distributions with increasing spill return ratios up to $RV = 0.4$, as shown in Fig. 1.5.10. For the highest value of spill return ratio investigated a change in spray characteristic occurs. Increase in the SMD near to the spray axis can be observed, combined with a decrease of SMD at the spray boundary. For these high spill return ratios the film thickness at the nozzle throat is smaller than 100 μm, therefore the effect of the boundary layer viscous friction for deceleration of the tangential velocity component must be taken into account. For an esti-

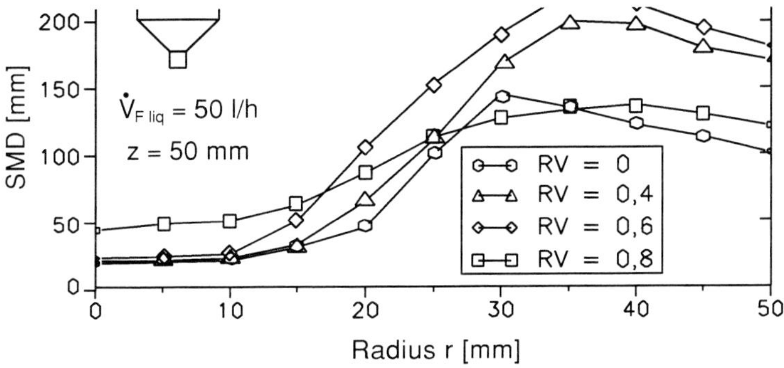

Figure 1.5.10: Radial distributions of SMD for different spill return ratios.

mation of the general spray characteristics of pressure swirl atomizers with spill return, the area and volume flux weighted integral SMD (SMD_{int}) is a better criterion as to practical nozzle qualification.

According to Fig. 1.5.11 the integral SMD increases only slightly with increasing spill return ratio up to $RV = 0.6$. Beyond this value, decrease from 190 µm ($RV = 0.6$) to 135 µm ($RV = 0.8$) takes place. At a spill ratio of $RV = 0$ the output volume flow rate ($\dot{V}$) is 50 l/h, whereas for $RV = 0.85$ it is only 7.5 l/h. When comparing this favorable result for the pressure-swirl nozzle with spill return with the behavior of a simple pressure-swirl nozzle it must be accentuated that is impossible to cover a control range of about 0.3 : 1 with nearly constant spray characteristics [4] for the latter nozzle type.

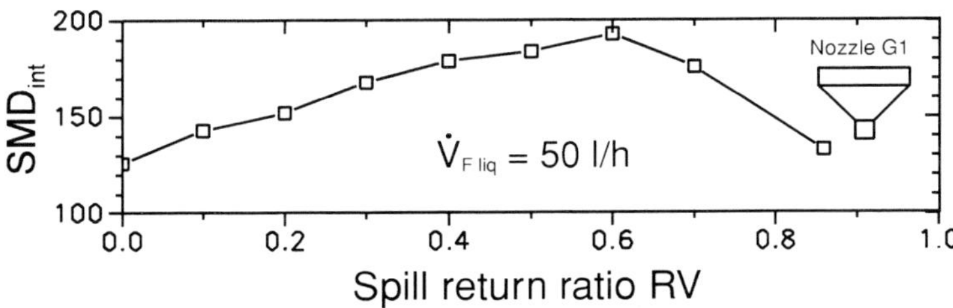

Figure 1.5.11: Integral SMD as a function of spill return ratio.

1.5.3 Results and Discussion for Internal-Mixing Air-Assist Atomization

With twin-fluid atomizers the energy required for the disintegration of the liquid is provided by a secondary fluid, usually compressed air or steam. The large differences of velocity between the slowly flowing liquid and the fast flowing surrounding air (or steam) generate aerodynamic shear forces. These shear forces initiate small disturbances at the liquid jet surface which grow continuously until disruption of the jet and formation of ligaments and finally droplets occur [1]. This process is commonly called primary atomization. Secondary breakup of ligaments and droplets in case of still high relative velocities between both phases may further contribute to the atomization process. It should be mentioned, however, that strict separation of primary and secondary atomization is always a simplification of the complex process of twin-fluid atomization.

The significant advantages of air-assisted internal-mixing atomizers are a homogeneous spray and low sensitivity as to the physical properties (surface tension, viscosity, particle content) of the liquid. Furthermore, by adjusting the atomizing gas flow rate to the liquid flow rate, it is possible to maintain the droplet size distributions approximately constant over a wide load range of 0.05 : 1.

In the past various kinds of air-assisted internal-mixing atomizers have been investigated [7, 8]. Published results are often very much dependent on the specific design of the atomizer and, furthermore, are valid for the investigated operating conditions only. There is still a lack of basic research on fundamental flow structures inside the atomizer and their correlation with resulting spray data. Such investigations should lead to better understanding of the atomization process. Furthermore, if combined with known correlations for physical phenomena like disintegration mechanisms [9] and two-phase flow patterns [10], they should allow the design of atomizers with required specific spray characteristics.

The effects of nozzle design and load on spray characteristics of internal-mixing twin-fluid atomizers were investigated using a nozzle with rotational symmetry (Fig. 1.5.12, left side) by employing a Phase Doppler Analyzer (PDA) equipment. The relevant flow structures were visualized by high speed photography in combination with an optically accessible planar nozzle (Fig. 1.5.12, right side). These investigations allowed to identify main parameters governing the atomization process and the correlation of spray data with flow structures. A detailed description of the test rig used can be found elsewhere [11].

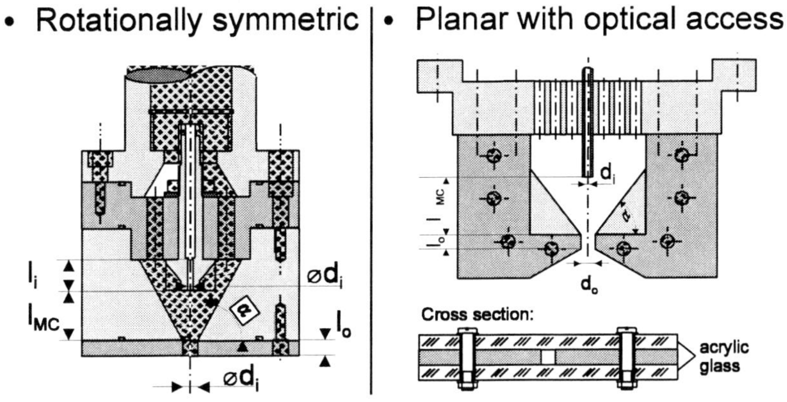

Figure 1.5.12: Internal-mixing air-assist nozzles, rotationally symmetric (left) and optically accessible (right) experimental versions.

1.5.3.1 Influence of Outlet Port Length

It was found that the dominant geometrical nozzle parameter influencing atomization quality is the length of the outlet port L [11]. Figure 1.5.13 (left) shows measured radial distributions of the local Sauter mean diameter (D_{32}) as a function of the length-to-diameter ratio. Corresponding photos of flow visualization with the optical accessible planar nozzle (Fig. 1.5.12, right side) are shown in Fig. 1.5.13 (right side). The liquid mass flow rate was $\dot{M}_{\text{liq}} = 40$ kg/h, and the air mass flow rate was $\dot{M}_{\text{air}} = 8$ kg/h, corresponding to an *ALR* (air to liquid ratio) of 0.2.

Strong influence of the length to diameter ratio *L/D* on the atomization quality can be seen from Fig. 1.5.13, left side. For the ratio $L/D = 1$ the atomization quality is poor over the entire cross-section of the spray. When looking at the flow structures inside the nozzle (Fig. 1.5.13a, right side) it becomes obvious that a great portion of the atomizing air stream leaves the nozzle as an annular co-flow around the central two-phase flow. The disintegration process starts in a small distance upstream of the beginning of the outlet port where air velocities due to the converging channel geometry are high enough to disturb the liquid jet. Hence, higher ratios of *L/D* increase time and path length available for disintegrating the liquid [12]. Such an extension leads to a reduction of the D_{32} mean droplet diameter across the entire cross-section up to *L/D* ratios of approximately 2. Flow visualization for $L/D = 2$ (Fig. 1.5.13b, right side) shows that the co-flow of air now disappears and nearly the whole cross section of the outlet port is covered by the two-phase flow now. Therefore, under those circumstances less air can escape from the nozzle as an annular co-flow without participating into the atomization process. However, according to Fig. 1.5.13 a further increase of *L/D* is linked with a pronounced increase of the D_{32} at the outer spray bound-

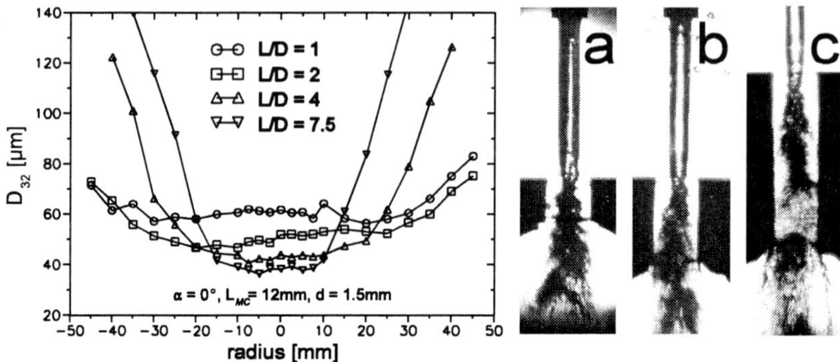

Figure 1.5.13: Influence of outlet port length L on local SMD and corresponding two-phase flow patterns.

ary. The improvement of droplet size in the center part of the spray by longer residence times and, therefore, better momentum exchange between air and liquid along the outlet port is small. As a reason for the distinct droplet size maximum at the spray boundary the development of an annular film in the outlet port for $L/D \geq 4$ (Fig. 1.5.13c, right side) could be identified. Droplets and ligaments start to accumulate on the wall of the outlet port and form an annular film that leaves the nozzle without being sufficiently atomized anymore. The slip velocity between air and liquid phase at the spray boundary is significantly smaller outside of the nozzle than inside the outlet port, which leads to significant coarser atomization and, hence, larger droplets in the outer regions of the spray cone. For $L/D = 7.5$ this flow condition was found even more pronounced than for $L/D = 4$.

The influences of other nozzle geometry parameters on the flow conditions inside the nozzle and their correlation with spray data, like air impingement angle α, length of the mixing chamber length L_{MC} and liquid orifice diameter d, have been described in detail in [13] and will not be discussed here.

1.5.3.2 Influence of Liquid Flow Conditions on Spray Characteristics

In Fig. 1.5.14 the pressure drop of the air flow is plotted against the volume flow rate of the liquid. The liquid flow Reynolds number (Re) is depicted in parallel as a second axis. The values plotted here are specific for constant air flow rate and a certain nozzle geometry. High speed photos of the disintegration process in- and outside the optically accessible nozzle – corresponding to the four most important liquid flow regimes – are assigned to the diagram.

For very low liquid flow rates (I) disintegration of the liquid jet is completed before it reaches the outlet port. The bulk of the liquid, however, impacts on the inner parts of the nozzle and enters the outlet as a wall

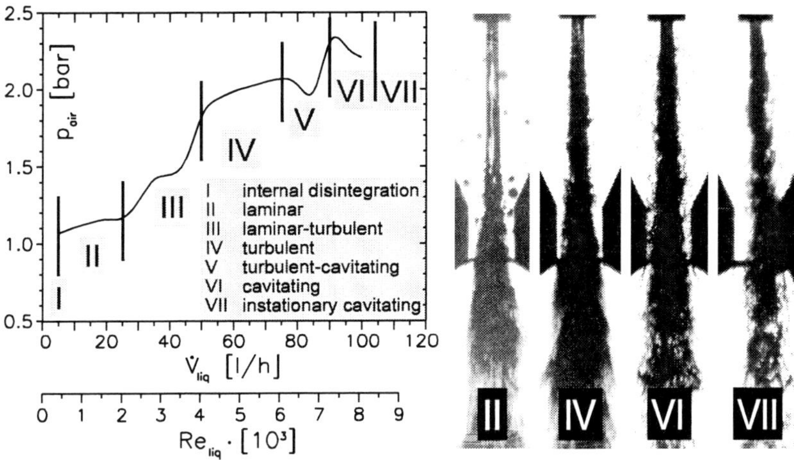

Figure 1.5.14: Pressure drop of atomizing air for varying liquid volume flow rate (left), corresponding phenomenological representation of internal nozzle flow (right).

film. Therefore, satisfactory atomization cannot be achieved. With increasing liquid flow rate (II) a regular and non-pulsating atomization process starts. In this regime the pressure drop weakly depends on the liquid flow rate. The disintegration of the laminar liquid jet starts closely upstream of the outlet port where the differences in velocity between the slowly flowing liquid jet and the fast flowing air are high enough to generate first visible disturbances at the liquid jet surface. At the end of the outlet port, marked by the horizontal lower edge of the window, the outlet port cross-section is still not homogeneously covered by two-phase flow. Therefore, part of the air passes the nozzle as an annular co-flow without really contributing to the atomization process.

Further increase of the liquid flow rate (III) causes strong increase of the pressure drop. This regime is characterized by sudden and oscillatory variations of spray characteristics and pressure drop. The atomization process becomes unsteady, accompanied by the emission of high-frequency acoustic waves. As a stationary atomization process cannot be achieved, this transition regime between the laminar (II) and the turbulent (IV) flow conditions of the liquid excludes its satisfactory technical application.

Like in the laminar case also for the turbulent liquid flow (IV) the pressure drop increases slowly with increasing liquid flow rate. The atomization process becomes stationary again. Disintegration now starts already in a short distance downstream the inlet port. The cross-section at the end of the outlet port is covered nearly completely by two-phase flow. Therefore, improved energy transfer from the gas to the liquid flow is now obtained as compared to the laminar case. For again higher liquid flow rates, non-steady transition (V) to cavitation (VI) takes place, related with further increase of pressure drop. The disintegration process in this cavitation regime is similar to the turbulent case. The corresponding improvement of the disintegration process is detectable, but not significant. For still higher volume flow rates (VII) cavitation becomes unstable. The liquid jet is now being deflected away from the nozzle axis. The pressure drop may decrease because the cross section is partly not covered, and in this case the atomization quality drastically deteriorates. For details concerning the liquid jet breakup with and without surrounding air, and as to the transition from laminar to turbulent jet behavior, see [14–16].

When investigating liquid jets in quiescent air, the regions of transition from laminar to turbulent are comparable to our findings here. It can be concluded, therefore, that transition (III) between two atomization regimes is caused by variation of the liquid jet characteristics – laminar to turbulent, respectively – which can be described by the liquid flow Reynolds number. On the other hand, the onset of cavitation strongly depends on the pressure inside the mixing chamber and cannot be described by the liquid flow condition only.

Since transition can be described by the Reynolds number it depends on the liquid flow rate, the diameter of the inlet port and the liquid viscosity. The inlet port diameter d_i was varied between 0.6 and 1.95 mm for constant air flow rate, outlet port length and diameter. Figure 1.5.15, left side, shows the pressure drop of the air flow as a function of the Reynolds number for four inlet port diameters ($l_i/d_i = 10$). There occurs a sudden increase in the pressure drop for Re $\geq$ 3000 except for the smallest diameter investigated. In this case a more gradual pressure increase could be observed. For d_i = 0.9 and 1.3 mm a transition regime from Re $\approx$ 2000 to Re $\approx$ 3000 occurs, whereas for d_i = 1.95 mm no transition regime was found. These differences of the appearance of transition can be explained by the formation of disturbances. The smaller the diameter the more important are external effects such as manufacturing imperfections. For the smallest diameter it was, therefore, impossible to generate a liquid jet with really laminar outlet conditions. This jet was transparent, but revealed a large scale helix structure. Therefore, these smallest inlet ports could not be investigated in detail.

Like mentioned before, transition from laminar to turbulent liquid flow is related to a drastic change of spray characteristics. Figure 1.5.15, right side, shows spatial D_{32} distributions (half profile) for two inlet port diameters with both flow conditions ($l_i/d_i = 10$). For both laminar cases maxima of D_{32} can be observed at the spray axis. At the spray boundary D_{32} values are far smaller. Like in Fig. 1.5.14-II air leaves the nozzle with high velocity and leads to disintegration of the liquid into fine droplets mainly in the outer zones of the spray. Complete disintegration of the ligaments near the spray axis is not achieved, however. In contrast to the laminar case, no ligaments are to be observed at the bottom of the photograph for the turbulent case (Fig. 1.5.14-IV). At the nozzle exit the disintegration process is nearly complete. This corresponds to the result that in both turbulent cases a nearly even profile of the D_{32} was obtained. When varying the viscosity of the liquid, these general findings of a dependence of the disintegration mechanism and

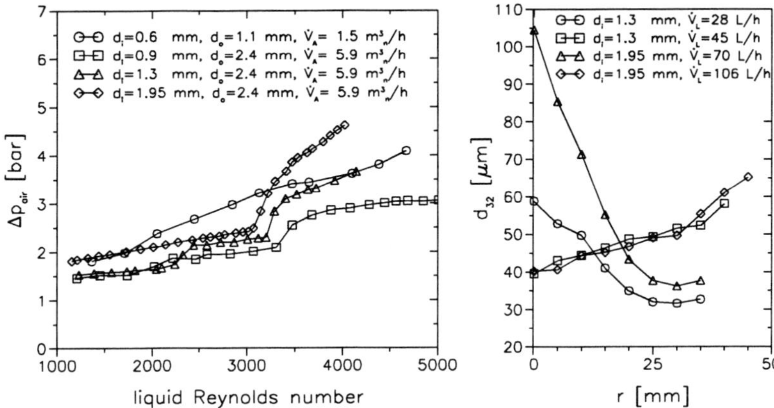

Figure 1.5.15: Pressure drop of atomizing air for different inlet port diameter values as a function of liquid flow Reynolds number (left), and corresponding local SMD for different liquid flow inlet conditions.

the spray characteristics on the liquid Reynolds number were confirmed as well [17]. The effects and the relevance of parameters apart from liquid flow conditions, like air mass flow rate and nozzle geometry dimensions, are described elsewhere [17] and will not be discussed here.

References

1. Chigier, N. A. (1991): The Physics of Atomization. Proc. ICLASS-91, 1–15, Gaithersburg, MD, U. S. A., pub NIST 813.
2. Lefebvre, A. H. (1998): Atomization and Spray. Hemisphere Publishing Corporation.
3. Horvay, M. (1985): Theoretische und experimentelle Untersuchungen über den Einfluss des inneren Strömungsfeldes auf die Zerstäubungseigenschaften von Drall-Druckzerstäubern. Dissertation, Universität Karlsruhe, Bereich Feuerungstechnik am Engler-Bunte-Institut.
4. Löffler-Mang, M. (1992): Düseninnenströmung, Tropfenentstehung und Tropfenausbreitung bei rücklaufgeregelten Drall-Druckzerstäubern. Dissertation, Universität Karlsruhe, Bereich Feuerungstechnik am Engler-Bunte-Institut.
5. Löffler-Mang, M. (1991): Atomization with Spill-Controlled Swirl Pressure-Jet Nozzles. Proc. ICLASS-91, 431–440, Gaithersburg, MD, U. S. A., pub NIST 813.
6. Löffler-Mang, M. (1991): Investigation of the Processes near to the Orifice of Spill-Controlled Swirl Pressure-Jet Nozzles. *Proc. Sprays and Aerosols* **91**, 72–77.
7. Nukiyama, S., Tanasawa, Y. (1938/1939): An Experiment on the Atomization of Liquid by Means of an Air Stream. *Trans. Soc. Mech. Engrs.* (Japan) **4/5**, 13–17.
8. Wang, G., Mao, C. P., Dietvorst, J., Chigier, N. (1987): An Experimental Investigation of Air-Assist Non-Swirl Atomizer Sprays. *Atomization and Spray Tech.* **3**, 13–36.

9. Faragó, Z., Chigier, N. (1992): Morphological Classification of Disintegration of Round Liquid Jets in a Coaxial Air Stream. *Atomization and Spray Tech.* **2**, 137–153.
10. Chin, J.S., Lefebvre, A.H. (1993): Flow Patterns in Internal-Mixing, Twin-Fluid Atomizers. *Atomization and Sprays*, **3**, 463–475.
11. Kufferath, A., Heyse, C., Leuckel, W. (1996): Influence of Outlet Port Length on Liquid Atomization by an Air-Assisted Internal-Mixing Nozzle. ILASS 96, Lund, Sweden, 179–184.
12. Pilch, M., Erdman, C.A. (1987): Use of Breakup Time Data and Velocity History Data to Predict the Maximum Size of Stable Fragments for Acceleration-Induced Breakup of a Liquid Drop. *J. of Multiphase Flow*. 741–757.
13. Kufferath, A., Leuckel, W. (1997): Experimental Investigation of Flow Conditions Inside an Air-Assisted Internal-Mixing Nozzle and their Correlation with Spray Data. Proc. ICLASS 97, Seoul, Korea, 262–269.
14. Schweitzer, P.H. (1937): Mechanism of Disintegration of Liquid Jets. *J. of Applied Physics* **8**, 513–521.
15. Wu, P.K., Faeth, G.M. (1993): Aerodynamic effects on primary breakup of turbulent liquids. *Atomization and Sprays* **3**, 265–289.
16. Faragó, Z., Chigier, N.A. (1992): Morphological Classification of Disintegration of Round Liquid Jets in a Coaxial Air Stream. *Atomization and Spray Tech.* **2**, 137–153.
17. Kufferath, A., Wende, B., Leuckel, W. (1998): Influence of Liquid Flow Conditions on Spray Characteristics of Internal Mixing Twin Fluid Atomizers. ILASS 98, Great Britain, Manchester, 179–184.

2 Flow and Combustion

Flow, Mixing, and Reaction in High Intensity Combustors

Bernhard Lenze*

Before the reaction of two and more components occurs these components have to be brought together and mixed very well. This is valid for every reaction scheme as well as for the combustion of gaseous, liquid, and solid fuels with oxidants like pure oxygen and/or air. The velocity of the reaction in combustion systems – in flames – depends on the fuel and its components, on the fuel concentration, on the temperature and on the pressure in the mixture of fuel and oxidant. The highest reaction velocities occur in nearly stoichiometric mixtures in which the highest reaction rate, combustion efficiency and flame temperatures are usually found with a complete burn out. The main question is, how to manage the fast and complete mixing with the result of a good, complete combustion with low concentration of incomplete, toxic and the atmosphere loading waste gas components like residual hydrocarbons, carbon monoxide, sulphur oxides, nitrogen oxides, soot, and other emissions. But before reaching that aim the stability of the flames has to be investigated, what means that flow, mixing, and temperature profiles should produce a very stable ignition zone or flame root near the burner outlet independent of the flow throughput at a given equivalence ratio, which is nearly stoichiometric in most cases.

The next aim is to ensure a good or proposed mixing in the flame zone after the stabilization zone with regard to sometimes a fast and complete or to sometimes a slower and, in the first stage, incomplete mixing and combustion with the aim of staged combustion for reducing thermal and/or fuel NO_x for instance. At the end a complete combustion is normally demanded in a certain time or length scale and with low emission concentrations in the flue gas. Beside these aims the demand for the heat exchange from the flame to the combustor walls and/or to the up heated products is to realize, which influences the overall efficiency of high intensity combustors. The heat transfer can influence the combustion process with its more or less high emission as

* Engler-Bunte-Institut, Bereich III – Verbrennungstechnik, Universität Karlsruhe, Kaiserstr. 12, 76128 Karlsruhe, Germany

well as the quality of the products and the lifetime of the burners, combustors, and furnaces.

In the frame of the experimental and theoretical work of the Collaborative Research Centre 167 one main topic was the stability of free and enclosed flames ensured by disc and/or swirl stabilized flames in the projects A1 and A2. The results obtained show how one can stabilize flames by inner and outer recirculation zones and how to ensure the right mixture of fuel and oxidant in the stabilization zone near the burner. Many experimental investigations and calculations by mathematical modelling have been done in the project A9, to describe the conditions for stable flames in dependence on fuel, turbulent flow field, and the stabilization methods like forming outer and inner recirculation zones by disc and/or swirl. Numerous in-flame measurements under different air flow rates and constant thermal throughputs help to understand the blow-off-mechanism. Based on this a Peclet number model could be constructed, to predict the stability of by disc and swirl stabilized flames in the dependence on fuel, geometric scales, and turbulent flow conditions. The central assumption of this model is an agreement of the time scales of the mean flow in the ignition zone with the chemical reaction time scale – according the Damköhler criteria – to describe the flame stability of different scaled burners using the input conditions only.

On the way to calculate the flow and concentration in the burner near field by mathematical modelling in project A9 difficulties arose, to describe the turbulent flow and mixing field of swirling jets and flames with the well defined k-ε-, or other (RSM-, ASM-, KLM-, ...) models. Therefore, an extended hot wire anemometry system with four or five hot wires and a fast compensated thermocouple was constructed, to measure the flow field of unswirled and highly swirled cold and heated streams. The measurements included all main velocities $\overline{u}$, $\overline{v}$, $\overline{w}$, the fluctuating velocities u', v', w' and the Reynolds stresses $\overline{u'v'}$, $\overline{u'w'}$ and $\overline{v'w'}$ as well as temperature fluctuations. The measurements in project A10 show unexpected results which explain the difficulties in modelling high intensive swirling flows and which lead to the so-called swirl induced turbulence at least. These results have been introduced into different mathematical model codes by varying Richardson and other dimensionless numbers. These so corrected or extended models give much better results for turbulent flow and mixing fields, for entrainment ratios as well as for extended intensities of inner and outer recirculation zones for enclosed swirling flows and flames. The found numerous results of experiments and calculations at first extracted from measurement on cold non-reacting systems have been translated to reacting flows. Then the mixing, reaction, and burn out of natural gas flames could be predicted with quite good agreement compared with the measurement results.

Combination of these adiabatic models with radiative heat transfer models leads at least to a prediction of the temperature field in high intensity combustors and enables the researchers to make first predictions regarding the NO_x-formation in different swirling flames depending on wall temperature, swirl parameter, and excess air factors. This is valid at first for gaseous fuels like natural gas, hydrogen, or higher hydro carbons.

But the described calculations and measurements are also important for the combustion of solid and liquid fuels, whereby liquid fuels – light oil and kerosene – will be burnt in gas turbines, power stations, and air craft gas turbines. Heavy fuel oils and industrial residual liquids will be fired in steam boilers of the refineries and waste incinerators of the chemical industry. For these burners and flames the atomization process is the first point of research, which has been done in numerous experimental projects on twin fluid (A11), swirl pressure (A3), shear driven liquid film (air blast) atomizers (A4) as well as in modelling projects as A12, A13, A14. After atomization process, which usually needs a short time or way in the near field of the burner only, the processes of evaporation, gasification, ignition, and mixing with the surrounding turbulent non-swirled or swirled combustion air will occur and should end – as for gaseous fuels – in a high stable, efficient, and complete combustion process with high turbulent exchange of mixture and heat transfer by radiation and convection and, at least, low emissions of soot, hydrocarbon, carbon monoxide, and nitrogen oxides. These problems have been investigated experimentally and numerically in the projects B1, B7, B8 and A6, A8, A9, and A11.

Heat transfer by radiation and/or convection combined with convective cooling systems has a high influence on the construction of burners and combustion chambers and their materials. These important influences have been investigated in some projects like B2, B7 and C1, C2/C3, C10 and C12.

Summarizing one can conclude, that turbulent flow restrains and influences the mixing and reaction rate in a flame which steers the stabilization, the heat transfer and emissions of high intensity combustion systems and influences the selection of geometry, construction and the lifetime of burners and combustion furnaces. On the other side the constructions and the sizes of burners and furnaces influence the heat transfer of the flame which causes altered temperatures and reaction rates and at least variations in turbulent flow structure, turbulent transfer coefficients of momentum, mass, and heat. All measurements in isothermal and reacting flow show influences of turbulence values on mixing and reaction and influences of reaction on turbulence behaviour. All investigations done in the Collaborative Research Centre 167 help to explain the differences between reacting and non-reacting turbulent flows and to develop better mathematical models for flames and ensure at least better designs and constructions of burners, combustors, furnaces, and high temperature industrial units with high efficiencies and low emissions.

References

Reports No. I (1986), II (1989), III (1992), and IV (1995) of the Collaborative Research Centre 167, University of Karlsruhe.

2.1 Stabilisation of Turbulent Concentric and Swirling Flames Based on Flow and Mixing Pattern Investigations

Peter Schmittel, Bernd Prade, Stefan Hoffmann, and Bernhard Lenze*

Abstract

In order to describe flame stability of turbulent, premixed, and diffusion flames, the flow-, mixing-, and reaction properties as well as the corresponding turbulence characteristics have been investigated in burners with inner by disk or swirl induced recirculation zones. Another aim of these research activities was to examine experimentally the mechanisms of ignition stability of these both types of flames, in order to generate and explain blow-off limits of differently scaled burners. Parallel to the experimental work, theoretical investigations were done to predict blow-off limits for both flames at different operating conditions as throughput, gas composition, burner geometry, and aerodynamic input conditions with mathematical models.

2.1.1 Introduction

The demands in highly loaded combustion systems – for example gas turbines, industrial furnaces, and boilers – for good stability and burnout characteristics, result in the frequent use of flames with inner recirculation zones. One of the essential problems associated with the design of these combustion systems is how to sustain a steady and stable flame at the burner exit over a wide range of operating conditions e. g. load, flow velocity and excess air. Because of their good stability, the characteristics of inner recirculation zones (storage of heat) induced by bluff-body or swirling flows are used as a flame stabilization technique in high speed flows.

The aim of this project is to examine the stability mechanism of turbulent swirling and concentric flames experimentally and to describe the

* Engler-Bunte-Institut, Bereich III – Verbrennungstechnik, Universität Karlsruhe, Kaiserstr. 12, 76128 Karlsruhe, Germany

blow-off with theoretical models. Hence the knowledge about the flow-, mixing- and reaction properties in the complete flame but specially in the ignition zone is needed. In this context experimental investigations of velocities with Laser-Doppler-Velocimeter, concentration and time resolved temperature measurements in the reacting flow were done. The experiments were performed for differently scaled concentric and swirl burners under free and enclosed conditions. Flame stability is the basic condition to do other fundamental investigations, e. g. of NO-formation in highly turbulent flames.

2.1.2 Flow- and Mixing Pattern

Inner recirculation zones are often used to stabilize highly turbulent flames. This zone can be induced with a bluff-body or with a swirl generator. The mechanism of stability improvement is based on the following positive conditions in the ignition zone even at high throughputs: near stoichiometric mixture (diffusion flames), low mean flow velocities, energy transport of heat and chemically active species by turbulent exchange from the reverse flow to ignite the fresh mixture and high turbulent burning velocities.

Basically, swirl and disk stabilized flames can be subdivided on the base of different mixing and reaction structure into two types of flames [1]. So-called type-1 flames are formed when the central fuel jet is able to penetrate the inner flow zone completely (see Fig. 2.1.1, left). Then the internal reverse flow zone builds a small annular region of recirculation zone between the fuel gas jet and the combustion air. Type-2 flames are formed when the inter-

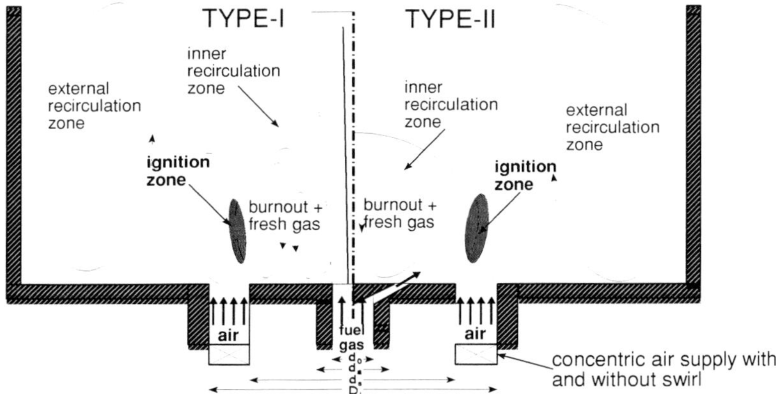

Figure 2.1.1: Flow- and mixing field of an enclosed type-I and type-II flame with inner recirculation zone formed by disk and/or swirl.

nal reverse flow strength prevents the forward progress of the fuel jet. Then the fuel flow is drifting radially away from the flame axis. Because of the rapid mixing in the shear layer the length of this flame type is very short. In the case of enclosure an external recirculation zone shown in Fig. 2.1.1 too is formed.

2.1.2.1 Disk Stabilized Flame

The flame stability characteristics of a bluff body burner are mainly determined by the aerodynamics of the flow around the body. Therefore intensive experimental investigations as well as calculations on the flow- and mixing patterns were done [8, 14, 16]. The used measurement techniques are described in Section 2.1.4. Figure 2.1.2 shows a typical stabilisation diagram of a type-I disk stabilized flame in dependence on thermal throughput $\dot{Q}$ and air ratio λ. The corresponding flow- and mixing field of different operating conditions with various ratios of the fuel/air-velocity at the burner exit is shown in Fig. 2.1.3. Obviously there is a remarkable influence of this ratio beside the disk diameter on the axial extension of the recirculation zone and therefore on the residence time in the stoichiometric zone with ignitable mixture. With higher air velocities (U_{L}) at the burner exit the near stoichiometric mixture is shifted into the forward flow (points C $\rightarrow$ B $\rightarrow$ A $\rightarrow$ Y in Figs. 2.1.2 and 2.1.3). More detailed investigations of this problem with different scaled burners have been done. In Fig. 2.1.3 the location of the stoichiometric mixture and the limits of flammability are shown. The critical cases for combustion stability are illustrated in Figs. 2.1.3E and 2.1.3F, whereby Fig. 2.1.3A shows a steady burning flame. In the case of small air flow the ignitable mixture is located inside the recirculation area. The heat transporting

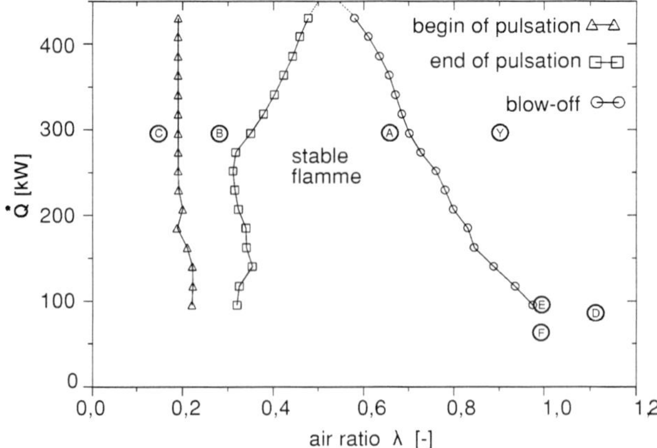

Figure 2.1.2: Stability diagram of a disk stabilized flame (D = 80 mm).

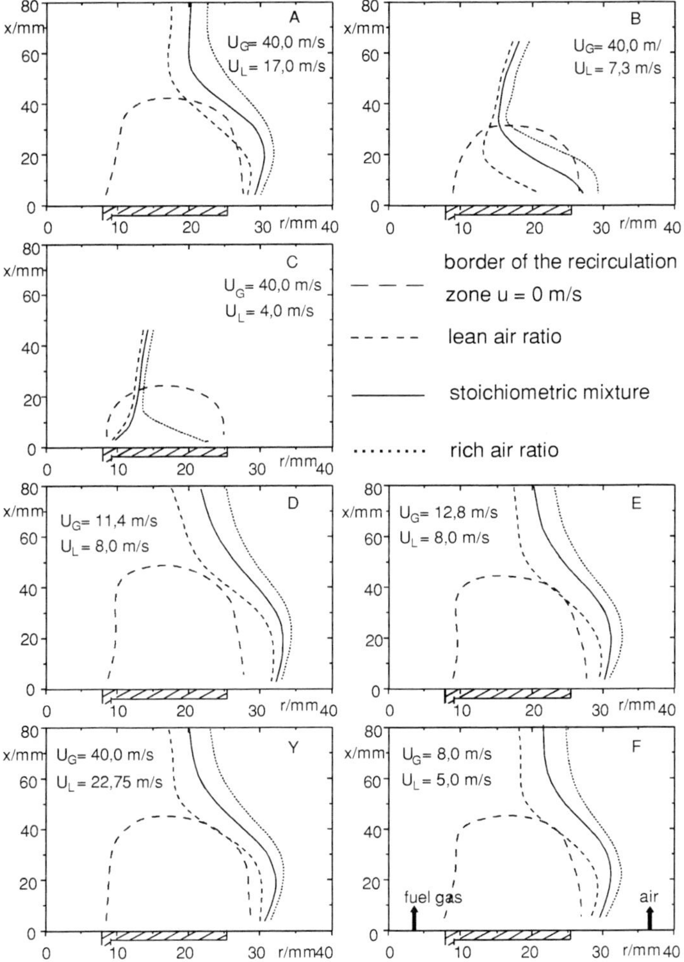

Figure 2.1.3: Location of the recirculation area and characteristic mixture fractions, variation of fuel (U_G)/air exit velocity (U_L).

streamlines in the forward flow will find an ignitable mixture in a distance far away from the burner tip, so that the main part of the fuel is burning as a lifted pulsating flame. The other critical case (Fig. 2.1.3Y) occurs at high air-exit momentum, because the ignitable mixture-area is then located completely in the forward flow with simultaneously high flow velocity, so that no balance of flow and turbulent flame velocity is possible and blow-off occurs.

2.1.2.2 Swirling Flames

Swirling flows develop low or actual reverse velocities near the jet axis after burner exit. This is a consequence of a large underpressure at the centre of the emerging jet, balancing the centrifugal forces. The development and intensity of the inner recirculation depends on the swirl intensity. The swirl level is indicated by the dimensionless swirl number S_0 defined as the ratio of angular momentum $\dot{D}$ and the axial momentum $\dot{I}$ multiplied with the throat radius at the burner exit:

$$S_0 = \frac{\dot{D}}{\dot{I} \cdot R_0} \tag{1}$$

To study the mechanism of swirling flame stabilization, detailed experimental and numerical [2, 3, 12] investigations of flow pattern, concentration, and temperature distributions for characteristic conditions were made near the blow-off limit of several typical swirling flames. Three examples of results of enclosed diffusion swirling-flames with different air ratio are shown in Fig. 2.1.4. The fat line (axial velocity $u = 0$) indicates the boundary between the forward flow and the inner recirculation zone. The area of near stoichiometric mixture (thin black line) contracts with increasing air flow rate. As a result of this, the size of high temperature zones gets smaller (dilution effect). In order to heat up the fresh combustible mixture to its ignition temperature, energy must be transported from the inner zone to the flame root. When the burner exit flow velocity is increased (increasing the air ratio at constant gas flow rate), the residence time of the gas in the reaction zone becomes shorter and in combination with lower temperatures in the ignition zone the burnout α (dotted line) in the downstream regions becomes more and more incom-

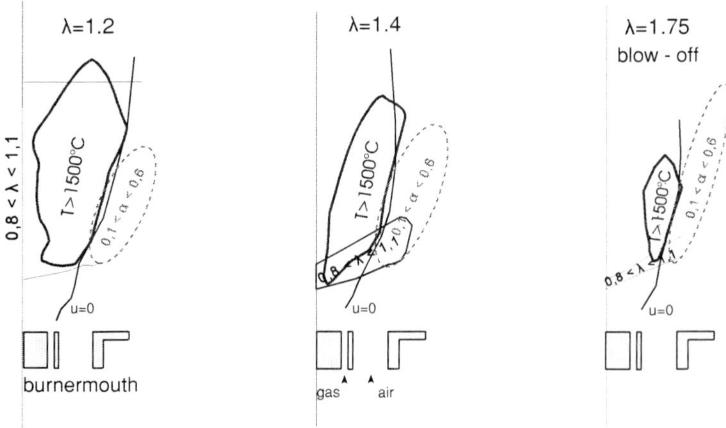

Figure 2.1.4: Location of the recirculation area, mixing (λ) and temperature (T) distribution in swirling flames for different air ratios ($P_{th} = 150$ kW; $S_{0,th} = 0.9$).

plete. At the stability limit, the low burnout causes that the heat input in the inner recirculation zone decreases to a level, where the mechanism of fresh gas ignition by hot burned gases is not able to keep the whole chemical reaction going. The result is a sudden "blow-out" of the flame, caused by too high velocity in the stabilisation zone.

Numerous investigations [2–4] about high turbulent flames show that aerodynamically controlled properties like the mean flow and the mixing field are almost Reynolds number independent. In this context investigations on premixed free swirling flames with different thermal loads and flow velocities were done. The dependencies of the reaction influenced parameters on different thermal loads are shown in Fig. 2.1.5. The CH_4-concentration as a characteristic mixing parameter is nearly independent for different throughputs. An increase of the throughput at constant air ratios leads to a decrease of the mean temperature, also the heat release is relative smaller because of the shorter residence time, what is pointed out by a decrease of the CO_2 concentration and lower burnout rates.

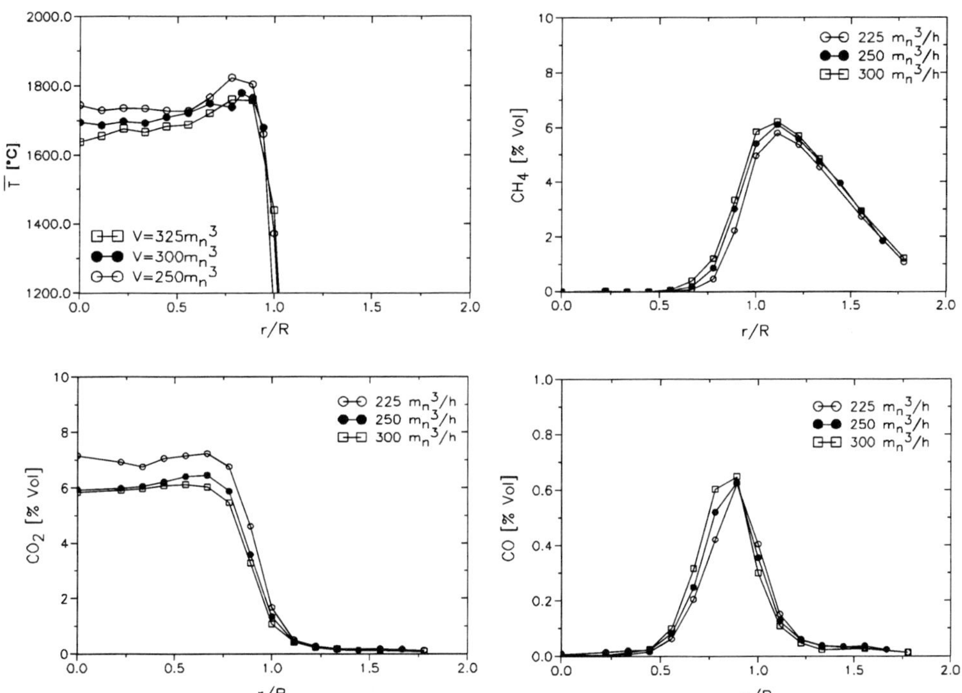

Figure 2.1.5: Radial temperature- and concentration distributions at $x/D = 0.5$ in premixed swirl flames ($D = 90$ mm, $\lambda_0 = 1.1$, $S_{0,\text{th}} = 1.5$) for different flowrates.

2.1.3 Flame Stability Models

Concerning the flame stability of laminar and turbulent premixed flames a lot of studies have been made in the past [4–7]. All these authors make different assumptions on the interaction of the flow and the reaction field to describe the flame stability with semi-empirical models. There are also numerical models in use for the prediction of flame stability [2]. The basic assumption of every stable flame is the balance between the flame velocity S and the opposite directed mean flow velocity $\bar{U}$ in the ignition zone.

$$S = -\bar{U} \tag{2}$$

Investigations [3, 8] about the reaction structure of highly turbulent swirling and disk stabilized flames showed that the reaction takes place at relative low Damköhler numbers (homogeneous-turbulent flame front in the Borghi Diagram [9]). This leads to the following expression for the turbulent flame speed.

$$\frac{S_t}{S_l} \sim \sqrt{\frac{u' L_t}{a}} \tag{3}$$

Using the geometrical resemblance of the burner/combustion chamber system with equal thermal boundary conditions (wall temperature) the aerodynamical proportionality leads to:

$$u' \sim U \quad ; \quad L_t \sim L_{char} \tag{4}$$

where U is a characteristic burner exit velocity and L_{char} a characteristic system length.

$$S_t \sim U \tag{5}$$

Equation (3) with (4) and (5) leads to:

$$\frac{U}{L_{char}} = K \cdot \frac{S_l^2}{a} \tag{6}$$

K is an empirical constant. Equation (6) can be transformed into a Peclet number based on the blow-off velocity U and the laminar flame speed S_L:

$$\frac{U \cdot L_{char}}{a} = K\left(\frac{S_l\, L_{char}}{a}\right)^2 \Rightarrow \mathrm{Pe}_U \sim \mathrm{Pe}_{S_l}^2 \tag{7}$$

This Peclet number based correlation can be used to predict the stability limits of different scaled premixed burners if U and L_{char} are adjusted on the respective flame system [7]. It is also possible to transfer this mode on non-premixed swirl and disk stabilized flames, because the results of field measurements show that the stabilization zone exists at stoichiometric mixture. The flame speed S_L has then been substituted with the amount of the stoichiometric laminar flame speed S_l.

2.1.3.1 Disk Stabilized Flames

As already mentioned, parts of fuel and oxidizer have been mixed at least partially in the recirculation zone, and therefore models based on the Peclet numbers can be used to correlate critical flow rates. Minx [10] submitted a Peclet number based model to describe the stability behaviour of concentric bluff-body flames. The mean assumptions of this model are:

- At the blow-off limit at maximum air flow u_{max} a stoichiometrically mixed area is located directly at the border of the inner recirculation area.
- The necessary kinetic time can be described using a/S_l^2.
- The length of the inner recirculation zone L_{RZ} is proportional to the burner geometry, leading to a characteristic residence time $\tau_{RZ} = L_{RZ}/u_{max}$.

The equivalence of residence and reaction time results in the correlation:

$$B \cdot \mathrm{Pe}_{\bar{u}_{max}} = C \cdot \mathrm{Pe}_{S_l}^{n} \tag{8}$$

with the blockage ratio B and the empirical constants; $C = 2{,}37 \cdot 10^{-5}$ and $n = 3.68$. The Pe-numbers are formed as:

$$\mathrm{Pe}_{\bar{u}_{max}} = \frac{\bar{u}_{max}\,(d_s - d_0)/2}{a} \quad ; \quad \mathrm{Pe}_{S_l} = \frac{S_l(d_s - d_0)/2}{a} \tag{9}$$

$$B = \frac{\text{blocked area (air exit)}}{\text{total area without disk (air exit)}} \tag{10}$$

The thermal diffusivity a [m^2/s] is calculated for a stoichiometric fuel/oxidizer mixture.

The plotted points in Fig. 2.1.6 represent the Pe-numbers with the burner-exit velocity of the combustion air at the point of maximum fuel flow.

Because of the small dimensions of the burner (24 mm $< D <$ 36 mm) Minx used for the experiments, the flow field can be classified in the laminar-turbulent transition range. This is the reason for the discrepancy between his experimental results and the results from investigations on larger burners (40 mm $< D <$ 80 mm) that were performed in this project [8].

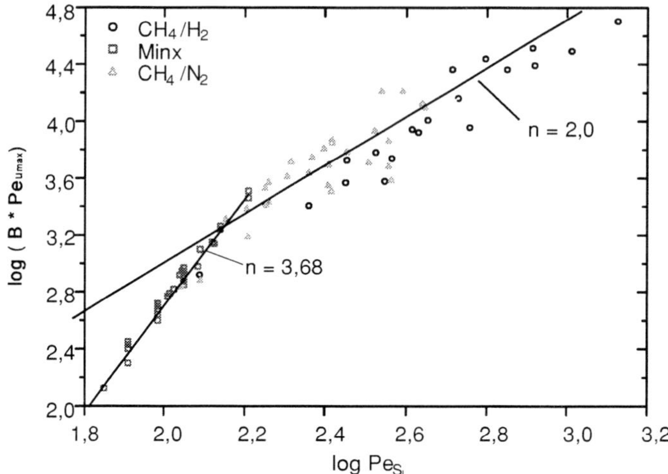

Figure 2.1.6: Flame blow-off maximum air flow, correlation between Pe-number formed with air exit velocity and Pe-number formed with the laminar flame speed.

The data show a quite good agreement with the suggested correlation, but the great disadvantage of this model is, that only one single point of the blow-off-limit can be predicted, at the point of maximum air flow. On the other hand, the empirical constant n is identical with the theoretical value of $n = 2$ for premixed flames, and premixed conditions in the stabilization zone were assumed to use this model. Therefore, a modification of this model is suggested here. The aim is to predict the trend of the complete blow-off-limit over the operating range of the burner. Therefore, it is necessary to observe the change of the flow field, especially in the ignition area, when the fuel/air ratio at the burner exit will be varied. The most important assumption remains, namely the stabilization directly before blow-off always takes place in a stoichiometric mixture located at the air side shape of the recirculation area, as has been proved here and in several investigations [10, 11].

Furthermore, the turbulence condition in this area is important for flame stabilization, followed by increasing exchange of heat and mass (resulting in an improved overall burning speed like the turbulent flame velocity S_t), whereby the intensity of turbulence obviously depends on the flow conditions at the burner exit. Thus, S_l is replaced by S_t in the suggested modification. The following correlation between S_l and S_t has been used (high flow and fluctuation velocities).

$$\frac{S_t}{S_l} \sim \left(\frac{u' \, L_t}{S_l \, \delta_l} \right)^{0,5} \tag{11}$$

Furthermore, the influence of the fuel/air ratio on the extension of the length of the recirculation area L_R can be taken into consideration, so that the correlation for the flame stability comes to:

$$\mathrm{Pe}_{u_l} = \frac{C_l}{B} \cdot \mathrm{Pe}_{S_l}^2 \cdot \frac{u' \cdot L_t}{S_l \cdot \delta_l} \tag{12}$$

where L_t is a characteristic turbulent length scale and δ_l the laminar flame front thickness.

2.1.3.2 Swirl Flames

Due to the experimentally proven Reynolds number invariance of the swirling flow field in fully turbulent flames, the length of the recirculation zone (as a characteristic length) is proportional to the burner diameter D [3, 11]. Furthermore, the tangential velocities are proportional to swirl number and to the exit velocities, so that a characteristic tangential velocity $w = u \cdot S_{0th}$ was considered as a good choice for a characteristic velocity as it represents the convective and dilution effects in **unconfined swirling flames** considered [3, 11]. The Peclet numbers can now be written as:

$$\mathrm{Pe}_w = \frac{u_0 \cdot S_{0th} \; D}{a} \sim \left(\frac{S_l \cdot D}{a}\right)^2 = \mathrm{Pe}_{S_l}^2 \tag{13}$$

This relation allows to predict stability limits of differently scaled premixed burners operating with different fuels and swirl intensities based on burner size, flow rate, and parameters of the cold gases only. The Peclet number model requires values of the laminar burning velocity as an input. Unfortunately, values of S_l published in literature are sometimes quite inconsistent and for many fuels there are hardly any data to be found. Moreover, errors in the flow rate measurements which lead to a certain amount of scatter of the stability measurements about a theoretical curve also cause errors in S_l which was derived according to the blow-off air ratio. In order to avoid those difficulties, the Peclet number expression can easily be transformed to obtain Eq. (13) and to calculate values of S_l by this equation using the results of the stability measurements.

If only one type of fuel is taken into account the laminar burning velocities and thermal conductivities depend on the initial air ratio only, so that they can be substituted by an empirical function λ_0:

$$\frac{u_0 \cdot S_{0th}}{D} \sim \mathrm{F}(\lambda_0) \rightarrow \mathrm{F} = \exp\left(\mathrm{A} + m \cdot \lambda\right) \tag{14}$$

In the last two formula, D/u_0 represents a typical residence time in the stabilization zone of the flame, and S_0 is included to take into account the effect of air entrainment by swirl. In the case of non-premixed flames the mixture field changes due to the varying momentum ratio of gas and air. As this effect is mainly caused by variations of the stoichiometric air ratio, the simplified model may also be able to predict stability curves of diffusion type-II swirling flames.

The stability limits of burners with exit diffusors which prevent the air entrainment in the initial part of the jet, show completely different swirl characteristics. In this case, the relation between blow-off limits and swirl strength has to be determined experimentally, or the models are limited to predictions at constant swirl levels.

In comparison with unconfined swirling flames the heat and mass transport **in enclosed flames** between the inner/outer recirculation zone and the forward flow depend more or less on the swirl intensity. The confinement-parameter B_{BR}/B_{BK} has a great influence on the size of the recirculation zones and with the swirl intensity a dominant influence on the mixing rates. This weakening effect on the flame stability can be considered in the confined swirling flames with the time mean velocity (geometrical assumption of the axial air velocity at the burner exit u_0 and the tangential velocity $w = u_0 \cdot S_{0th}$). This leads with Eq. (6) to:

$$u_0 \cdot \sqrt{\frac{1+S_{0th}^2}{D}} = K \cdot \frac{S_1^2}{a} \quad \text{with } K \sim \mathrm{P}_{th}^{0.75} \tag{15}$$

The constant K considers the influence of increasing throughput.

2.1.4 Experimental Setup and Measurement Technique

Figure 2.1.7 shows the concentric burner with central, axial fuel pipe and unswirled air flow with an annular blockage which is used to investigate flame stability at various geometries (D = 40–80 mm; B = 0.2–0.76). The whole air flow is split into six single flows and redirected two times and ducted through a honeycomb grid to get a symmetrical incoming flow in the burner mouth. The **swirling experiments** were performed in swirling free and enclosed flames generated by a geometrically similar burner with a "movable-block swirl generator" [1]. This swirl generator allowed a continuous adjustment of the theoretical swirl number between S_{0th} = 0.0–2.0. The diameter of the cylindrical exit nozzle was varied between D = 60 mm and D = 100 mm. The fuel gas was injected in diffusion mode concentric with different kinds of gas nozzles. For premixed flames fuel gas was injected radially upstream of the swirler through eight nozzles in order to provide

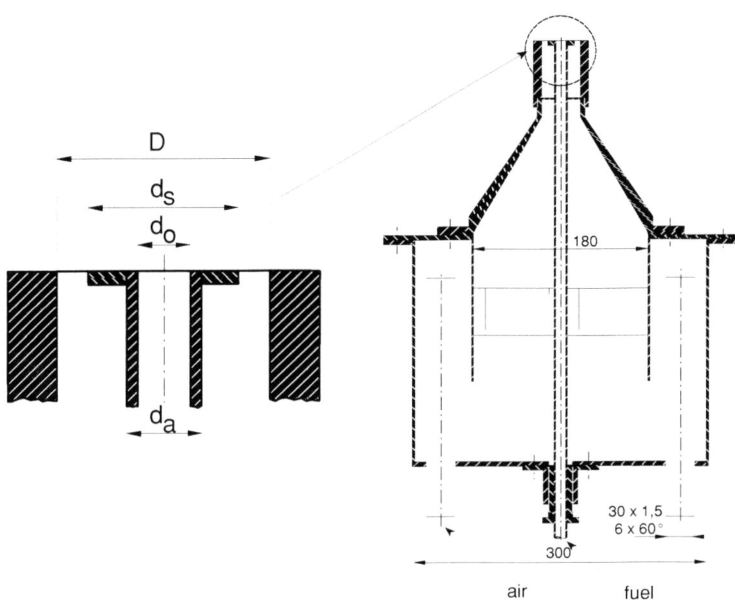

Figure 2.1.7: Bluff-body burner.

complete mixing in the burner exit plan (see Fig. 2.1.8). To get comparable flow input conditions with the diffusion flames, a cylindrical bluff-body was installed instead of the gas nozzle.

The experiments with enclosed swirling flames were carried out using a water cooled combustion chamber (inner diameter 0.5 m, length 1.7 m, $D_{BR}/D_{BK} = 0.12$). The burner ($D = 60$ mm) can be moved axially in the combustion chamber to perform field measurements. The boundary and inlet conditions during the experiments were controlled and kept constant to guarantee the reproducibility of the stability measurements.

The measurements in the near-field of the burner covered the determination of the turbulent velocity and temperature field as well as volumetric concentrations of relevant stable species.

Velocity measurements were carried out using a counter-type two colour Laser-Doppler-Velocimeter. Magnesia oxide particles ($d_P = 1$–5 microns) brushed off a massive cylinder served as tracer material for the combustion gases and the entrained air. Time-mean and fluctuating temperatures were measured using thermocouples with diameters of 50 and 100 microns. After preamplification, the thermal inertia of the thermocouples is compensated electronically, so that temperature fluctuations up to about 1 kHz can be detected. After digitising the signals, the data are stored and processed in a computer. Concentrations of stable species were determined by commercial gas analysers working on the basis of infrared absorption (CH_4, CO, CO_2, H_2O), paramagnetism (O_2) and thermal conductivity (H_2). The gas samples

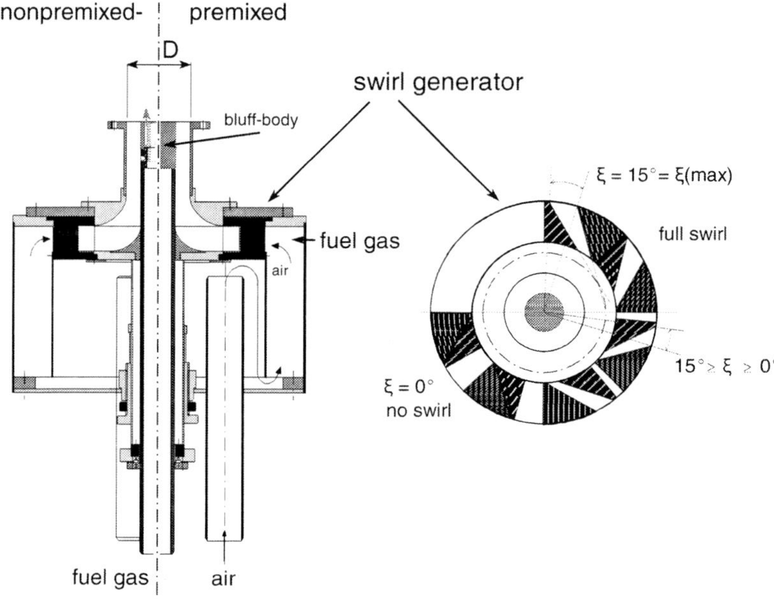

Figure 2.1.8: Swirl burner for premixed and non-premixed operation (60 mm $\leq D$ $\leq$ 100 mm).

were taken isokinetically with a small water cooled suction probe, whereby the mouth was turned into the mainstream direction. More information about the used burner-/combustion-systems can be found in [3, 8, 12].

2.1.5 Results and Discussion

The confinement influence on the flame stability of type-1 disk stabilized flames is negligible if the confinement parameter is low [8, 13]. Figure 2.1.9 shows the stability limits of free and enclosed disk stabilized type-I flames. The reason for this is that the flame will be stabilized usually by the inner recirculation zone and from that the ignition zone of the flame is isolated from the outer recirculation zone by the air flow. If the mixing rates upstream from the burner increase (type-II swirling flames) the effect of confinement rises up. Because of this the majority of investigations on disk stabilized flames were done without confinement.

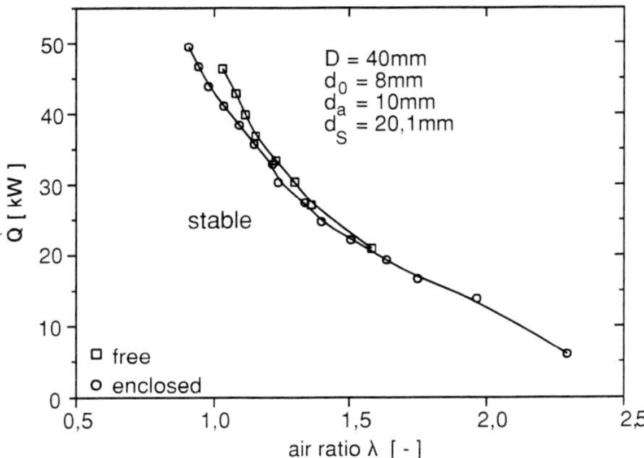

Figure 2.1.9: Comparison of the lean stability limits of an enclosed and a free disk stabilized flame.

2.1.5.1 Disk Stabilized Flames

In Fig. 2.1.10 the blow-off curves are shown for different fuel compositions or flame speed (fuel mixture of methane and nitrogen or hydrogen) in dependence on excess air ratio. The results point out the high influence of the laminar flame speed expressed by hydrogen concentration on the stability limit. Figure 2.1.11 shows the influence of various burner geometries (blockage ratio) on the stability limits at constant fuel concentration. As a consequence of higher air velocities (constant gas flow) with increasing blockage ratio B

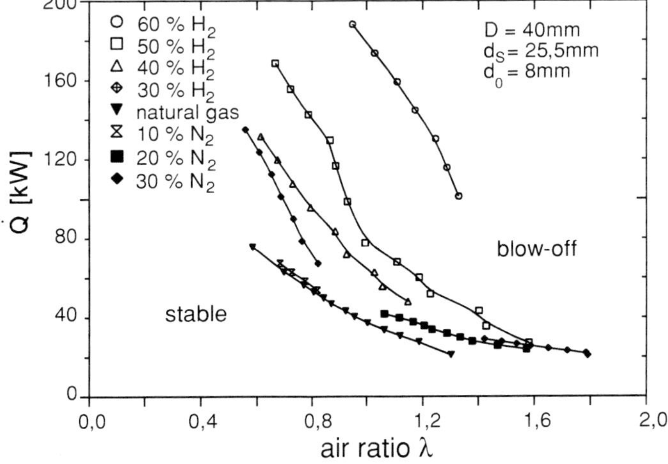

Figure 2.1.10: Influence of the laminar flame speed on the flame stability of a disk stabilized flame (D = 40 mm).

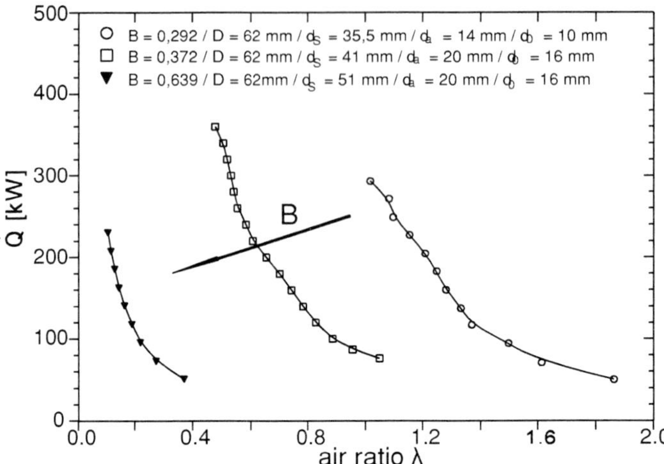

Figure 2.1.11: Influence of the blockage ratio (*B*) on the stability limit of a disk stabilized flame.

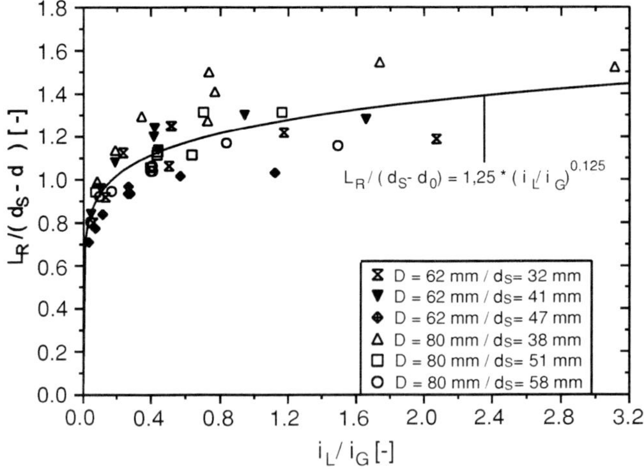

Figure 2.1.12a: Normalized length of recirculation area versus weighted air/fuel momentum ratio.

the blow-off occurs at a lower overall air ratio. Figure 2.1.12a shows the correlation found in the described burner systems that are of interest for the use of the modified Pe-model. The characteristic length on the recirculation zone normalized with $(d_S - d_0)$ is correlated with the momentum ratio of fuel and air at the burner exit. Assuming that fuel and air jet were independent free jets with the same half angle, the boundaries of both jets would intersect at $L_R/(d_S - d_0) = 1.5$. In the case of high ratio of i_{air}/i_{fuel} the fuel jet will no

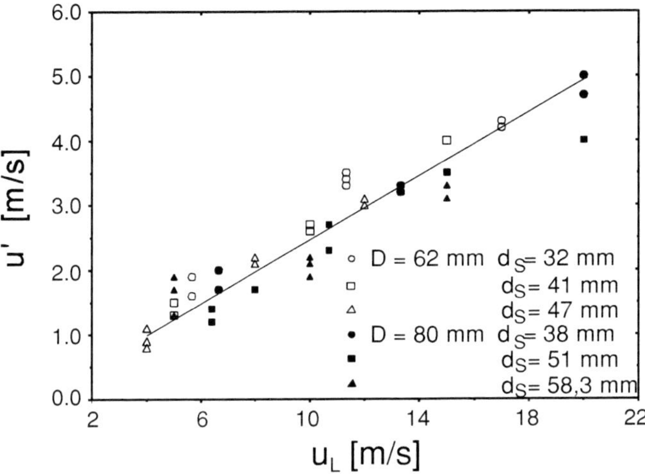

Figure 2.1.12b: Fluctuation velocity in the stabilization zone correlated with mean burner exit air velocity.

longer penetrate the recirculation zone (type-II flame). In this case of transition the recirculation zone becomes again strongly dependent of the ratio of the exit moment and then should be normalized with the disk diameter d_S. The correlation $L_R/(d_S - d_0) = 1{,}25\ (i_{air}/i_{fuel})^{0{,}125}$ is used to determine the length of the recirculation area. The second modification of the original Pe-model (Eq. (8)) concerns the substitution of the laminar burning velocity by the turbulent. Therefore, the fluctuation velocity u' in the stabilization zone has to be known. Figure 2.1.12b shows the well known correlation between u' and the main air velocity U_{air}. The experiments were done under isothermal conditions and it could be shown that this results are comparable with flame measurements [14].

With these correlations the modified model (Eq. (12)) can be tested, only using conditions as geometry, burner exit flows, density of air, and fuel and laminar burning velocity at the burner exit at blow-off condition. As before, the thermal diffusivity is calculated for a stoichiometric fuel-air ratio. All the measured values corresponding to the blow-off limits at different geometries and fuel-compositions of the not enclosed flames are included in Fig. 2.1.13. Although there are still deviations from the regression line, which probably depend on the flame temperatures in the different systems and the corresponding influence on the reaction kinetics, the correlation is acceptable, rather somewhat better than the original model. The most important result from the modification is that the shape of the blow-off limits can be described by this correlation.

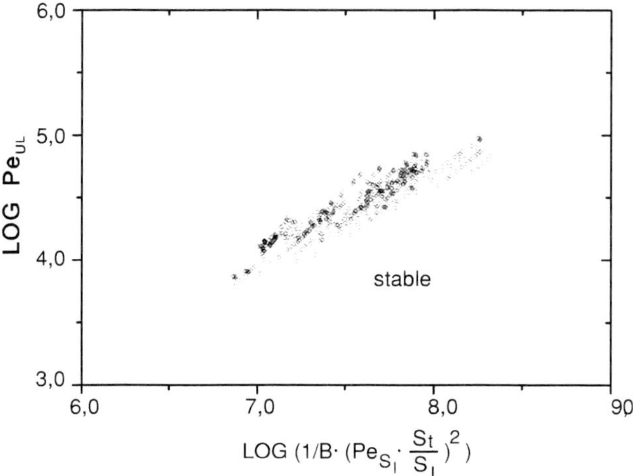

Figure 2.1.13: Stability limits of disk stabilized flames, described by modified Pe-model (Eq. (12)).

2.1.5.2 Swirling Flames

Figure 2.1.14 shows the results of stability measurements on premixed swirling flames at different swirl numbers and throughputs (150 kW = P_{th} = 550 kW). Stable combustion is possible for flame conditions below and to the left of the given data points. At a given swirl number and increasing thermal load (i. e. decreasing residence times), the regime of stable combustion is shifted towards smaller initial stoichiometric air ratios where stable combustion can take place due to the higher kinetic rates of the mixture. As the

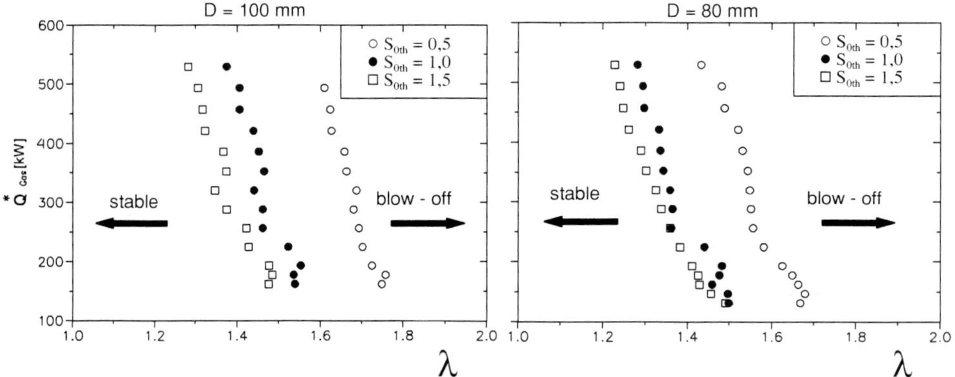

Figure 2.1.14: Stability limits of free premixed swirl flames as a function of air equivalence ratio λ and swirl intensity $S_{0,th}$ for different burner diameters (D = 100 mm/ 80 mm).

amount of entrained surrounding air increases proportionally to the swirl number, increasing swirl leads to a substantial shift of the local air ratio towards higher values resulting in lower flame temperatures and flame speeds. Therefore, increasing swirl decreases the limits of stable combustion. The same behaviour was found for turbulent swirling diffusion flames with ring and multihole fuel nozzles as shown in Figs. 2.1.15a and 2.1.15b but the flames are much more stable compared with premixed flames. The reason for that is that the mixture in the ignition zone consists of stoichiometric

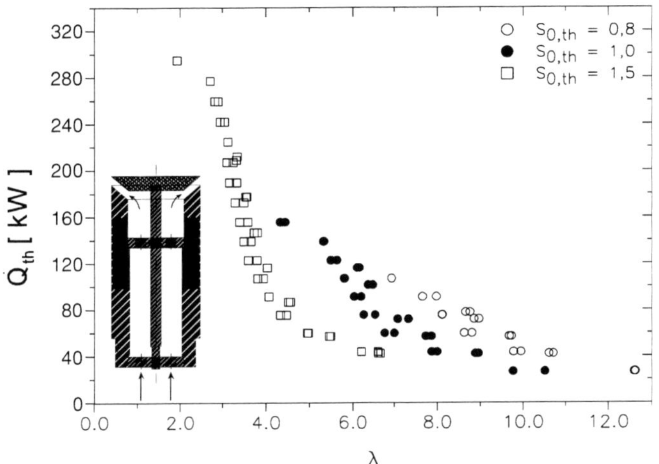

Figure 2.1.15a: Lean stability limits of diffusion swirling flames with a ring gas nozzle (D = 100 mm).

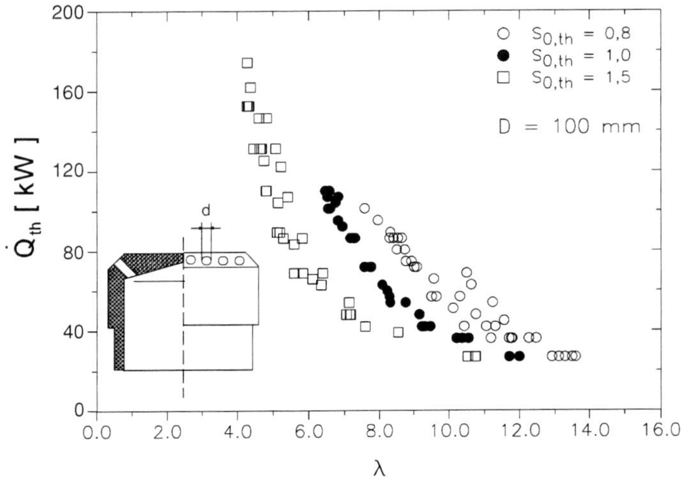

Figure 2.1.15b: Lean stability limits of diffusion swirling flames with a multihole gas nozzle (D = 100 mm).

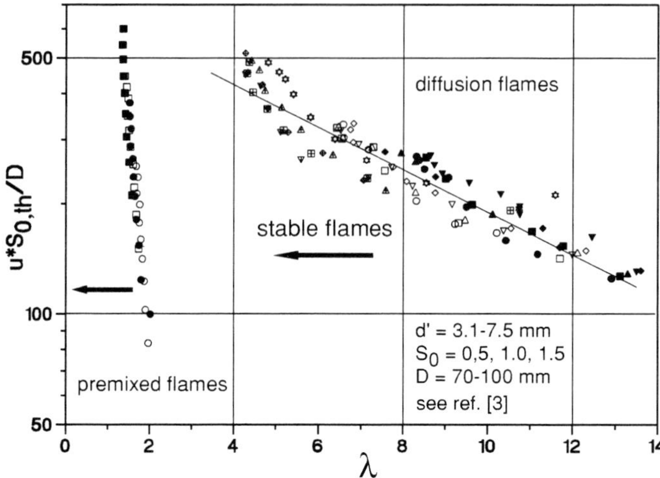

Figure 2.1.16: Correlation of stability data based on the simplified relation for swirling flames in dependence on λ.

mixture. Figure 2.1.16 shows the comparison of the stability points of premixed and diffusion flames in the same diagram from which one can see that diffusion flames are much more stable over a wider range of excess air factor and load.

Figure 2.1.17 shows stability data evaluated according to the Peclet number relation (13), based on flame speeds determined on a Bunsen flame [15]. The Peclet number dependency correlates the measurements for three different swirl numbers and four burner sizes rather satisfactorily, however, for decreasing values of $Pe^2_{S_l}$ (i. e. increasing air ratios) the measurements show increasing scattering. This is most probably due to the above

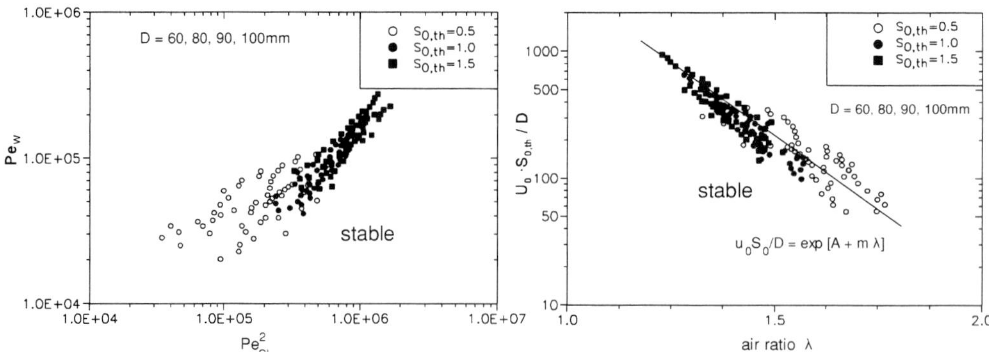

Figure 2.1.17: Correlation to stability data based on the Pe-number (Eq. (13)), for premixed swirling flames (right: simplified Eq. (14)).

mentioned uncertainties about the correct value of S_l. In order to avoid this source of errors, the simplified relation (14) can be used as long as only one type of fuel is taken into account. In the right side of Fig. 2.1.17, the stability data of the left side are presented according to Eq. (14) (with the values $A = 13.4$ and $m = 5.35$) in a semi-logarithmic diagram. It can be seen that the inverse characteristic residence time $u_0 \cdot S_{0,th}/D$ is well reproduced by an exponential function of the form and that the accuracy of the predictions is increased. The stability data for a natural gas flame are again represented by a single straight line, whereby the empirical constants of the exponential function had to be adapted for this other fuel.

Figure 2.1.18 shows the influence of different thermal loads and swirl intensities on the lean blow-off limits of enclosed swirling flames. When the thermal input increases, the mean velocity in the reaction zone becomes higher. This is the reason, why the air ratio at the stability limit decreases with higher throughput.

Equation (15) for enclosed diffusion flames can be fitted on an arbitrarily measured blow-off point. Then it is possible to calculate the stability limits under different operation conditions (swirl number, thermal input $K \sim P_{th}^{0.75}$). The comparison between the values calculated with the model and the measured blow-off limits shows good agreement. The maximum difference between measurements and calculated air ratio at the stability limit is 4.1 %. The influence of higher wall temperature ($T_{wall} = 700\,°C$) on the lean blow-off limit is shown in Fig. 2.1.18, right. The lean blow-off limits are extended caused by shorter chemical time scales by a higher ignition zone temperature at constant air ratio in combination with higher values of heat input in the flame root.

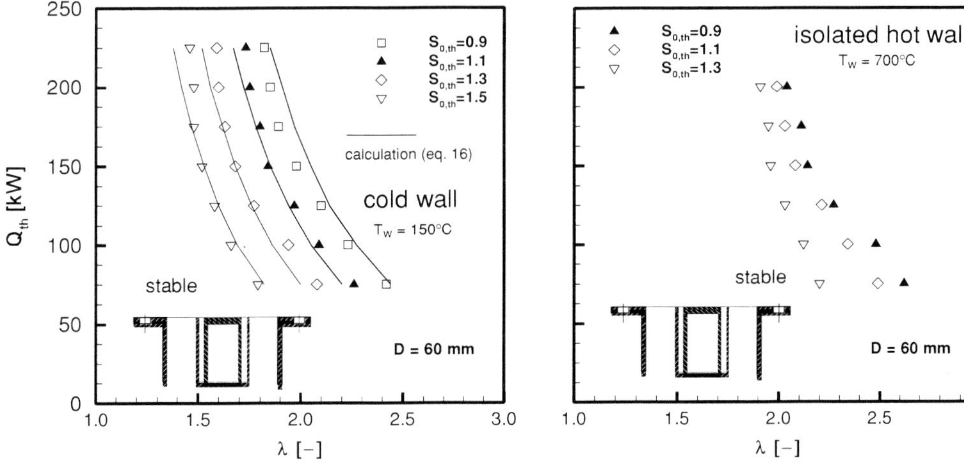

Figure 2.1.18: Blow-off points of a confined swirling flame (calculation Eq. (15), cold and hot wall.

References

1. Leuckel, W., Fricker N. (1976): The Characteristics of Swirl-Stabilized Natural Gas Flames, Part 1: Different Flame Types and their Relation to Flow and Mixing Patterns, *J. Inst. of Fuel* **49**, 103–112.
2. Phillip, M., Hoffmann, S., Habisreuther, P., Lenze, B., Eickhoff, H. (1992): Experimental and Numerical Study Concerning Stabilization of Strongly Swirling Premixed and Nonpremixed Flames. 24th Symposium (International) on Combustion, The Combustion Institute, Pittsburgh, 361–368.
3. Hoffmann, S. (1994): Untersuchungen des Stabilisierungsverhaltens und der Stabilitätsgrenzen von Drallflammen mit innerer Rückstromzone, Ph. D. Thesis, Universität Karlsruhe (TH).
4. Hillemanns, R. (1988): Das Strömungs- und Reaktionsfeld sowie Stabilisierungseigenschaften von Drallflammen unter dem Einfluß der inneren Rezirkulationszone, Ph. D. Thesis, Universität Karlsruhe (TH).
5. Zukowski, E. E., Marble, F. E. (1956): Experiments Concerning the Mechanism of Flame Blow-Off from Bluff-Bodies, Proceedings of the Gas Dynamics Symposium on Aerothermochemistry, Evanston, Northwestern University, 205–210.
6. Putman, A. A., Jensen, R. A. (1949): Application of Dimensionless Numbers to Flash-Back and other Combustion Phenomena, 3rd Symposium (International) on Combustion, Flame and Explosion Phenomena, p. 89.
7. Kremer, H. (1971): Kennzahlen zur Beurteilung der Stabilität von Vormischflammen, GWI **20**, No. 3, 101–106.
8. Prade, B. (1993): Experimentelle und theoretische Untersuchung zum Abblaseverhalten von turbulenten Stauscheibendiffusionsflammen, Ph. D. Thesis, Universität Karlsruhe (TH).
9. Borghi, R. (1985): On the Structure and Morphology of Turbulent Premixed Flames, Recent Advances in the Aerospace Science, Ed. C. Casci, Plenum Publ. Corp., 117–138,
10. Minx, E. (1969): Über die Stauscheibenstabilisierung turbulenter Diffusionsflammen an Brennern mit zentraler Brennstoff- und ringförmiger Luftzufuhr, Ph. D. Thesis, RWTH Aachen.
11. Rawe, R., Kremer, H. (1981): Stabiliy Limits of Natural Gas Diffusion Flames with Swirl.18th Symposium (International) on Combustion, The Combustion Institute, 667–677.
12. Schmittel, P., Lenze, B., Leuckel, W. (1997): Messungen zur Stabilität turbulenter, eingeschlossener Drallflammen unter Variation der Einflussgrößen, Deutsch-Niederländischer Flammentag, *VDI Bericht* **1313**, 121–126.
13. Schefer, R. W., Namazian, M., Kelly, J., Perrin, M. (1996): Effect of Confinement on Bluff-body Burner Recirculation Zone Characteristics and Flame Stability, *Combust. Sci. Tech.*, **120**, 185–211.
14. Prade, B., Lenze, B. (1990): Stability of Concentric Diffusion Flames in Dependance on Flow Properties and Burning Velocity, Joint Meeting of the Soviet and Italien Section of Combustion, Pisa, Italy, 6.3.
15. Leisenheimer, B. (1989): Messungen der laminaren Flammengeschwindigkeit von CH_4-H_2-Luftgemischen an der Bunsenflamme, Seminararbeit, Universität Karlsruhe (TH).
16. Prade, B., Lenze, B. (1992): Experimental Investigation in Extinction of Turbulent Nonpremixed Disk-Stabilized Flames, 24th Symposium (International) on Combustion, The Combustion Institute, Pittsburgh, 369–375.

2.2 Velocity-Fields, Reynolds Stresses, and Swirl-Induced Intermittency in Free and Enclosed Rotating Flows

Frank Holzäpfel, Klaus Döbbeling, and Bernhard Lenze*

Abstract

Swirling flows are used very often in industrial application to mix better and faster, to build recirculation zones, to enhance residence time, and to stabilize and intensify reaction in cyclone combustors, flames, and high intensity combustion chambers. This work shows the results of detailed measurement of the three mean velocities and the complete Reynolds stress tensor for swirling free jets with and without vortex breakdown and gives a reliable and detailed database, which serves for validation purposes of mathematical models and, in particular, turbulence closures. It also contributes to a better understanding of the nature of free and enclosed, turbulent swirling flows. Swirl induced intermittency (SII) was found and introduced, whereby entrained fluid elements are accelerated towards the center of the free jet, and, hereby, considerably increase the radial turbulent transport. A quantitative analysis of SII was performed by means of measured conditioned JPDFs of the Reynolds shear stresses realized by a four- to six-wire technique, designed for the simultaneous measurement of the instantaneous velocity vector and temperature. It was shown that effects which are based on mechanisms similar to SII can also occur in low-confined swirl flows. The experimental and modeling work leads to the conclusions that SII gives an additional explanation for the outstanding mixing characteristics and can be used to optimize technical swirling flows. With swirl-induced intermittency, the discrepancies between measurements and numerical predictions can be explained to a large extent, and for good prediction of swirling free jets, turbulence models have to be complemented by a model which considers SII.

---

* Engler-Bunte-Institut, Bereich III – Verbrennungstechnik, Universität Karlsruhe, Kaiserstr. 12, 76128 Karlsruhe, Germany

2.2.1 Introduction

Swirling flows are frequently encountered in many engineering applications, including swirl atomizers, cyclone separators and different reactor systems. In the field of combustion, they are used in numerous types of devices e. g. gas turbine combustors, industrial burners, or pulverized coal burners, because the formation of the internal recirculation zone enables excellent flame stabilization, high heat release rates, and good burnout characteristics.

The generic swirling flow is that of a single, axisymmetric swirling jet discharging into stagnant surroundings, whereby the swirling free jet without vortex breakdown is commonly considered as an ideal benchmark test for the assessment of turbulence closures, since it is determined by turbulent transport more than by pressure effects, but reliable flow data including mean velocities and turbulence quantities are scarce. Still, flow data based on single hot-wire measurements by Morse [1] or data made available by Sislian and Cusworth [3] are used for current validations of turbulence models by for example Younis et al. [2]. Unfortunately, these data, according to Fu et al. [4], are not as consistent as desirable. Due to this obvious lack of reliable data, Cheng et al. [5] recently performed Laser Doppler Velocimetry measurements in swirling free jets but without seeding the ambient air. The first objective of the current work is to provide a complete detailed database comprising the three mean velocities and the complete Reynolds stress tensor for the isothermal, turbulent swirling free jet in stagnant surroundings with and without vortex breakdown. These measurements were performed with the quintuple, a further development of the by Döbbeling [10] designed quadruple-hotwire measurement technique, which was developed by the authors and is described in detail in [6]. This five-wire technique was designed for fast measurements of the instantaneous velocity vector in highly turbulent flows. Its main benefits are increased accuracy and an expanded uniqueness domain covering $\pm 90°$ angle of attack, as compared to the four-wire techniques which are restricted to about $\pm 40°$.

The predictive qualities of turbulence models for swirling flows discussed in literature are quite contradictory. Especially, conflicting reports arise from the different predictive capabilities of identical turbulence models being applied to free and confined swirling flows. Although high-order turbulence models like the Reynolds Stress or the Algebraic Stress model quite successfully predict confined swirling flows [7], they have highlighted serious defects in the prediction of free swirling jets [4]. The turbulent radial transport is underestimated, leading to smaller spreading angles and longer internal recirculation zones. Several researchers have developed different model corrections, (pressure-strain correlations) [2]. The confusion is increased even more by the fact that the k,ε-model, which sometimes fails completely in predicting confined swirl flows [8], yields partially satisfactory results in swirling free jets. The last work of this long term project shows that swirl-induced intermittency considerably modifies the momentum transport and can explain discrepancies between measurement and prediction of free swirling jets.

In the second part of the paper, the physical mechanism of swirl-induced intermittency (SII) is introduced and subsequently qualitatively proven by measured joint probability density functions (JPDFs) of the Reynolds shear stresses [9]. Beyond that, it is quantified based on conditioned JPDFs measured in the slightly heated jet. These measurements were performed with an extension of the quintuple method for fast, simultaneous measurements of temperature and velocity.

2.2.2 Experimental Apparatus

Figure 2.2.1 shows the movable blocks swirl generator used in [10] and the present experiments. Its annular ring nozzle with an outer diameter of 107 mm is surrounded by a horizontal front plate to ensure that the ambient air far away from the swirl jet boundary is entrained almost radially. The swirl strength is characterized by the outlet swirl number S_0 defined by the ratio of angular to axial momentum flux in the nozzle divided by the outside nozzle radius R as given by

$$S_0 = 2\pi \int_0^R \varrho(\overline{uw}+\overline{u'w'}r)r\mathrm{d}r / 2\pi R \int_0^R \left((p - p_{\mathrm{ref}})+\varrho(\overline{u}^2+\overline{u'^2})\right) r\mathrm{d}r \tag{1}$$

In the current paper, the theoretical swirl number $S_{0,\mathrm{theo}}$ calculated according to Leuckel, [11], as a function of the swirl generator geometry is used, which is approximately equal to S_0. The letters, $\overline{u}$, $\overline{v}$, $\overline{w}$ denote the mean velocities in the axial, radial, and tangential direction in cylindrical coordinates (x, r, φ) and u', v', w' their respective fluctuating portions.

The measurements in cold air were conducted with the quintuple measurement technique [6]. For the purpose of this work, the directional calibration area was expanded – compared to the earlier probe of Döbbeling [12] – from $\pm 45°$ to $\pm 90°$ and the angular sensitivity functions were calibrated on three velocity levels instead of one. For conditional sampling, the quintuple probe was supplemented by a platinum wire 1 µm in diameter running in the constant current mode for fast measurements of temperature. The resulting probe geometry, shown in Figure 2.2.2, is a result of the following – to some extent contradictory – requirements: The measuring volume should be as small as possible, whereas the temperature wire must not sense fluid heated by the hot-wires. Furthermore, the flow sensed by the hot-wires should be disturbed as little as possible by the prongs supporting the cold wire.

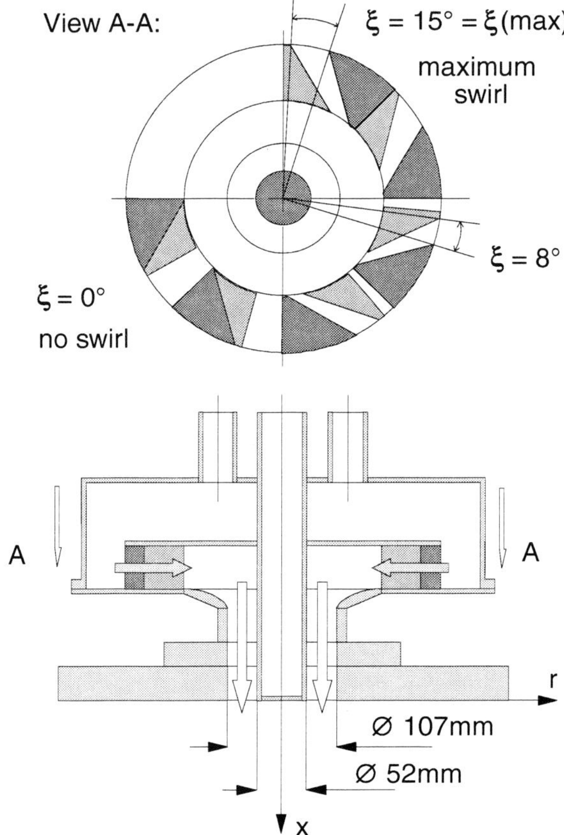

Figure 2.2.1: Movable blocks swirl generator (top view and cross section).

For temperature correction of the hot-wire voltages, the method of Meyer [13] was adopted. The resulting six-wire method allowed the simultaneous measurement of temperature and velocity with good accuracy for frequencies up to 2 kHz [14].

2.2.3 Mean Velocities and Turbulence Quantities

All measurements with the quintuple method were conducted with cold air (21 °C) at the theoretical swirl numbers of $S_{0,\text{theo}} = \dot{D}/\dot{I}R_0 = 0.4$, 0.95, and 2.53 and a volumetric flow rate of $\dot{V} = 418\ \text{m}^3/\text{h}$. The measurements were performed at 18 axial locations including the nozzle exit, and up to 38 radial

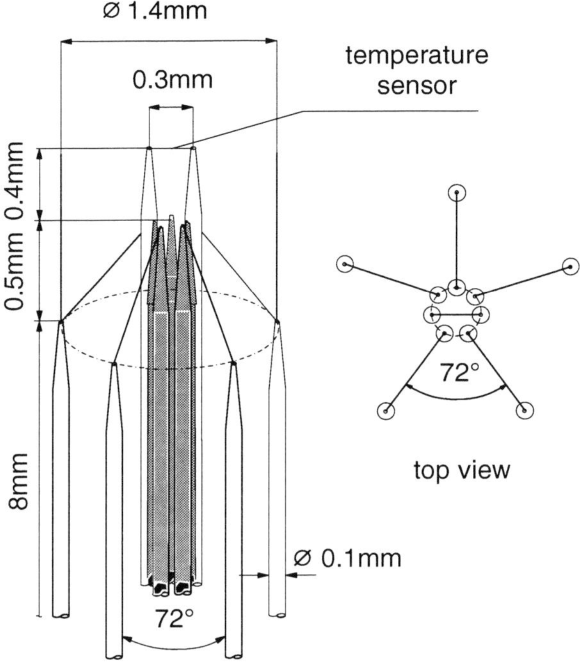

Figure 2.2.2: Wire arrangement of the six-wire probe for simultaneous measurements of temperature and velocity.

positions. At each location, 4096 samples of the instantaneous velocity vector recorded with a data rate of 50 Hz were used to calculate, subsequently, the three mean velocity components and all six Reynolds stresses. Figures 2.2.3 and 2.2.4 show the mean axial ($\overline{u}$) and tangential velocity ($\overline{v}$) profiles in seven selected axial locations for the swirl numbers 0.4 and 0.95. Detailed results for $S_{0,\text{theo}} = 2.53$ are not presented because almost all features of the flow fields (including the Reynolds stress distributions) at $S_{0,\text{theo}} = 0.95$ and 2.53 and their respective interpretations are very similar (Figs. 2.2.5 and 2.2.6). Assuming that the flow-field will not change basically for intermediate swirl numbers, the current work is representative for a wide range of swirl numbers, covering swirling free jets with and without vortex breakdown. For $S_{0,\text{theo}} = 0.4$, the $\overline{u}$-profile in the annular ring nozzle displays an almost constant value, whereas for $S_{0,\text{theo}} = 0.95$, the axial mean velocity component increases linearly with the radius. For the $\overline{w}$-profiles, a similar tendency is observed, which leads to a stronger accentuation of the forced vortex region with increasing swirl number. For the smaller swirl number no internal recirculation zone (IRZ) is formed. The radial position of the $\overline{u}$-maximum, which is located in the first axial measuring position at $r = 45$ mm, is shifted up to $x = 80$ mm towards even lower radii, to reach in the last measuring plane ($x = 400$ mm) a radial position of $r = 60$ mm. The center-line velocity

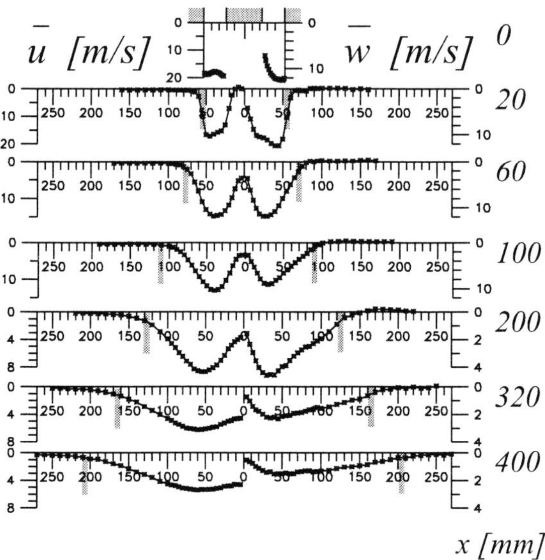

Figure 2.2.3: Mean axial and tangential velocities; $S_{0,\text{theo}} = 0.4$ (area of maximum instability is marked dotted).

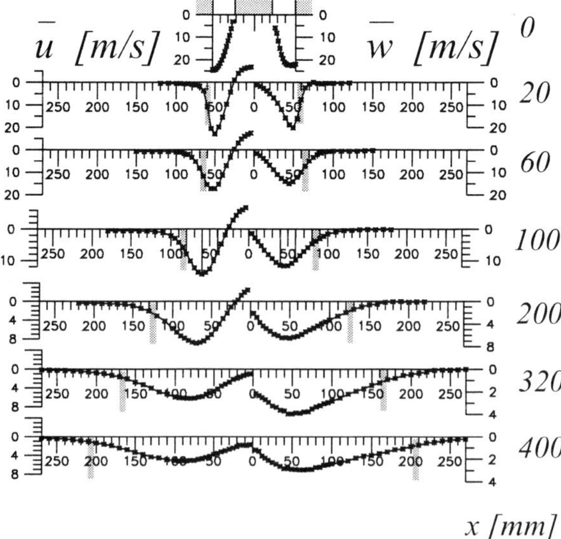

Figure 2.2.4: Mean axial and tangential velocities; $S_{0,\text{theo}} = 0.95$ (area of maximum instability is marked dotted).

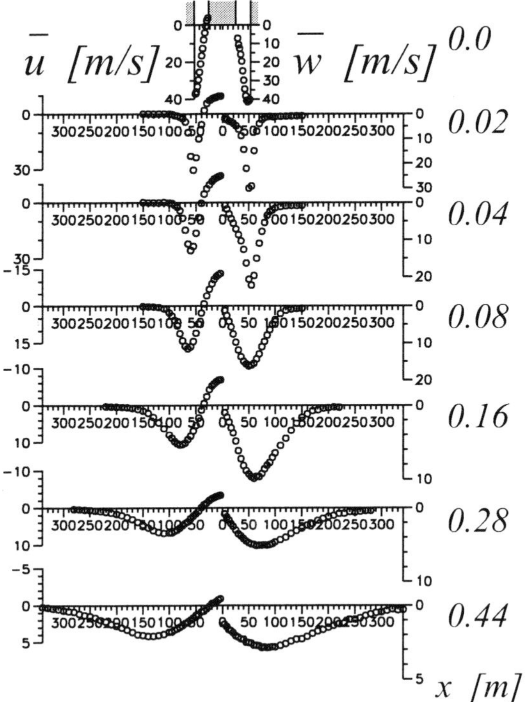

Figure 2.2.5: Mean axial and tangential flow velocities of a free swirl jet at $S_0 = 2.53$.

increases from $\bar{u} = 3.5$ m/s at $x = 40$ mm to 4.6 m/s at $x = 400$ mm where the annular jet-type of the $\bar{u}$-profiles has almost disappeared. For $S_{0,\text{theo}} = 0.95$ and along the first 160 mm of axial length, the width of the reverse flow in the IRZ corresponds to the diameter of the central bluff body, thus resembling a Taylor-Proudman column [15]. Since the IRZ does not close before $x = 280$ mm, the annular jet-type flow persists up to the last measured location. The corresponding $\bar{u}$-maximum remains at $r = 50$ mm in the first three measuring planes and then gradually moves up to $r = 90$ mm in the last measuring plane. The tangential velocity profiles, $\overline{w}(r)$, consist of the combined forced-free vortex type for both swirl numbers. The corresponding maxima are located almost at a fixed radial position in the whole flow domain under investigation. For $S_{0,\text{theo}} = 0.95$, the radial position of the $\overline{w}$-maximum ($r = 45$ mm) in the swirl generator coincides with those of the flow-field. For $S_{0,\text{theo}} = 0.4$, however, this coincidence is not given. Here, the radial position of the $\overline{w}$-maximum, which is located in the swirler nozzle at $r \approx 50$ mm, is shifted to the position of $r = 30$ mm, which is maintained up to $x = 200$ mm. Downstream of $x = 200$ mm, the maxima slowly begin to shift towards increasing radius for both swirl numbers. For $S_{0,\text{theo}} = 0.95$, the IRZ influences the shape of the forced vortex region. To understand this, the conservation of

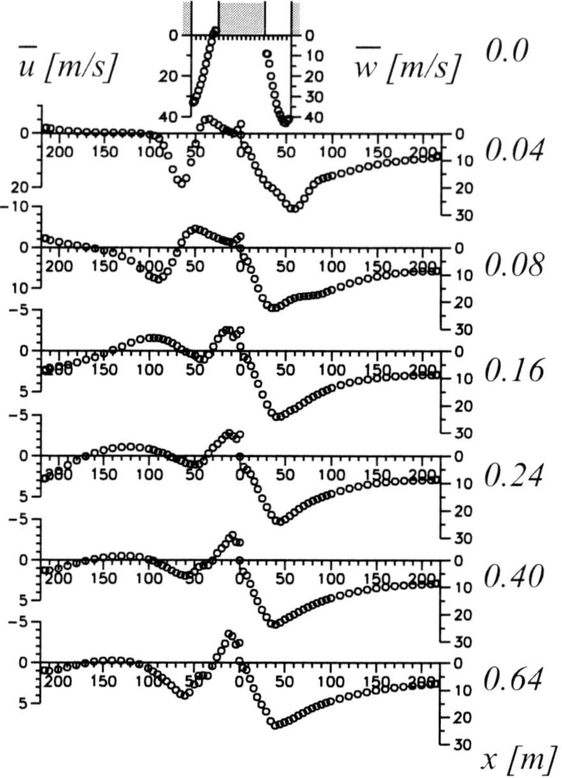

Figure 2.2.6: Mean axial and tangential flow velocities of an enclosed swirl jet at $S_0 = 2.53$.

angular momentum of the recirculating fluid has to be considered. The forced vortex part is concave-shaped up to $x = 40$ mm, where the recirculating fluid with little angular momentum moves to increasing radius. In the downstream part of the IRZ ($x = 100$ mm), the fluid moves with high angular momentum towards the axis of rotation, so that the forced vortex part becomes convex-shaped.

In contrast to confined swirl flows [10, 14, 23], the free vortex part of the tangential velocity profiles cannot be approximated by a superposition of a forced and a free vortex as expressed by $\overline{w}(r) = a \cdot r + b/r$. This suggests that the distribution process of angular momentum is not predominantly governed by the partial conservation of angular momentum, as shown by Hirsch [9] for confined swirling flows. Obviously, an additional effect, probably SII, determines the almost linear decrease of the tangential velocity in the swirling free jet, which is better elucidated by the Reynolds stress profiles shown in Figs. 2.2.7 and 2.2.8. In the first measuring planes, the peaks of all Reynolds stresses close to the swirl jet boundary almost coincide with the radial position of maximum instability (dotted areas), which is indicated by the minimum of the circulation gradient ($\min(\partial(\overline{w} \cdot r)/\partial r)$) in the free vortex. The instable situation at the jets boundary originates from the inequality of the radial pressure gradient and the centrifugal forces acting on fluid elements which are radially displaced by turbulent fluctuations [16]. Further downstream, these peaks begin to widen and merge into other peaks of different origin. It is interesting to note that now the radial positions of maximum instability and the maxima of the Reynolds stresses are displaced. The position of maximum instability moves faster towards the swirl jet boundary, indicating the existence of a predominant transport mechanism. The positions and magnitudes of the inner maxima originate either from the convective transport of the stresses produced in the swirl generator or from turbulence production by intense shearing processes in the IRZ. For $S_{0,\text{theo}} = 0.95$, these maxima soon merge into the peaks close to the jet boundary as a result of the displacement effect of the vortex breakdown bubble. Figures 2.2.7 and 2.2.8 also demonstrate that the decay of turbulence is damped in the forced vortex region in agreement with the observations of Jacquin et al. [17]. Further discussion of the Reynolds stress distributions can be found in [14]. In order to assess the quality of the measurement technique, the entrainment rate $(\dot{M}_X - \dot{M}_0)/\dot{M}_0$ which was calculated by the integration of the radial profiles of the mean axial velocity, is shown in Fig. 2.2.9a.

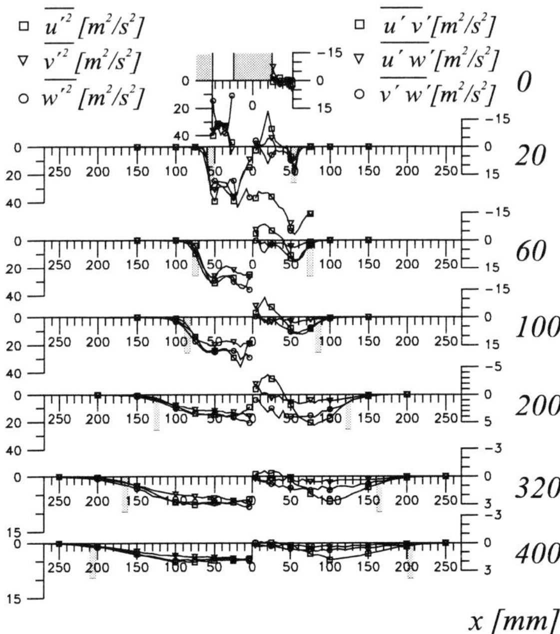

Figure 2.2.7: Reynolds stress; $S_{0,\text{theo}} = 0.4$ (area of maximum instability is marked dotted).

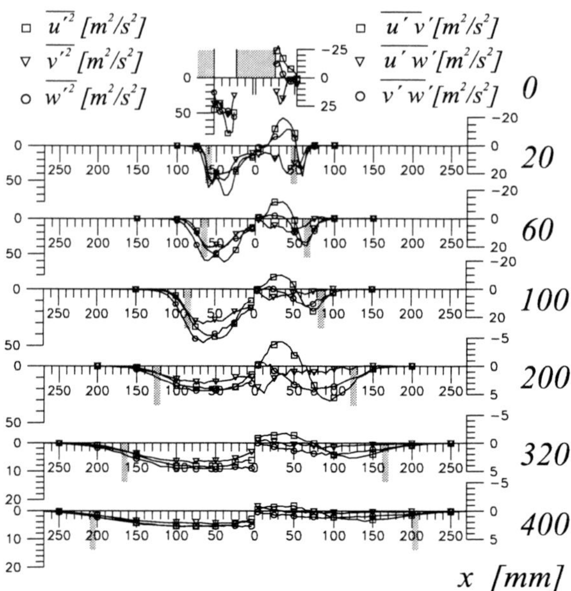

Figure 2.2.8: Reynolds stress; $S_{0,\text{theo}} = 0.95$ (area of maximum instability is marked dotted).

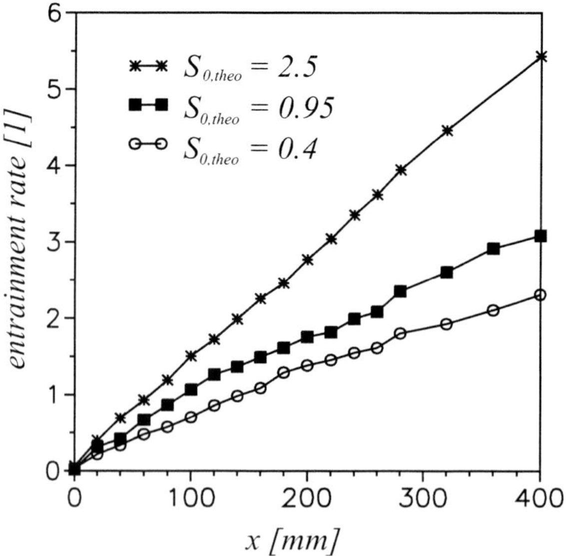

Figure 2.2.9a: Non-dimensionalized entrainment rates.

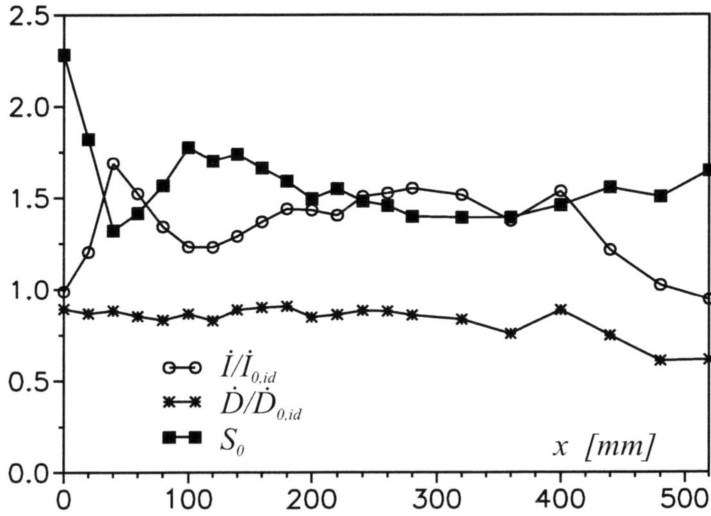

Figure 2.2.9b: Axial and tangential integral momentum-flow and swirl parameter over axial distance of a free swirling jet.

The entrainment rate increases as the swirl number increases. In contrast to non-swirling free jets, it varies almost linearly with the axial distance even in the initial region. Beneath the consistent velocity profiles shown, the lack of any scattering in the entrainment rate indicates the reliability and accuracy of measurements. The values of conservative ratios of axial $\dot{I}/\dot{I}_0$ and swirl momentum $\dot{D}/\dot{D}_0$ as well as of the swirl parameter S_0 are shown in Fig. 2.2.9b and present good results for swirl but only a reasonable result for the axial momentum, for which the static pressure was calculated by the tangential velocity distribution.

2.2.4 Swirl-Induced Intermittency

In turbulent free jets, the entrainment process is mainly governed by the larger-scale motions of turbulent eddies enclosing irrotational fluid [19]. The entrained fluid is sheared, stretched, and accelerated by the pressure forces of the surrounding eddies. During this process, the entrained fluid elements become rotational themselves and are turbulent according to the classical definition. In the non-swirling case, the alternating occurrence of rotational and irrotational fluid, commonly called intermittency, is limited to the boundary region of the jet.

However, in swirling free jets, an additional effect occurs, which even reaches to the axis of the jet. We name this effect swirl-induced intermittency (SII), which initiates as soon as irrotational fluid has been entrained by the swirling jet. The irrotational fluid will sense the applied volumetric force F_{dc}, which is formed by the difference of the centrifugal forces acting on the entrained (denoted by the asterisk) and its surrounding fluid as given by F_{dc} accelerates the irrotational fluid towards the axis of rotation as long as it has not yet reached the local mean tangential velocity level. On its way towards the center of the jet, the entrained fluid is accelerated, deformed, and diluted by the jet flow until it loses its identity finally. The measurements reveal that the irrotational fluid may still be accelerated towards the axis in regions with positive circulation gradients (local disturbances are already damped here), provided they have a lower tangential velocity than their local surrounding. Due to their inertia, the entrained fluid elements can even reach the axis of rotation, where the force F_{dc} already decelerates the entrained fluid.

$$F_{dc} = \frac{\varrho \cdot \overline{w}^2}{r} - \frac{\varrho \cdot w^{*2}}{r} \tag{2}$$

In the following, the consequences of SII on the turbulent momentum transport are demonstrated by means of measured joint probability density functions (JPDF) of the Reynolds shear stresses. Figure 2.2.10 schematically shows the expected manifestation of SII in JPDFs of $v'w'$. For local turbulence production (Fig. 2.2.10a) an elliptic JPDF without skewness can be expected according to Da Costa Ribeiro [20]. Figure 2.2.10c shows a JPDF where SII dominates local turbulence production. The maximum of this skewed JPDF is situated in the third quadrant. This position means that the corresponding fluid elements move with the velocity v_p towards the center of the jet and have a tangential velocity deficit Δw relative to the local mean tangential velocity level (which corresponds to $w = 0$). According to Eq. (2), Δw causes a further acceleration towards the center of the flow. An equal superposition of both effects, local turbulence production and SII, leads to a bimodal JPDF as can be seen in Fig. 2.2.10b.

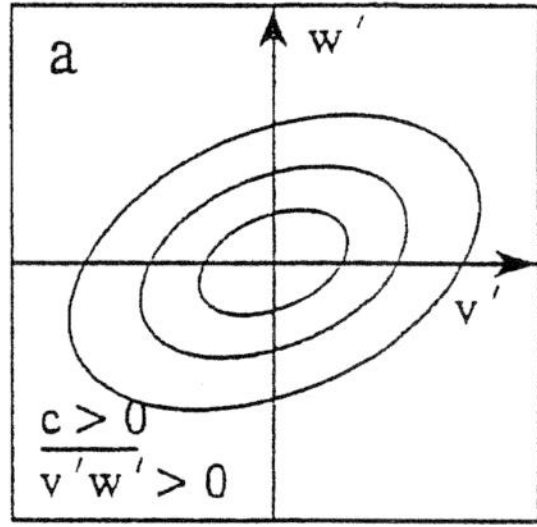

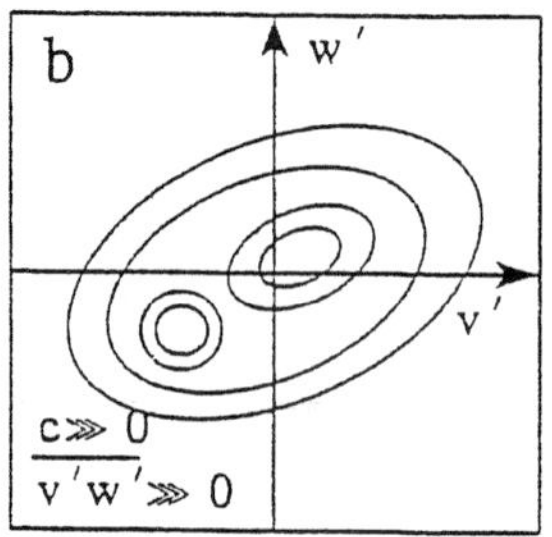

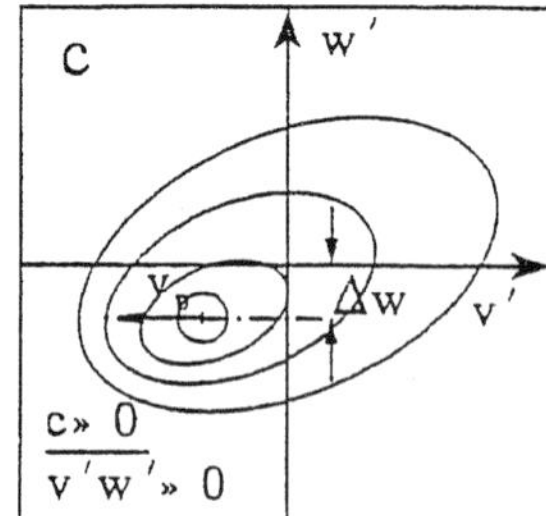

Figure 2.2.10: To the interpretation of JPDFs regarding SII.

Actual measurements of JPDFs of all three shear stresses are shown in Fig. 2.2.11. These JPDFs were established in one axial location (x = 100 mm) on three different radial positions for $S_{0,\text{theo}} = 0.95$. They are representative of different measuring positions and different swirl numbers. Every JPDF was calculated from 102,400 samples taken with the quintuple measurement technique and comprises the value of the corresponding shear stress $\overline{u'_i u'_j}$ and correlation coefficient $c = \overline{u'_i u'_j}/(\overline{u_i'^2} \cdot \overline{u_j'^2})^{1/2}$. The value of the lowest isoline is plotted in the diagrams and also corresponds to the increment of the equidistant isolines. The position of the origin of the coordinates for mean velocities is marked by a cross. This allows us to determine

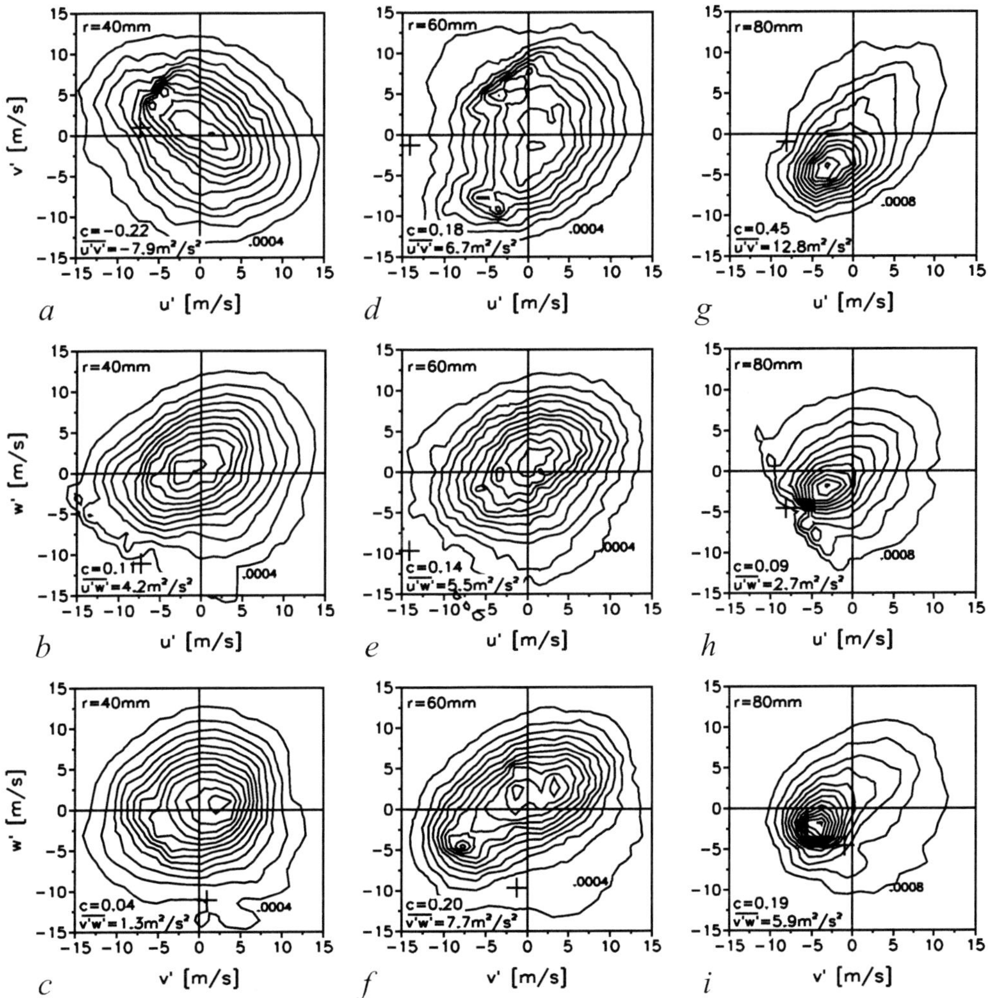

Figure 2.2.11: JPDFs of the Reynolds shear stresses, x = 100 mm, $S_{0,\text{theo}} = 0.95$.

the mean velocity levels of individual parts of the JPDFs. All three types of JPDFs which were described in Fig. 2.2.10 for $v'w'$ are found by measurements: local turbulence production (c), dominating SII (i), and the bimodal JPDF (f).

Further structural information is made available by the JPDFs. Since SII is equivalent to fluid elements with little axial and tangential momentum moving towards the axis of rotation, the maxima of the JPDFs for $u'v'$ (g) and $u'w'$ (h) at $r = 80$ mm are also located in the third quadrant. These particular locations together with their high probabilities substantially modify the value of the corresponding shear stresses. At $r = 60$ mm, $u'v'$ (d) exhibits a distribution with three maxima. Beneath the central maximum (local production), two further maxima exist, which indicate transport of low axial momentum from the center (second quadrant: $v > 0$) and the boundary (third quadrant: $v < 0$, SII) of the jet. Since $\overline{u'v'}$ is positive, SII obviously also dominates the $u'v'$ distribution. The JPDF for $v'w'$ (f) clearly shows two maxima. The maximum in the third quadrant denotes the irrotational fluid which has meanwhile picked up more axial and tangential momentum and moves at about $\overline{v} = -6$ m/s towards the axis. Due to its location, this maximum contributes substantially to the high value of the corresponding shear stress and to the radial transport of angular momentum. SII also considerably increases $\overline{u'w'}$. This can only be proved by the following conditioned measurements, since the two characteristic peaks of local production and SII melt into one another. In contrast to the JPDFs for $u'v'$ and $v'w'$, they are not separated by the high penetration velocity of the irrotational fluid.

Now, the ambition to obtain a more comprehensive understanding of SII by separating SII from the jet flow is near at hand. This is achieved by slightly heating the swirling free jet and by measuring the temperature and the velocity vector simultaneously. The main benefit of this approach is the possibility of using the value of the instantaneous temperature to decide whether the coincident measured velocities should be attributed to the entrained fluid (cold) or the turbulent flow (heated). This enables us to quantitatively assess the contribution of SII to the entire momentum exchange. Figure 2.2.12 illustrates this method with the typical development of temperature and a velocity-proportional hot-wire voltage as observed on the oscilloscope. When the temperature signal falls under the threshold value T_{thr}, the high-frequency velocity fluctuations cease, which corresponds to irrotational fluid moving through the measuring volume.

From these measurements, PDFs of temperature, temperature-velocity JPDFs and conditioned JPDFs of the Reynolds shear stresses were calculated. Furthermore, characteristic time and length scales as well as velocities of the entrained irrotational fluid were extracted. A drawback of this method is the decrease of the force F_{dc} caused by the density difference of the cold entrained and the heated flow according to Eq. (2). The heating of the jet-air by 25 K above the ambient temperature leads to a maximum decrease of F_{dc} by 8 % and was neglected in the interpretations.

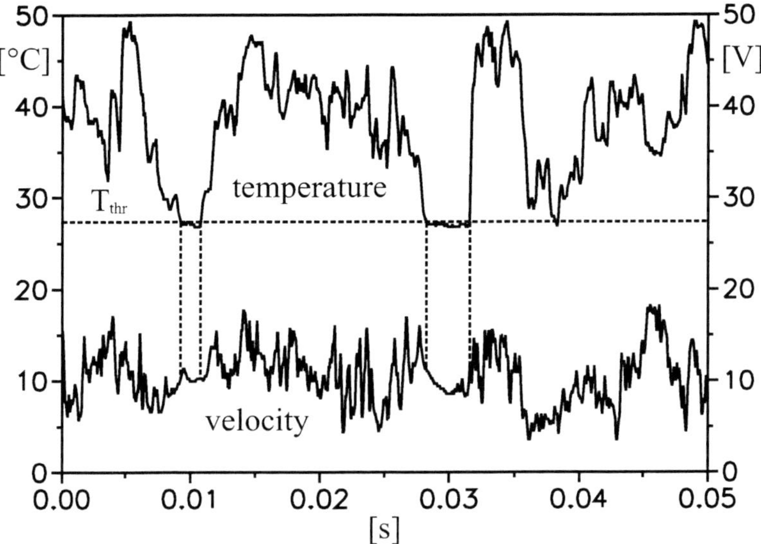

Figure 2.2.12: Typical development of temperature and a hot-wire voltage in the heated swirling free jet.

Figure 2.2.13 displays conditioned JPDFs (CJPDF) taken at $S_{0,\text{theo}} = 0.95$ and $x = 100$ mm at the radius of $r = 80$ mm, which correspond to the JPDFs shown in Fig. 2.2.11g–i. The results of the conditioned measurements are highly dependent on a properly chosen threshold value. The temperature-velocity JPDFs (Fig. 2.2.13a–c) suggest choosing a threshold value of $T'_{\text{thr}} = -5$ K. This value clearly separates the distributions of the heated and the cold flow. At this threshold level, the probability of the CJPDFs for the entrained fluid (Fig. 2.2.13d–f) is 14 %. Because they are completely situated in the third quadrant, these CJPDFs have considerably higher covariances and correlation coefficients than the jet fluid (Fig. 2.2.13g–i). The entrained fluid has reached an average of 60 % of the local mean axial velocity and 43 % of the tangential velocity and moves with about $\overline{v} = -3.0$ m/s towards the axis of rotation. Its mean residence time in the measuring volume is 2.8 ms and its mean diameter is 21 mm. Due to the chosen threshold value, the CJPDFs of the jet fluid (Fig. 2.2.13g–i) are still skewed and their maxima are still entirely situated in the third quadrant, i. e. SII is not completely separated from the jet flow. When the threshold value is increased to $T'_{\text{thr}} = 0$ K, CJPDFs with little skewness are obtained (Fig. 2.2.13p–r). In that case, the already stronger heated boundary layer of the enclosed irrotational flow is attributed to the entrained fluid as well. This layer also originates from the jet boundary and features velocity levels similar to those in the irrotational core, thus contributing considerably to the momentum transport attributed to SII. Then, SII contributes substantially to the turbulent momentum transport: $\overline{u'v'}$ is increased by 70 % and $\overline{v'w'}$ by 64 % whereas $\overline{u'w'}$ even changes

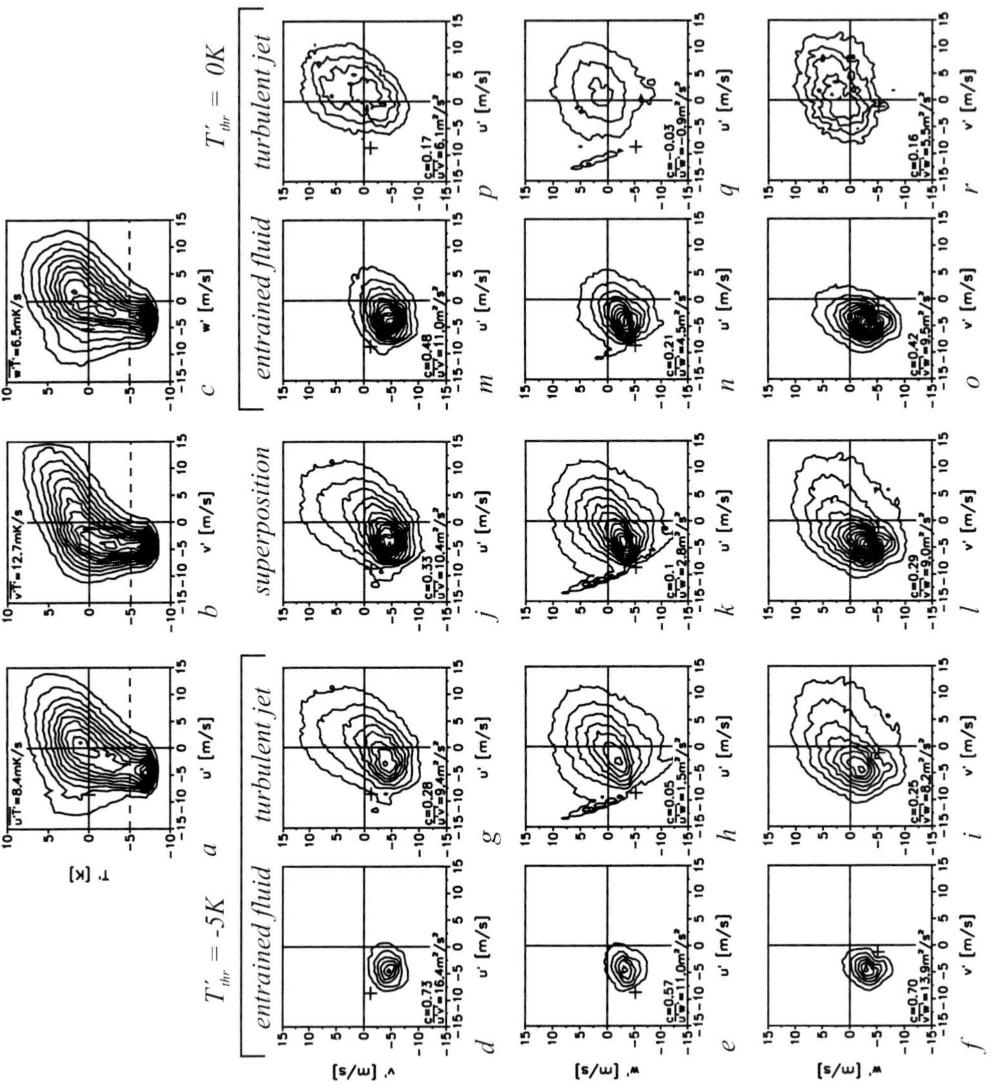

Figure 2.2.13: Temperature-velocity JPDFs and conditioned JPDFs of the Reynolds shear stress, $x = 100$ mm, $r = 80$ mm, $S_{0,\text{theo}} = 0.95$, $T_{\text{thr}} = -5$ K, resp. 0 K.

its sign. When the threshold value is fixed at $T_{thr} = -5$ K, which corresponds to an underestimation of the effect under consideration, $\overline{u'v'}$ and $\overline{v'w'}$ are increased by about 10 % whereas $\overline{u'w'}$ is doubled.

The JPDFs described and their qualitative interpretations are representative for different axial locations in the free jet and also for the other swirl numbers investigated. Further JPDFs are found in [14]. However, in regions where the turbulent momentum transport is small due to vanishing gradients of the mean velocities, SII can enhance the turbulent transport up to 800 %. On the other hand, there are regions where the effects are partially reversed. Close to the axis of rotation, SII can slightly reduce the shear stresses.

A short outline of the current investigations for the numerical prediction of swirling free jets will conclude the current section. Since the CJPDFs (Fig. 2.2.13p–r) are determined mainly by local turbulence production and not by swirl-induced radial transport, they approximately correspond to JPDFs which are modeled in conventional calculations of swirling free jets without consideration of SII. This statement is confirmed by calculations which Younis et al. [2] performed for the swirling free jet with a Reynolds stress closure. They predict negative values of $\overline{u'w'}$ (corresponding to the CJPDF, Fig. 2.2.13q), whereas their measured data display positive values (corresponding to the JPDF, Fig. 2.2.11h). Furthermore, they identify the negative values as the main reason for the defects of their predictions since $\overline{u'w'}$ contributes additionally to the rate of production of the shear stress $\overline{u'v'}$, which is responsible for the radial transport of axial momentum. Younis et al. [2] tried to overcome these defects with modifications of the model for the pressure-strain correlations without obtaining consistent predictive agreement.

Finally, the apparent contradiction that the k,ε-model often fails to predict confined swirl flows, whereas it provides satisfactory results in free jets can be exposed. The overestimation of the angular momentum transport, while sometimes leading to unacceptable results in the confined swirl flows, it partially compensates the missing contribution of SII to the overall momentum transport in the free jet.

2.2.5 Is Swirl-Induced Intermittency Periodic?

Park and Shin [21] performed detailed investigations of the entrainment characteristics of swirling free jets. They found that the entrainment process is considerably enhanced by large-scale periodic motions in the jet boundary regions near the nozzle exit. From these findings, the following questions arise: Do their results complement our investigations by describing the manifestation of SII in the free jet boundary region? Is SII periodic and eventually caused by the precessing vortex core which Park and Shin [21] identify as the source of the periodic motions?

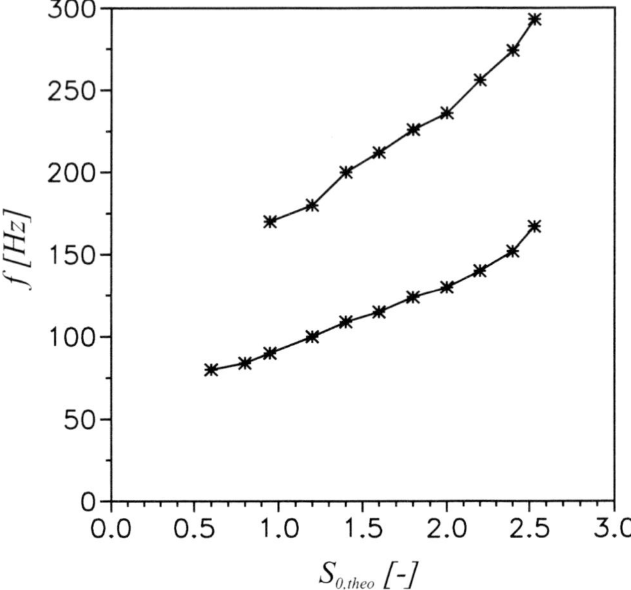

Figure 2.2.14: Dominant frequencies of spectrum analyses, $x = 10$ mm, $r = 40$ mm.

To answer these questions, we established power spectra from linearized single-hotwire measurements. Figure 2.2.14 shows the dominant frequencies in the vicinity of the nozzle exit. These frequencies correspond exactly to those occurring in confined swirling flows [14]. The maximal amplitudes of the peaks in the power spectra appear inside the nozzle and are strongly dampened downstream in the free and the confined flows. At $x = 100$ mm, the frequencies have completely disappeared. However, at the same location, SII was not only proven to exist, but even partially dominates the momentum transport. For $S_{0,\text{theo}} = 0.4$, we could not find any dominant frequencies, although SII is manifested for this swirl number just as for the stronger swirling cases. These results suggest that SII is not periodic and the frequencies occurring for swirl numbers exceeding $S_{0,\text{theo}} = 0.6$ are not related to SII. Altogether, our results do not allow for a consistent interpretation with all aspects of the results of Park and Shin [21].

2.2.6 Similar Effects in Confined Swirling Flows

After SII was shown to modify the momentum transport in swirling free jets further investigations were performed to see whether similar effects occur in low-confined swirling jets, which have been examined regarding JPDFs of the shear stresses in a cylindrical model of a combustion chamber of 500 mm diameter and 1000 to 2000 mm length in many detailed measurements [10, 14, 18]. In fact, JPDFs featuring the characteristics of SII were established in the jet's boundary in the vicinity of regions with negative circulation gradients. Figure 2.2.15 shows JPDFs for $v'w'$ measured on two radial positions at $x = 40$ mm and $S_{0,\text{theo}} = 0.4$. A comparison of Fig. 2.2.15a,b (confined flow) and Fig. 2.2.11c,i (free jet) illustrates the similarity of the respective JPDFs. Both JPDFs measured in the confined flow are located in stable regions. This means that the effects seen in Fig. 2.2.15b originate in the adjacent unstable region, which is extended from $r = 65$ to 80 mm. Here, the parts of the turbulent flow, which have an angular momentum deficit against the local mean value, are accelerated towards the center of the flow. Then, they penetrate into the stable region ($r \leq 60$ mm) due to their inertia. However, the measured penetration velocity $\overline{v} \approx -1.7$ m/s is much smaller than in the free jet. This fact is supported by the different radial distances of the JPDFs with similar characteristics. The distances, which are proportional to the penetration depth of the fluid elements, amount to $\Delta r = 40$ mm in the free jet and $\Delta r = 10$ mm in the confined flow. It is not surprising that in unstable regions of swirling flows fluid elements tend to separate according to their individual angular momenta. This effect is well-known and leads to

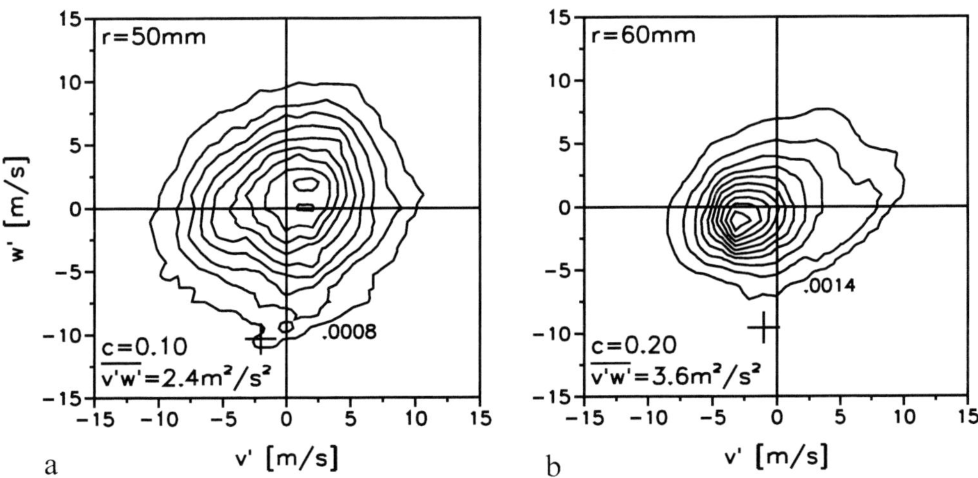

Figure 2.2.15: JPDFs of $v'w'$ in the confined swirling flow, $x = 40$ mm, $S_{0,\text{theo}} = 0.4$.

a promotion of the turbulent transport which is completely described by the Reynolds stress transport equations [22]. However, the penetration of such accelerated fluid elements into stable regions and the related enhancement of the momentum transport is not represented by the stress transport equations. On the other hand, negative circulation gradients usually only appear in very limited regions of confined swirl flows, because these tend to form similar as well as stable tangential velocity profiles along the whole length of the confinement. Therefore, we conclude that the effects described are of secondary importance for the overall momentum transport in common confined flows. This judgment is confirmed by the superior prediction qualities of Reynolds stress models in confined swirl flows as opposed to swirling free jets.

References

1. Morse, A. P. (1980): Axisymmetric free shear flows with and without swirl, Thesis, Fac. of Engineering, University of London.
2. Younis, B. A., Gatski, T. B., and Speziale, C. G. (1994): On the prediction of free turbulent jets with swirl using a quadratic pressure-strain model, Institute for Computer Applications in Science and Engineering, *NASA ICASE, NASA Contractor Report* 194964, Hampton.
3. Sislian, J. P., Cusworth, R. A. (1984): Laser Doppler velocimetry measurements in a free isothermal swirling jet. University of Toronto, Institute of Applied Science, Rep. No. 281, CN ISSN, 0082–5255.
4. Fu, S., Launder, B. E., and Leschziner, M. A. (1987): Modelling strongly swirling recirculating jet flow with Reynolds-stress transport closures. *Proc. 6th Symp. Turb. Shear Flows*, Toulouse, 6171–6176.
5. Cheng, T.-C., Hassel, E. P., and Janicka, J. (1995): Some velocity measurements in a swirling jet and a swirling flame with strong recirculation. *Eng. Research* **61**, 1–5.
6. Holzäpfel, F., Lenze, B., and Leuckel, W. (1994): Assessment of a quintuple hotwire measurement technique for highly turbulent flows. *Exp. Fluids* **18**, 100–106.
7. Weber, R., Visser, B. M., and Boysan, F. (1990): Assessment of turbulence modeling for engineering prediction of swirling vortices in the near burner zone. *Int. J. Heat Fluid Flow* **11**, 225–235.
8. Hirsch, C. and Leuckel, W. (1996): A curvature correction for the k-ε model in engineering applications. *Proc. 3rd Symp. Eng. Turb. Model. Meas.*, Heraklion-Crete, 71–80.
9. Hirsch, C. (1995): Ein Beitrag zur Wechselwirkung von Turbulenz und Drall. Dissertation, Universität Karlsruhe.
10. Döbbeling, K. (1990): Experimentelle und theoretische Untersuchungen an stark verdrallten, turbulenten isothermen Strömungen. Dissertation, Universität Karlsruhe.
11. Leuckel, W. (1967): Swirl intensities, swirl types and energy losses of different swirl generating devices. IFRF Doc. Nr. G02/a/16, Ijmuiden.
12. Döbbeling, K., Lenze, B. (1990): Computer-aided calibration and measurement with a quadruple hotwire probe. *Exp. Fluids* **8**, 257–62.

13. Meyer, L. (1992): Calibration of a three-wire probe for measurements in nonisothermal flow. *Exp. Therm. Fluid Sci.* **5**, 260–267. Morse, A. P. (1980): Axisymmetric Free Shear Flows With And Without Swirl. Thesis, Fac. of Engineering, University of London.
14. Holzäpfel, F. (1996): Zur Turbulenzstruktur eingeschlossener und freier Drehströmungen. Dissertation, Universität Karlsruhe.
15. Taylor, G. I. (1921): Experiments with rotating fluids. *Proc. Roy. Soc.* **A100**, 114.
16. Rayleigh, Lord, O. M. (1917): On the Dynamics of Revolving Fluids. *Proc. Roy. Soc.* **A93**, 148–154.
17. Jacquin, L., Leuchter, O., and Geffroy, P. (1990): Homogeneous turbulence in the presence of rotation. *J. Fluid Mech.* **220**, 1–52.
18. Holzäpfel, F., Lenze, B., and Leuckel, W. (1996): Quintuple hot-wire measurements of the turbulence structure in confined swirling flows. Submitted to *J. Fluids Eng.*
19. Hinze, J. O. (1975): Turbulence. McGraw-Hill, New York.
20. Da Costa Ribeiro, M. M. (1976): The turbulence structure of free jet flows with and without swirl. Ph. D. thesis, Fac. of Eng., University London.
21. Park, S. H., Shin, H. D. (1993) Measurements of entrainment characteristics of swirling jets. *Int. J. Heat Mass Transfer* **36**, 4009–4018.
22. Hirai, S. and Takagi, T. (1995): Parameters dominating swirl effects on turbulent transport derived from stress-scalar-flux transport equation. *Int. J. Heat Mass Transfer* **38**, 2175–2182.
23. Kitoh, O. (1990): Experimental study of turbulent swirling flow in a straight pipe. *J. Fluid Mech.* **225**, 445–479.

2.3 Mathematical Modeling of Turbulent Swirling Flames

Peter Habisreuther, Matthias Philipp, Heinrich Eickhoff, and Wolfgang Leuckel *

Abstract

Important aims of flame predictions are the flame ignition stability and NO_X-emissions. As swirling flames usually comprise highly turbulent flows, the interaction of turbulence and oxidizing reactions have to be considered carefully, e. g. by using a coupling-model of the "presumed-pdf"-type. Due to its strong dependence on temperature and oxygen atom concentration, a further application of such a model has been performed concerning thermal formation of NO. For this purpose an extension to a "presumed-pdf"-model has been developed and tested thoroughly.

2.3.1 Introduction

For modeling kinetically controlled phenomena like blow-out and pollutant emissions, proper representation of the flow- and mixing fields in swirling flames is a crucial prerequisite. Here, modeling of the non-reacting (isothermal) swirling flow is considered usually as a first step before taking into account also the more sensitive chemical equations. However, predictions of isothermal swirling flows are connected with large uncertainties, and sometimes reveal unrealistic results, when using the simple standard k,ε-model for turbulence closure [1–3]. Therefore, a comparison of results obtained using different turbulence closure schemes is shown first in the next section. Flame stability and pollutant emissions are the most important properties when designing industrial combustion devices. Therefore, numerical investigations of these phenomena for swirling flames were performed. The model used for turbulence/reaction coupling [4] in order to predict blow-out (or flame stability) was extended afterwards for the prediction of

* Engler-Bunte-Institut, Bereich III – Verbrennungstechnik, Universität Karlsruhe, Kaiserstr. 12, 76128 Karlsruhe, Germany

thermal NO_X-formation. This extension is explicitly described in Section 2.3.5.3.

2.3.2 Combustion Systems

Experimental investigations were carried out with swirling premixed and non-premixed flames, either confined in cylindrical combustion chambers (e. g. Fig. 2.3.1, premixed configuration) or unconfined in free environment. Swirl was generated by devices of the *movable-blocks*-type [5], allowing to adjust swirl intensity in a wide range [6]. Details on operational conditions and geometrical parameters are given in [2, 6, 7] for the premixed case, in [8, 6] for the non-premixed flame configuration, and in [4] for the flame

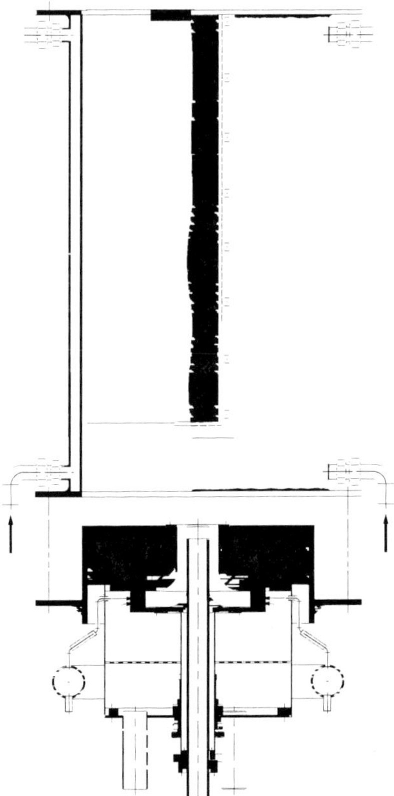

Figure 2.3.1: Overview over burner and water-cooled combustion chamber for premixed operational mode.

stability experiments shown below. Besides mixing conditions, the most important differences between the first two (confined) configurations refer to the thermal boundary conditions, as the premixed flame was operated in a fiber-insulated burning-chamber, while the non-premixed flame was strongly cooled by bare water-cooled walls ($T_W \approx 80\,°C$).

2.3.3 Swirling Flow Modeling

The poor accuracy often experienced when using the standard k,ε-model for turbulence closure in swirling flows is a result of wrong predicted turbulent momentum exchange by this model. In particular, the exchange of circumferential momentum is over-predicted, leading to tangential velocity profiles of the forced vortex type [1]. As a backlash, these tangential velocities often result in unrealistic axial velocities, i. e. flow field patterns, such as much too long central recirculation zones. A remedy to this problem is offered by higher order turbulence schemes such as a Reynolds-stress-model (RSM), or simpler models especially designed for swirling flows. The latter ones are e. g. *Richardson-type* corrections [9, 4], *Renormalization Group* (RNG) based corrections, or an algorithm which was developed by C. Hirsch [10] (MKE = modified k,ε-model) within the Collaborative Research Centre 167. This correction algorithm itself was tested extensively for isothermal swirling flows and is described in [10], so that a detailed description will not be given here.

The focus of this work is modeling of swirling flames, one aim being to prove the applicability of relatively simple correction schemes and to compare their results to more sophisticated models such as the RSM.

In a reacting surrounding, effective swirl intensity is reduced due to the radial expansion of flue-gas. This is one of the primary explanations found in literature for simulation results often being in relatively good agreement to experimental results under this special conditions, even when using the standard k,ε-model.

Holzäpfel [1] offers a detailed discussion of experimental results found on this topic in literature. Summarizing his conclusion, the good agreement found using the simple k,ε-model is often a result of global geometry parameters, especially the form of the combustion chamber outlet. When using a centrally blocked outlet geometry, differences between k,ε-model and higher order turbulence closure results are found small. This is due to the global flow pattern which enforces tangential velocity profiles of the solid body type vortex, e. g. [10]. In contrast, when using a circumferential blockage of the chamber outlet, forcing the flow to exit centrally, distinct differences are found, e. g. [9, 11]. On the other hand Weber [12, 3] found that even for a central burner outlet geometry application of the k,ε-model yields reasonable good accuracy.

Because of these somehow contradictory results, comparisons of predictions of the flow field under hot conditions will be discussed below.

2.3.3.1 Numerical Method and Turbulence Modeling

For the following predictions a 2-d TEACH-based code was used for solving the balance equations of momentum and mass. The MLU high order discretisation scheme [13] and the SIMPLEC algorithm for pressure coupling served for the numerical realization.

Three models were applied for turbulence closure. The first one was the standard k,ε-model (STKE), which is widely used, extensively discussed in literature and was applied in this work with the standard modeling constants. This approach is based upon Boussinesq's approximation, assuming an isotropic turbulent viscosity which is a very crude assumption in swirling flow situations [10, 9]. The most complex model used was a Reynolds-stress-model (RSM), which is discussed in detail for combustion e. g. in [11, 14].

The third model compared below is a modified k,ε-model (MKE) by C. Hirsch [10], corrected on the basis of the local time-mean acceleration vector of the flow. The basic approach was derived on the basis of comparison of the *algebraic stress model* (ASM) with the equations for the Boussinesq' stresses. Assuming the central difference to be the *added convection* terms, Hirsch defines a correction of the turbulent shear stress $\overline{\varrho v'w'}$. Details of this model are explained in [15, 10].

2.3.3.2 Turbulent Reaction Model for Heat Release

Heat release was calculated solving balance equations for the time-mean concentrations of the species CH_4 and CO, and one equation for the time-mean mixture fraction f. The concentrations of the species O_2, N_2, CO_2, and H_2O then were computed from mass-balance according to mixture fraction. These species concentrations contribute to the specific heat capacity, resulting in the local temperature value due to the local values of enthalpy.

Due to the non-linear dependence of reaction rates on temperature and species concentrations, the influence of turbulent fluctuations of these values on the time-mean reaction rates have to be modeled. This was done by applying a model concept of the "presumed-pdf"-type which was developed by M. Philipp [4] and will be described in Section 2.3.4.1. This turbulent reaction model was extended for predicting thermal NO_X, as will be discussed in Section 2.3.5.3.

2.3.3.3 Flow Predictions

Premixed Configuration

To give an impression of the flow field of the premixed swirling flame, the normalized stream function $\Psi/\dot{m}_0$ is shown in Fig. 2.3.2 with $\dot{m}_0$ as total the mass-flow. The flow calculation was performed using the RSM for turbulence closure. The flow field shows two separated recirculation zones, a central one (IRZ) and another in the bottom edge of the combustion chamber (ARZ). It should be pointed out that the IRZ does not close and extends up to the central outlet-blockage (right). This result was obtained for each of the turbulence-closure models mentioned above. Looking more into detail, predictions are compared with measured data in the near burner zone in Fig. 2.3.3 for all three turbulence models. First, a relatively good accuracy can be stated for the predictions, despite of a slightly under-predicted recirculation intensity in the ARZ. Second, the differences between the individual calculation results are small, supporting the result of Peters and Weber [12]. In addition, the tangential velocity profiles of all turbulence models are compared in Fig. 2.3.4: The best accuracy is achieved using the RSM, but the differences are small. Because the outlet-geometry in this combustion chamber was of the central blockage type, this result supports Holzäpfel's interpretation [1]. Although similar, a characteristic difference exists: The maxima of the tangential velocity profiles show the smallest radial shift using the RSM. This result can also be observed for the axial velocity profiles which are not plotted below, and corresponds to stronger recirculation at the ARZ as shown in Fig. 2.3.3.

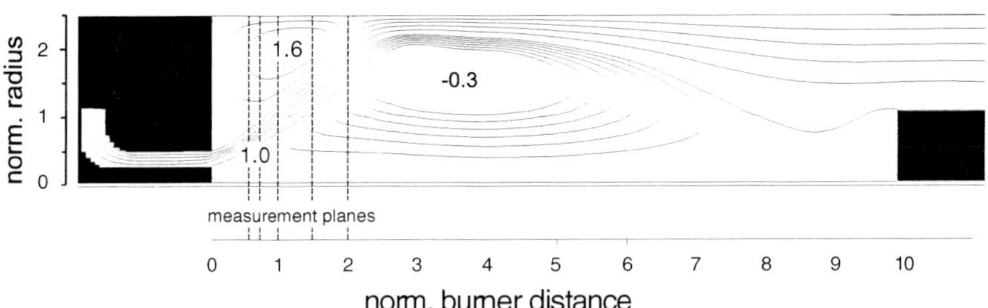

Figure 2.3.2: Normalized stream function (RSM-model) in a meridian plane of the combustion chamber for premixed swirling flame (D_0 = 90 mm; $S_{0,\text{th}}$ = 1.5; λ = 1.1; $\dot{V}$ = 308 m_N^3, T_w = 80 °C).

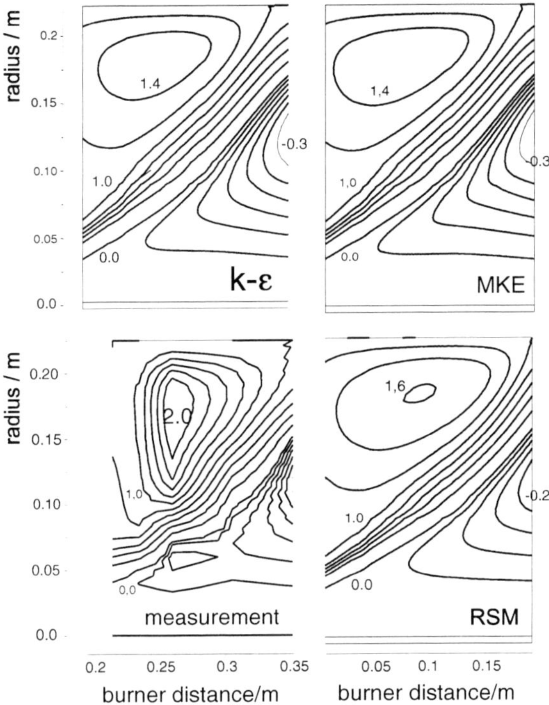

Figure 2.3.3: Iso-lines of the normalized stream function in the near burner zone for premixed flame; comparison of calculated with measured data (D_0 = 90 mm; $S_{0,th}$ = 1.5; λ = 1.1; $\dot{V}$ = 308 m_N^3, T_w = 80 °C [10]).

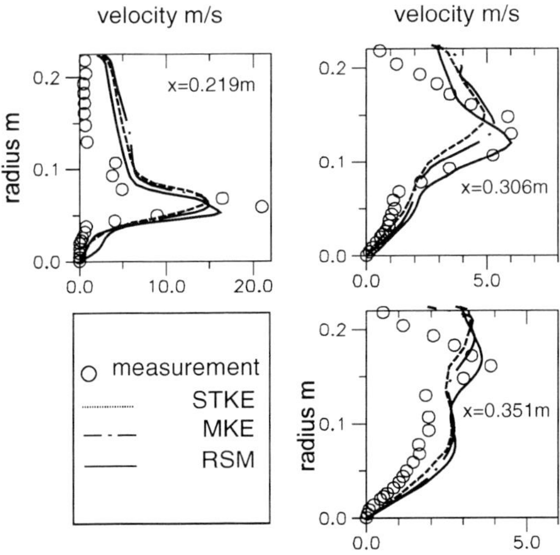

Figure 2.3.4: Comparison of calculated with measured tangential velocity profiles [10] for premixed swirling flame.

Non-Premixed Configuration

For a non-premixed configuration, a comparison of results obtained with different turbulence closures was performed accordingly [16]. Summarizing the results shown there, the following conclusions may be drawn: The differences in flow field predictions are found small too, supporting the results of Weber et. al., e. g. [3]. Because of the outlet geometry being of the central blockage type in combination with a relatively narrow combustion chamber, this also supports the results of Holzäpfel [1, 15].

2.3.4 Modeling Combustion Stability

2.3.4.1 Turbulent Reaction Model

The experimental investigations to which the following is referring to are reported in [6]. For predicting stability limits, the proper modeling of heat-release is a fundamental prerequisite which includes the effect of turbulent fluctuations of temperature and concentrations on the mean fuel consumption rate. Therefore, a turbulent reaction model, basically comprising a "*presumed-pdf*" approach, was developed by M. Philipp [4] for calculating the time-mean source terms of species balance equations.

In the following section a brief overview will be given. A *joint-probability density function* (JPDF) is **constructed** by means of two statistical moments of two parameters, one describing the local stoichiometry (mixture fraction f), the other one the reaction progress (e. g. normalized temperature Θ or rate of burnout a_r). For these statistical moments (mean value and variance), transport equations have to be solved, assuming these parameters to be mass bound (see Fig. 2.3.5). While time-mean values for mixture fraction and temperature are usually available, the transport equations of variances, explicitly discussed in [4], have to be solved additionally. Assuming statistical independence of reaction progress from mixture yields the joint pdf by simply multiplying the single pdf's:

$$\mathrm{jpdf}(f,\Theta) = \mathrm{pdf}(f) \cdot \mathrm{pdf}(\Theta) \tag{1}$$

The construction of the single pdf's is done for each control volume of the calculation domain, *assuming* a principal shape of the single pdf's and using the adiabatic flame temperature for normalizing the temperature to $\Theta = T/T_{ad}$ (Fig. 2.3.5). Subsequent calculation of the chemical rates for every point of the JPDF with an adequate global mechanism [17, 2] and integration of the probability weighted rates yield the time-mean reaction rates.

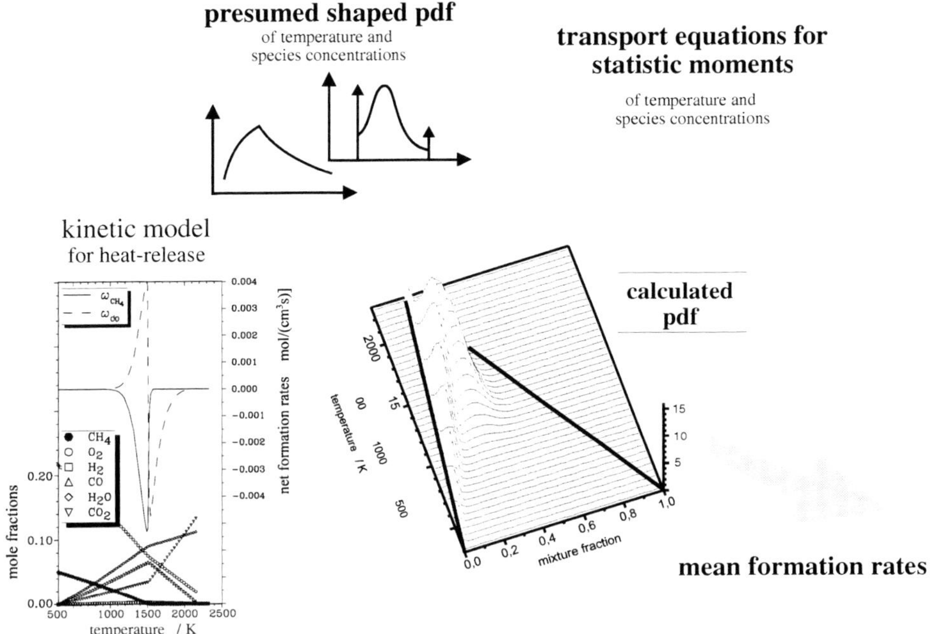

Figure 2.3.5: Schematic view of the JPDF-reaction model.

Further details of this turbulent reaction model are explained in greater detail in [4].

2.3.4.2 Results

Calculations were performed for unconfined premixed and non-premixed swirling flames. In order to predict stability limits, several calculations were made, varying the thermal load of the flame. As a first result of these calculations we found an approximate independence of the flow pattern and mixing-field on the thermal load. This *Reynolds-independence* is shown in Fig. 2.3.6 where on the left the stream function, normalized with the mass-inflow, and on the right the mixture fraction representing the local stoichiometry has been plotted for two thermal loads in the vicinity of the burner exit. The central recirculation-zone can be identified with negative normalized stream-function values. As can be seen from this figure, the stoichiometry as well as the flow pattern remains approximately unchanged despite the fact that the volume flow rate of the $\dot{V}$= 350 m_N^3 case is already very close to blow-off.

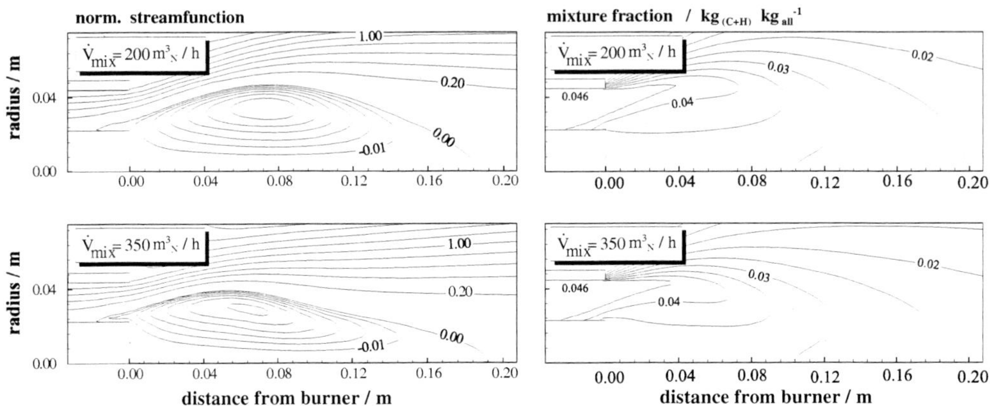

Figure 2.3.6: Iso-lines of the normalized stream function ψ/ψ_0 and mixture fraction in the near burner zone of the premixed flame-configuration ($S_{0,th}$ = 1.5, λ = 1.2) at different thermal loads.

Stability Limits

Stability limits were calculated by increasing the thermal load until blow-off occurred. Blow-off was identified by situations where no converging solution could be achieved anymore. In Fig. 2.3.7 a comparison of calculated stability limits with measured data is shown in terms of the highest thermal load achievable for different operational parameters. The dimensionless para-

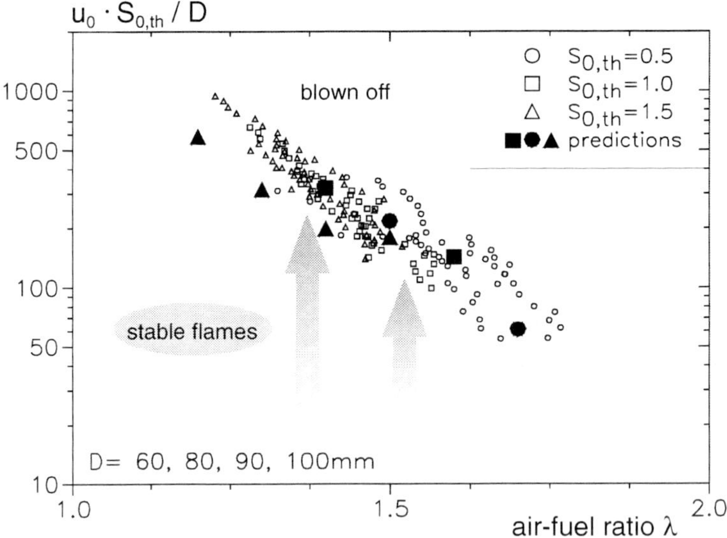

Figure 2.3.7: Calculated and measured stability limits of the unconfined, premixed, swirling flames operated at different swirl parameters $S_{0,th}$.

meter $u_0 \cdot S_{0,\text{th}}/D_0$, was found to uniquely define flame stability as a function of the air fuel ratio, and that over a range of burners and flames of different values of mean nozzle velocity u_0, nozzle diameter D_0 and even different swirl-numbers [6, 18, 19]. The good accuracy shown in this diagram confirms that the model is able to predict the effect of swirl on stability.

Iso-lines of the time-mean values of temperatures, for the two thermal loads already shown in Fig. 2.3.6, are plotted in the near-burner zone on the left side of Fig. 2.3.8. Increasing the thermal load leads to an isolated spot of high temperatures in the shear zone between forward flow and recirculation zone, monotonously decreasing in size with increasing load until blow-off occurs. Corresponding to the temperature field, the fuel-consumption rates, shown on the right side of Fig. 2.3.8, decrease too.

The experiments on confined flames with different cooling rates have demonstrated a strong thermal influence on blow-out [6]. If the power is being increased, flame temperature increases as well, resulting in a counter effect of decreasing residence time and increasing kinetic reaction rate on stability. Hence, for special conditions, stability may even increase with decreasing residence time.

The influence of the kinetic time scale on stability can be separated, by applying preheating of the mixture and maintaining constant residence time with reduced thermal power. Then the air ratio λ at blow-out becomes a function of preheat temperature alone.

It has been shown that this function is identical for swirling flames and a *perfectly stirred reactor* (PSR) [20]. Hence, if the air ratio is known at blow-out for a flame at one preheat temperature, all other values can be evaluated from PSR-calculations.

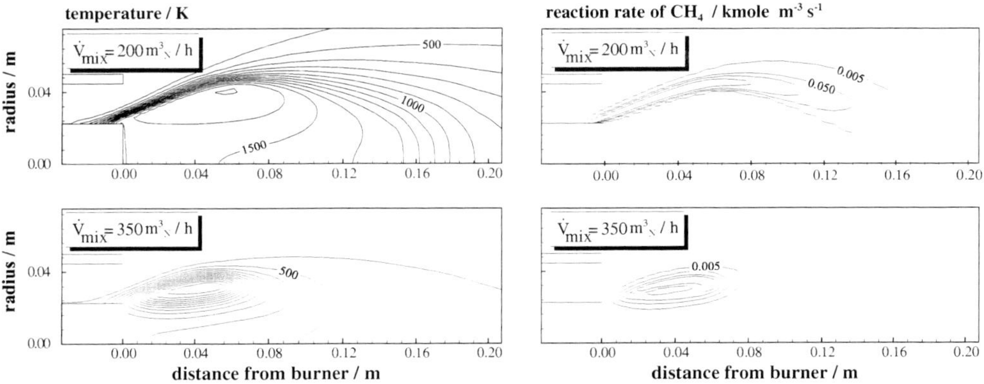

Figure 2.3.8: Iso-lines of time-mean temperature and fuel-consumption rate in the near burner zone of the premixed flame configuration ($S_{0,\text{th}} = 1.5$, $\lambda = 1.2$) at different thermal loads.

2.3.5 Modeling Thermal NO-Formation

The good prediction quality obtained within the flame stability calculations, by taking the influence of turbulent fluctuations of temperature and mixture into account, encourages the use of this approach in the context of modeling thermal formation of nitrogen oxide, well known to be very sensitive to the combustion temperatures and, therefore, its fluctuations too. Additionally, as the influence of the concentrations of oxygen atoms becomes crucial in many situations, an extension to the approach used for prediction of heat-release, was developed and will be discussed below.

2.3.5.1 Kinetics

Despite the fact, that the chemical mechanism for the so-called thermal formation of nitrogen oxide is well-known since several years, calculations of this pollutant species often reveal unrealistic results in highly turbulent flames as applied e. g. in gas turbine combustion. An analysis of the chemical process gives an explanation for this behavior.

According to Zeldovich [21] and to an addition of Fenimore [22] for fuel-rich conditions, the thermal formation rate of NO can be calculated employing three elementary reactions; the first and rate-controlling step describing the attack on the stable N_2 molecules by oxygen atoms, the second much faster step forming NO while recycling the oxygen atoms, and a third step forming NO with the use of OH-radicals:

$$\begin{aligned} N_2 + O &= NO + N \\ N + O_2 &= NO + O \\ N + OH &= NO + H \end{aligned} \tag{2}$$

Assuming the nitrogen atoms to be in steady-state and neglecting the reverse rates, as justified because being far from NO equilibrium concentration, yields the well-known simple expression for the NO-formation rate using coefficients from literature [24]:

$$\dot{\omega}_{NO} = 2 \cdot 1.8\ 10^{14} \cdot \exp\left\{\frac{-319\ \text{kJ/mol}}{RT}\right\} \cdot [O] \cdot [N_2] \tag{3}$$

As can be seen, besides the strong dependency on temperature ($E_A = -319$ kJ/mol), the formation rate of NO is a function of the concentrations of the oxygen atoms, too (N_2 is approx. constant for constant local stoichiometry). While temperature and N_2 are known by experiments or can be calculated using a global reaction scheme for describing heat-release, the concentration of oxygen atoms usually is not available and, therefore, has to be modeled. Such models usually comprise partial equilibrium assumptions. While the most often applied approach, assuming the very slow recombination reaction of oxygen atoms to be in partial equilibrium, is only valid for high temperatures and, even more restrictive, for long residence times, a slightly more complex model assumes an O/H subsystem to be in partial equilibrium [7]. With coefficients from [23], the second assumption yields the following expression for the concentration of oxygen atoms:

$$[O] = 3.0502 \cdot [H_2]\,[O_2] \;/\; [H_2O] \cdot \exp\{-589.37/(T/K)\} \tag{4}$$

With this additional equation the thermal formation rate of NO can be calculated on the basis of the concentrations of N_2, H_2, O_2, and H_2O, which have to be supplied by experimental data or calculation of heat-release.

2.3.5.2 Turbulence/Reaction Coupling

The success of a model applying the *"presumed pdf"* approach strongly depends on the numerical representation of the pdf, i. e. the discretisation used. However, the discretisation requirements themselves strongly depend on the ratio between the time scales of mixing and of the reaction progress: If reaction is very fast with respect to turbulent mixing the pdf degrades to a double-peak shape, one corresponding to the non-mixed part and the other one for the mixed and reacted part; if reaction is very slow the pdf becomes single-peak shaped, yielding time-mean formation rates depending on the local residence time, analogous to a well-stirred reactor. Due to discretisation efforts, both of these cases are hardly possible to describe with a pdf approach, at least the accuracy of model predictions is rather low. As will be shown below, another problem arises choosing the reaction-progress variable as one independent parameter of the pdf. Therefore, as a first step, the characteristics of the thermal NO-formation were investigated in a 1-dim. laminar premixed flame.

On the left side of Fig. 2.3.9 a typical progress of the formation rates for lean conditions is plotted against flame temperature. Three characteristics can be seen: 1.) Significant formation rates are restricted to temperatures above approx. 1700 K; 2.) In the subsequent temperature range formation is mainly controlled by the temperature rise, due to the exponential dependence in Eq. (3); 3.) A sharp decrease of formation rates can be seen for very high temperatures, despite the fact that NO-concentrations are far from equilibrium values. This fact is due to the super-equilibrium concentrations of the oxygen atoms, e. g. found at 2000 K ($\lambda = 1$). The right hand diagram of Fig. 2.3.9 shows characteristic time-scales of the NO-formation for different operational parameters, calculated according to [25]. Two additional

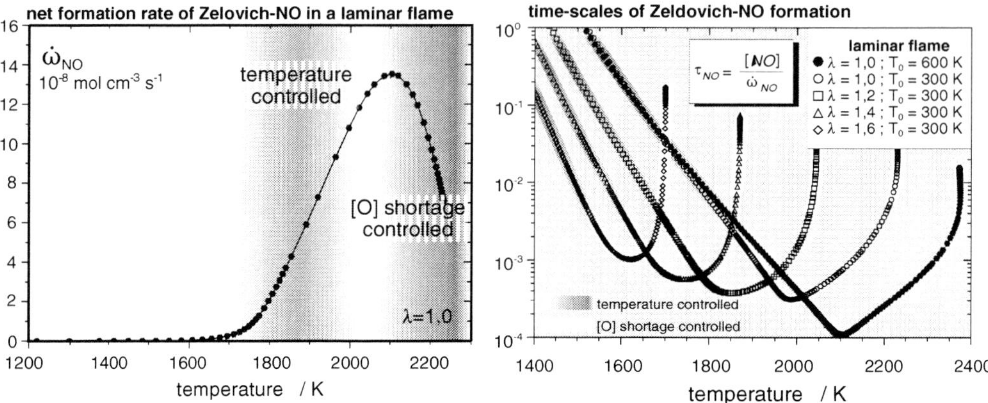

Figure 2.3.9: Characteristics of Zeldovich-NO formation: formation rate (left) and characteristic time-scales (right) in laminar flames.

aspects can be seen: 1.) The time-scales vary in a wide range comprising up to three orders of magnitude; 2.) The time-scales (and therefore the formation rates, too) are nearly **independent** of temperature at highest temperature values. Summarizing the results, with respect to the implementation of the chemical scheme in the frame of a "presumed-pdf" approach, two domains with different behavior can be distinguished:

1. One domain of reaction progress, where the temperature is distinctly apart from the highest temperature. In this domain NO-formation rates are mainly dominated by their temperature dependence. Temperature or some burnedness variable may be used as the independent variable for the NO-formation progress. Here, the oxygen atoms are related with the progress variable due to Eq. (4). Because the time-scales of NO-formation are of similar order as those of turbulent mixing (approx. 0.1 to 1 ms), the impact of turbulent fluctuations of stoichiometry and reaction progress has to be taken into account, e. g. by the aid of statistical models.

2. For the highest temperatures NO-formation is very slow and varies mainly due to the recombination of oxygen atoms. The time-scales of NO-formation are far higher than those of turbulent mixing. Because the temperature is approximately constant (variation occurs only due to heat-losses), the turbulent fluctuations of temperature may be neglected. Interaction with mixing can be described similar to a "perfectly-stirred" reactor system, where the residence time can be used substituting a progress variable.

2.3.5.3 Two-Domain Model for Thermal NO-Formation

When describing thermal NO-formation in highly turbulent swirling flames with variable operational conditions, one faces the fact that both of the domains mentioned above may significantly contribute to the total formation of NO. For this purpose a turbulent NO-formation model, comprising both sub-models, has been developed.

"Presumed-pdf" Model for the Main Reaction Zone

As a fundamental basis, the "presumed-pdf" model already discussed above, was applied with mixture fraction and burnedness serving as pdf-variables, where the mean-value of burnedness is calculated on the basis of O_2-consumption as discussed in [2]. For the variance of burnedness a transport equation is solved accordingly.

As stated above, the presumed-pdf can be constructed using these mean-values and variances. For each discrete point of the pdf the composition of stable species can be calculated as a function of burnedness and mixture fraction, and by assuming a global 1-step reaction scheme [2]. Subsequent application of Eqs. (3) and (4) for each pdf-point and integration of the calculated rates over the pdf yields the time-mean NO-formation rate. As the temperature, which can be calculated at each point of the pdf as corresponding to the value of burnedness, is an adiabatic one, an additional correction accounting for heat-losses has to be applied: According to [25] an approximation for the reaction velocity coefficient of NO-formation under variation of temperature can be given as follows:

$$k_{NO}^{*} = k_{NO,ad} \cdot (\exp(-T^{*}/T_{ad}) \cdot \exp(T^{*}(T_{ad}-T^{*})/T_{ad}^{2})) \quad (5)$$

where T_{ad} denotes the adiabatic temperature corresponding to burnedness, T^{*} stands for the temperature with account to heat-losses, and $k_{NO,ad}$ is the NO-formation coefficient calculated above. As shown in Fig. 2.3.10, the temperature T^{*} is calculated by assuming the mean-value of heat-losses being distributed over the pdf-temperatures equally:

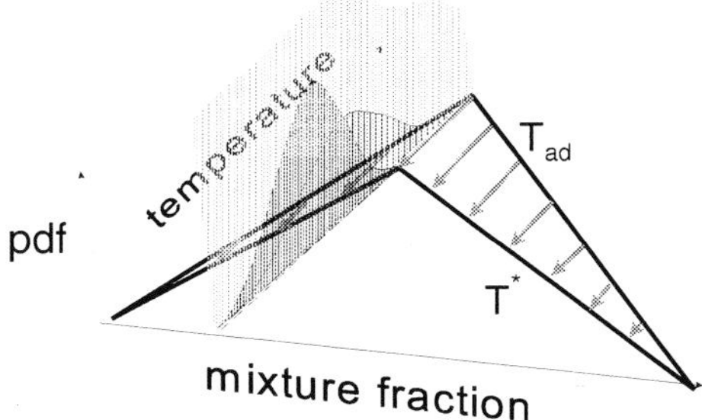

Figure 2.3.10: Temperature correction of adiabatic pdf-values due to heat-losses.

$$T^* (a,f,\overline{q}_v) = T_{ad} (a,f) \cdot C_q (\overline{q}_v) \tag{6}$$

"Perfectly Stirred Reactor" Model for the Burned Zone

In order to avoid separate models and the difficulties arising when dividing the calculation domain into different zones manually, the "perfectly stirred reactor" (PSR) sub-model was integrated into the "presumed pdf" sub-model by simply over-writing the results of the pdf-model for all discrete pdf-points, complying with the condition $a > 0.98$. For this pdf-points the concentration of oxygen atoms was chosen according to values precalculated by solving the differential equations for the ideal PSR system. Further calculation of the NO-formation rates is performed by applying the Zeldovich formula (Eq. (3)). For a fast look-up of the precalculated (PSR) oxygen atom concentrations, they were expressed as a polynomial function of residence time (τ), stoichiometry (λ) and the heat-loss per mass of fuel (ΔH^*):

$$\begin{aligned} X_O &= X_O^* \cdot X_{O,ref}^* \\ X_O^* &= \tau_{norm}^{m(\lambda)+1} \\ X_{O,ref}^* &= m_{ref} (\lambda) \cdot \Delta H^* + c_{ref} (\lambda) \end{aligned} \tag{7}$$

with the polynomial coefficients given in Tab. 2.3.1. The local stoichiometry and heat-loss is calculated as discussed already in the pdf sub-model section. The local mean residence time is determined by solving an additional transport equation which reads in cylindrical coordinates:

$$\frac{\partial(\varrho U \overline{\tau})}{\partial x} + \frac{1}{r} \frac{\partial(r \varrho V \overline{\tau})}{\partial r} - \frac{\partial}{\partial x} \left(\Gamma_\tau \frac{\partial \overline{\tau}}{\partial x} \right) - \frac{1}{r} \frac{\partial}{\partial r} \left(r \Gamma_\tau \frac{\partial \overline{\tau}}{\partial r} \right) = S_\tau \tag{8}$$

with Γ_τ calculated analogous to mass transport. This equation integrates the transport terms by setting the source term S_τ to 1 (seconds per second) times the mean density ϱ, for dimensional reasons. As a result, local mean residence time may be regarded as the mean value of the residence times of small fluid elements which are transported and exchanged according to the turbulent motion.

Table 2.3.1: Polynomial coefficients for [O]-concentration fit.

$f_{polynom} (\lambda) = b \cdot (a_0 + a_1 \cdot \lambda - a_2 \cdot \lambda^2 + a_3 \cdot \lambda^3 - a_4 \cdot \lambda^4)$						
	a_0	a_1	a_2	a_3	a_4	b
$m(\lambda)$	−16.2691	44.9667	−50.2799	24.6992	−4.5181	1.0
$m_{ref}(\lambda)$	−2.4739	8.0850	−6.9762	2.6612	−0.3883	10^{-8}
$c_{ref}(\lambda)$	−4.8725	12.8453	−11.1927	4.0247	−0.5026	10^{-2}

2.3.5.4 Results of the Turbulent Thermal NO-Formation Model

The validation of the NO formation model comprises calculations of two swirling flame configurations, one being a premixed flame burning in the combustion chamber shown in Fig 2.3.1, with an additional insulation yielding higher fluid temperatures, and the non-premixed swirling flame operated in a water-cooled combustion chamber. The main differences between these configurations with respect to NO-formation are the local combustion chamber temperatures (premixed: 1850 K; non-premixed: 1400 K) and, of course, the local stoichiometry values, which are variable and subject to turbulent fluctuations in the non-premixed case. Both combustion chambers comprise cylindrical segments and a swirl burner fixed to their bottom (nozzle diameter premixed: D_0 = 90 mm; non-premixed: D_0 = 60 mm). The burner exit geometry in the premixed case employed in addition a diffusor. Table 2.3.2 shows the operational conditions applied in both cases.

The calculation of the NO concentrations was performed as a postprocessing step solving the mean species transport equation and using calculation results in the premixed case [2] as the basic data for calculating the NO-formation rates. For the non-premixed case a combination of measured data (velocities and turbulent kinetic energy ; mean temperatures and composition) and calculation results (turbulent dissipation rate, fluctuations of burnedness and mixture fraction, local mean residence time) was applied. Details on the measured data are given in [16, 8].

As a result of the premixed case calculation, Fig. 2.3.11 shows a comparison with measured NO concentrations in four axis normal planes, investigated experimentally. As can be seen, despite of a small deviation of the profiles at $x/D_0 = 1.33$, the agreement is very good, which – to a large extent – is a result of the good representation of the mean local temperatures obtained using a correction scheme that accounts for the contribution of radiative heat losses [16]. Additionally, several calculations were carried out varying the global air to fuel λ ratio. For this variation, the comparison of measured with calculated outlet concentrations of NO as shown in Fig. 2.3.12 also yields good agreement.

With the non-premixed configuration, the complexity of the calculation increases by the local variation and the turbulent fluctuation of local stoichiometry. Additionally, the flue gas temperature is very low, allowing NO-formation only to take place near the main reaction zone. This results in very

Table 2.3.2: Operational conditions of the investigated flames.

	Premixed flame	Non-premixed flame
air to fuel ratio λ	1.1, 1.2, 1.3, 1.5	1.2
swirl-parameter $S_{0,\text{th}}$	0.8	0.9
thermal load P_{th}	300 kW	150 kW

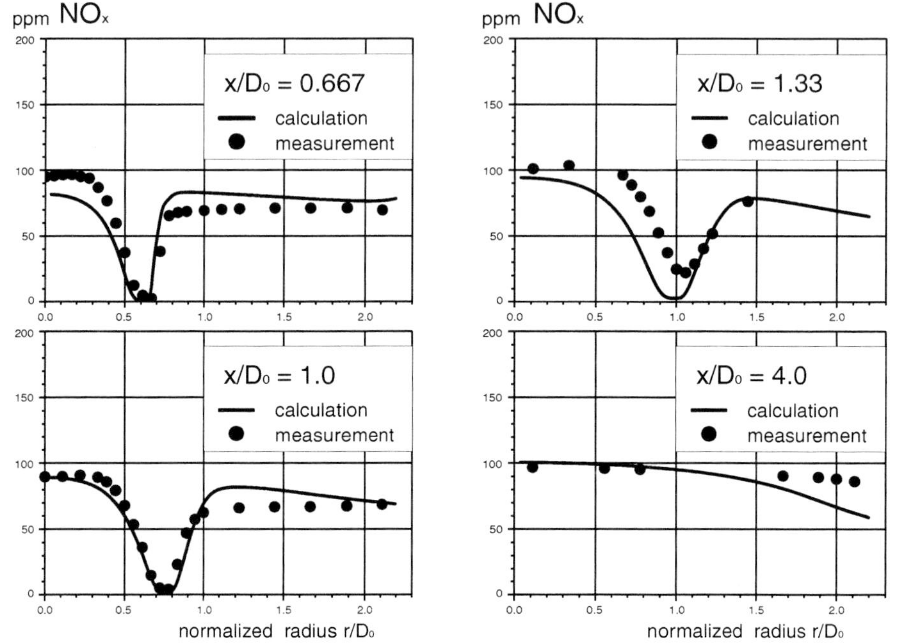

Figure 2.3.11: Comparison of measured NO-concentrations [8] to calculated data at four measurement planes.

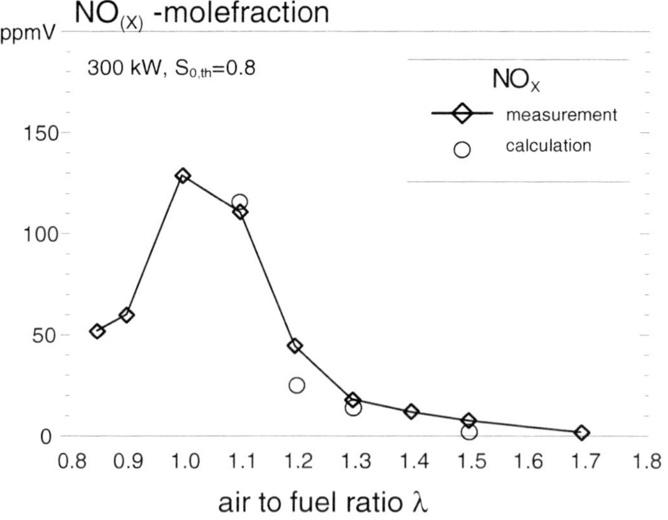

Figure 2.3.12: Calculated and measured $NO_{(x)}$ emissions of the premixed configuration for varying global stoichiometry.

low experimental NO_X-emissions of approx. 14 ppm at the chamber outlet and maximum concentrations experimentally observed of approx. 20 ppm. Looking at Fig. 2.3.13, the measured distribution of NO-concentrations is compared with calculated values. The calculated concentrations are predicted too high near to the chamber axis, and somewhat too low towards the chamber walls. Looking into details NO_X is plotted against the radius of the combustion chamber at four cross sections, in Fig. 2.3.14. First, a good representation of the local minima of the curves can be observed, as regards position and value. At zones near to the walls, the measured data show an approximately constant shape, while predictions exhibit a slight increase. As the calculated NO-formation rates in this zone are negligible, this deviation indicates only an under-prediction of turbulent mixing rates. In contrast, the over-prediction of NO-concentrations near to the chamber axis may result from the NO-model. But, as the predicted concentrations show (in contrast to the measured data) local maxima near the burner axis (e. g. $x/D_0 = 1.5$), the over-prediction may again be an effect of mixing rates, which are calculated too low. Generally, a very good prediction of emissions (see $x/D_0 = 5$) can be observed with predicted values of 18 ppm against measured data of 14 ppm. Summarizing the results of the NO-calculations in swirling flames, the turbulent formation model shows the ability, to calculate NO-formation in two very different configurations, and thus offers the possibility of predicting NO-emissions with good accuracy even in flame configurations from industrial type burners.

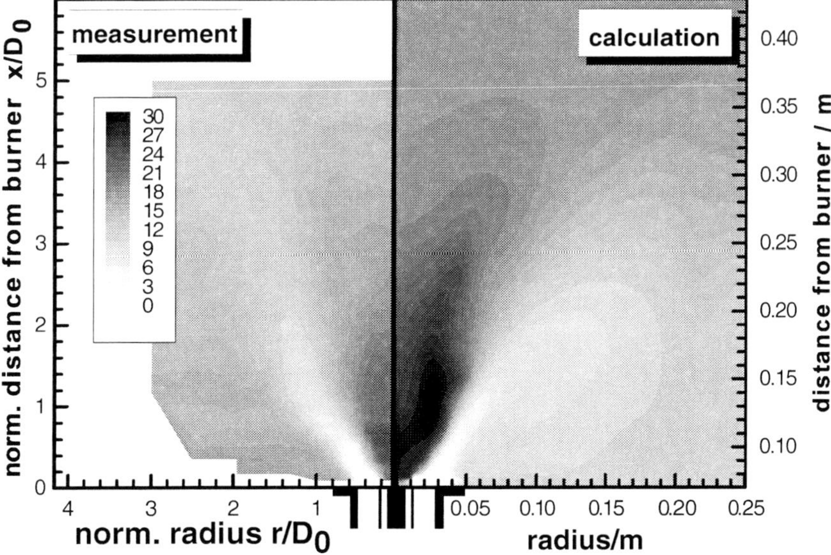

Figure 2.3.13: Comparison of the calculated NO concentration distribution (ppm) to measured values in the near-burner zone of the non-premixed configuration.

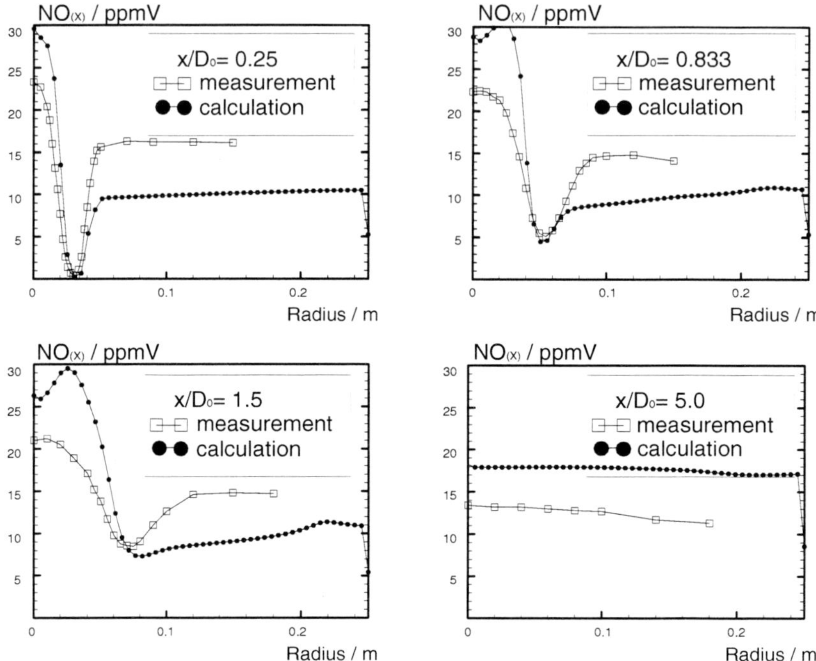

Figure 2.3.14: Comparison of measured and calculated cross sections of NO-concentration of the non-premixed configuration.

References

1. Holzäpfel, F. (1996): Zur Turbulenzstruktur freier und eingeschlossener Drehströmungen, Dissertation, Universität Karlsruhe (T. H.).
2. Habisreuther, P., Leuckel, W., Eickhoff, H. (1995): Modellierung und experimentelle Validierung von eingeschlossenen Vormisch- und Diffusionsdrallflammen. Forschungsbericht 1993–1995, Sonderforschungsbereich 167, Universität Karlsruhe, 143–171.
3. Weber, R., Dugu, J. (1992): Combustion Accelerated Swirling Flows in High Confinements. *Prog. Energy Combust. Sci.* **18**, 349.
4. Philipp, M. (1991): Experimentelle und theoretische Untersuchungen zum Stabilitätsverhalten von Drallflammen mit zentraler Rückströmzone, Dissertation, Universität Karlsruhe (T. H.).
5. Leuckel, W. (1967): Swirl intensities, swirl types and energy losses of different swirl generating devices, IFRF Doc. No. G 02/a/16, Ijmuiden.
6. Prade, B., Schmittel, P., Lenze, B. (1998): Stabilisation of turbulent concentric and swirling flames based on flow and mixing pattern investigations, Abschlußbericht, Sonderforschungsbereich 167, Universität Karlsruhe.
7. Stapf, D. (1998): Experimentell basierte Weiterentwicklung von Berechnungsmodellen der NO_X-Emission technischer Verbrennungssysteme, Fortschritt-Berichte VDI, Reihe **6**, Nr. 395, Dissertation, Universität Karlsruhe (T. H.).

8. Schmittel, P., Lenze, B., Leuckel, W. (1997): Messungen zur Stabilität turbulenter, eingeschlossener Drallflammen unter Variation der Einflußgrößen, Verbrennungen und Feuerungen – 18. Deutsch-Niederländischer Flammentag, Delft/NL, 121.
9. Döbbeling, K. (1990): Experimentelle und theoretische Untersuchungen an stark verdrallten, isothermischen Strömungen. *Dissertation, Universität Karlsruhe (T.H.).*
10. Hirsch, C. (1995): Ein Beitrag zur Wechselwirkung von Turbulenz und Drall, Dissertation, Universität Karlsruhe (T. H.).
11. Weber, R., Visser, B. M., Boysan, F. (1990): Assessments of turbulence modelling for engineering prediction of swirling vortices in the near burner zone. *Int. J. Heat and Fluid Flow* **11**, No. 3, 225.
12. Peters, A. A., Weber, R. (1995): Mathematical Modeling of a 2.25 MW_t Swirling Natural Gas Flame. Part 1: Eddy Break-up Concept for Turbulent Combustion; Probability Density Function Approach for Nitric Oxide Formation. *Comb. Sci. and Technol.* **110**, 67.
13. Noll, B. (1992): Evaluation of a bounded high resolution scheme for combustor flow computations, *AIAA Journal* **30**, No. 1, 64.
14. Jones, W. P. (1994): Turbulence Modelling and Numerical Solution Methods for Variable Density and Combusting Flows, in: *Turbulent Reacting Flows*, Academic Press.
15. Döbbeling, K., Holzäpfel, F., Hirsch, C., Lenze, B. (1998): Experiments on free and enclosed swirling flows with regard to Reynolds stress tensors and swirl induced intermittency, Abschlußbericht, Sonderforschungsbereich 167, Universität Karlsruhe.
16. Habisreuther, P., Schmittel, P., Idda, P., Eickhoff, H., Lenze, B. (1997): Experimentelle und numerische Untersuchungen an einer eingeschlossenen Drall-Diffusionsflamme, Verbrennungen und Feuerungen – 18. Deutsch-Niederländischer Flammentag, Delft/NL, 127.
17. Westbrook, C. K., Dryer, F. L. (1981): Simplified Reaction Mechanisms for Oxidation of Hydrocarbon Fuels, *Comb. Sci. Tech.* **27**, 31–43.
18. Philipp, M., Hoffmann, S., Habisreuther, P., Lenze, B. Eickhoff, H. (1992): Experimental and Numerical Study Concerning Stabilization of Strongly Swirling Premixed and Nonpremixed Flames, *24th Symp. (Int.) on Combustion*, The Combustion Institute, Pittsburgh, 361–368.
19. Hoffmann, S., Habisreuther, P., Lenze, B. (1994): Development and Assessment of Correlations for Predicting Stability Limits of Swirling Flames, *Chem. Eng. Proc.* **33**, 393–400.
20. Eickhoff, H., Braun-Unkhoff, M., Frank, P., Merkle, K., Zarzalis, N. (1998): Analysis of Lean Blow out for Swirl Stabilized Combustion, WIP Report W2 EO2, 27th Symp. (Int.) on Combustion, Boulder.
21. Zeldovich, Ya. B., Sadovnikov, P. Ya., Frank-Kamenetzkii, D. A. (1947): Oxidation von Stickstoff bei der Verbrennung (übersetzt von Ina Nickel). Akademie der Wissenschaften der UdSSR, Moskau.
22. Fenimore, C. P. (1970): Formation of Nitric Oxide in Premixed Hydrocarbon Flames, 13th Symp. (Int.) on Combustion, 373.
23. Maas, U., Pope, S. B. (1992): Simplifying Chemikal Kinetics: Low-Dimensional Manifolds in Combustion Space, *Combust. Flame* **88**, 239–264.
24. Warnatz, J. (1984): in *Combustion Chemistry*, (ed. Gardiner, W. C.), Springer Verlag.
25. Correa, S. M. (1992): A review of NO_X formation under gas-turbine combustion conditions, *Comb. Sci. Tech.* **105**, 357–375.
26. Weber, R., Peters, A. A., Breithaupt, P. P., Visser, B. M. (1995): Mathematical Modelling of Swirling Pulverized Coal Flames: What Can Combustion Engineers Expect from Modeling. *ASME J. Fluid Eng.*, **17**, 289.

2.4 Stability and Burnout of Swirling Flames with Wastewater Injection

Karsten Ehrhardt and Wolfgang Leuckel*

Abstract

The incineration of wastewater streams by the "in-flame evaporation burner" was investigated in a pilot scale furnace with regard to burnout and flame stability. With this burner type a wastewater spray is injected into the swirling flame of an auxiliary gaseous or liquid fuel. Two requirements must be satisfied. First, complete burnout of carbon monoxide (CO) and of possibly toxic hydrocarbons (HC) must be ensured, and second, the amount of auxiliary fuel must be minimized. Consequently, a combined optimum for burnout and flame ignition stability must be found. To achieve this, a sound understanding of burnout and of the flame stabilization process is required.

Two regimes of limitation were identified in the burnout controlling chain of mixing, evaporation, and gas phase oxidation. At temperatures below approximately 850 °C burnout is limited by chemical kinetics, which in principle can be described accurately by detailed reaction schemes. At temperatures higher than approximately 900 °C the evaporation of big droplets limits burnout. In this latter regime the mixing type (solution/emulsion) and the volatility of the wastewater hydrocarbons exhibit strong influences on the evaporation processes. Rapid mixing of the wastewater spray with the hot flue gases of the auxiliary fuel flame is, on one hand, advantageous for complete burnout but; on the other hand, fast mixing impairs flame stabilization because of extraction of sensible heat from the flame by evaporation. Therefore, an optimal balance between rapid mixing and aerodynamic separation of the stabilization zone of the gas flame, which is located at the outer edge of the internal recirculation zone, from the evaporating spray which penetrates through the internal recirculation zone, must be achieved. The performed investigations, referring to the influence of the major system parameters on flame stability, allow to find this balance, and they indicate that the mixing process between the spray and the internal recirculation zone is crucial for flame stability.

* Engler-Bunte-Institut, Bereich III – Verbrennungstechnik, Universität Karlsruhe, Kaiserstr. 12, 76128 Karlsruhe, Germany

2.4.1 Introduction

Thermal incineration is one way of industrial wastewater treatment and meets broad application. Incineration is favorable if the organic fraction is relatively high and consists of a wide variety of different compounds which are of too little value to be separated for recycling into a production process. Other aspects, like the non-availability of a wastewater purification plant or the biological toxicity of the waste, apply as well. In some cases incineration can even be used as an unit operation of a production line, e. g. as a production stage of salts like Na_2CO_3 or of HCl from highly chlorinated hydrocarbons to be recirculated into the process or re-used in another process. Burning wastewater streams is, in general, typical for chemical processes and oil refineries.

A special burner system of great practical importance is the so called "in-flame evaporation burner" [1] shown in Fig. 2.4.1. This burner type is a swirl stabilizing burner with axial injection of the liquid waste and radial injection of auxiliary fuel. Air-assist atomizers are widely used in those systems, as they provide fine sprays with high jet momentum and can be applied even if liquid wastes are contaminated by abrasive particles. A pilot scale burner of this type has been investigated in the present study. Another common and competitive industrial system separates auxiliary fuel combustion and waste injection, i. e. the wastewater spray is being injected into the hot flue gas of the auxiliary fuel burnt in a primary combustor. This latter setup avoids problems with flame stability which could arise from the former sys-

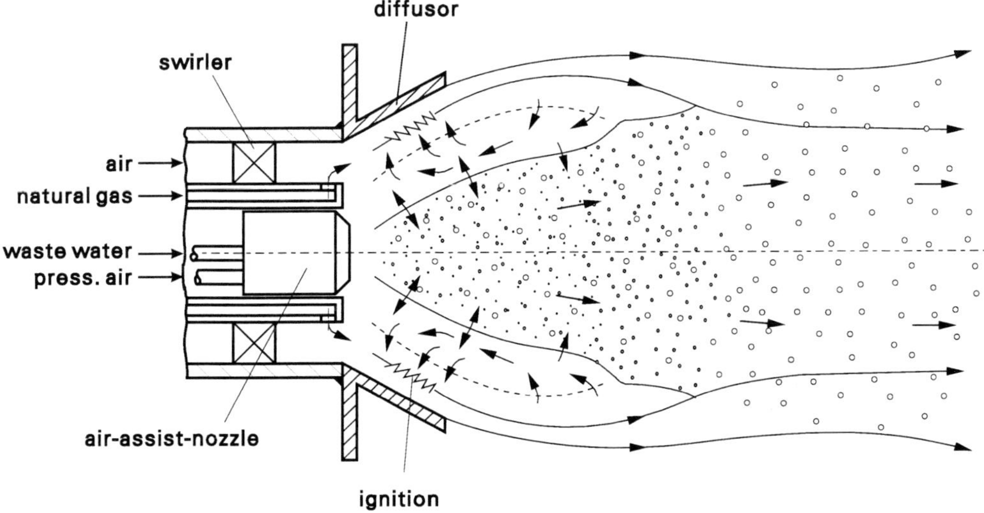

Figure 2.4.1: Scheme of "in-flame-evaporation burner" [2].

tem in the way that quenching of the auxiliary fuel flame by the evaporating wastewater spray might occur. The major advantages of the in-flame evaporation burner, on the contrary, are rapid and thorough mixing of the waste stream with the hot flue gas with the consequence of smaller incinerator volume, no need for an additional primary combustor, and due to lower peak temperatures reduced thermal NO_X-formation as well as less thermal load onto the refractory lining [2].

Incineration of wastewater must be regarded as an industrial process for waste destruction which needs a high-caloric auxiliary fuel. Therefore, two requirements must be satisfied by the process. First, complete burnout of hydrocarbons (HC) and carbon monoxide (CO) must be ensured, and low emission limits must be guaranteed as wastes might contain highly toxic compounds. Second, the amount of auxiliary fuel must be minimized due to economic but also ecological demands. Flame stability must, therefore, be optimized. Optimization of the entire process requires searching for a combined optimum to reconcile burnout and flame stability as they depend – with opposite tendency – on some parameters of combustion, e. g. swirl intensity and spray angle [3]. To achieve this, sound and detailed quantitative understanding of the occurrence of incomplete burnout and of flame stabilization are required.

2.4.2 Experimental

The pilot scale furnace which was designed for the experiments is shown in Fig. 2.4.2. The furnace is water cooled and well insulated by refractory fibers and bricks. Therefore, heat losses are approximately 30 kW of the thermal input of 150–350 kW only and, by this, comparable to industrial applications (5–10 %). The furnace has an inner diameter of 600 mm and a variable inner height of 2.0 to 2.5 m. Two arrays, of three windows each, allow in combination with the vertically adjustable burner optical access to the flame. The burner of the described type is mounted for down-firing at the top of the furnace. The swirl number of the cold burner air flow could be adjusted between 0 and 1.87 by means of a moveable swirl generator, and air flow rates between 150 and 500 m^3/h ensured fully turbulent flow. A 45 mm long diffusor with a 30° half quarl angle was used as a burner mouth exit nozzle to enhance flame stability. The diffusor was mounted at the end of the annular duct for the main combustion air stream with an outer diameter of 90 mm. The origin of the coordinate system (axial: x, radial: r) was defined at the center of the outlet plane of the diffusor. The fuel lance (diam. 42.4 mm) was located inside the diffusor, with its tip placed usually 40 mm upstream of the outlet plane of the diffusor ($x = -40$ mm). The fuel lance consists of an annular duct for the auxiliary fuel (natural gas) and an exchangeable air-assist atomizer for

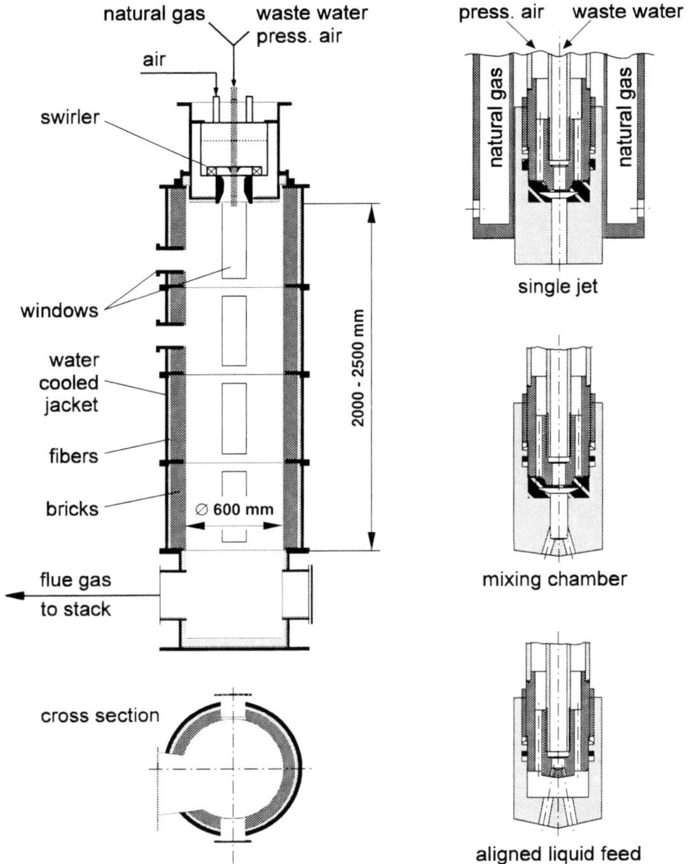

Figure 2.4.2: Schemes of pilot scale furnace (left) and of air-assist nozzles (right); top: single-jet nozzle within annular gas lance; middle: multi-jet mixing chamber nozzle; bottom: multi-jet nozzle with individual and aligned liquid feed of outlet ports.

wastewater injection (Fig. 2.4.2, top right side). Figure 2.4.2 also shows the three basically different types of atomizers used in this work. Most results were obtained with radial injection of the natural gas through 16 holes of 5 mm diameter. The natural gas stream usually provided approx. 90 % of the total thermal input. Water flow rates between 40 and 150 kg/h corresponded to adiabatic flame temperatures between 700 and 1300 °C. Temperatures were measured by a fine-wire (100 µm) Pt/PtRh10 thermocouple, and suction probes were used for sampling of flue gas. The concentrations of O_2, CO, CO_2, CH_4, and H_2 were measured on dry basis. A FID was used to measure the sum of unburnt hydrocarbons, whereas selective detection of methane and of the hydrocarbons fed with the waste stream was accomplished by FTIR diagnostics.

2.4.3 Results and Discussion

In the first part of this section the results concerning the occurrence of incomplete burnout will be presented, whereas flame stability data will be presented in the second part.

2.4.3.1 Incomplete Burnout

It seems useful to start with some theoretical consideration of the combustion process which is illustrated in Fig. 2.4.3, and the resulting principal sources of incomplete burnout. At first, the wastewater is being atomized to generate a large evaporation surface. In the second step, the spray mixes with the hot flue gases which provide the sensible heat needed for droplet evaporation. Evaporation itself represents the third step. Before the vapor can be oxidized in the gas phase (fifth step) it must mix with oxygen on a molecular basis (fourth step). Heat and mass transfer (second and fourth step) can, however, be treated as one step, namely mixing, since they take place simultaneously by turbulent and molecular exchange. The limiting process of the two is determined by the ratios between demand and supply of heat and oxygen, respectively. For usual wastewater compositions the demand for heat usually exceeds the demand for oxygen, as the water fraction must be vaporized but not oxidized. Since the in-flame evaporation burner injects the wastewater into the primary flame of the auxiliary fuel, the primary flame and its interaction with the evaporating spray must be considered as an essential mechanism as well. Again, hydrocarbons and oxygen have to mix before they can react. Each step in the described chain may be rate limiting, hence, incomplete burnout may result either from poor mixing or from slow evaporation due to poor atomization, and/or from slow oxidation kinetics. In case of poor mixing, cold and/or fuel rich pockets may prevent complete burnout [4]. This source of incomplete burnout is avoided with the in-flame evapora-

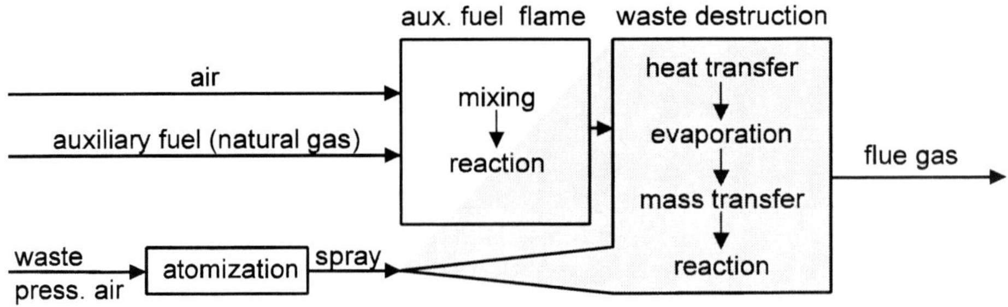

Figure 2.4.3: Illustration of the combustion process.

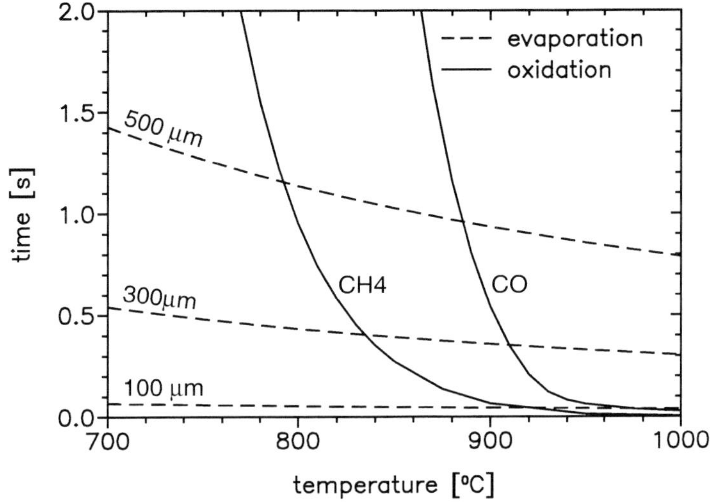

Figure 2.4.4: Calculated temperature dependence of times required for the evaporation of water droplets and for oxidation of CO and CH_4 from 1000 to 10 ppm.

tion burner as mixing of all streams – wastewater spray, auxiliary fuel and combustion air – is fast in the highly turbulent flame jet [2, 5]. Figure 2.4.4 illustrates the dependence of droplet evaporation [5] and of oxidation kinetics [6] on flame temperature. It can be seen that reaction kinetics depends much stronger on the temperature than evaporation does. Obviously, depending on the temperature and droplet diameter, there exists an evaporation controlled as well as a reaction controlled regime. For temperatures higher than approximately 900 °C chemical kinetics is fast and burnout is limited only by evaporation, whereas at temperatures lower than approximately 850 °C burnout is controlled by reaction. In order to generate truly reaction controlled conditions evaporation must be fast, hence, atomization must be fine, since droplets must evaporate before the contained hydrocarbons can be oxidized in the gas phase. Both regimes were realized by suitable experimental conditions and investigated in a detailed manner.

First it was attempted to find a correlation between atomization quality and the occurrence of incomplete burnout under evaporation controlled conditions [5]. In other words, a measure was sought to assess atomization quality with respect to burnout. For this purpose, three different single-jet air-assist nozzles were used to inject a mixture of 3.8 kg/h glycol and 40 kg/h water into a 200 kW flame. Glycol was selected for its high boiling point of 196 °C compared to water, because it should chiefly remain in the liquid phase until the end of the droplet lifetime. So the end of droplet/spray evaporation could be detected by FID. In those tests, significant differences of the emissions of unburnt hydrocarbons (HC) and carbon monoxide (CO) were detected as can be seen in Fig. 2.4.5 (right side, open symbols). Parallel

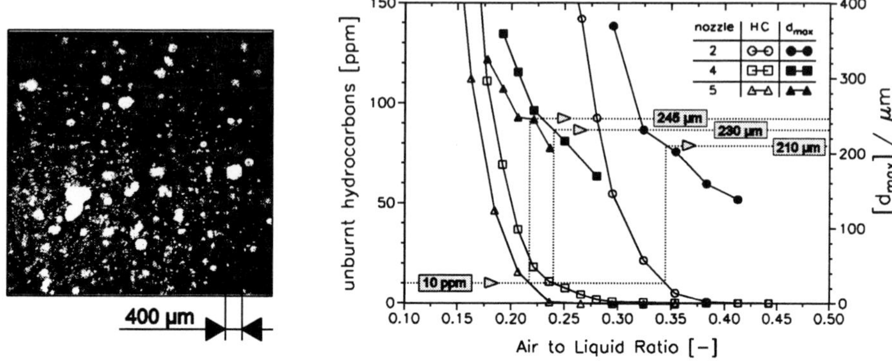

Figure 2.4.5: Direct sizing of large droplets by high-speed photography (left). HC-emissions and diameter of the largest droplets in dependence of ALR for three air-assist nozzles (right).

to this, the sprays of the three nozzles were investigated by PDA at ambient temperature with the same load by PDA. The obtained data showed that the upper end of the droplet size distribution determined the onset of incomplete burnout. It was not possible, however, to find an appropriate measure to assess atomization quality on the basis of the PDA data, as the end of the size distribution could not be detected with sufficient accuracy. This, on the one hand, is a consequence of statistical limitations, as less than 1 % of total liquid mass could cause significant emissions but may be contained in a few big droplets only. Those are hard to detect statistically within thousands of small droplets. On the other hand, random and systematic errors, due to e. g. surface reflection, limited dynamics of photo-multipliers or non-spherical droplet shape, are disastrous in regimes of size distributions where droplets occur seldom. To overcome this problem, high speed photography was used for direct sizing of the largest droplets in the sprays (Fig. 2.4.5, left side). When comparing the size of the biggest droplets in the sprays with the corresponding HC-emissions, good correlation was found between the onset of incomplete burnout and a critical droplet size. Recent results prove that the basic requirement for finding good correlation was the existence of comparable initial velocities of the big droplets. For the three atomizers those big droplets originate from the same process, namely insufficient breakup of the liquid jet near to the nozzle axis. Nozzles with long outlet ports avoid incomplete breakup on the axis, but generate big droplets at the spray boundary due to the formation of a film at the wall of the outlet duct [7]. This class of droplets can still evaporate within the experimental furnace even if the droplets are significantly larger (approx. 350 µm), as their initial velocity (approx. 20 m/s) is much smaller than the initial velocity of big droplets on the liquid jet axis (approx. 60 m/s). For the investigated system, the distance a droplet travels until complete evaporation occurs was determined approximately by calculation of the simultaneous deceleration and evaporation of a single droplet in infinite environment flowing with the given space velocity, using appropriate Nu- and c_W-correlations for evaporating droplets [8, 9]. It was found sufficient to consider an isolated droplet, since the length of the flame (< 0.25 m) is much shorter than the burnout zone (approx. 2 m) where still profiles were detected. Otherwise more detailed computations would have been necessary [10].

For industrial application the termination of spray evaporation is the key question with respect to burnout, as industrial wastewaters usually contains a varying and broad composition of light and heavy organics which is often not even known exactly. Nevertheless, the influences of volatility and of the type of mixture were investigated as well [11], since these are of scientific interest and in some cases of practical relevance too. To simulate dissolved organic loads with different volatility, C_1- to C_4-alcohols and glycol were added to a water flow of 60 kg/h. The mass flows of the alcohols were set to equal the heat input of a glycol mass flow of 1 kg/h (4.7 kW), they varied, therefore, according to their heating value between 0.86 kg/h for methanol and 0.52 kg/h for 1-butanol. Emulsions of *n*-hexane, *n*-heptane, and fuel oil EL (comparable to fuel oil N° 2) in water were generated in a continuous process atomizing oil into water [12]. Again the hydrocarbon mass flows were set to equal 4.7 kW. A coarse emulsion (SMD = 90 μm for fuel oil) and a fine emulsion (5 μm) could be generated by using different pressure-swirl nozzles. The emulsions were generated directly upstream of the single-jet air-assist nozzle (Fig. 2.4.2, top right side), and they were stable between generation and atomization due to the short residence time of 0.5 s within the liquid feed through the lance. Evaporation was completed approximately 1.5 m below the burner, whereas the natural gas flame was shorter than 0.25 m. The natural gas (18.1 m^3/h = 180 kW) covered 97.5 % of the total heat input. Due to this high fraction, the temperature distribution and the flow field were almost unaffected by differences of the speed of evaporation and of subsequent heat release. The adiabatic temperature ($\lambda = 1.15$) was 1110 °C, and the temperature of the flue gases in the burnout zone was higher than 1000 °C which ensured fast chemical kinetics.

Figure 2.4.6 shows on the left side the radial concentration profiles of the dissolved hydrocarbons measured 0.83 m below the burner, which were normalized to account for the different initial molar concentrations. The reference concentrations $[HC]_{REF}$ were calculated by assuming that the hydrocarbons were fully vaporized and homogeneously mixed with the flue gas, but not yet oxidized; the values were 2030, 1050, 700, 530, and 1220 ppm for methanol, ethanol, 1-propanol, 1-butanol, and glycol, respectively. Obviously, evaporation and burnout depend on volatility. Significant differences were detected between the four alcohols, but especially between all others and glycol for which $[HC]/[HC]_{REF}$ has a peak value of unity. The observed volatility of the alcohols in aqueous solution rises from methanol to butanol, hence, contrary to the volatility of the pure alcohols. This "reversal of volatility" may be attributed to mixture effects in aqueous solution, in other words, the reversal is a consequence of the rise of the Henry coefficient

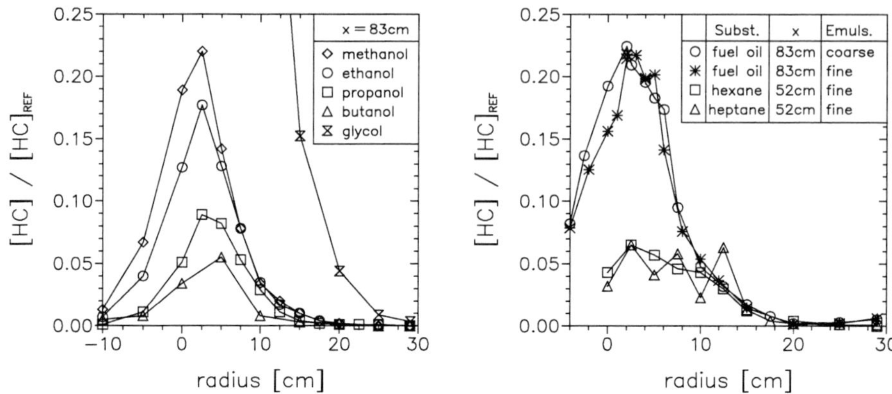

Figure 2.4.6: Normalized concentrations of unburnt hydrocarbons for alcohols dissolved in water (left) and oil-in-water emulsions (right).

from methanol to butanol. For the four alcohols the Henry constants at 100 °C are 7.9, 11.6, 16.6, and 29.1 bar, respectively [13]. The detected high glycol concentrations are a consequence of its very small Henry constant, at 100 °C only 0.014 bar. On the right side of Fig. 2.4.6 the normalized concentrations of the emulsified hydrocarbons are shown ($[HC]_{REF}$ = 336 and 290 ppm for hexane and heptane, respectively, and 1.35 g/m^3 ≈ 135 ppm for fuel oil). It seems surprising that the disperse oil phase of an emulsion evaporates significantly faster than a dissolved hydrocarbon with equal volatility. Fuel oil evaporates as fast as methanol, and hexane and heptane could not even be detected anymore 0.83 m below the burner. Their normalized concentrations are only 0.06 at the axis for $x = 0.52$ m where the investigated alcohols have peak values of the normalized concentration between 0.20 and 0.60. No difference could be observed between evaporating a fine (5 μm) and a coarse (90 μm) emulsion of fuel oil, which was also the case for heptane. The fast evaporation of fuel oil, which does not correspond to the thermodynamic data, could be explained by phase separation during droplet evaporation, like observed by Chung and Kim [14] for the vaporization of emulsion droplets on a hot surface. The disperse oil phase is accumulated at the droplet surface where it evaporates quickly.

As mentioned above, reaction kinetics becomes rate limiting at lower temperatures, but as the liquid must still be vaporized before the hydrocarbons can react, atomization must be very fine to rule out an influence of evaporation within the post flame zone of the experimental furnace. The key problem during the experiments was to realize sufficiently fast evaporation and mixing without flame extinction. For realizing this challenging aim an atomizer was applied which apportions the liquid mass onto six jets, developing along a 20° cone to accelerate mixing, and which at the same time ensures very fine atomization and avoids the generation of big droplets issuing from a wall film. It was proved that evaporation of the sprays was completed 0.5 m below the burner for the pressurized air flow of 10 m^3/h (ALR = 21 mass-%). This distance is short compared to the furnace length and, therefore, allows investigation of oxidation rates in the gas phase without overlaid evaporation. A high swirl number of 1.5 enhanced mixing, and a lance for radial injection of the natural gas with low momentum provided wide stability limits. The temperature of the flue gas was usually varied by reducing the gas flow and leaving all other settings constant. Doing so, air excess rises, but mixing remains fairly constant as the momentum of the gas can be neglected compared to the momentum of the spray and of the burner air. At low flame temperatures, however, quenching of the gas flame by the evaporating spray is too strong to allow complete oxidation of the natural gas in the hot primary flame. Therefore, unburnt hydrocarbons and CO do not originate from the liquid waste only, which is the case for evaporation controlled conditions, but stem from the auxiliary fuel as well. The thermal stability of unburnt species decreases in the order of their appearance with decreasing temperature. The experiments confirm the well known order: Methane is significantly more stable than the alcohols, and the slowest reacting compound, CO, can be used to monitor burnout. It was easily possible to determine the oxidation rates, since the concentration and temperature profiles were flat in the furnace for $x > 0.5$ m. The data were evaluated by assuming that oxidation rates in the burnout zone can be described by a global reaction of first order and, therefore, plotted in an Arrhenius diagram (Fig. 2.4.7). The influence of oxygen concentration was neglected in this evaluation as oxygen was present in large excess ($[O_2] > 5$ vol-% in the dry flue gas). The data for CO and CH_4 were compared with other global rate expressions and with the predictions of detailed reaction schemes [11]. The schemes (e. g. [6]) agree satisfactorily with the experimental data, their usage can, therefore, be recommended for system layout. On the contrary, some of the

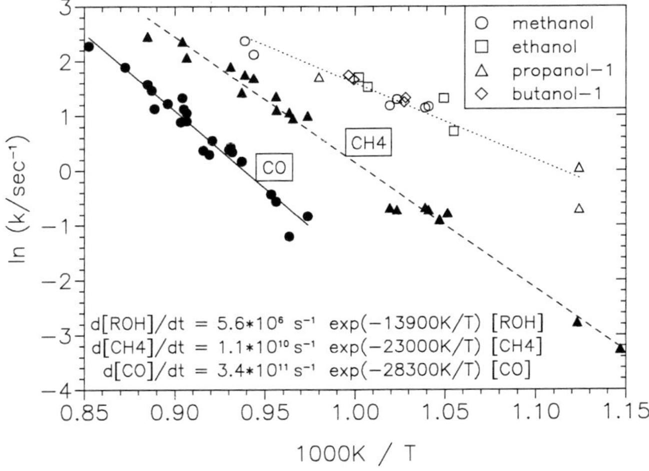

Figure 2.4.7: Arrhenius-diagram for oxidation of CO, CH_4, and alcohol vapors.

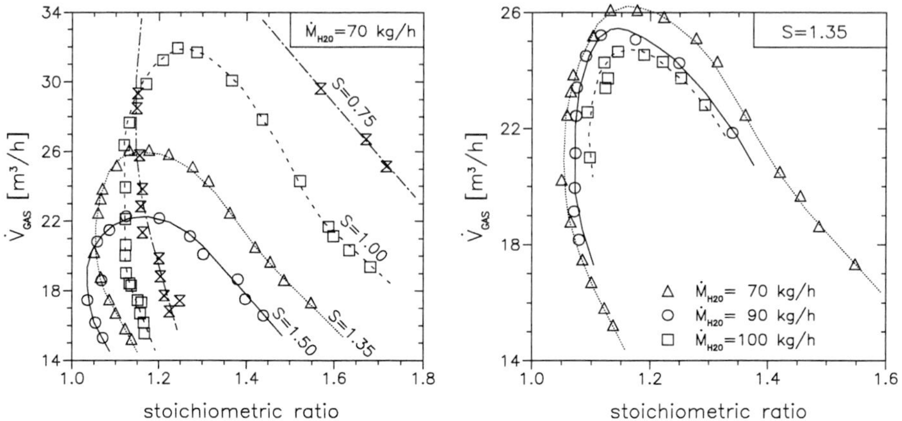

Figure 2.4.8: Flame stability limits as a function of swirl number (left) and of injected water flow (right). $\dot{M}_{GLYCOL}$ = 3.8 kg/h; $\dot{V}_{P.AIR}$ = 10 m^3/h.

global expressions give unreasonable results for wastewater incineration, probably as a consequence of being fitted only for a certain experimental system and for the resulting reaction conditions (stoichiometry, reaction intensity, etc.) which have a strong influence on reaction rates. Their application for conditions much different from the fitted ones may, therefore, lead to incorrect results. It was further attempted to investigate the oxidation kinetics of benzene which is known to have a high thermal stability, but it was not possible to detect benzene in the burnout zone even under conditions for which the thermally unstable alcohols were detected. Probably, benzene never reaches the cold burnout zone in detectable concentrations, but evaporates entirely in the hot flame zone where it is oxidized rapidly. Such a rapid evaporation is conceivable when regarding the hexane and heptane concentrations in Fig. 2.4.8 for evaporation controlled conditions, hence poor atomization.

2.4.3.2 Flame Stability

The aim of the investigations regarding flame stability was – after realization of reaction controlled conditions as described before – the general characterization of flame stability behavior, since only little information, e. g. [1, 15] can be found in the literature on this phenomenon. Figure 2.4.1 [2] schematically shows the structure of a swirling gas flame with wastewater injection. The swirled burner air induces an internal recirculation zone which is typically enforced by a diffusor. Radial injection of the natural gas leads to rapid mixing with the air and, therefore, to high reaction density. Without wastewater injection, radial injection would lead to type-II flames [16], which have been investigated in numerous studies, e. g. [17–19]. Air-assist atomizers are being applied for the wastewater stream, as they generate fine (⇒ burnout) sprays with high jet momentum. The spray can, therefore, penetrate through the internal recirculation zone, like the fuel jet of a type-I flame, and mixing of the spray with the flue gas of the gas flame in the root zone of the flame is being delayed. Rapid mixing of the auxiliary fuel with the combustion air and the delayed mixing of the spray lead to aerodynamic staging of combustion, which allows the establishment of a high temperature zone being significantly hotter than the final adiabatic flame temperature. Note, that the aerodynamic staging would not be possible if an atomizer with low jet momentum, e. g. a pressure swirl atomizer, were applied.

The flame stabilizes directly at the burner at the outer edge of the hot internal recirculation zone. There, the criterion for flame stabilization – turbulent flame speed equals the opposing flow velocity – will be satisfied, since flow velocity is relatively low, whereas turbulent flame speed is high. The latter is a consequence of the presence of near stoichiometric mixture, recirculated hot flue gases and high turbulent mixing rates. The modification of flow, temperature, and stoichiometry patterns by the spray is of special interest, as this flame type could be regarded as a swirling gas flame modified by the central spray. The flow field is modified by the momentum transfer from the spray to its surroundings, the temperature is lowered by spray evaporation, and the pressurized air and the vaporized organic fraction change the stoichiometry of the gas phase. Unfortunately, the three mechanisms are very much coupled with each other, which complicates the interpretation of experimental results.

In the following, the influence of some important parameters of flame stability behavior will be presented (Figs. 2.4.8–2.4.10). If the gas flow rate is plotted against the stoichiometric ratio, the boundaries of the stable area can generally be approximated by two straight lines. The rich limit (left) hardly depends on the stoichiometric ratio, whereas lower throughputs allow higher stoichiometric ratios before blow-off occurs at the lean limit (right). The shape of the area of stability is similar to swirling flames without water injection [19] (only lean limit, as the rich limit was not investigated) and also to flames stabilized by bluff bodies [20]. It was shown [18, 20] that the rich limit is set by mixing, which means that at this limit conditions at the outer edge of the internal recirculation zone get too rich to allow flame stabilization. This rich limit does not depend on the stoichiometric ratio, as the flow does not depend on the Reynolds number (throughput) for fully turbulent flow. The lean limit, on the contrary, is determined by reaction kinetics [18–20]. If the throughput is increased, the residence time in the stabilization zone gets shorter, which has to be compensated by less air excess corresponding to higher laminar flame speed. The end points of the boundary line in Figs. 2.4.8–2.4.10 do not indicate flame extinction but the onset of incomplete burnout ([CO] $\approx$ 100 ppm, $x = 2$ m) because of too low temperatures of the flue gas ($\approx$ 900 °C). No experiments were done under such conditions.

A key parameter of flame stability is swirl intensity (Fig. 2.4.8, left side). With increasing swirl number the stable region got smaller and was shifted towards richer conditions. This shift was also observed for swirling gas flames [18, 19]. The effect of swirl on stability could be a consequence of intensified mixing. By this, the entrainment of cold gas from the environment into the flame and of spray into the inner recirculation zone gets stronger, which cools the flame and deteriorates stability. The rich limit also shifts, since faster mixing decreases the concentration of natural gas close to the fuel lance where it is injected. Hence, conditions get leaner at the outer boundary of the inner recirculation zone for constant overall stoichiometry. The effect upon the lean limit is obviously stronger due to the non-linear dependence of reaction on temperature.

The mass flow rate of the injected water affected both stability limits only slightly, although the final adiabatic flame temperature drops significantly as can be seen from the upward shift of the end points of the boundary lines. The relatively small influence the composition of liquid (not the spray, see below!) was confirmed by varying the heat input from the liquid waste (Fig. 2.4.9). To do this, the flow of heating oil, which was emulsified with constant water flow of 70 kg/h, was varied between 0 and 6.0 l/h (0–60 kW). The stability limits were shifted with increasing oil flow towards richer conditions (Fig. 2.4.9, left side), hence towards lower stoichiometric ratios λ_{TOTAL}, which was the formal consequence of the additional demand of oxygen for complete oxidation. This becomes clear if the oxygen demand for the heating oil is not included into the calculation of the stoichiometric ratio λ_{GAS} (Fig. 2.4.9, right side). Now, the rich limits coincide for all oil flows, whereas a small extension of the stable area on the lean side due to the oil becomes obvious. This obser-

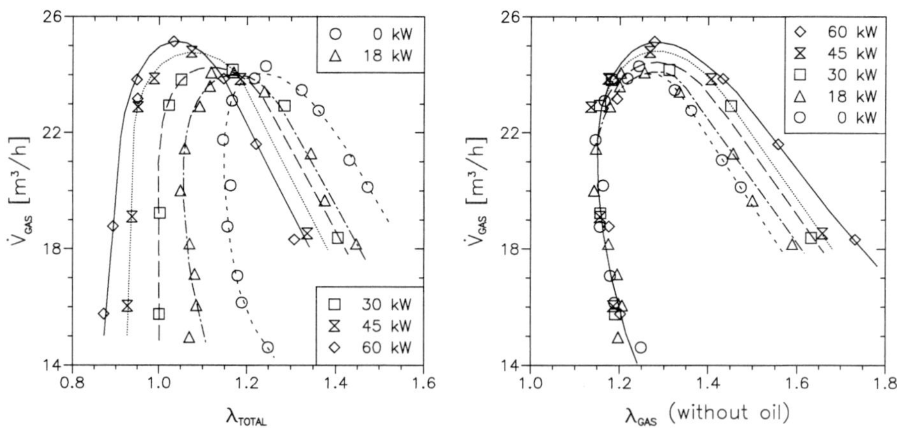

Figure 2.4.9: Flame stability as a function of heat input from emulsified heating oil, shown for two different abscissa. The demand of oxygen for oxidation of the heating oil is considered by the stoichiometric ratio, λ_{TOTAL} (left side), but not considered by λ_{GAS} (right). $S = 1.35$; $\dot{M}_{H_2O} = 70$ kg/h; $\dot{V}_{P.AIR} = 10$ m³/h.

vation agrees with the governing processes described above. The relatively small change of the liquid flow (water + oil) from 70 to 76 l/h hardly changes the aerodynamic properties of the spray, and this should, consequently, not affect mixing of natural gas with air and, therefore, not affect the rich stability limit. On the other hand, heat input from the oil reduces cooling by spray evaporation and, therefore, allows higher air excess. It can be concluded that the liquid flow asserts only a minor influence on stability limits if the flows of air and auxiliary fuel only are considered in the diagram plots. This characteristic allows, from the point of view of application, on the one hand to turn conditions fuel rich, e. g. to avoid excessive formation of nitric oxides, and on the other hand to rise the temperature level in the furnace without increasing the gas flow, which could cause flame extinction. An interesting phenomenon is that the water flow could be maximized for an "optimal" swirl number if the air and the gas flow were maintained constant. For the optimal swirl number, the stoichiometric ratio was situated approximately in the middle between the rich and the lean limit, but tended slightly towards the rich side as the lean limit depends stronger on the swirl number than the rich limit. The optimal swirl number got larger if spray momentum was increased, which confirms the importance of momenta and their ratios as is well known from usual burner scale-up.

Figure 2.4.10 shows the influence of some nozzle design parameters on flame stability for a constant nozzle load. Nozzles of the two basically different designs shown in Fig. 2.4.2 were used: nozzles with individual and aligned liquid feed of each outlet port, and also nozzles with a mixing chamber characteristic realized by de-alignment of the liquid feed. The left side shows the variation of the number of jets per nozzle N, which were inclined

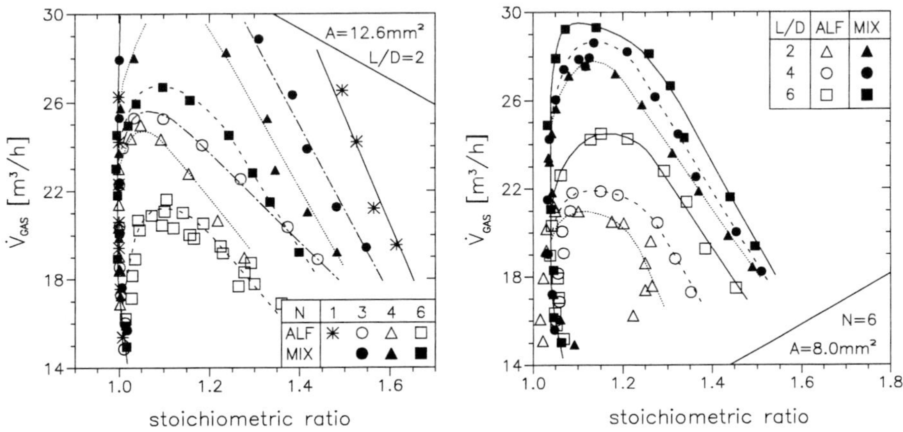

Figure 2.4.10: Flame stability for different multi-jet mixing chamber nozzles (MIX) and multi-jet nozzles with aligned liquid feed (ALF). Left: variation of number of jets per nozzle N. Right: variation of length L of the six outlet ports. $S = 1.35$; $\dot{M}_{H_2O} = 70$ kg/h; $\dot{M}_{GLYCOL} = 3.8$ kg/h; $\dot{V}_{P.AIR} = 10$ m³/h.

by 10° against the axis ($N \neq 1$) and which had a constant total cross sectional area $A = N \cdot \pi/4 \cdot D^2$. Atomization did not change as the liquid feeds and the outlet ports were geometrically similar (e.g. $L/D = 2$), but mixing between the spray and its environment – the internal recirculation zone – was intensified with increasing number of jets due to the decreasing diameter of the outlets, D. The consequence of accelerated mixing, again, was enhanced cooling of the internal recirculation zone which reduced the lean blow off limit. The rich limit, on the contrary, remained unchanged, as mixing between air and natural gas was not affected (Fig. 2.4.10, left side). The reason for flame stability to be worse with the aligned liquid feed compared to the de-aligned feed as well as with shorter outlet ports (Fig. 2.4.10, right side) was again the resulting intensified exchange of spray with its hot gas environment. This was proven by measurement of the entrainment of air into the sprays.

The results elucidated how demanding the task is to complete burnout, favorably within a short distance from the burner to keep the furnace dimensions small, and at the same time to obtain good stability which allows wide control ranges, and furthermore to minimize the demand of auxiliary fuel. The most serious problems arise for atomization, as burnout and stability showed a strong dependence on N, L/D, and the nozzle type (and others) which will make it difficult to select a suitable nozzle without any experimental testing. Similar problems arise for other parameters, e.g. burnout improves for intensified swirl while ignition stability gets worse. Due to the diversity of industrial wastes and of the required incineration conditions (temperature and stoichiometry) no general recommendations can be given, but the results should enable and support to deduce a strategy for system optimization in any particular case of application.

References

1. Hess, K., Leuckel, W., and Minx E. (1972): Aufarbeitung von organisch verunreinigten Abwässern durch Verbrennungsprozesse, *Chemie-Ing.-Techn.* **44**, 325–332.
2. Leuckel, W. and Römer, R. (1977): Die thermische Vernichtung heizwertarmer flüssiger und gasförmiger Prozessrückstände unter Anwendung eines Mehrstoffbrenner-Systems, *VDI-Bericht*, **286**, 141–150.
3. Ehrhardt, K., Schöbel, A., and Leuckel, W. (1997): Optimierung der Flammenstabilität eines Mehrstoffbrenners für heizwertarme Industrieabwässer, *VDI-Bericht*, **1313**, 151–156.
4. Seeker, W.M.R. (1990): Waste combustion, Twenty-Third Symposium (Int'l) on Combustion, The Combustion Institute, Pittsburgh, PA, 867–885.
5. Ehrhardt, K., Kufferath, A., and Leuckel, W. (1998): Assessment of atomization quality with respect to burnout for the incineration of organically contaminated waste waters, *Combust. Sci. and Technol.* **136**, 333–347.

6. Glarborg, P. and Hadvig, S.(1991): Development and Test of a Kinetic Model for Natural Gas Combustion, Nordic Gas Technology Centre, ISBN 87-89309-44-8.
7. Kufferath, A. and Leuckel, W. (1997) Experimental investigation of flow conditions inside an air-assisted internal-mixing nozzle and their correlation with spray data, ICLASS'97, Seoul, Korea, 262–269.
8. Yuen, M.C. and Chen, L.W. (1976): On drag of evaporating liquid droplets, *Combust. Sci. and Technol.* **14**, 147–154.
9. Renksizbulut, M. and Yuen, M.C. (1983) Experimental study of droplet evaporation in a high-temperature air stream, *J. Heat Transfer* **105**, 384–388.
10. Mulholland, J.A., Srivastava, R.K., Wendt, J.O.L., Agarwal, S.R., and Lanier, W.S. (1991): Trajectory and incineration of rogue droplets in a turbulent diffusion flame, *Combust. Flame* **86**, 297–310.
11. Ehrhardt, K., Ehret, A., and Leuckel, W. (1998): Experimental study on the dependence of burnout on the operation conditions and physical properties in wastewater incineration, Twenty-Seventh Symposium (Int'l) on Combustion, The Combustion Institute, Pittsburgh, PA, in press.
12. Heyse, C. (1996): Erzeugung von Öl-in-Wasser Emulsionen mittels flüssig-flüssig Zerstäubung, Internal Report, Universität Karlsruhe, Engler-Bunte-Institut.
13. Gmehling, J., Onken, U., and Arlt, W. (1977): Vapor-Liquid Equilibrium Data Collection, DECHEMA Chemistry Data Series, Frankfurt, M.
14. Chung, S.H. and Kim, J.S. (1990): An experiment on vaporization and microexplosions of emulsified fuel droplets on a hot surface, Twenty-Third Symposium (Int'l) on Combustion, The Combustion Institute, Pittsburgh, PA, 1431–1435.
15. Steinebrunner, K., Fassoth, N., and Seifert, H. (1993): Einsatzmöglichkeiten eines Mehrstoff-Brennersystems zur thermischen Entsorgung von heizwertarmen flüssigen Prozeßrückständen, *VDI-Bericht* **1090**, 451–459.
16. Leuckel, W. (1972): Einfluß der Verdrallung auf die Eigenschaften turbulenter Diffusionsflammen, *Archiv für das Eisenhüttenwesen* **43**, 189–200.
17. Syred, N. and Ber, J.M. (1974): Combustion in swirling flows: a review, *Combust. Flame* **23**, 143–201.
18. Rawe, R. (1978): Über die Drallstabilisierung frei brennender turbulenter Diffusionsflammen mit zentraler radialer Gaszufuhr, PhD Thesis, Ruhr-Universität Bochum.
19. Hillemanns, R. (1988): Das Strömungs- und Reaktionsfeld sowie Stabilisierungseigenschaften von Drallflammen unter dem Einfluß der inneren Rezirkulationszone, PhD Thesis, Universität Karlsruhe.
20. Prade, B. (1993): Experimentelle und theoretische Untersuchung zum Abblaseverhalten von turbulenten Stauscheibendiffusionsflammen, PhD Thesis, Universität Karlsruhe.

3 Pollutant Formation

Formation of Pollutants in Combustion

W. Leuckel*

Apart from the efficiency of energy transformation, minimizing of pollutants in the flue gases has become the leading goal of research and development in the area of industrial combustion systems since about two decades. As far as gaseous or liquid fuel combustion is concerned, pollutants of general importance are residual hydrocarbons and carbon monoxide due to incomplete gas phase burnout, soot and, possibly, coke particles, and particularly nitrogen oxides (NO, NO_2). Legal standards of pollutants emission limitation, being quite low already, are becoming more and more stringent in all industrialised countries with the consequence that very intense efforts are required in research and development in order to meet or even undercut those limits without making allowance for essential disadvantages like reduced ignition stability or the occurrence of combustion pulsations.

All the above mentioned types of pollutants have been addressed in the joint research programme of the Collaborative Research Centre 167, although main emphasis has been on nitrogen oxides and soot. For gas turbine combustors and industrial high temperature furnaces, steam generators, and chemical process heaters, thermal NO_x formation represents the main mechanism responsible for nitrogen oxide emissions. However, in case of fuels with fuel-bound nitrogen like heavy fuel oil, off-gases from chemical or refinery processes, and producer gas from coal or solid waste gasification, also fuel-N may heavily contribute to NO_x emissions.

For thermal NO_x formation, the basic chemical scheme, the so-called "Generalized Zeldovich-Mechanism", is well established and reveals that high temperature flame zones (beyond about 1500 °C), the local oxygen/fuel stoichiometry, and the flame gas residence time inside those zones are the influencing parameters. In order to reduce thermal NO_x it is essential, therefore, to avoid the coincidence of high temperature and excess oxygen conditions in the flame by matching the flow, mixing, and oxidation reaction patterns in the appropriate way. For heat generating combustors in industrial

* Engler-Bunte-Institut, Bereich III – Verbrennungstechnik, Universität Karlsruhe, Kaiserstr. 12, 76128 Karlsruhe, Germany

or domestic installations, furnace interior recirculation of already partially cooled flue gas may be engaged to lower the temperature as well as the oxygen partial pressure in the burner-near flame region for efficiently minimizing thermal NO_x formation. With gas turbine combustors, however, the limitation of turbine inlet gas temperatures to about 1200–1250 °C for stationary gas turbines and around 1400 °C for air craft engines offers the possibility to apply lean premixed (LP) combustion with comparatively high air equivalence ratios (in the order of 1.6 to 2.0), thus avoiding peak temperatures beyond adiabatic in general. Difficulties and complications arising in this context from ignition instability and/or combustion pulsations are not easy to overcome, so that also air-staged (RQL) combustion is being considered as an alternative to LP-combustion (see below).

Whenever fuel-N nitrogen oxide formation must be restrained, flue gas recirculation is not an appropriate measure since temperature dependence of this chemically quite complex process is only moderate. Instead, oxygen deficient combustion is a very effective way to avoid conversion of fuel nitrogen to NO_x, and even to reduce NO and NO_2 formed in preceding flame zones. This led to the concept of "staged combustion" in the early 1980th, either in the form of air-staged (2 stage) or as fuel-staged (3 stage) combustion. In a well designed staged system conversion of fuel nitrogen to NO can be brought down to about 5 % for fuel gases containing C-H-N compounds with e. g. NH, NO or CN bonds (for example NH_3, amines, HCN etc.).

For combustion systems which serve the purpose of heat generation, it is essential to extract sufficient thermal energy (heat) already before adding the final combustion air stream downstream of the reducing stage in order to achieve complete burnout, so that peak temperatures in this burnout stage are limited to below 1500 °C. With gas turbine combustors, however, heat extraction should not occur, and therefore the secondary air has to be injected directly into the hot gas stream from the reducing stage, such that it mixes extremely fast (within a very few milliseconds) with it, thus avoiding extended near-stoichiometric high-temperature zones with intense NO formation. Essential questions to be answered by research in the context of staged combustion were the optimal choice of the reducing stage combustion parameters (air equivalence ratio, temperature, minimum residence time), and in how far the NO_x formation/reduction behaviour will be different for different types of nitrogen bonds.

In the Collaborative Research Centre 167, the before mentioned problems and goals regarding the NO_x formation and NO_x reduction processes in turbulent diffusion flames, as well as the described concepts of NO_x emission reduction (i. e. delayed mixing, flue gas recirculation, staged combustion, in-flame air staging) have been investigated experimentally in the project A 7 over three granted periods. Natural gas has been applied as the basic fuel gas at a thermal power level of between 100 and 300 kW, and doped with ammonia when investigating NO formation from fuel-bound nitrogen. However, also other chemical species like amines and organic CN- and NO-species have been used for comparison. Axial-jet as well as swirl-type burners were applied in order to vary mixing intensity as well as the basic burner-near flow mixing and reaction patterns. The chief results of this project A 7 are being reported and interpreted in one of the following detailed papers, also including results from extensive experiments on fuel-N reaction chemistry in an isothermal plug flow reactor as well as corresponding mathematical modelling of the fuel NO_x chemistry. Another paper contains, amongst others, first results from the A 8 project of the final granting period of the Collaborative Research Centre 167, referring to NO_x formation modelling in turbulent flames.

Soot formation from hydrocarbons in flames is a phenomenon favourable for many heat generating industrial combustion plants, because of intensified heat transfer from soot particles suspended in high temperature gases. However, soot particles must then burn completely in the tail zone of those flames in order to avoid soot emission in the flue gas. For other applications, in particular gas turbine combustors, domestic burners, and waste gas incinerators, soot formation is not desirable at all, and combustion should be controlled such that soot concentrations in the flame are being at least minimised. In order to meet all those different goals, basic and quantitative knowledge is required about soot formation and oxidation processes in turbulent diffusion flames.

Different from nearly all fundamental soot research work reported in literature which refers to premixed (mostly laminar) flames, project A 6 of the Collaborative Research Centre 167 dealt with soot formation from various hydrocarbons – methane, propane, and acetylene – injected into a hot flue gas turbulent plug flow inside a 100 mm diameter insulated refractory tube. By varying the excess oxygen content and the temperature of the flue gas stream it was possible to simulate the actual local soot formation conditions in technical turbulent diffusion flames, i. e. in high temperature flame zones (above 1150 up to 1600 °C) with mixtures of burnt flue gas, HC-fuel gas, and limited (sub-stoichiometric) concentrations of residual oxygen.

One of the following papers describes the plug flow reactor experiments of the A 6 project and their very interesting and application relevant results in more detail. In addition, soot concentration patterns measured in free burning turbulent axial-jet diffusion flames are being reported and compared with predicted concentrations from a semi-theoretical soot formation model. This model has been deduced from the reactor experiments and, coupled with a successful turbulent flame prediction model, seems to be able to deliver realistic soot pattern prediction for practical flames of the fuel gas types investigated in the project.

Carbon monoxide and hydrocarbon emissions from flames caused by incomplete combustion have been addressed particularly in the project A 11 which was run over two granted periods of the Collaborative Research Centre 167. These investigations referred to a special type of process wastewater incinerator, which has proved in practice to be very efficient and cost saving as regards plant investments and operation expenditures. The process is called "In-Flame Vaporizing Combustion" and is characterised by the injection of (gaseous or liquid) auxiliary fuel and wastewater separately through a three-fluid or four-fluid nozzle mounted in the centre of the divergent exit quarl of a swirl burner.

Main problems of this burner concept are to achieve good ignition stability in certain ranges of specific wastewater load (or adiabatic flame temperature, respectively), thermal power, and excess air ratio, as required by the particular application, and also satisfactory burn-out in terms of low CO and HC emissions in the flue gas. With this burner, the degree of burn-out depends on three partial processes in the flame, i. e. wastewater atomisation and droplet evaporation, spray and gas phase mixing, and chemical kinetics of HC und CO oxidation. As regards the kinetics, it must be considered that adiabatic flame temperatures are chosen to be low (between 980 and 1100 °C) in order to save auxiliary fuel and to avoid thermal NO_x formation. The detailed paper on the A 11 project analyses these mechanisms and interprets their relevance under practical plant conditions. It also discusses how to design the wastewater spray, the fuel gas injection and the swirl intensity of the combustion air in order to ensure good flame stability.

3.1 Formation and Reduction of Thermal and Fuel Nitrogen Oxides in Flames

Dieter Stapf, Peter Jansohn, Stefan Koger, and Wolfgang Leuckel*

Abstract

This work focuses on analysing and modelling the combustion chemistry of the formation and reduction of nitrogen oxides in practical flames and combustors. The influence of several parameters on NO_x formation in turbulent swirling flames has been investigated systematically. Swirl-stabilized turbulent diffusion flames can be switched easily from a short type-II flame with rapid and effective mixing to a longer type-I flame with slower mixing of fuel and combustion air by means of changing the fuel nozzle geometry. Local measurements applying the FTIR technique showed that, in case of a type-I flame, fuel-N is converted mainly under fuel rich conditions. This results in significantly lower in-flame NO_x production from fuel nitrogen with a type-I flame compared to a type-II flame. Premixed combustion, as a further version, results in the highest NO_x emission from fuel nitrogen. Besides this in-flame staging technique of type-I flames, staged combustion with two spatially separated combustion zones of different air equivalence ratios λ is most effective against NO_x formation. In those systems, NO_x formation from fuel-N species, independent of their chemical structure, is being reduced drastically by optimising the stoichiometric ratio of the first stage which is operated in fuel rich mode (air equivalence ratio between 0.75 and 0.8).

In order to develop chemical kinetic models for computing these phenomena, a plug flow reactor was set up for kinetic experiments under practical reaction conditions. A kinetic model from literature was modified on the basis of the experimental data, by introducing better kinetic parameters of those elementary reactions which are important under practical conditions of reburning, and not just by fitting the complete model to specific experimental data. All modelling calculations were performed using the CHEMKIN code as a computing tool.

To simulate the entire process of NO_x reduction by reburning in an industrial furnace, a hybrid model has been developed. It couples the calculation of the turbulent reacting flow by a CFD-code with the detailed

* Engler-Bunte-Institut, Bereich III – Verbrennungstechnik, Universität Karlsruhe, Kaiserstr. 12, 76128 Karlsruhe, Germany

chemical kinetic model of NO_x reduction by a postprocessor code which uses the CFD-code results as available input data. Agreement of modelling and detailed measurements was found very satisfactory.

Furthermore, thermal NO formation in a turbulent swirling flame has been calculated using a simplified three step reaction mechanism. The calculation has been performed again by a postprocessor on the basis of pdf methods to consider the influence of temperature fluctuations. Computation results and experimental data agreed satisfactorily, too.

3.1.1 Introduction and Motivation

Despite a variety of efforts, environmental problems from fossil fuel based power generation have been increasing steadily in the past. Although NO_x emission from power generation and industrial furnaces has been reduced in Germany to about 25 % within the past two decades, nitrogen oxide emission (contributing to the greenhouse effect, acid rain, smog, and ozone problems) has been a continuous problem because of increasing traffic.

In other countries like China, with economical growth rates of about 10 % per year, traffic is increasing rapidly, and the energy demand will be covered mainly by available nitrogen containing coal.

To achieve low NO_x emissions from high performance combustors like gas turbines or industrial incinerators, preferably the combustion processes themselves should be modified, rather than applying more expensive end-of-pipe gas cleaning technologies. The major nitrogen oxide sources are thermal NO formation (Zeldovich NO [1]) and NO_x formation from fuel nitrogen. Modern NO_x reduction technologies have been developed during the past two decades in order to fulfil legislative demands efficiently, e. g. lean premixed combustion in gas turbines to reduce thermal NO, or three-staged combustion (NO_x reburning) for industrial residues with high nitrogen content [2].

Therefore, a major subject of research activities in this area must be the understanding and the theoretical description of partial processes and the development of comprehensive models. With those models it will then be possible to design combustion processes without performing extensive and quite expensive semi technical scale and scale-up experiments.

Most practical flame combustors have one feature in common: The bulk of combustion takes place in highly turbulent swirl-stabilized flames. In these flames, NO_x formation and reduction depend upon the interaction of complex chemistry and turbulent flow, with the consequence that designing a comprehensive mathematical model is difficult and time consuming.

3.1.2 Fundamentals

3.1.2.1 Nitrogen Oxide Formation in Combustion

In combustion processes nitrogen oxides are formed and destroyed by reactions with radicals or stable species of the C/H/O-system. Therefore, conversion of N-species is directly dependent on the quality of the combustion process. This means that mathematical modelling of hydrocarbon pyrolysis and oxidation is a prerequisite to describe NO_x formation correctly.

The principle of NO formation and destruction is shown in Fig. 3.1.1. Nitrogen sources for NO formation are either N_2 from the combustion air or fuel-bound nitrogen. During fuel decomposition the latter is converted by several reaction steps to NH_i species. At this point of the mechanism, two competitive reaction pathways exist, depending on the local stoichiometry expressed by the air equivalence ratio λ: Under lean conditions, oxidation to NO is the dominant path, whereas lack of oxidation partners (under fuel rich conditions) mainly leads to the formation of molecular nitrogen, N_2.

Decomposition of NO by the attack of CH_i or HCCO radicals is called reburning. Reaction products of the reburning path are in case of CH_i radicals HCN, and in case of HCCO radicals HCN and NH_i species. Subsequent reactions to NH_i and N_2 occur according to the mechanism described above.

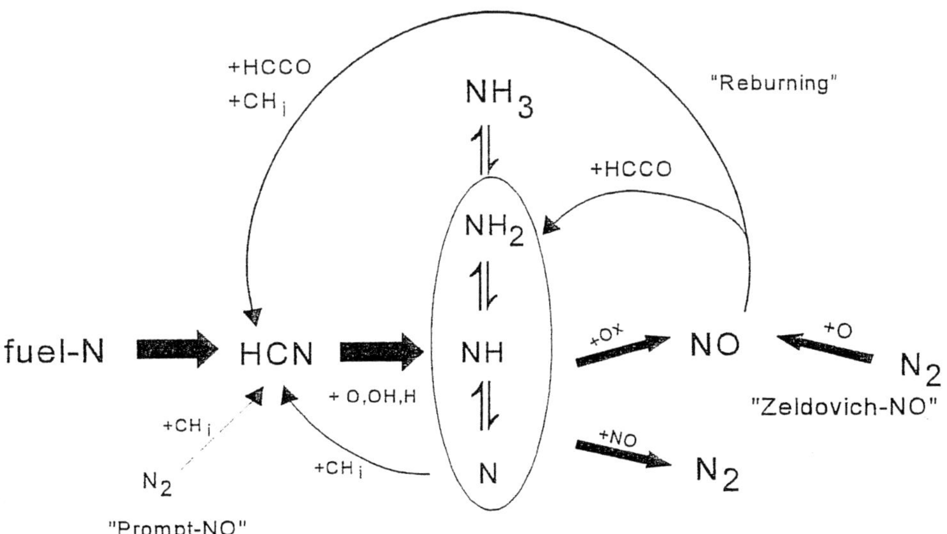

Figure 3.1.1: Main reaction pathways in the fuel-N reaction mechanism of NO-formation.

NO formation by reaction of small hydrocarbon radicals with N_2 (prompt NO formation) is negligible in most cases.

The major portion of worldwide NO_x emissions is due to a reaction process with molecular nitrogen from the combustion air according to the mechanism experimentally derived by Zeldovich et al. [1]:

$$N_2 + O \rightleftharpoons NO + N \qquad (1)$$

$$O_2 + N \rightleftharpoons NO + O \qquad (2)$$

Zeldovich NO formation is, however, important only in the higher temperature range $T > 1500\,°C$ (thermal NO) because of the high stability of the N≡N triple bond and the resulting high activation energy of reaction (1). Under fuel rich conditions the following reaction gets important, too.

$$OH + N \rightleftharpoons NO + H \qquad (3)$$

Assuming several justifiable simplifications under technical reaction conditions, the following NO production rate equation $\dot{\omega}_{NO}$ [mol/(cm^{3}s)] can be derived:

$$\dot{\omega}_{NO} \cdot \frac{p}{R \cdot T} = \frac{d[NO]}{dt} = 2 \cdot k_1 \cdot [N_2] \cdot [O] \qquad (4)$$

with

$$k_1 = 1.8 \cdot 10^{14} \cdot \exp\left[\frac{-319\ kJ/mol}{R \cdot T}\right] \cdot \frac{cm^3}{mol \cdot s} \qquad (5)$$

3.1.2.2 Primary Measures for NO_x Reduction

For technical applications the influence of stoichiometry on NO formation from fuel nitrogen as well as from molecular nitrogen is of particular interest. By lowering the air equivalence ratio λ of the first (reducing) combustion stage it is possible to decrease NO formation drastically. But if λ is too low, NH_3 and HCN are being preserved in this stage and will be oxidized almost quantitatively to NO in the following combustion stage necessary to guarantee the burnout of H_2 and CO formed in the first stage. Therefore, optimising of a staged combustion system means minimizing the sum of all N-species ([TFN] = [HCN] + [NH_3] + [NO]) in the reburn zone. Nowadays, staged combustion is widely used for most combustion processes in power generation and industry in order to reduce NO_x emission by this inexpensive primary measure instead of expensive downstream application of catalysers.

3.1.2.3 Characteristics of Type-I and Type-II Flames

In Fig. 3.1.2 typical flow patterns of type-I and type-II flames are shown [3]. Type-I flames are slim and long shaped. Their flow and reaction structure is mainly influenced by the axial fuel jet enclosed by an annular reverse flow zone. For generating a type-I flame, a sufficiently high axial momentum of the fuel jet to push through the inner recirculation zone formed at swirl numbers $S_{0,\text{th}} > S_{0,\text{th,krit}} \approx 0.4$ is a prerequisite. The jet characteristics of type-I flames enable the heat release to be distributed over the whole flame length, hence resulting in not very high peak flame temperatures ($T_{\max} \leq 1800$ K), which is a desired feature for certain industrial furnace applications avoiding very inhomogeneous heat transfer.

Type-II flames, on the contrary, have a short and compact shape with a central inner recirculation zone. Rapid mixing of fuel and air is favoured by spraying the fuel into the shear zone between the air jet and the inner recirculation vortex. Recirculation of hot gases back to the root of the flame and high reaction intensity result in high flame temperatures ($T \approx 2000$ K).

As to the production and removal of N-species, the described flow patterns of the two flame types are particularly interesting, since their reaction conditions for pyrolysis and oxidation of the fuel are quite different.

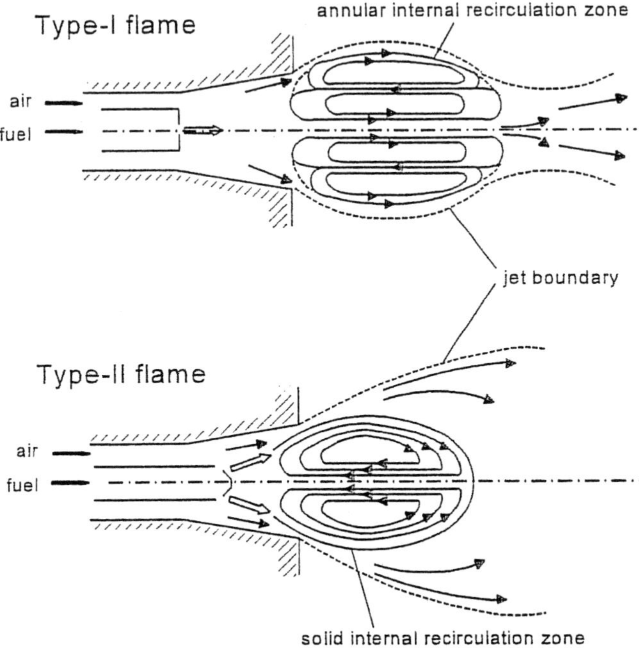

Figure 3.1.2: Mixing and reaction field patterns of type-I and type-II swirling flames.

3.1.3 Experimental

3.1.3.1 Swirl Burner

The combustion facility to investigate formation and reduction of N-species in turbulent swirling diffusion flames is shown in Fig. 3.1.3 [4]. The swirl burner (moveable block principle) is mounted on the top of the main combustion chamber, and the flame is burning downwards. It is possible to vary the burner exit geometry and the fuel nozzle geometry in order to produce different flow fields. The total thermal input of the flames investigated was 0.35 MW. The fuel was natural gas (NG) doped with nitrogen containing compounds for simulating fuel-bound nitrogen.

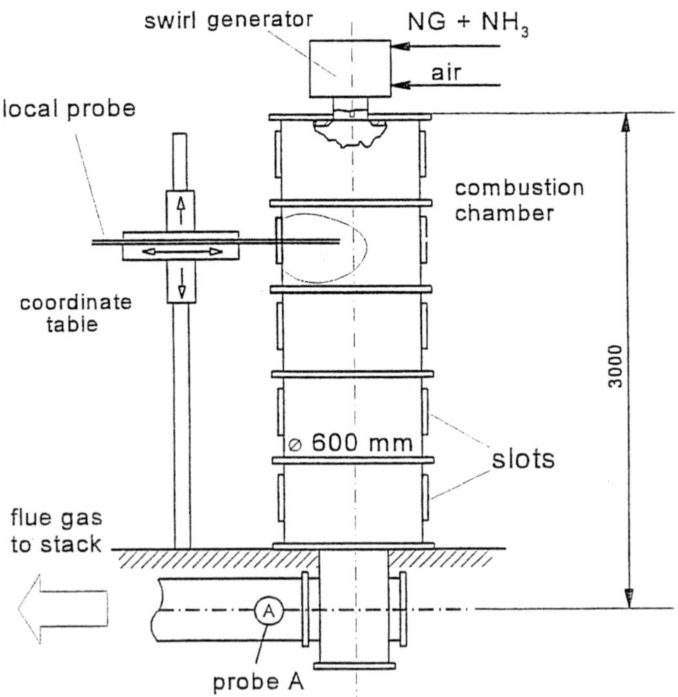

Figure 3.1.3: Combustion chamber and swirl burner arrangement.

3.1.3.2 Plug Flow Reactor

A plug flow reactor has been set up to study NO_x reburning kinetics under conditions comparable to those met in industrial furnaces. This type of reactor was chosen because it can generate diluted reburning conditions with low radical concentrations along the whole combustion process. Figure 3.1.4 shows a schematic of the flow reactor [5, 6] which consists of the primary combustion chamber, the mixing system, and the plug flow reaction tube of 100 mm diameter and about 2000 mm length.

A quarter circle shaped nozzle between the primary combustion chamber and the reaction tube accelerates the hot product flow smoothly up to velocities of about 15 m/s inside the ceramic tube. At its entrance section, natural gas is being injected radially through a multihole nozzle (Fig. 3.1.5). Number and flow momentum of the near sonic injector jets have been optimised for very fast mixing. In order to maintain fast mixing for all operating conditions it was important to add nitrogen to the reburn natural gas stream which varies depending on the equivalence ratio Φ in the primary combustion chamber and in the flow tube. By varying the equivalence ratio in the primary combustion chamber as well as the natural gas flow through the injector nozzle it was possible to control initial oxygen content and equivalence ratio in the reaction tube independently of each other. The reaction tube is heated electrically to avoid significant heat losses, hence the temperature of the reacting flow was almost constant. Temperature variations were within a range of $\pm 30\,°C$ only.

Both combustion facilities were set up in a semi technical scale rather than in a laboratory scale in order to enable applying probe sampling techniques and to avoid wall effects. Furthermore, the chosen scale provides

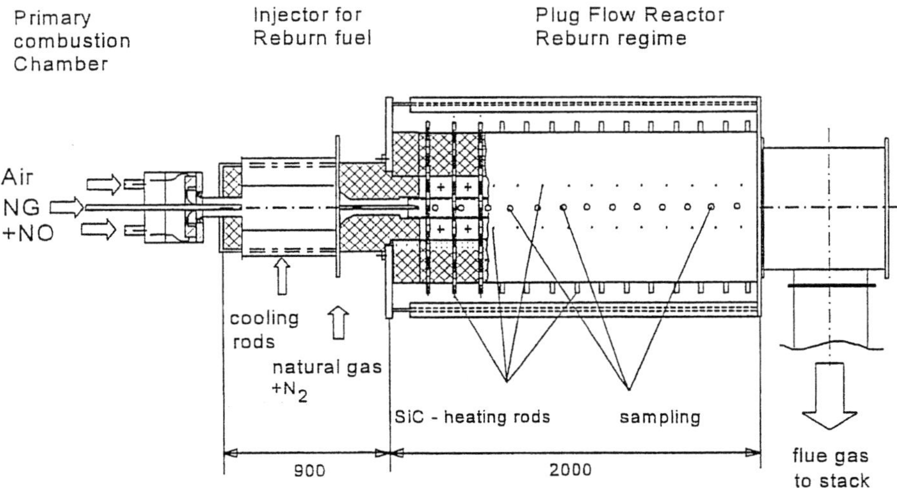

Figure 3.1.4: Schematic of the plug flow reactor.

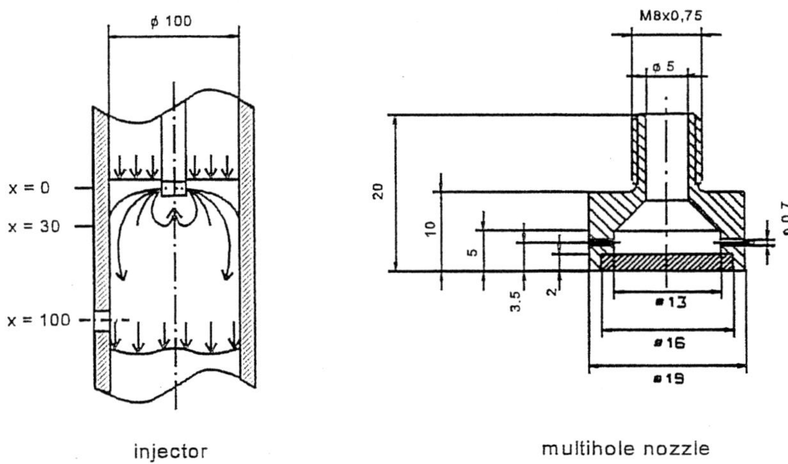

Figure 3.1.5: Injector and injector nozzle geometry.

well defined turbulent flow conditions and high spatial resolution of the reaction progress. Temperature and concentration profiles in both combustion systems have been measured by means of water cooled stainless steel probes introduced through side stream ports. Apart from conventional analysers for O_2, H_2, CH_4, CO, CO_2, and NO/NO_x, an FTIR has been used to determine the concentrations of HCN, NH_3, N_2O, NO_2, C_2H_2, and C_2H_4 in the wet flue gas.

3.1.4 Results

3.1.4.1 Minimization of NO_x Emission of Swirling Turbulent Diffusion Flames

The effects of the following experimental variables on NO_x emission have been investigated [4]:

- air equivalence ratio
- fuel nitrogen content
- chemical bound of fuel nitrogen (NH_i, CN, NO)
- swirl intensity of the combustion air
- fuel nozzle geometry

Figure 3.1.6 gives an overview of NO_x emission variations achieved only by modification of burner parameters. Fuel nozzle geometry and swirl intensity of the combustion air show significant effects on NO_x emission. This is due to their influence on flow and mixing patterns of swirl-stabilized diffusion flames, i. e. the formation and extension of the different flow zones (internal and external reverse flow, jet mixing zone of air and fuel, see Fig. 3.1.2) as well as heat and mass exchange rates between those zones. Especially at high fuel-N contents, fuel nozzle geometry had a great impact on NO_x emission. The emission level with divergent multihole nozzles producing type-II flames was considerably higher than the level with axial jet nozzles generating type-I flames. This fact could also be observed in flames without any fuel nitrogen content, i. e. with thermal NO production only. Minimum NO_x emission could be achieved with axial jet nozzles by minimizing the fuel jet momentum flux $\dot{I}_{0,g}$ such that just a type-I flame was produced, i. e. near to

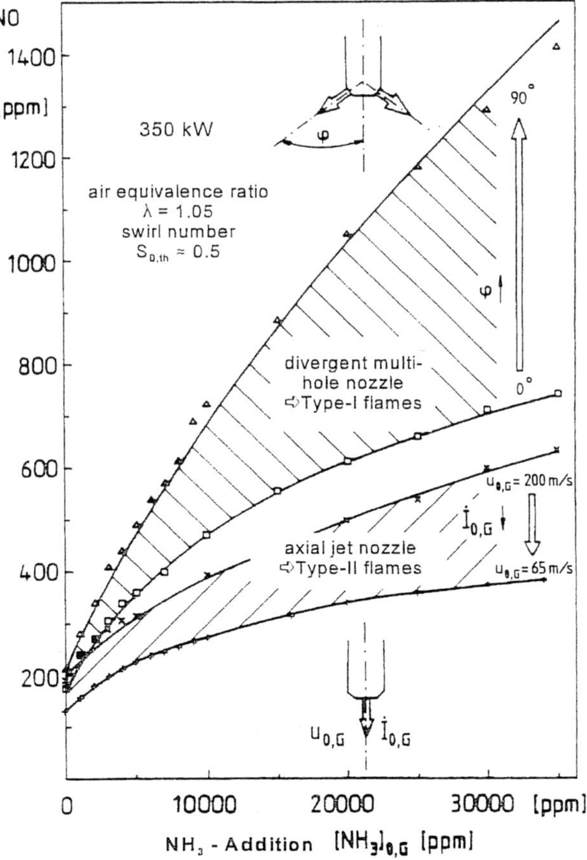

Figure 3.1.6: NO_x emission concentrations as a function of diverse burner parameters and fuel nitrogen content.

transition to type-II. Under those conditions maximum residence time of fuel nitrogen released under reducing conditions was realized ($\lambda_{loc} < 1$) and, therefore, conversion to molecular nitrogen (N_2) was favoured. The results show that modifications of flame patterns by varying burner parameters is a very simple and efficient method to reduce NO_x formation in practical burner applications.

Local measurements in different flow zones of type-I and type-II flames have been performed to obtain detailed knowledge of NO formation and reduction occurring in the combustion zone of technical flames. HCN and NH_3 have been considered to be important fuel nitrogen sources and intermediate reaction species from released fuel bound nitrogen as well.

The measured concentration field (Fig. 3.1.7) and the evaluated exhaust molar flow rates of NO, NH_3, and HCN from NH_3 doped type-I flames (fuel nitrogen content: $[NH_3]_{0,g} = 30\,000$ ppm) show that fuel nitrogen is being transformed mainly to N_2. Only small portions of fuel nitrogen are being transformed to NO in air rich zones. In the fuel rich natural gas jet ($\lambda_{loc} = 0.7$) formation of HCN occurs, however, being destroyed towards the end of the flame under slightly fuel rich conditions without additional formation of NO. The local measurements confirm, as discussed before, the existence of efficient sub-stoichiometric conditions in type-I flames, leading to less NO formation and, therefore, explain the only small conversion rates of fuel nitrogen to NO in such flames.

In NH_3-doped type-II flames, however, rapid decomposition of the fuel nitrogen compound NH_3 under near stoichiometric conditions ($0.9 \leq \lambda_{loc} \leq 1.1$) is observed. Here, the reactions of N-species lead mainly to the formation of NO. Formation and destruction of HCN are restricted to the narrow fuel rich jet area and to the part of the internal recirculation zone near to the burner exit ($\lambda_{loc} \geq 0.7$).

Premixed flames with air equivalence ratios above unity have the highest NO_x emission, especially at high fuel nitrogen content, because of their overall super-stoichiometric conditions. Moreover, all flames, independent of flame type, show a degressive increase of NO_x emission, i. e. smaller fuel nitrogen conversion rate to NO, with increasing fuel nitrogen content.

In order to investigate the effect of the type of fuel nitrogen bound on NO_x emission the natural gas was doped alternatively with 1 vol-% of ammonia, acetonitrile, pyridine, or nitromethane [7]. Figure 3.1.8 shows the N-species emissions of those four additives measured for variable flame stoichiometry. Starting from near stoichiometric conditions, the total fixed nitrogen (TFN) emission which is almost entirely NO decreases with decreasing λ_{prim} because of the increasing lack of oxidizing species. This leads to the well-known minimum emission at $\lambda \approx 0.75$. Under very fuel rich conditions ($\lambda < 0.7$), however, the TFN concentration increases again. This is due to the emission of the intermediates HCN and NH_3, because of their extremely slow oxidation kinetics in this stoichiometry range.

The TFN emission curves of all N-species are obviously very similar, apart from nitromethane which produces somewhat higher TFN(NO)-values

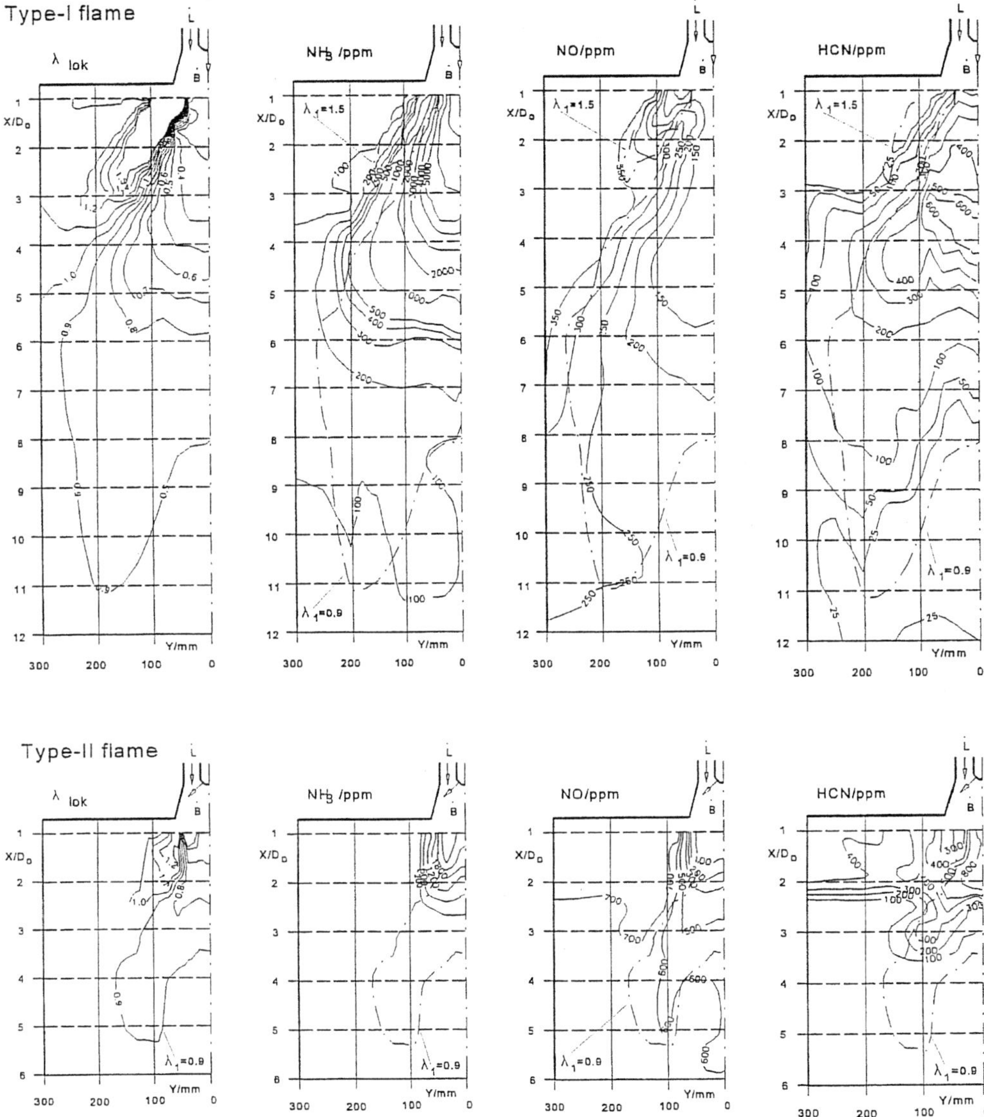

Figure 3.1.7: Field measurements of type-I and type-II flames doped with 3 % ammonia.

between $\lambda_{\text{prim}} = 0.8$ and 1.0. In general, the influence of the fuel nitrogen bound on TFN emission is negligible compared to the influence of the stoichiometry. This is due to the fact that all nitrogen compounds decompose to species of the NH_i pool. At this crucial point of the NO formation mechanism only the actual stoichiometry is important for NO production.

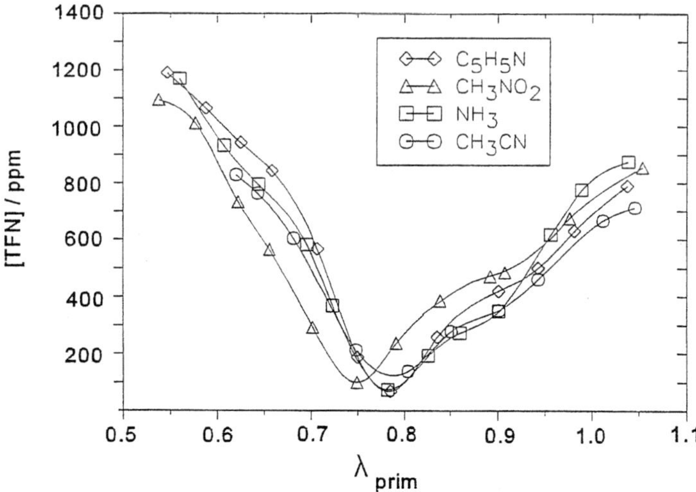

Figure 3.1.8: TFN concentrations from different fuel-N additives as a function of the air equivalence ratio.

3.1.4.2 Development of Chemical Kinetics Modelling

In this part of the project, detailed measurements of temperature and concentrations as a function of residence time were performed using a plug flow reactor [5, 6]. These measurements have been compared with theoretical calculations applying the CHEMKIN code [8–10], which has been established and improved during the past two decades as a standard tool for chemical kinetics modelling in combustion. The calculations were performed using different reaction mechanisms from recent literature and setting the measured initial conditions as input parameters. Also, the model calculation results have been checked with measured data from radical-rich flame front systems (i. e. own measurements in laminar premixed Bunsen flames and data from literature of a laminar premixed flat flame and a stirred reactor [5]).

Comparison of measurement and modelling results showed a considerable underestimation of the reaction time scales at an initial oxygen content of $[O_2] > 1\,\%$ in the hot gas mixture from the precombuster (Fig. 3.1.9). Furthermore, the reaction of NO with HCCO to HCNO followed by HCN formation as the most important NO reburning reactions have been very much overestimated by calculations. This was concluded from the reaction rate analysis presented in Fig. 3.1.10, which shows that, according to the literature models [11], removal of NO occurs almost entirely from attacks of HCCO radicals, HCCO being an intermediate in C_2H_2 decomposition under fuel rich conditions. In this case HCNO species are formed (see Fig. 3.1.9) as an intermediate.

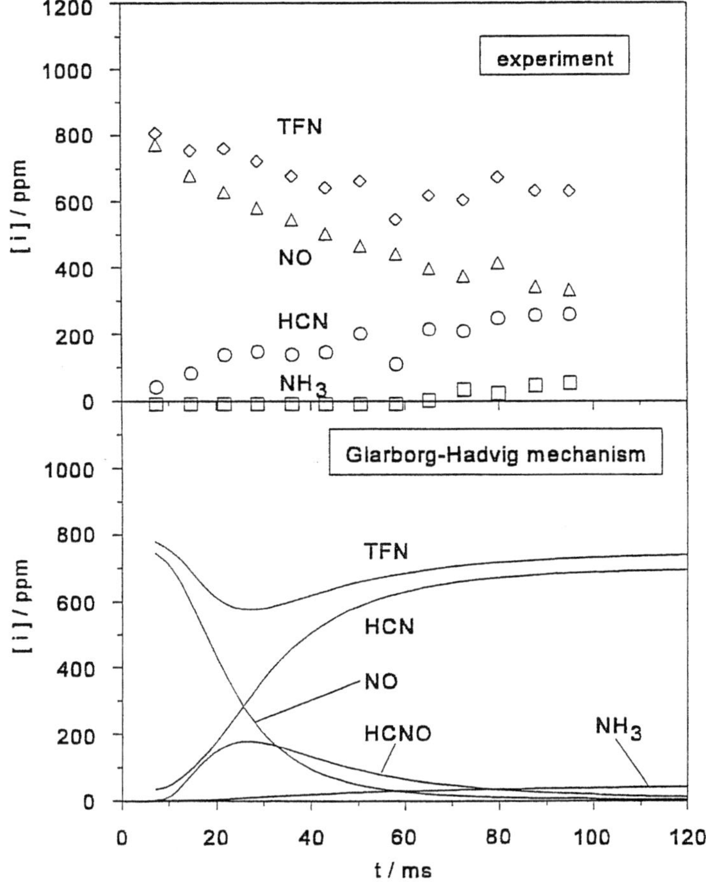

Figure 3.1.9: Calculated and measured concentration of the major N-species in a plug flow reactor under typical reburn conditions (λ_{reb} = 0.8, 1300 °C) as a function of residence time.

Based on reaction rate equations and a sensitivity analysis for the dominant reactions in the C/H/O/N-system, more recent chemical kinetic data have been introduced into the overall most reliable mechanism of Clarborg and Hadvig. Those modifications [5] led indeed to improved agreement of measured and calculated reaction time scales and NO destruction rates under fuel rich conditions, as can be seen from the comparison of experimental data and those from the modified mechanism in Fig. 3.1.11. However, HCN formation is still being overestimated, to the disadvantage of N_2 formation. Under principally different radical-rich flame front conditions at $\lambda < 1$, improved agreement of the modified mechanism with measured data has been achieved as well. Also, laminar burning velocities at 1 bar total pressure have been calculated with all of the fore-mentioned mechanisms applying

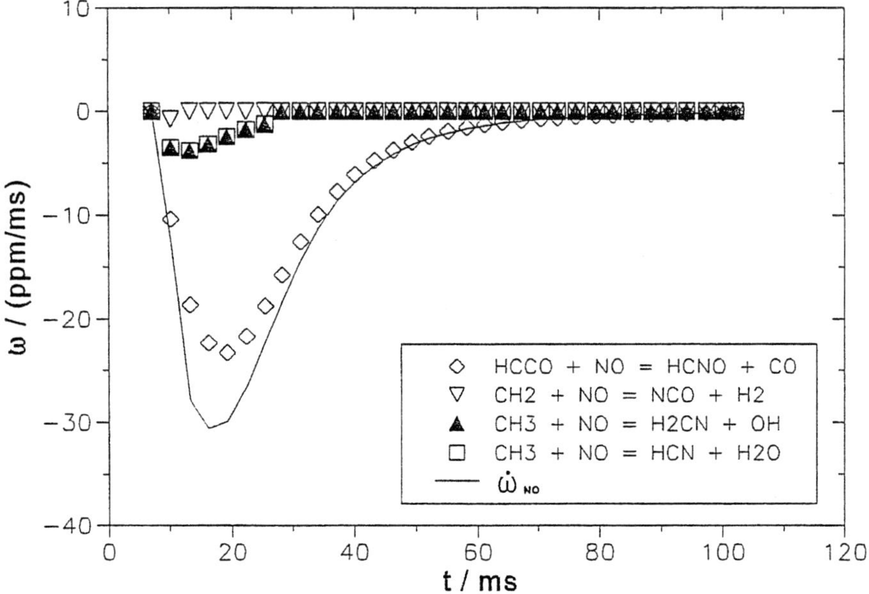

Figure 3.1.10: Analysis of the NO destruction rate (calculation with the Clarborg-Hadvic mechanism) for the experiment in Fig. 3.1.9.

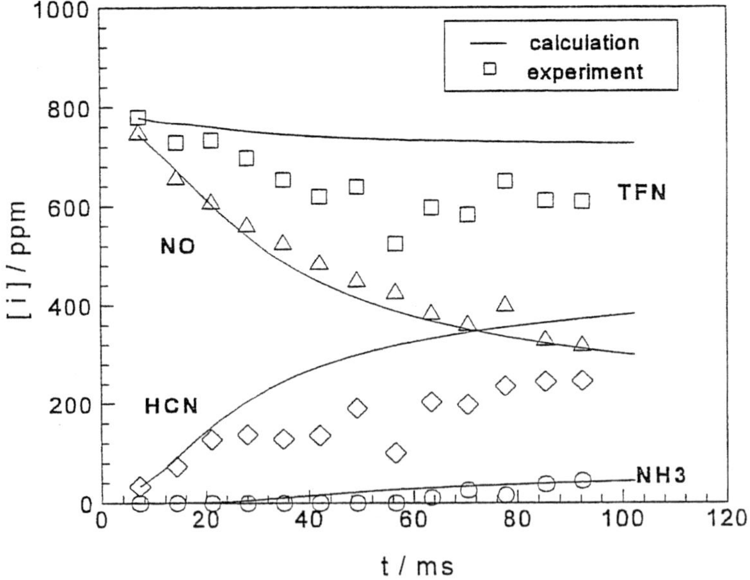

Figure 3.1.11: Calculated (modified mechanism) and measured concentrations of the major N-species with the reaction conditions of Fig. 3.1.9.

the CHEMKIN mass transfer model and the appropriate thermochemical data base. Whereas the models from literature, according to the lower reaction time scales deduced from the plug flow reactor calculations, calculate too high values of burning velocity, the modified C/H/O mechanism reproduces the experimental data quite satisfactorily for all the reburn stage air equivalence ratios investigated.

Considering the structures of laminar premixed flames at $\lambda < 1$, calculated with the diverse mechanisms, in terms of concentration profiles reveals the reasons for the differences under technical reburning conditions: In radical-rich flame fronts, even at $\lambda = 0.8$, the consumption of NO is dominated by reactions with CH and CH_2 radicals. Their reaction rates are relatively high compared to the reaction of NO with CH_3, because the concentrations of CH and CH_2 are some orders of magnitude higher than they are under technical reburning conditions. C_2-chemistry and HCCO are less important in those premixed flame fronts, since in contrast to fuel rich technical flames, in laboratory laminar flame systems there exist sufficient oxygen containing chain propagators resulting from molecular diffusion to keep the concentrations of HCCO and CH_3 low. Therefore, inaccurate reaction rate constants become obvious just under technical reburning conditions, where the reactions of NO + CH_3 seem to be dominant. The role of the reaction path HCCO + NO is in any case not yet clear and has been omitted in the modified mechanism.

Figure 3.1.12 represents the reaction pathways for NO_x reburning. Under technical conditions, i. e. under a deficiency of oxygen and radicals, the reactions with CH_3 are dominant. Allowing the reaction pathway NO + HCCO $\rightleftarrows$ HCNO + CO, the HCN formation from the following HCNO decomposition is significantly overestimated. Under these conditions HCN is relatively stable because of the lack of oxygen containing chain propagators. If, on the other hand, HNCO is an important reaction product, subsequent NH_3 formation and higher N_2 production under fuel rich conditions may occur.

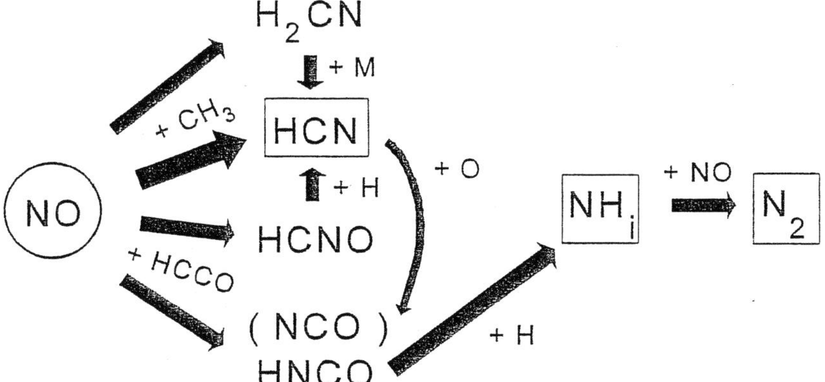

Figure 3.1.12: Reaction pathways of NO_x reburning under technical conditions.

Figure 3.1.13 shows a comparison of experimental data and of concentrations of the main N-species calculated from detailed and reduced mechanisms under technical reburning conditions. The "enhanced modified mechanism" additionally contains the NO reburning path with HCCO and parallel diminution of the reaction path NO + CH_3. The "reduced mechanism" contains only 15 from 48 species of the detailed mechanism and was derived by sensitivity and reaction rate analysis as well as introduced

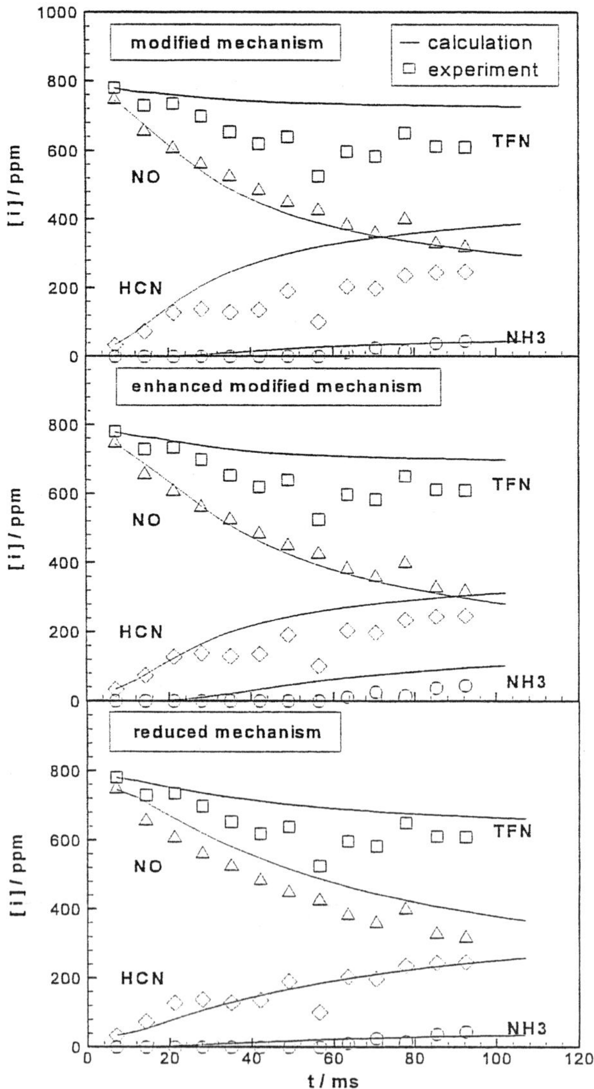

Figure 3.1.13: Comparison of detailed and reduced mechanisms under reaction conditions as in Fig. 3.1.9.

steady state assumptions. It will be presented in detail in a further publication.

3.1.4.3 Coupling of Chemistry and Turbulence in Process Modelling of NO

3.1.4.3.1 Zeldovich NO in Turbulent Swirling Flames

Calculations of NO formation in technical turbulent swirl-stabilized natural gas flames at 1 bar have been performed. In this case effective mechanisms are NO formation from molecular nitrogen at high temperatures according to Zeldovich (with great influence of turbulent temperature fluctuations on NO formation) as well as prompt NO formation under fuel rich conditions. The aim of these investigations was to develop a calculation method as simple as possible for practical applications, but considering the most important physical and chemical influencing variables without "fiddle factors".

Prompt NO formation has been investigated by means of a comparison of measurements and calculations for laminar premixed natural gas flames. This system can be treated using the CHEMKIN code with the application of detailed chemistry. At 1 bar and $\lambda > 1$ the prompt NO emission is significantly lower than 10 ppm. It increases with decreasing air equivalence ratio. The generally known reaction mechanisms showed, amongst each other, at $\lambda < 1$ significant discrepancies in their results for prompt NO, the main reason obviously being uncertainties of kinetic data of the NO formation from small CH_i radicals.

The Zeldovich NO mechanism is controlled by the three step mechanism described before. The required concentrations of the radicals are calculable from the concentrations of the stable species H_2, O_2, and H_2O, by assuming partial equilibrium of the following three reactions in the O/H system:

$$H + O_2 \rightleftharpoons OH + O$$

$$O + H_2 \rightleftharpoons OH + H$$

$$OH + H_2 \rightleftharpoons H_2O + H$$

This mechanism is relatively simple and free from essential influences on the real hydrocarbon combustion mechanism in turbulent flames. On the basis of the partial equilibrium, the Zeldovich-kinetics is strongly dependent on the local H_2, O_2, H_2O concentrations and the local fluctuations of the temperature in the flame front, because of the highly non-linear temperature dependence of the first Zeldovich reaction. It is advisable, therefore, to apply post processor methods, based upon the data resulting from preceding modelling of the turbulent reacting flow.

The computational analysis of detailed reaction mechanisms supplied the following results: For natural gas combustion at 1 bar the known partial equilibrium of the O/H system is fulfilled above 1700 K, independent of the air equivalence ratio. Below this temperature, the required radical concentrations cannot be calculated this way, but in this range the Zeldovich NO formation is anyway negligibly small. To compute the local temperature-averaged NO production rates $\overline{\dot{\omega}}$, clipped Gauss distributions have been constructed using mean temperatures $\overline{T}$ and their fluctuations T_{rms}. This probability-density-function (pdf) is typical for the NO_x producing main reaction zone of turbulent swirling flames. The alternative application of β functions supplied no advantages as to improved accuracy. The local mean production rate of NO is then calculated from local mean concentrations and temperature values and from rms data of temperature fluctuations:

$$\overline{\dot{\omega}_{NO}} = \overline{\dot{\omega}_{NO}}\left(\overline{[O_2]}, \overline{[H_2O]}, \overline{[H_2]}, \overline{[N_2]}, \overline{T}, T_{rms}, \overline{\lambda}\right) \tag{6}$$

This method has been checked by using measured data from confined turbulent swirling natural gas flames ($\lambda = 1.1$). The local production rates of NO have been integrated along the main stream lines, and the obtained NO concentrations have been compared with measured NO_x data. This quasi one-dimensional approach is, however, only justified in convection dominated flame zones. NO influx by cross streamline turbulent exchange with the reverse (NO containing) flow is being neglected.

For this reason, in a premixed flame the NO concentration has been calculated 30 % too low. It was found that with this flame approximately half of the Zeldovich NO production takes place already in the flame front below the maximal mean combustion temperature because of super-equilibrium O-concentrations. Furthermore, about half of the total NO emission results from temperature fluctuations above the local time averaged value.

The experimental data of a diffusion flame showed, however, reasonable agreement with the calculations (Fig. 3.1.14), as the neglect of stoichiometry fluctuations has only small effects in this case. The NO production rate is, in contrast to the temperature dependence, only slightly non-linear dependent on λ, so that positive and negative λ-amplitudes compensate each other with regard to NO production. This justifies the neglect of λ- and concentration-rms parameters in Eq. (6).

3.1.4.3.2 NO_x Reburning in an Axial Jet Diffusion Flame

Employing detailed reaction mechanisms for the calculation of NO_x reduction by reburning in furnaces requires mathematical coupling of detailed chemical kinetics with fluid-dynamics. Numerical codes for the calculation of turbulent reacting flows (CFD) are commonly available, but the implementation of a detailed reaction mechanism would lead to prohibitive demands of CPU

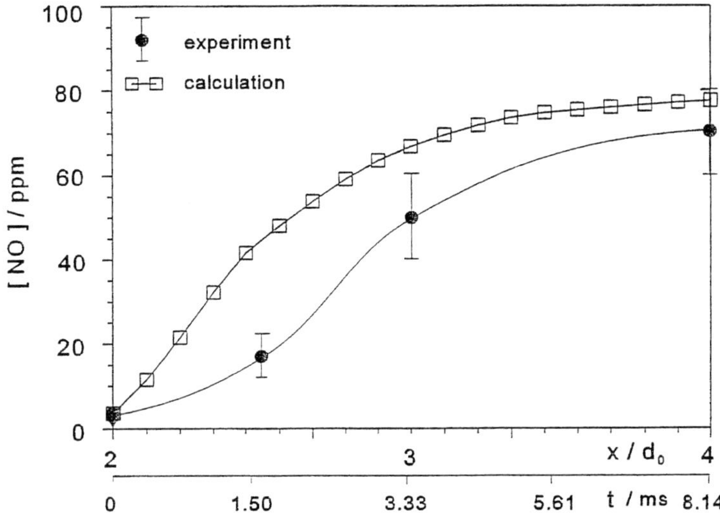

Figure 3.1.14: Comparison of measured and calculated NO concentration along the 40 % carbon stream line.

time, due to the large number of reactions and species to be considered. A significant reduction in number of considered species and reactions which would be required for the implementation in a CFD-code, but which still must allow an adequate description of NO_x reduction by reburning, has not been accomplished yet. A different approach, based on the decoupling of fluid-dynamics and pollutant formation, has been followed in this work [12, 13]. In a first step, the turbulent reacting flow is calculated considering mainly the heat release by fuel oxidation. In a separate second step NO_x reduction in the previously determined hot gas flow field is calculated by "post-processing". The decoupling is justified since species like NO or HCN exist in small concentrations only, and their effects on local heat release by reaction and on the flow pattern can be neglected entirely.

The postprocessor divides the flow field into a series of plug flow reactors which follow the stream line pattern. Their size is significantly larger than the size of the CFD-grid, in order to keep the calculation time moderate. Reaction progress is then calculated in each reactor using a detailed reaction mechanism. Subsequently, concentrations are corrected for turbulent exchange between reactors, which provide the values for the next iterative step in the calculation.

For validating of the model detailed measurements of temperature and species concentration distributions were performed in the fuel rich second stage of a pilot scale furnace. The total thermal input was 335 kW. The stoichiometric air ratios were 1.11 in the first stage, 0.93 in the investigated second (reburn) stage and 1.05 in the third stage. The reburn fuel was injected axially to allow accurate calculation of the flow and to simplify the code of the postprocessor. The NO_x concentration in dry flue gases before fuel injection was 1200 ppm.

Figure 3.1.15 shows a comparison of measured (left hand side) and, applying the modified mechanism, calculated (right hand side) distributions of NO, HCN, and NH_3. In the shown case the reburn fuel was doped with 3 % NH_3 to simulate chemically bound nitrogen. Therefore, NH_3 is predominantly fed with the reburn fuel and is to a much smaller extent an intermediate of NO-reduction.

The agreement between experiment and calculation is overall satisfying, although some discrepancies remain. The predicted NO reduction is restricted to a somewhat too narrow core where it is overestimated. Parallel to this, the predicted HCN formation is somewhat low and the HCN-peak is shifted downstream. NH_3 is in good agreement, however. The observed differences are not necessarily resulting from deficiencies in the detailed reaction mechanism, but seem to be a consequence of weaknesses in the preceding flow simulation by the CFD-code.

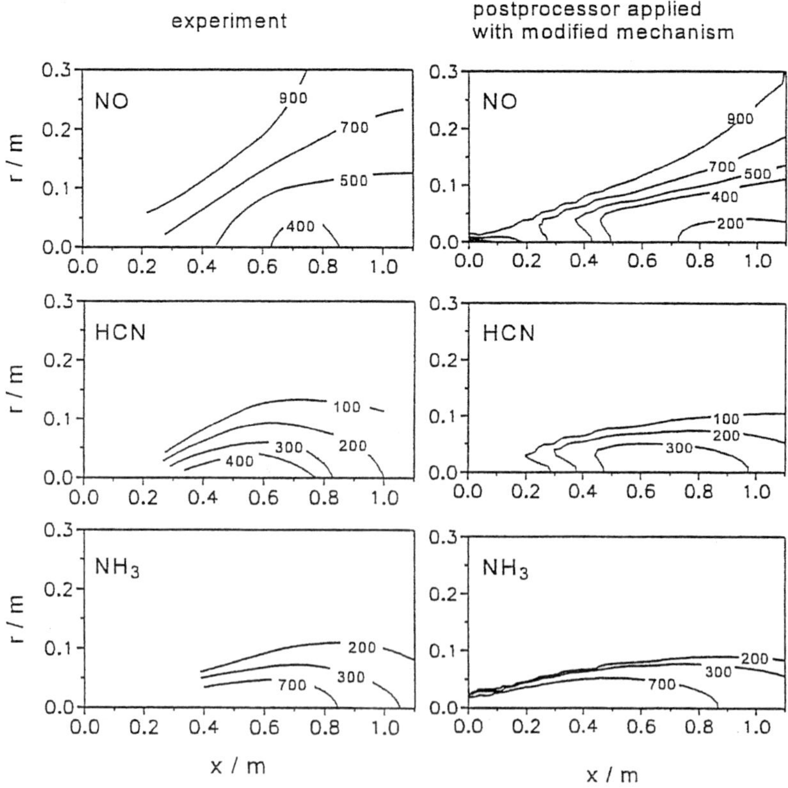

Figure 3.1.15: Comparison of the predicted and the measured NO reduction in the reburn stage of a pilot scale furnace for axial injection of NH_3-doped reburn fuel.

The overall agreement of models predictions (when using the modified reaction mechanism) with the experiments seems to be sufficiently good to allow prediction of NO_x reduction by reburning with technical accuracy.

3.1.5 Conclusions

The main objectives of the A8 project concerned the analysis and modelling of formation and reduction of nitric oxides in industrial-type flames and combustors. This was accomplished by providing detailed flame data from experiments in a semi technical scale, by applying laboratory scale derived chemistry models from literature on practical combustion conditions, and by validating the resulting models for practical low NO_x process design. During the project phase I until the early 1990s, detailed parametric studies and field measurements of swirling turbulent diffusion flames were carried out at a 350 kW thermal scale with applying the new FTIR gas species concentration measurement technique to investigate NO_x formation and reduction chemistry. Afterwards during phase II, detailed and reduced NO_x chemistry models were developed for these practical conditions using a large data basis from systematic plug flow reactor experiments. Finally, those detailed and reduced chemistry models were introduced into modelling of the turbulent reacting flow for a NO_x reburn combustor and for thermal NO formation in a swirling turbulent premixed flame, respectively.

With our present knowledge, computation of fuel NO_x formation within a turbulent swirling flame, based upon existing complex kinetic schemes, is not yet possible because of limited computer capacity. The complex fuel NO_x model cannot yet be reduced to much less reactions and species without loosing significant information. However, NO_x reburning in an axial jet diffusion flame has been calculated in good agreement with experimental data, applying the developed detailed mechanism in a postprocessor.

Furthermore, a reduced Zeldovich NO model was set up in the frame of the A8 project. It has been coupled with the JPDF model developed in the frame of the A9 project. By joining the A8 and A9 projects in the final phase of the Collaborative Research Centre 167, thermal NO formation in confined premixed swirling flames has been simulated successfully [14].

References

1. Zeldovich, J. B., Sadovikov, P. J., Frank-Kamenetskii, D. A. (1947): Oxidation des Stickstoffs bei der Verbrennung (Transl.: Ina Nickel), Akademie der Wissenschaften der UdSSR, Moskau, Leningrad.
2. Kolb, T., Jansohn, P., Leuckel, W. (1988): Reduction of NO_x Emissions in Turbulent Combustion by Fuel-Staging/Effects of Mixing and Stoichiometry in the Reduction Zone, 22nd Symposium (International) on Combustion/The Combustion Institute, 1193–1203.
3. Leuckel, W., Fricker, N. (1976): The characteristics of swirl stabilized natural gas flames, *J. Inst. of Fuel* **49**, 103–112.
4. Jansohn, P. (1991): Bildung und Abbau N-haltiger Verbindungen, insbesondere von HCN, NH_3 und NO, in turbulenten Diffusionsflammen, Dissertation, Universität Karlsruhe (TH).
5. Stapf, D. (1998): Experimentell basierte Weiterentwicklung von Berechnungsmodellen der NO_x-Emission technischer Verbrennungssysteme, Dissertation, Universität Karlsruhe (TH).
6. Stapf, D., Leuckel, W. (1996): Flow Reactor Studies and Testing of Comprehensive Mechanisms for NO_x Reburning, 26th Symposium (International) on Combustion/ The Combustion Institute, 2083–2090.
7. Stapf, D., Leuckel, W. (1993): NO_x Formation from Fuel Nitrogen in Technical Flames, 2nd International Conference on Combustion Technologies for a Clean Environment, Lisbon, Portugal.
8. Lutz, A. E., Kee, R. J., Miller, J. A. (1988): SENKIN: A Fortran Program for Predicting Homogeneous Gas Phase Chemical Kinetics with Sensitivity Analysis, Sandia Report SAND87-8248.
9. Kee, R. J., Grcar, J. F., Smooke, M. D., Miller, J. A. (1992): PREMIX: A Fortran Program for Modelling Steady Laminar One-Dimensional Premixed Flames, Sandia Report, SAND85-8240.
10. Kee, R. J., Rupley, F. M., Miller, J. A. (1989): CHEMKIN-II: A Fortran Chemical Kinetics Package for the Analysis of Gas-Phase Chemical Kinetics, Sandia Report SAND89-8009.
11. Glarborg, P., Hadvig, S. (1991): Development and Test of a Kinetic Model for Natural Gas Combustion, Nordic Gas Technology Centre, DK, ISBN 87-89309-44-8.
12. Ehrhardt, K., Toqan, M., Jansohn, P., Teare, J. D., Beer, J. M., Sybon, G., Leuckel, W., (1998): Modelling of NO_x Reburning in a Pilot Scale Furnace Using Detailed Reaction Kinetics, *Combust. Sci. and Tech.* **131**, 132–146.
13. Stapf, D., Ehrhardt, K., Leuckel, W. (1998): Modelling of NO_x Reduction by Reburning, *Chem. Eng. Tech.* **21**, 412–415.
14. Habisreuther, P., Stapf, D., Eickhoff, H., Lenze, B. (1997): Validierung eines PDF-Modells für die thermische NO_x-Bildung in eingeschlossen brennenden turbulenten vorgemischten Drallflammen, *Gaswärme Int.* **46**, Heft 2, 115–121.

3.2 Soot Formation from Gaseous Hydrocarbons in Turbulent Combustion

Wolfgang Leuckel, Michael Huth, and Bernd Bartenbach*

Abstract

Soot formation and soot oxidation in technical hydrocarbon flames are essential processes for many industrial combustion systems, due to the fact that soot concentrations may largely contribute to radiative heat transfer. Under ecological aspects, however, soot particle emission from the flue gas to the atmosphere has to be avoided by complete burnout in the flame. Soot prediction in flames is, therefore, an important task of combustion research.

This contribution reports on a systematic and extended experimental study on soot formation from methane, propane, and – for fundamental reasons – also from acetylene in a plug flow reactor (PFR). This reactor setup offered the possibility to realise well-defined fuel gas concentration, temperature, and residence time conditions within ranges relevant for soot formation and oxidation in industrial flames. On the basis of the experimental data obtained, a simple autocatalytical mechanism approach and a superimposed soot surface deactivation step are shown to be adequate for interpreting the results in wide ranges of flame temperature (1200 °C till 1600 °C), residence time (up to 100 ms), and local stoichiometry (i. e. the air (or oxygen) equivalence ratio during the soot formation process).

Under approximately isothermal conditions as realised in the PFR, soot growth rates increase with residence time up to a maximum, and then decrease asymptotically to zero. As a consequence, there exist final asymptotic soot concentration values which depend on the [C/O]-ratio and on the reaction temperature, the increase of which results in a decrease of the final soot concentration. Also, the soot formation kinetics depends mainly on the temperature and can be expressed in terms of an Arrhenius equation with the activation energy 192 kJ/mol, independent of the soot generating fuel gas (methane, propane, and acetylene). As a result of those quite complex interrelationships, there occurs a very distinct maximum of the soot concentration as a function of temperature at about 1350 °C, for constant residence time and [C/O]-ratio.

* Engler-Bunte-Institut, Bereich III – Verbrennungstechnik, Universität Karlsruhe, Kaiserstr. 12, 76128 Karlsruhe, Germany

The physico-chemical model developed on the basis of those interrelationships enables to predict local soot formation (i. e. local soot mass growth) rates as a function of the local stoichiometry, the local flame gas temperature, and the local residence time increment in flames. This modelling capability has been shown to be in agreement with results from systematic and extensive gravimetric soot concentration field measurements in free axial-jet type turbulent diffusion propane and methane flames. The computed data compare favourably with the measuring results and prove the applicability of the developed kinetic soot formation model, which now may be implemented into comprehensive gas phase combustion CFD-codes for turbulent hydrocarbon flames.

3.2.1 Introduction

3.2.1.1 Physico-Chemical State of the Art

Prediction of soot concentration fields in technical combustion systems like boilers, production furnaces e. g. in the metallurgical, ceramic and chemical industry, waste incinerator systems, gas turbine combustors, and piston engines is still an area of quite limited reliable results and assured experience. In spite of a large number of laboratory type experiments performed over many years [1–4], there is hardly any validated soot formation or oxidation scheme available to simulate soot concentrations under really high turbulence industrial combustion conditions in satisfactory agreement with experimental data from large scale flames. This is true even for gaseous hydrocarbon combustion, and is the more so for liquid fuel or pulverized bituminous coal flames.

The reason for this situation is the extremely complex nature of many interdependent and interacting individual steps and physico-chemical processes leading from high-molecular hydrocarbon species, which originate from gas phase reactions of fuel gas or fuel vapour components, to the final soot particle suspensions in high temperature flames [1, 3, 5, 6]. These fundamental processes have been studied in great detail mainly in laminar premixed, but also in laminar diffusion flame fronts by applying various experimental concepts and systems as well as different gas species, soot nuclei/particle, and temperature diagnostics. In this way, a fairly convincing and satisfying qualitative and semi-quantitative basic understanding of soot formation has been achieved [1, 3–6], which represents the necessary platform to attack the again more complex conditions of soot forming in turbulent flames of technical combustion systems [7–11].

3.2.1.2 Soot Formation in Turbulent Combustion

At least the final step of the total soot forming process, namely soot particle growth which contributes predominantly to total soot mass concentrations, is a fairly slow (10 to 80 or 120 ms) process compared to most C-H-O-gas phase reactions in flames. For high-turbulence flames, even local gas concentration and temperature turbulent fluctuation times are in general shorter ($\tau_t \approx 1 \div 10$ ms) than the soot formation time scales (which leads to $Da_t < 1$). However, because of strong non-linearity of the soot particle growth dependence on gas concentrations (hydrocarbons, especially acetylene and PAH's, and oxygen) and gas temperature, fluctuations of these factors are still essential for predicting soot formation rates in industrial flames [7, 10].

Beyond that, soot formation in turbulent diffusive combustion – quite different from laminar flame front conditions usually investigated – preferably occurs in fuel-gas/air/hot-burnt-gas mixtures. The presence of higher or lower burnt gas fractions in soot forming flame zones is due either to fuel injection into near-burner internal recirculation zones (induced by bluff body or swirl stabilisation of flame ignition), or simply to high unmixedness in large burning turbulent eddies. Therefore, the influence of burnt gas mixture fraction upon soot formation (i. e. soot particle growth) requires thorough systematic further experimental research.

3.2.1.3 Experiments Performed

The objective of the present investigations performed in project A 6 of the Collaborative Research Centre 167 [12, 13] was to develop and to validate a descriptive model for soot formation in flames which – combined with an adequate model of the turbulent transport, and also with a soot oxidation equation – would enable to predict soot concentration patterns in industrial gaseous hydrocarbon flames. For reasons outlined above, a turbulent plug flow reactor (PFR) has been used, with a pre-combustor generating a hot flue gas stream into which different hydrocarbons were injected, and providing well-defined but variable conditions of gas temperature, stoichiometry, and residence time for the soot formation process to be investigated.

Parallel with the plug flow reactor experiments, extensive field-distributed measurements have been made in free-burning axial-jet turbulent diffusion flames, in order to examine whether the correlation model deduced from the PFR-experiments applies to non-homogeneous combustion conditions as well. The final correlation was intended to allow the prediction of soot concentrations in gaseous fuel turbulent diffusion flames with an accuracy adequate for engineering applications.

3.2.1.4 Practical Relevance

Process quality and energy efficiency of high temperature processes, which in most cases are combustion heated systems, depend to a large extent on the overall heat transfer intensity and its spatial distribution along the flame(s) or the furnace. Since thermal radiation is by far the dominating heat transfer mechanism at high or very high temperatures, radiative properties of flames are very important. Non-luminous flame radiation, which is radiation emission from flame gas species (mainly CO_2, H_2O), may be considerably enhanced – even by more than one order of magnitude – with soot in the (now luminous) flame, and so are heat radiation fluxes [11, 14, 15]. An additional consequence of soot enhanced radiation is lowering the upper temperature level of a flame in its main reaction zone, thus reducing thermal NO_x emission (e. g. in metallurgic melting furnaces or glass tanks).

Another important aspect, related to soot, of industrial and engine combustion as well as of waste incineration is the ecological requirement to avoid even minor soot emissions in the flue gas stream; in cases where NO_x-reducing catalyser units are installed in the flue gas line, they may suffer losses of activity due to soot particle deposits onto their active surface.

In all those cases – whether soot in the flame offers advantages or would be of disadvantage, and when soot as a flue gas emission component must be avoided – a reliable soot prediction method would be an extremely valuable tool for practical burner, combustor and furnace design.

3.2.2 Experimental

3.2.2.1 Plug Flow Reactor Investigations

In the project A 6 of the Collaborative Research Centre 167 basic systematic measurements of soot growth rates from lower hydrocarbon gases (methane, propane, and acetylene) have been performed under well-defined, although variable conditions of mixture stoichiometry, temperature, and residence time using a plug flow reactor setup (Fig. 3.2.1) [16–18]. Hot flue gas, generated by natural gas/air combustion in a precombustor located at the lower end of the reactor system, entered the actual reactor tube (int. diam. 100 mm) at temperatures variable between 1100 °C and about 1600 °C. This temperature of the "primary mixture" was controlled by applying variable air preheating versus variable precombustor chamber cooling by means of water-charged cooling tubes, and thus could be shifted within the indicated range, independent of the primary mixture stoichiometry. The pre-combustor burner was designed to operate with air equivalence ratios between 0.5 and

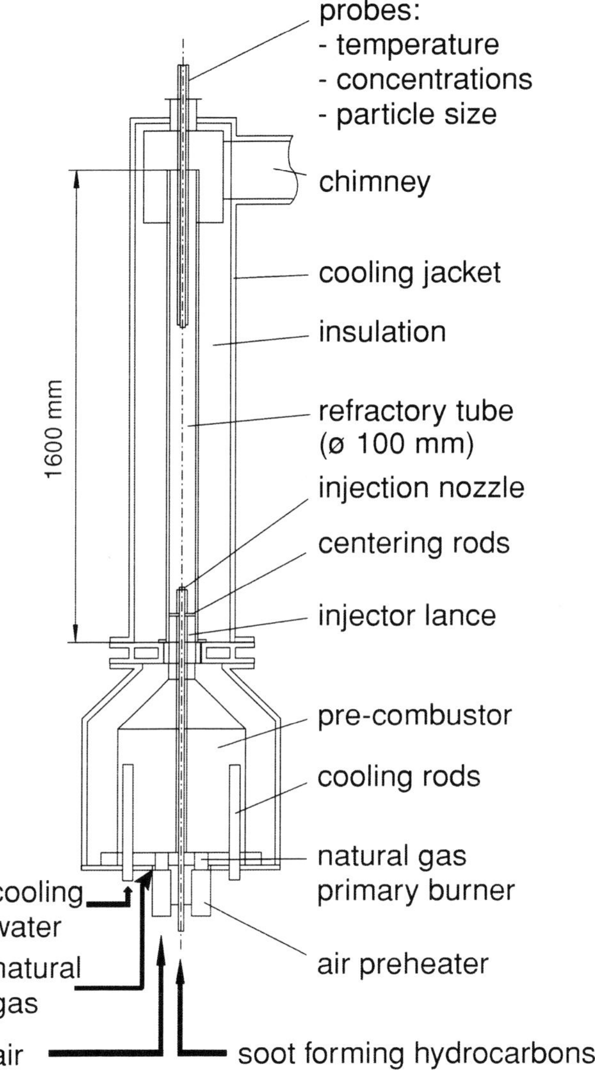

Figure 3.2.1: Plug flow reactor design.

2.0, thus rendering very different primary mixture stoichiometry conditions for the soot formation process in the subsequent plug flow reactor tube.

The soot-forming hydrocarbons, i. e. methane, propane, and acetylene, were injected with high velocity radially through a multihole nozzle (15 holes, 0.7 mm hole diam.) placed in the centre of the reactor tube near to its lower end. Due to extensive preliminary mixing investigations, the hydrocarbon-gas/primary-flue-gas mixing process has been optimised in order to ensure residence times less than 10 ms for achieving homogeneous mixture prior to main soot formation which was to be analysed. Soot formation conditions provided in this way obviously correspond to those occurring in turbulent diffusion flames within large turbulent eddies of high unmixedness, where soot particle growth (the dominant soot formation step) takes place by thermal hydrocarbon decomposition in hot oxygen-deficient eddy regions.

The initial hydrocarbon molar fraction in the injection section has been raised up to 6 % for methane, 3 % for propane, and 4 % for acetylene. The corresponding overall stoichiometry downstream of the injection covered an air equivalence ratio range from 0.4 till 0.9. Temperatures up to 1600 °C could be generated – and maintained fairly constant – along the reactor tube.

Using different sampling probes (Fig. 3.2.2) at various distances downstream of the injector tube, residence time resolved measurements of soot and gas species concentrations, soot particle size distributions (as described in [18, 19]), and gas temperatures were performed for residence times ranging from 10 to about 100 ms. Soot concentrations were determined by weighing soot deposits sampled inside a sintered bronze filter, which was incorporated into the tip of the sampling probe as shown in Fig. 3.2.2; the soot mass capacity of the bronze filter could be extended by adding a glass fibre filter. The soot mass collected during a certain sampling time was then related to the suction flow gas volume during the same time interval, in order to determine soot concentration data. Gas temperature was measured by thermocouple suction pyrometry. Gas species concentrations were determined by miniature probe gas sampling and subsequent physical gas analysis as well as gas chromatography [19].

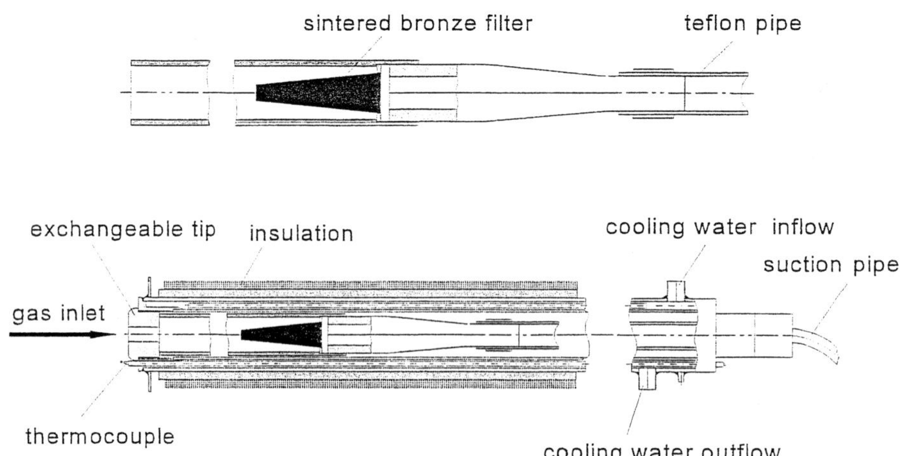

Figure 3.2.2: Gas species and soot sampling probe with sintered bronze soot filter.

3.2.2.2 Turbulent Diffusion Flame Investigations

Local measurements of gas and soot concentrations and gas temperature were performed also in several free-burning turbulent diffusion flames, with the same hydrocarbons as fuel gases like injected into the plug flow reactor tube [17, 19]. In addition, flow velocity fields, as required for analysing convective and turbulent diffusive transport rates, were determined by Prandtl-probe and LDV diagnostics.

Simple axial jet burners equipped with nozzle diameters 4, 10, and 15 mm were used, the flames being stabilised by small hydrogen or hydrogen/oxygen mixture streams through concentric annular thin slots at the nozzle rim. The program of investigation comprised measurements with methane and propane turbulent diffusion flames in a Reynolds number range between $Re = 25\,000$ and $Re = 150\,000$ defined at the nozzle exit.

Simultaneous determination of soot volume fraction and mean soot particle size by means of the 3-wavelength extinction ratio technique [19, 20] was carried out, as well as soot concentration measurement by the gravimetric technique described above. Satisfactory agreement of soot concentration data was achieved by assuming a suitable soot particle density (namely 1.1 g/cm^3) for converting mass concentrations to volume fractions. Figure 3.2.3 presents an example of comparing soot volume fractions along the axis of a propane diffusion flame from both methods; data from gravimetric sampling have been integrated over the corresponding flame cross section.

3.2.2.3 Experimental Parameters of the PFR Tests

From laminar premixed flame experiments it had been found [16, 21] that soot formation starts when the [C/O]-ratio in the flame feed gas exceeds a critical value $[C/O]_{crit}$, which is far below the value one deduced from thermodynamic equilibrium and depends strongly on temperature (Fig. 3.2.4).

Data obtained from the plug flow reactor strongly indicate that the influence of gas mixture stoichiometry on soot particle growth comprises two basically independent parameters: the air equivalence ratio of the precombustor flue gas approaching the injection nozzle of the secondary fuel gas (called "primary air equivalence ratio" $\lambda_I \geq 1$); and the air equivalence ratio after secondary fuel injection (called "total air equivalence ratio" $\lambda_g < 1$) which relates the total molar amount of oxygen introduced by the precombustor air feed to the amount of oxygen required for complete burn-off of the total (primary and secondary) fuel streams. By varying both these stoichiometry parameters over a wide range, a novel application-orientated interpretation of the soot formation processes has been achieved, as described and discussed later.

Figure 3.2.5 shows results from preliminary PFR experiments demonstrating that the soot formation limit, expressed as the critical minimum secondary fuel gas concentration for soot generation in the PFR reactor tube,

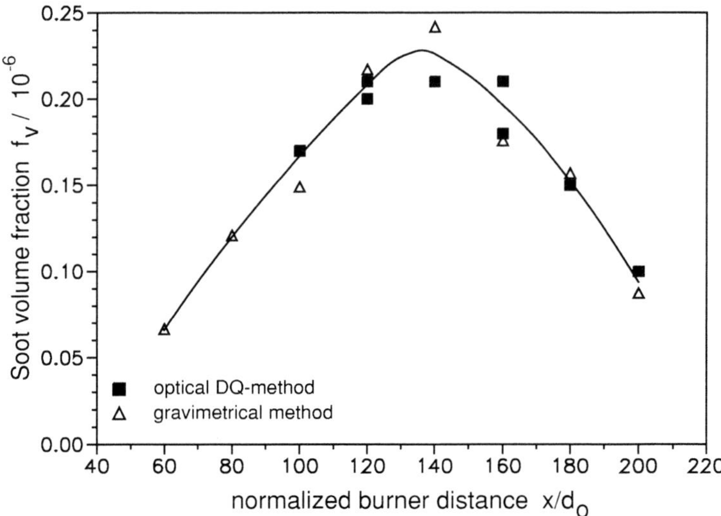

Figure 3.2.3: Soot volume fraction data measured at the axis of a free burning flame. Comparison between gravimetry and 3-wave-length extinction optical method (DQ-method).

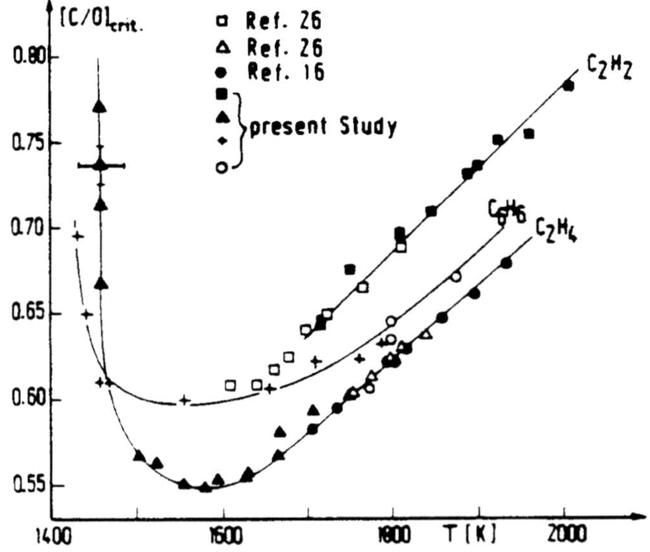

Figure 3.2.4: Critical [C/O]-ratio at the soot formation limit as a function of temperature [21].

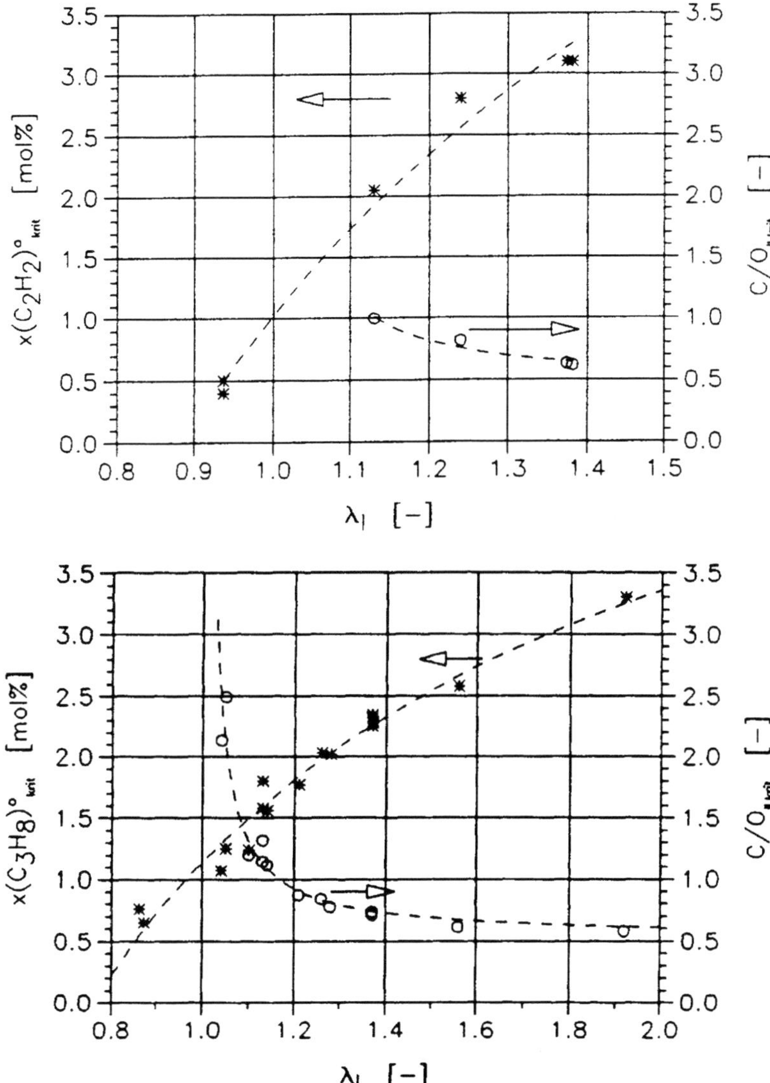

Figure 3.2.5: Critical fuel concentration in air and critical [C/O]-ratio at the soot formation limit versus the primary air equivalence ratio, for propane (lower diagram) and acetylene (upper diagram) reaction.

increases with the primary air equivalence ratio λ_I, as would be expected [16]. The corresponding $[C/O]_{crit}$ value, calculated on the basis of the molar carbon content in the secondary fuel and the free molar oxygen content in the primary flue gas at the reactor tube entrance, drops with increasing λ_I (i. e. soot formation is being enhanced), finally assuming values in the range 0.55 till 0.6 as known for premixed flames under favourable flame temperature conditions (see Fig. 3.2.4). These results prove that the primary air equivalence ratio is a main parameter for the PFR experiments.

3.2.3 Results

3.2.3.1 Experimental Results from the PFR

In a first step of presenting the measuring data, soot concentration results have been plotted versus gas residence time after secondary fuel injection, with the PFR temperature and the air equivalence ratios λ_I and λ_g as the essential process parameters. Figure 3.2.6 shows an example of those plots for soot formation from **methane** as secondary fuel. From the diagrams it is obvious that, for all conditions of stoichiometry and temperature, soot concentration increases with gas residence time along the reactor [13, 17].

Apart from a comparatively high primary air equivalence ratio ($\lambda_I = 1{,}14$) combined with high temperatures (range 1350 °C till 1450 °C) for which the opposite tendency was observed, soot formation was strongly enhanced by higher gas temperature. As regards the primary air equivalence ratio, values near to or slightly beyond stoichiometric (i. e. $\lambda_I = 1.0$ and 1.05 in the diagrams) seem to lead to maximum soot formation at the same gas temperature. Those tendencies can be seen even clearer from a different plot of the same data in Fig. 3.2.7.

The influence of the total air equivalence ration λ_g as the second main stoichiometry parameter is shown – again for methane – in Fig. 3.2.8 where, with increasing total air equivalence ratio in the substoichiometric range ($\lambda_g = 0.64$ till 0.83), a sharp decrease of soot concentration occurs down to zero. The curves apply for constant gas residence time ($t_v = 42$ ms), constant primary air equivalence ratio, and three different gas temperatures.

Similar dependencies of soot concentration as a function of residence time have been found for propane and even acetylene as soot generating secondary hydrocarbon fuels. Figure 3.2.9 presents an example for **propane**, and Fig. 3.2.10 for **acetylene**, both corresponding to the previous diagrams in Fig. 3.2.6 for methane. Figures 3.2.11 and 3.2.12 demonstrate the soot con-

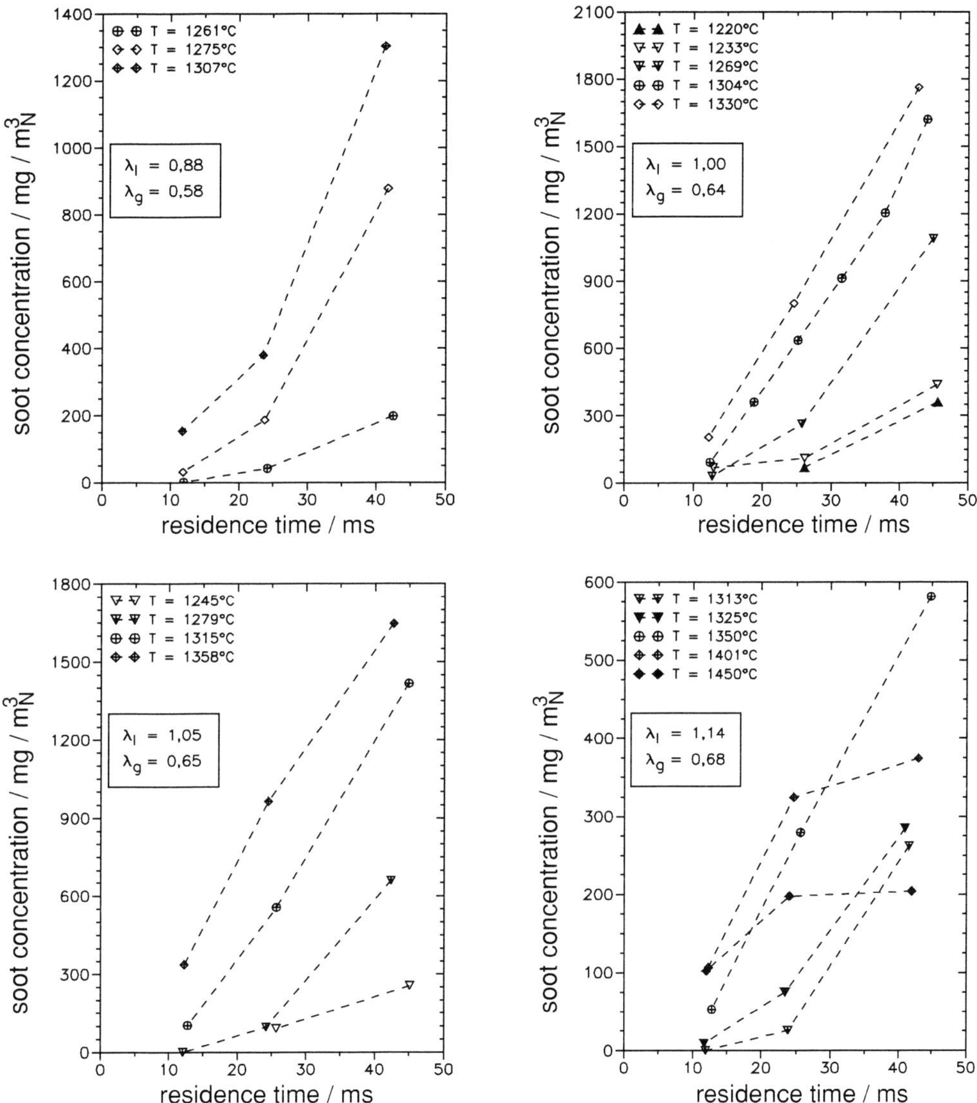

Figure 3.2.6: Soot concentration from methane reaction versus reactor gas residence time. Parameters: reactor gas temperature, primary and total air equivalence ratio.

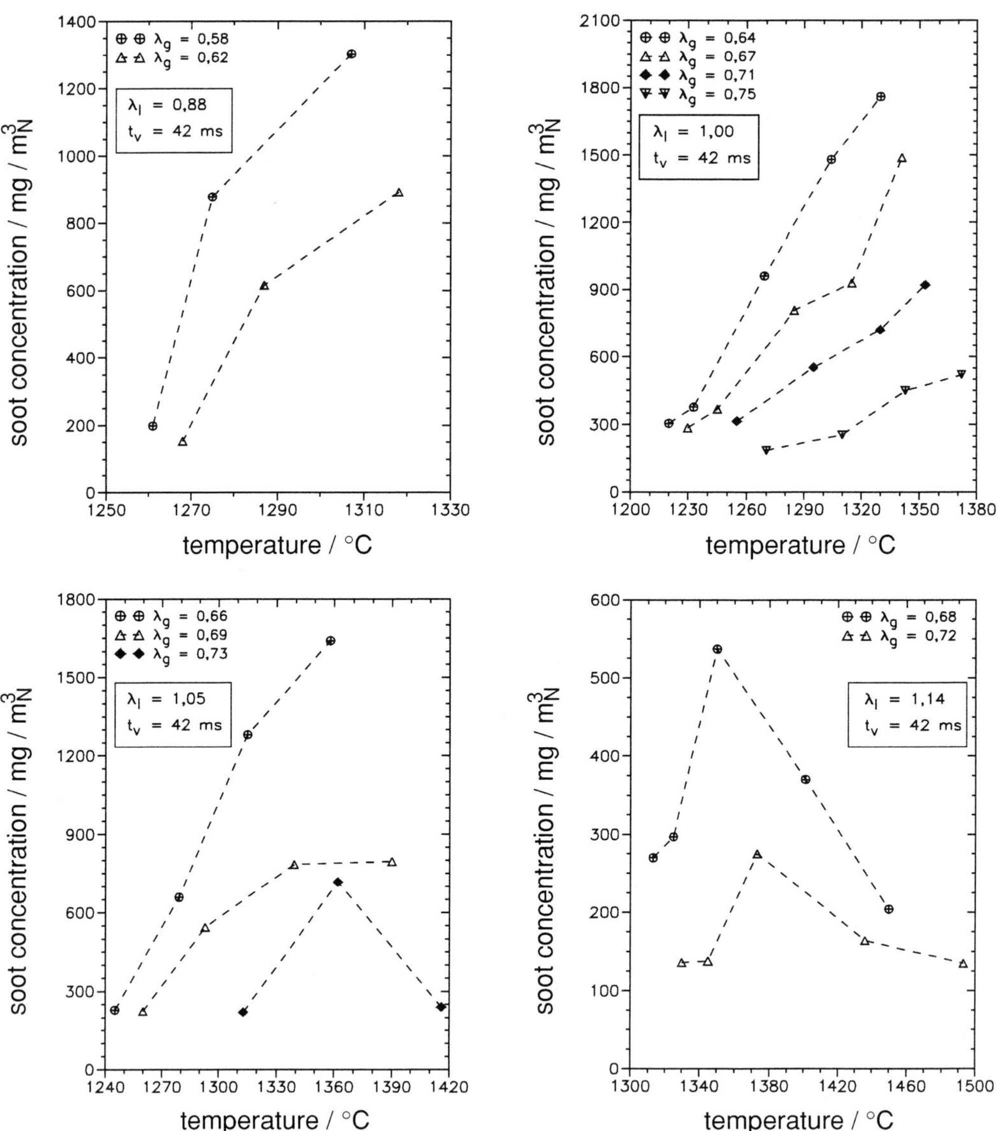

Figure 3.2.7: Soot concentration from methane reaction versus reactor gas temperature. Parameters: total and primary air equivalence ratio; constant: gas residence time.

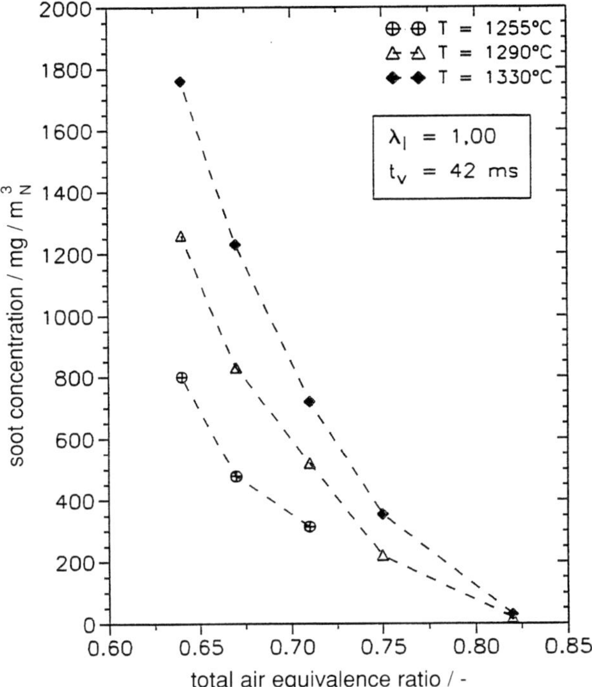

Figure 3.2.8: Soot concentration from methane as a function of the total air equivalence ratio. Parameter: reactor gas temperature; constant: primary air equivalence ratio and gas resicence time.

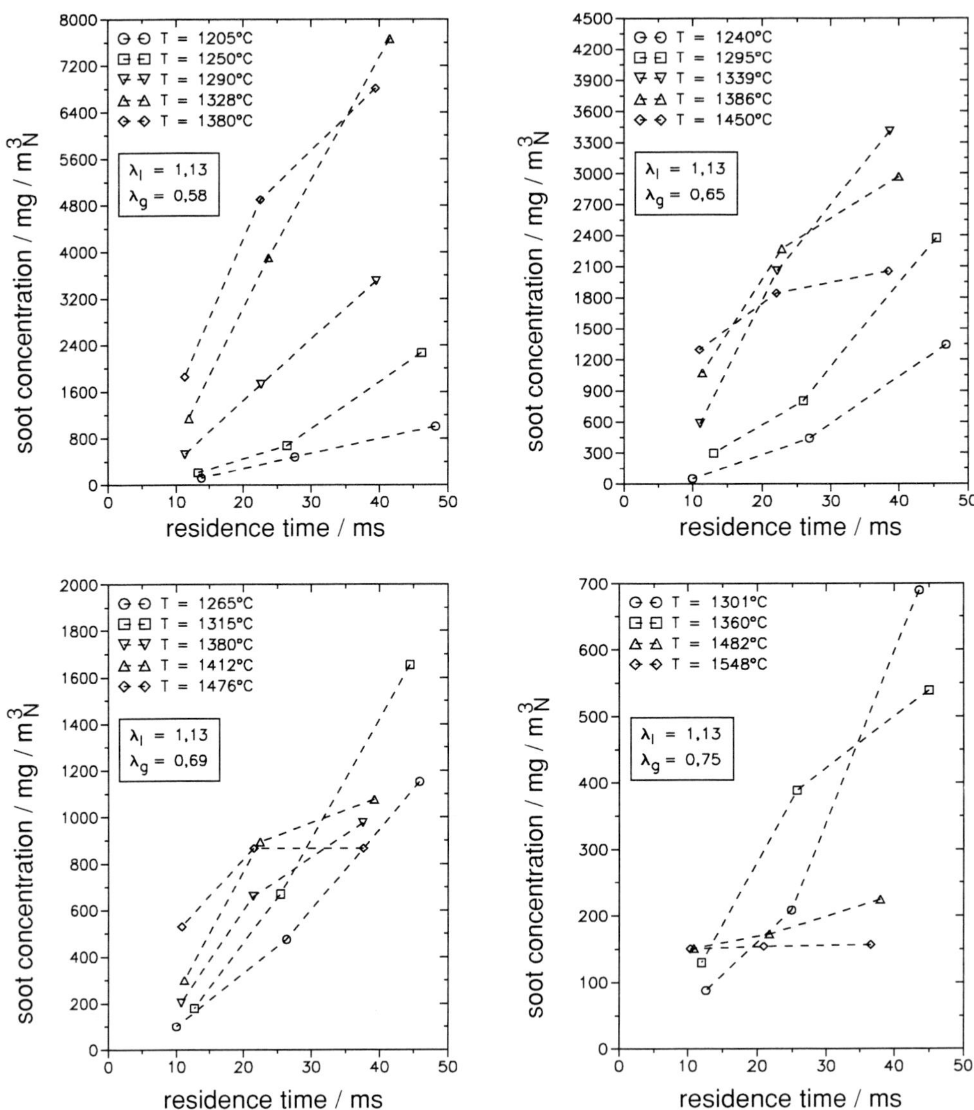

Figure 3.2.9: Soot concentration from propane reaction versus reactor gas residence time. Parameters: reactor gas temperature and total air equivalence ratio; constant: primary air equivalence ratio.

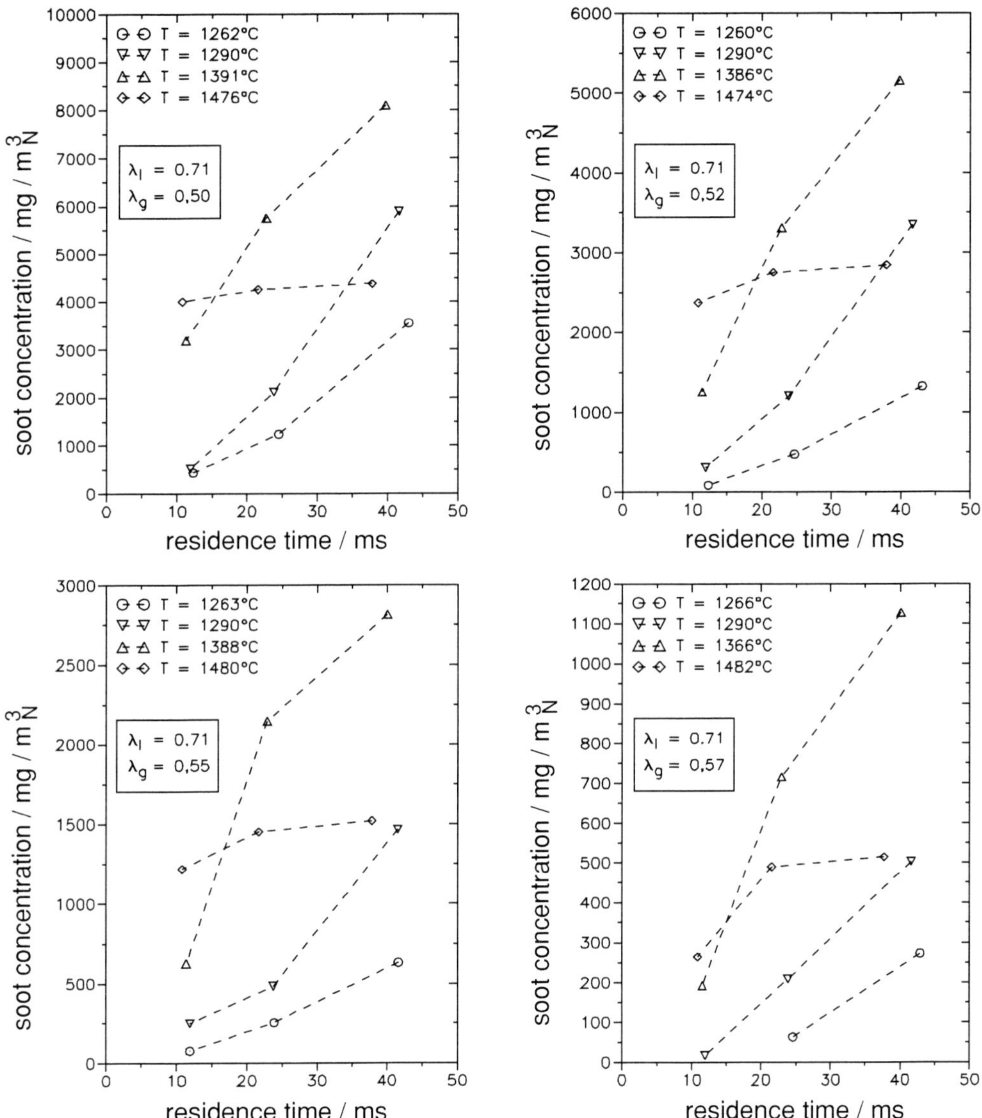

Figure 3.2.10: Soot concentration from acetylene reaction versus reactor gas residence time. Parameters: reactor gas temperature and total air equivalence ratio; constant: primary air equivalence ratio.

centration dependencies with regard to the total air equivalence ratio λ_g for propane and acetylene, respectively, corresponding to Fig. 3.2.8 for methane.

A very essential question refers to the gas phase level precursors of soot formation or, more precisely, of the soot particle growth by carbon condensation from particular gas phase hydrocarbons. Part of the fundamental literature assigns to acetylene a dominant soot precursor role in aliphatic hydrocarbon flames; other authors support different hypotheses [22]. In the present investigations acetylene and other C_2 and C_3 hydrocarbons have been measured in parallel with soot in order to contribute in clarifying this question.

Indeed, acetylene has been found with concentrations in the order of 1 vol-% along the reactor, the acetylene concentrations showing an obvious correlation with the soot formation rates. Apart from acetylene, ethylene was detected but in much lower concentration. Figure 3.2.13 presents an example of a constant acetylene concentration at about 0.9 vol-%, corresponding to also constant soot formation rates (i. e. linear increase of soot concentration) measured for **methane** as secondary fuel. In Fig. 3.2.14, again for methane reaction, data of acetylene and soot concentration are being compared, showing very clearly higher soot concentration levels correlated with higher acetylene concentrations (which themselves go along with higher reactor temperature). The constant temperature curves of both diagrams also exhibit enhanced soot formation (corresponding to progressively increasing soot concentrations) along the reactor, being correlated with increasing acetylene concentrations. It can be concluded, therefore, that acetylene is indeed the essential primary soot precursor for soot mass generation in the soot particle growth process when burning aliphatic hydrocarbons like methane or propane under substoichiometric oxygen conditions.

Using **acetylene** directly as a secondary fuel, basically the same parameter effects upon soot formation were observed. With acetylene, however, as may be expected, soot formation intensity is shifted towards higher values of the total air equivalence ratio, as is obvious from Fig. 3.2.12 compared to Figs. 3.2.8 and 3.2.11 and can be attributed to the lower H/C molar ratio of C_2H_2. There are also differences in the sense that with acetylene the saturation of soot particle growth due to progressing deactivation of the particle surface occurs faster, resulting in final soot concentrations not (much) higher than from propane (e. g. Figs. 3.2.11 and 3.2.12 to be compared with each other, or Fig. 3.2.9 in comparison with Fig. 3.2.10).

Investigations on **soot oxidation** have also been made, performing a limited number of experiments in the plug flow reactor setup. The diagram Fig. 3.2.15 gives an example of the effects of air equivalence ratio and temperature upon the decay of soot concentration versus residence time. Nearly irrespective of the temperature within the range above 1300 °C, the final asymptotic values of soot concentration were found to be correlated only with the total air equivalence ratio λ_g (Fig. 3.2.16). Complete soot burn-off was achieved above $\lambda_g = 1.10$ or 1.15. Based upon those results, an empirical

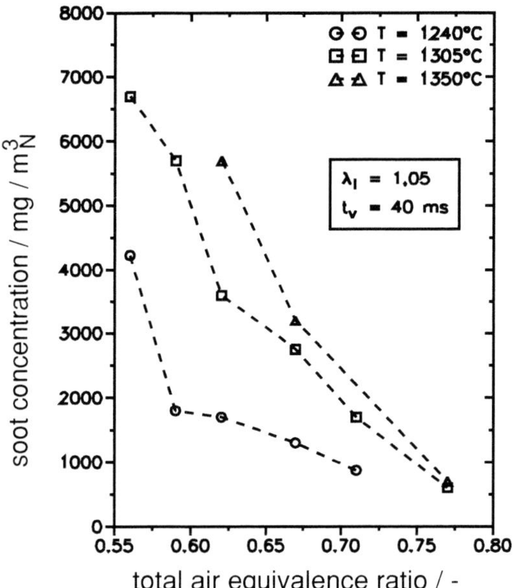

Figure 3.2.11: Soot concentration from propane reaction as a function of the total air equivalence ratio. Parameter: reactor gas temperature; constant: primary air equivalence ratio and gas residence time.

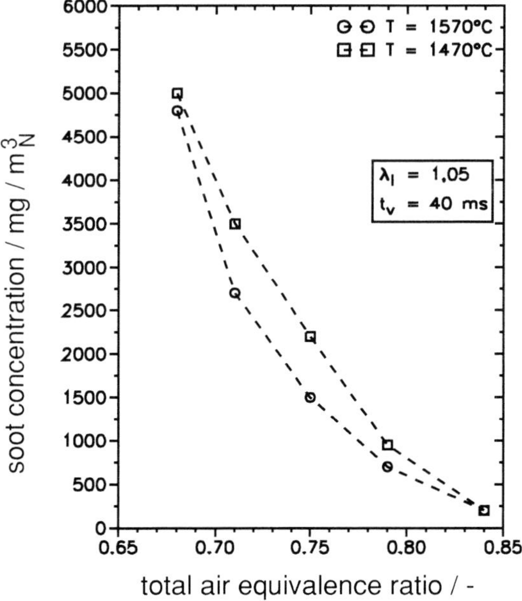

Figure 3.2.12: Soot concentration from acetyle reaction as a function of total air equivalence ratio. Parameter: reactor gas temperature; constant: primary air equivalence ratio and gas residence time.

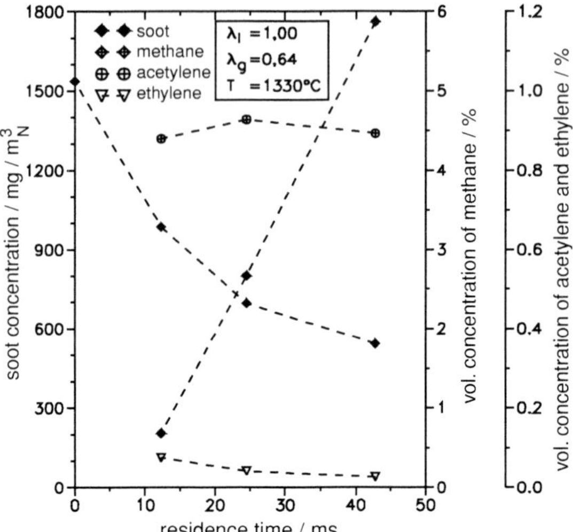

Figure 3.2.13: Intermediate hydrocarbon species and soot concentration from methane reaction as a function of reactor gas residence time.

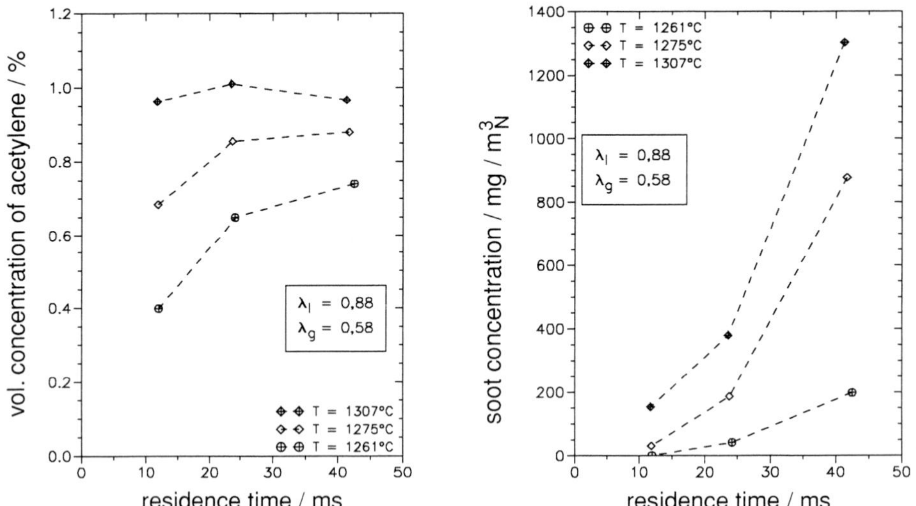

Figure 3.2.14: Comparison of acetylene and soot concentration versus gas residence time profiles for methane reaction. Parameter: reactor gas temperature; constant: primary and total air equivalence ratio.

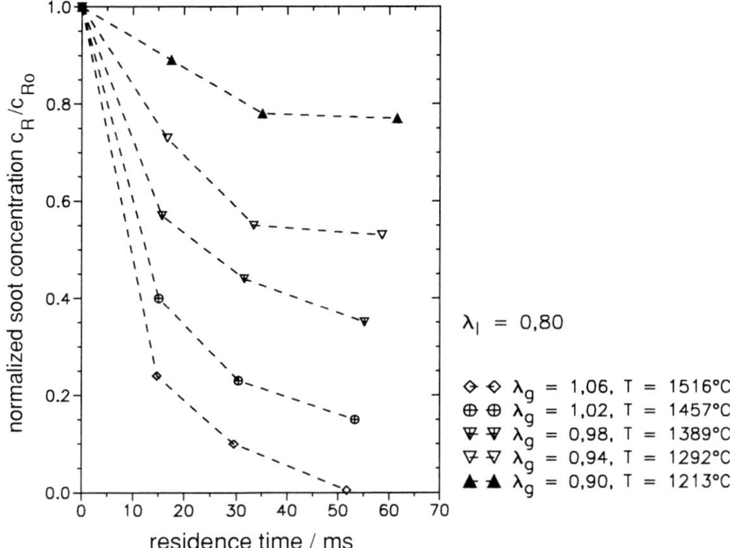

Figure 3.2.15: Decay of the normalised soot concentration by oxidation versus reactor gas residence time. Parameters: total air equivalence and reactor gas temperature; constant: primary air equivalence ratio in the soot generating precombustor.

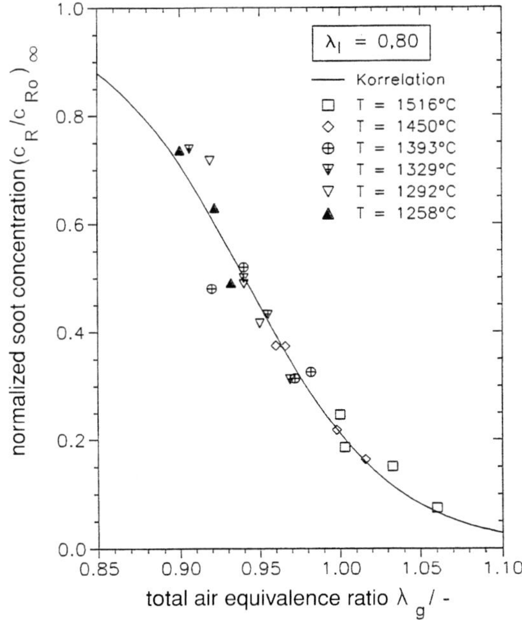

Figure 3.2.16: Decay of the final (infinite residence time) normalised soot concentration by oxidation versus total air equivalence ratio. Parameter: reactor gas temperature; constant: primary air equivalence ratio in the soot generating precombustor.

relationship has been established for the soot decay ratio after long oxidation time:

$$\frac{C_{R\infty}}{C_{Ro}} = \frac{b_1 \cdot \exp(-b_2 \cdot \lambda_g)}{1 + b_1 \cdot \exp(-b_2 \cdot \lambda_g)} \tag{1}$$

with parameters b_1, b_2 being dependent on the type of soot present, and in particular on whether the soot has been formed under very or only moderately rich mixture conditions; burning of soot from very rich mixture flame zones occurs comparatively faster, i. e. this type of soot has got higher degree of activation with respect to surface oxidation.

3.2.3.2 A Soot Formation Model Derived from the Experiments

From the experimental data it was concluded that for the soot particle growth process, which represents by far the major part of soot mass generation from hydrocarbon combustion, one may distinguish three domains characterised by the parameters temperature, total air equivalence ratio, and residence (soot formation) time:

1. Start-up domain with soot formation rates increasing with residence time (e. g. left-side upper diagram in Fig. 3.2.6); the "induction time delay" of soot formation may be considered negligible in this context.
2. Transition domain with constant soot formation rates, i. e. linear soot concentration increase; (e. g. curves for 1304 °C, 1330 °C in the right-side upper diagram and for 1315 °C, 1358 °C in the left-side lower diagram of Fig. 3.2.6); each of those constant soot formation rates represents also the maximum formation rate for the corresponding data of temperature and air equivalence ratio.
3. Saturation domain with soot formation rates decreasing with increasing residence time (e. g. curves for 1401 °C/1450 °C in the right-side lower diagram of Fig. 3.2.6; or curves 1474 °C till 1482 °C of Fig. 3.2.10). This saturation effect may be due to impoverishment of precursor species (acetylene), radicals (H-atoms) or/and reactive sites at the soot particle surface.

Based upon this knowledge it seemed justified to try to correlate all the data in a diagram of the type shown in Fig. 3.2.17, with the final soot concentration $C_{R\infty}$ (after infinite residence time) and the residence time value at the turning-point of the curve t_{wp} as reference parameters of both ordinates. The sigmoidal shaped curve corresponds to increasing soot formation rate in the short-time domain I, approximately constant rate in the intermediate domain II, and decreasing soot generation in the final domain III which leads asymptotically to the final (constant) soot concentration $C_{R\infty}$. The mathematical formulation of this general soot formation characteristic described

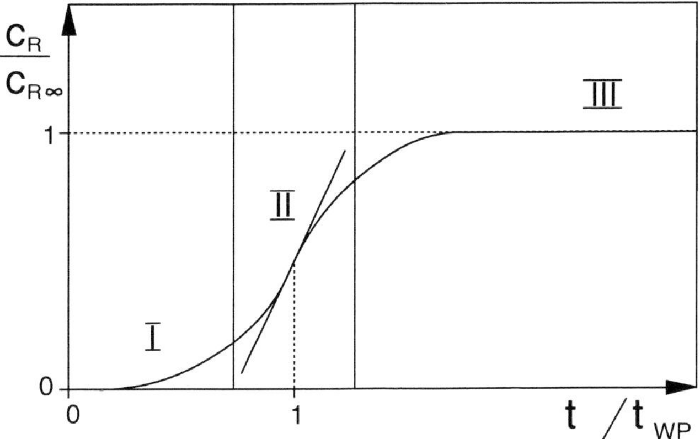

Figure 3.2.17: Schematic of the normalised soot concentration/reactor gas residence time profile.

by the domains I to III, which as a whole corresponds to the soot particle growth process by carbon condensation at the particle surface, follows the equation of an autocatalytic reaction [17]:

$$\frac{dC_r}{dt} = k \cdot C_A \cdot C_R \tag{2}$$

with the solution:

$$\frac{C_R - C_{RO}}{C_{R\infty}} = \frac{\exp\,(g \cdot t) - 1}{\exp\,(g \cdot t) + q} \tag{3}$$

wherein:

$C_A(t) = C_{AO} + C_{RO} - C_R\,(t)$
(acetylene concentration decreasing with time)

$g = \mathrm{K} \cdot C_{R\infty}$
(temperature-dependent kinetic exponential factor)

$q = \dfrac{C_{AO}}{C_{RO}}$
(start concentration ratio)

Assuming $C_{RO} << C_R$ and $C_{A\infty} = 0$,

so that $C_{AO} = C_{R\infty}$

the final equation of the sigmoidal curve reads:

$$\frac{C_R}{C_{R\infty}} = \frac{\exp(g \cdot t) - 1}{\exp(g \cdot t) + q} \tag{4}$$

This soot formation model equation has been applied to all measuring data obtained from the PFR test runs, by fitting (via regression analysis) the parameters $C_{R\infty}$, g, q to the $C_R(t)$-dependencies of the type shown in Figs. 3.2.6, 3.2.9, and 3.2.10 before. Furthermore, the final soot correlation $C_{R\infty}$ has been expressed – following previous investigations of Baumgaertner et al. [23] – as:

$$C_{R\infty} = \delta \cdot ([\mathrm{C/O}] - [\mathrm{C/O}]_{crit})^n \tag{5}$$

with the temperature-dependent pre-factor δ and the exponent n again having been determined from the total set of experimental $C_{R\infty}$-data.

Figure 3.2.18, as an example, presents a comparison between fitted curves from the equation above and the measuring points and shows good agreement. The kinetic exponential factor g could be well correlated with temperature, and was found independent of the kind of secondary fuel, as illustrated by Figs. 3.2.19 and 3.2.20 and represented by the function:

$$g = 2.04 \cdot 10^8 \cdot \mathrm{s}^{-1} \cdot \exp\left(-\frac{192\,\mathrm{kJ/mol}}{R \cdot T}\right) \tag{6}$$

The concentration ratio q turned out to be fairly constant at $q = 20$. The final (asymptotic) soot concentration values $C_{R\infty}$, as evaluated from the experimental data for propane as a secondary fuel, are plotted in Fig. 3.2.21 versus the [C/O]-ratio with the temperature as the second parameter. As to be expected, the diagram reveals a strong influence of the [C/O]-ratio. Furthermore, there is a distinct decrease of $C_{R\infty}$ with increasing temperature. Further evaluation with $[\mathrm{C/O}]_{crit} = 0.305$ then supplied plots like Fig. 3.2.22 for the factorial coefficient $\delta(T)$ in the empirical equation:

$$C_{R\infty} = \delta \cdot ([\mathrm{C/O}] - [\mathrm{C/O}]_{crit})^n \tag{7}$$

As regards the influence of the temperature level upon soot formation, the evaluation as a whole finally leads to the conclusion that increasing temperature accelerates the kinetics of soot particle growth, but also lowers the final soot concentrations obtained after long residence times. These opposite effects lead to the result that, for constant gas residence time (in the range 10 till 80 ms) and [C/O]-ratio, maximum soot yields occur for a gas temperature level of about 1350 °C, independent of the type of secondary fuel, as schematically illustrated in Fig. 3.2.23. The domains denoted by I, II and III correspond to those in Fig. 3.2.17.

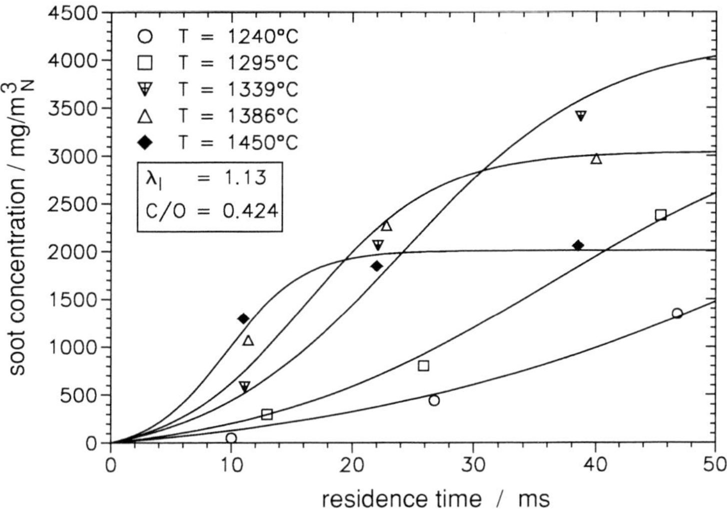

Figure 3.2.18: Soot concentration data from propane reaction versus reactor gas residence time. Comparison of measured data (symbols) with the profiles calculated from the model (curves) with the reactor gas temperature as a parameter.

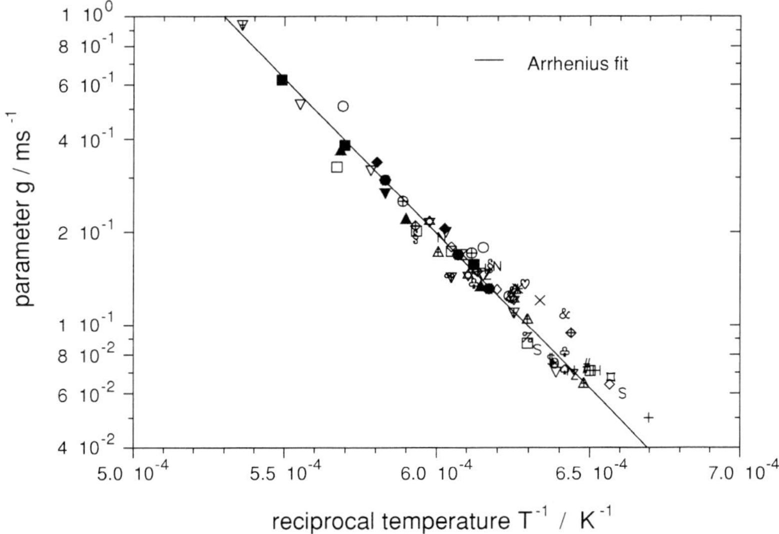

Figure 3.2.19: Exponential coefficient g as a function of the reciprocal reactor gas temperature. Comparison of all measured data from propane for different primary and total air equivalence ratios (symbols) with an Arrhenius equation (full line).

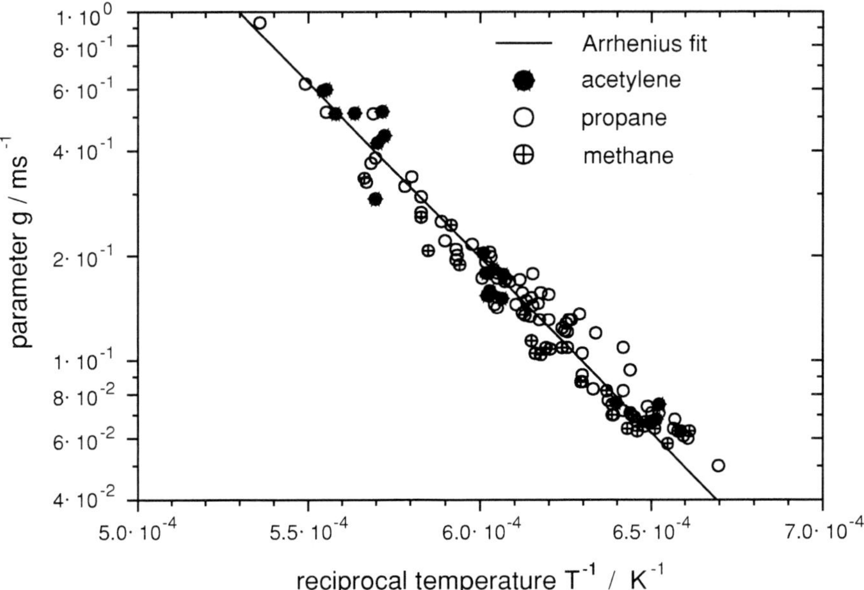

Figure 3.2.20: Exponential coefficient g as a function of the reciprocal gas temperature. Comparison of all measured data from methane, propane, and acetylene (symbols) with an Arrhenius equation.

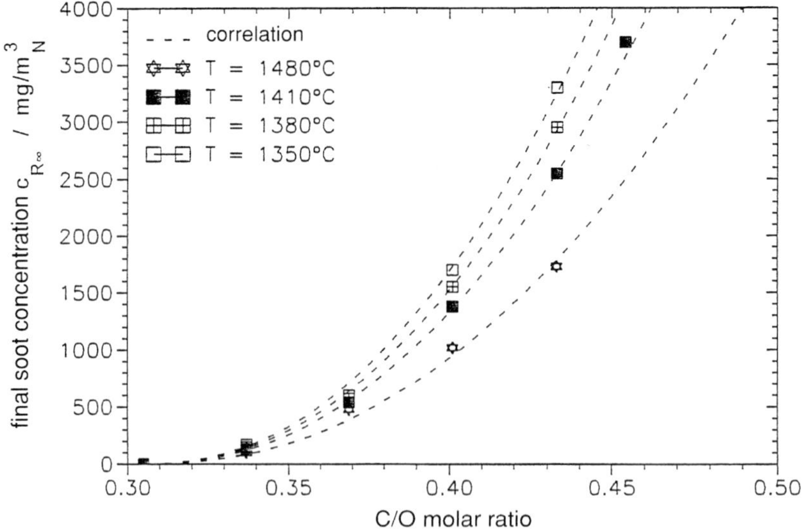

Figure 3.2.21: Final (infinite residence time) soot concentration from propane reaction versus the [C/O]-ratio according to the overall stoichiometry. Comparison of measured data (symbols) with the empirical correlation (n = 2,2; dotted curves) for different gas temperatures.

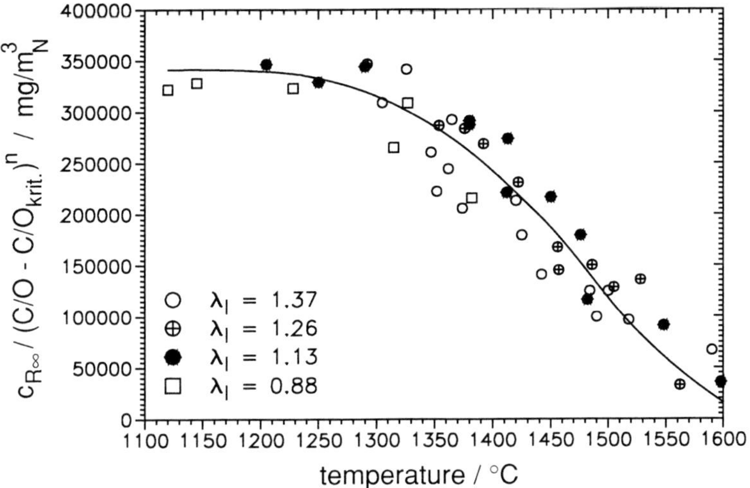

Figure 3.2.22: Reduced expression of the final soot concentration versus reactor gas temperature, with all measured data from propane reaction. Parameter: primary air equivalence ratio.

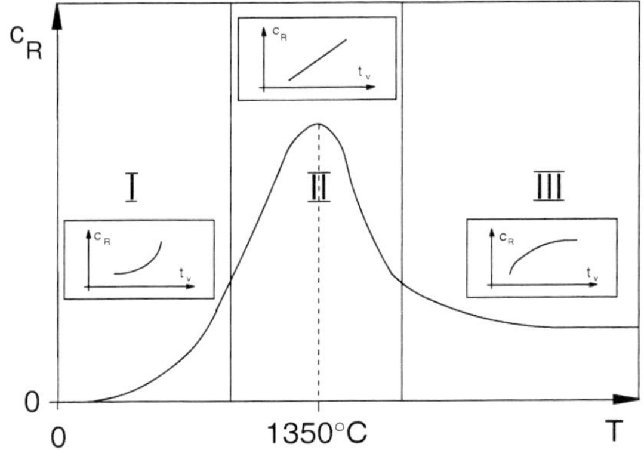

Figure 3.2.23: Schematic of the soot concentration generated during a fix residence time, as a function of reactor gas temperature, with soot formation domains I, II, and III.

3.2.3.3 Results from Turbulent Diffusion Flame Measurements

Unconfined axial-jet type turbulent diffusion flames have been measured in the range between 24 000 and 107 000 of the Reynolds number defined by burner nozzles exit data [17, 24]. At least those high upper values guarantee fully turbulent flow conditions all over the flame volume, in spite of the sharp increase of the kinematic gas viscosity and, hence, decrease of the local turbulent Reynolds number for elevated flame temperature. In most literature references, however, this condition is not sufficiently considered, so that the flames investigated there were not really turbulent and cannot be referred to for reliable data comparison. Fuel gases used for the present work were again methane (CH_4) and propane (C_3H_8). In the data evaluation, buoyancy effects had to be accounted for, in particular for the lower exit velocity flames.

Flow velocity, local equivalence ratio, and temperature distribution along and across the jet flames are the chief influencing parameters of soot formation and (in the flame tail zone) also soot oxidation. Amongst others, stable gas species like shown in Fig. 3.2.24 for a propane flame have been measured by gas sampling and subsequent physical analysis, which enable to evaluate the field distribution of the local stoichiometry and degree of reactedness as time-mean values. The formation of CO and H_2 as intermediate species is evident. Corresponding hydrocarbon species concentrations at the flame axis, again as functions of the dimensionless burner distance x/d_0 (d_0 = burner nozzle diameter), can be picked off from Fig. 3.2.25. Propane, as the fuel gas, drops with monotonic tendency, whereas methane, ethylene, and acetylene, being intermediate reaction products, appear and vanish towards the burn-out end of the flame like CO and H_2 in Fig. 3.2.24. Maximum concentrations of those species are quite high (between 2.5 and 3.6 % vol. at x/d_0 = 100 to 120). Of course, acetylene, as a soot precursor, amongst these intermediates is of particular importance in this context.

Fig. 3.2.26 compares measured soot concentrations along the flame axis for all **three propane flames**. With increasing Reynolds number, i. e. increasing flame jet velocities and, hence, shorter gas residence times for a given burner distance x, the soot concentration maxima shift downstream. Furthermore, the soot concentration maximum in the highest Reynolds number (highest flow velocity) flame amounts to only half of the maximum soot concentrations in the lower Reynolds number flames.

These findings are to be explained by the fact that soot formation (or more precisely: the soot particle growth process) is kinetically slow compared to the gas phase flame reactions. Figure 3.2.27 verifies that the acetylene concentration and temperature profiles of the three flames obviously cannot be responsible for their different soot formation intensity, although a moderate kinetic influence upon acetylene formation as well as decomposition becomes obvious from the curves. As a very clear statement it can be deduced from Figs. 3.2.26 and 3.2.27 that – for all three flames – the acetylene concentration maximum coincides with the x/d_0-range of the main soot concentration increase, i. e. the main soot formation region. As regards the

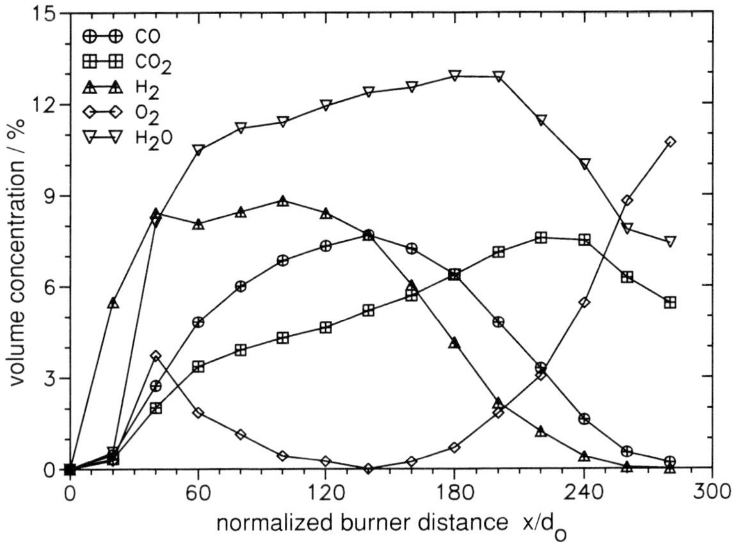

Figure 3.2.24: Stable inorganic gas species concentrations measured on the flame axis at various burner distances x for the propane flame Re = 60 000.

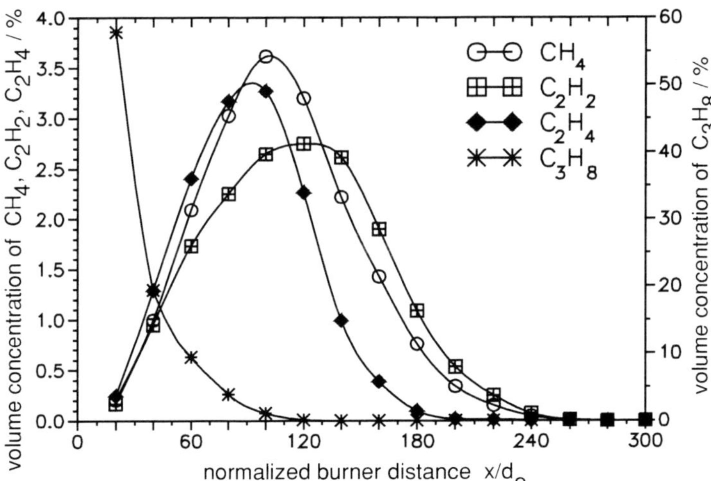

Figure 3.2.25: Propane and intermediate hydrocarbon species concentrations, measured at the flame axis, versus burner distance (propane flame Re = 60 000).

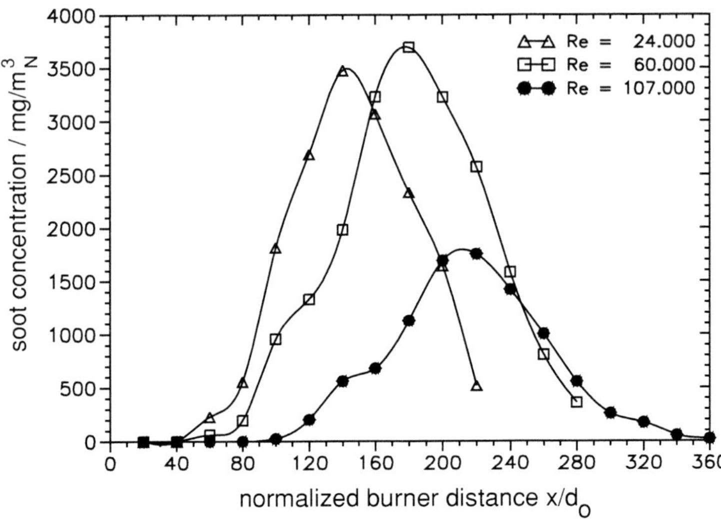

Figure 3.2.26: Soot concentrations measured at the flame axis, versus burner distance (propane flames with Re = 24 000, 60 000, and 107 000).

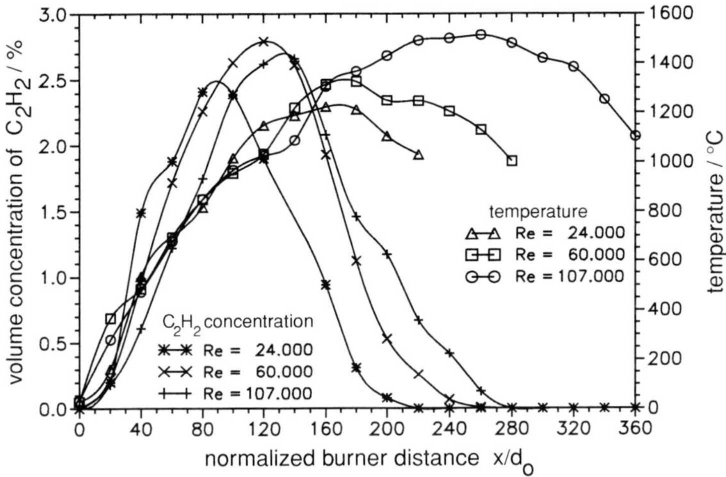

Figure 3.2.27: Acetylene concentrations and temperatures measured at the flame axis, versus burner distance (propane flames with Re = 24 000, 60 000, and 107 000).

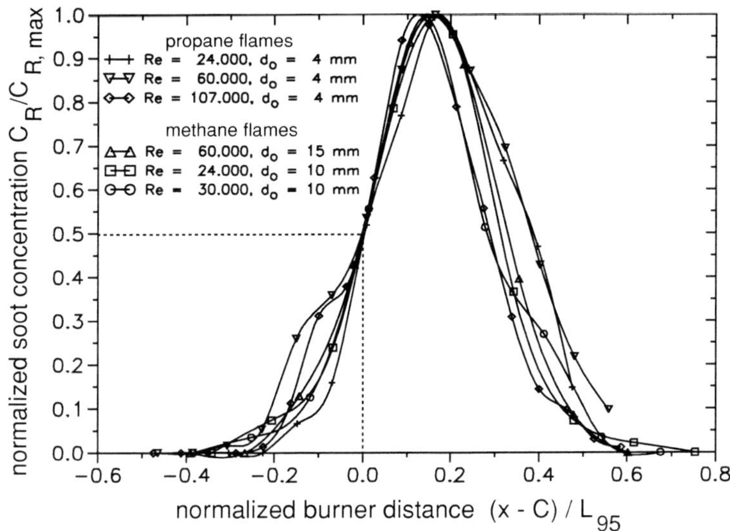

Figure 3.2.28: Measured axial profiles of soot concentration at the flame axis, plotted in reduced mode for comparison (three propane and three methane flames, with different Reynolds numbers).

temperature decline for larger burner distances, this is to be attributed to the thermal off-radiation from the flame, the radiation energy rates being relatively smaller for the higher velocity flames (Fig. 3.2.27). It should be mentioned that the maximum soot concentrations measured of about 3.5 g/m^3_n correspond to a secondary fuel carbon conversion to soot ("soot yield") of about 4.5 %.

Finally, Fig. 3.2.28 illustrates that the axial soot concentration profiles, in a reduced form, of all six propane and methane flames have a quite similar shape and yield a narrow scatter plot; reducing parameters for the *x*-scale are the special burner distance *C*, where half the maximum soot concentration is present, and the 95 %-burnout flame length. This similarity of profiles confirms the similarity of processes determining soot formation in gaseous hydrocarbon diffusion flames, independent of the particular type of fuel gas.

3.2.3.4 Modelling Soot Concentrations in Axial-Jet Type Turbulent Diffusion Flames

In the project A 6 of the Collaborative Research Centre 167, a much simplified model of soot formation and oxidation in unconfined jet-type turbulent diffusion flames, based upon the obtained experimental results and their phenomenological interpretation, has been developed. The basic local parameters of the flame field, as required for applying the soot formation and oxidation equation deduced from the PFR-experiments and described and discussed above, were the local air equivalence ratio or the mixing fraction, the local [C/O]-ratio, and the free molecular oxygen concentration, apart from the local flame temperature. Oxygen concentration and mixture fraction showed to be strongly correlated, as demonstrated by Fig. 3.2.29.

One problem of calculating local soot oxidation rates is the lack of precise information on the local soot particle size distribution or the specific surface area of the soot, respectively. Particle size is the result of soot particle growth and coagulation of primary soot particles [25], so that predicting soot particle size in flames would be very difficult. Instead, a start value of particle size in the flame under consideration has been estimated from experimental experience.

As regards the calculation of local soot formation rates, another problem arises, that is how to account kinetically for the local "soot age", i. e. the total preceding formation time of the soot present at a given position in the flame. Furthermore, since the formation time has to be weighted by the preceding temperature distribution (higher temperature accelerates aging for the same soot residence time), the problem becomes even more difficult. This was overcome by integration of the Arrhenius type function $g(T)$, introduced before as the kinetic parameter of soot formation, over the average flame jet residence time as a function of burner distance x. Turbulent transport (diffusion) processes have been neglected in this residence time estimation. They could be taken into account only by implementing the soot formation/oxidation laws into a complete turbulent gas phase reaction CFD-code, but this has not been realised in the present work.

In a first step [17], the present calculation has been restricted to the axis streamline of the propane flame with Re = 107 000, and has been based upon measured data of flow velocity, gas temperature, and gas species concentrations as input data for the soot formation and oxidation rate equations. Turbulent diffusion rates of soot in the axial and radial directions were taken into account in this evaluation. Figure 3.2.30 compares the soot formation rates predicted in this way with those evaluated from the measured soot concentration data. For the exponent $n = 2.0$ there is good agreement along the entire soot formation region, whereas soot oxidation rates obviously are being overpredicted, resulting in a too sharp soot concentration decrease between $x/d_0 = 220$ and 300.

Another realised modelling procedure [17, 24] consisted in applying a 2-dimensional k,ε-based CFD-code to calculate turbulent flow, mixing and reaction patterns of the propane flames, but again implementing the measured temperature distributions in order to overcome the difficulty of achieving good radiative heat release prediction in the presence of soot for closing the energy balance. Free oxygen concentrations were deduced from the mixture fraction values according to the correlation in Fig. 3.2.29. The soot formation and oxidation kinetics was implemented into the code, but with the

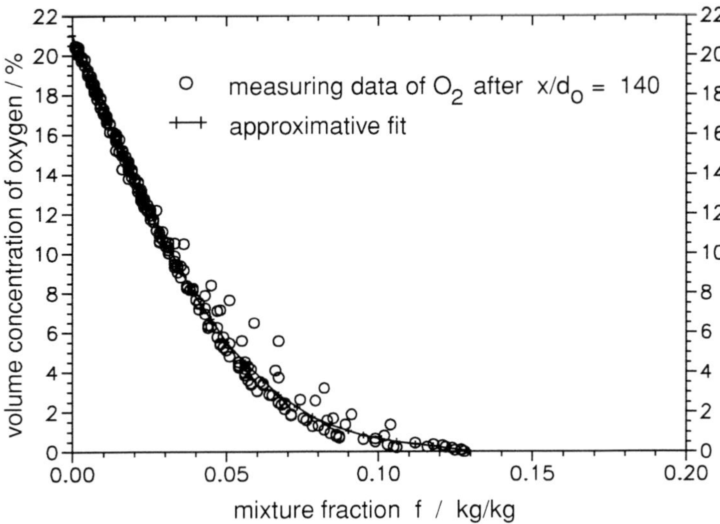

Figure 3.2.29: Molecular oxygen concentration versus mixture fraction f measured in the burnout region ($x/d_0 > 140$) of the propane flame Re = 60 000.

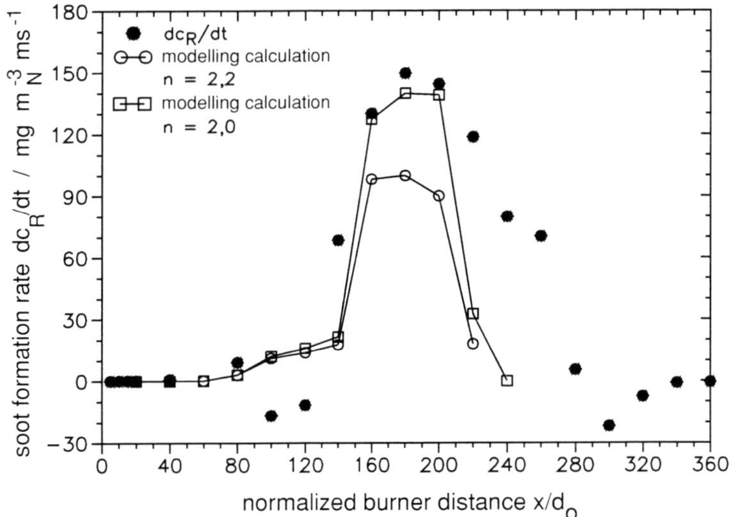

Figure 3.2.30: Soot formation rate at the axis of the propane flame Re = 107 000, evaluated from soot concentration measurements (full symbols) and from the modelling calculation (open symbols).

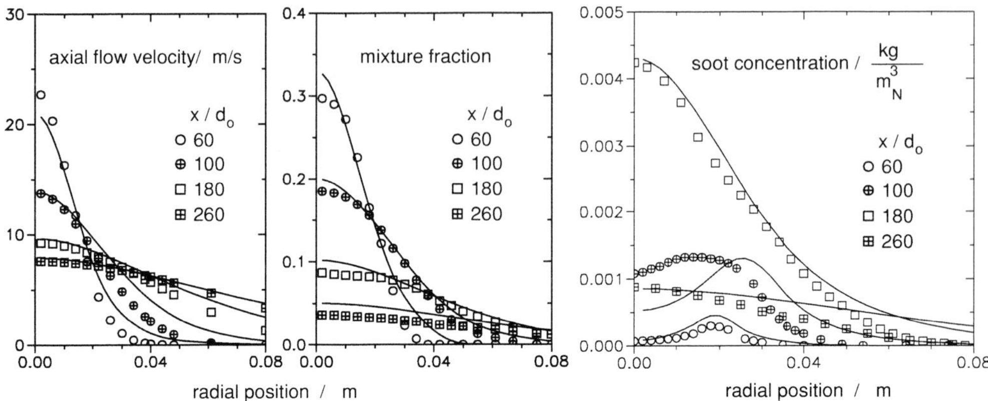

Figure 3.2.31: Radial profiles of axial velocity, mixture fraction, and soot concentration at four different burner distances in the propane flame Re = 60 000. Comparison of measured data (symbols) and modelled profiles (full line curves).

severe restriction of neglecting the soot aging effect in this calculation as described in [17]. Results of the calculation, plotted as four radial traverses of axial velocity, mixture fraction and soot concentration, are shown in Fig. 3.2.31. Considering the strong simplification applied for the soot formation mechanism, predicted soot concentration profiles are in fair agreement with the measured ones. Improvements are, of course, desirable, and also seem promising to be achievable on the basis of the complete soot formation model developed from the PFR-results and well-predicted flow and mixing patterns as shown in Fig. 3.2.31.

Beyond that, the major goal of future soot formation modelling for practical flames should be to include acetylene, as the chief precursor, into the gas phase reaction prediction and into the formulation of soot formation rates.

3.2.4 Conclusions

- The plug flow reactor (PFR) has proven to be a very effective experimental system for investigating soot formation under a great variety of flame conditions, ranging from nearly premixed to turbulent substoichiometric combustion by staging the fuel/oxygen stoichiometry.
- In nearly isothermal substoichiometric reaction as realised in the PFR, soot particle growth rates increase with residence time, adopt a maximum, and after this decrease asymptotically towards zero. At higher temperature, the maximum growth rate (i. e. the maximum soot concentration gradient) shifts to shorter reactor residence time.

- The temperature effect upon soot formation can be described in terms of an Arrhenius formulation. The activation energy turned out independent of the soot generating fuel gas to be 192 kJ/mol.
- Oxygen/fuel stochiometry (i. e. the excess of the actual [C/O]-ratio against the $[C/O]_{crit}$-ratio for the first appearance of soot traces) affects the final soot concentration appearing after very long residence time.
- The asymptotic decrease of the soot formation rate with long residence times is being interpreted as a deactivation effect of the active sites for carbon condensation at the particle surfaces. With gas temperature increasing from 1200 °C till 1600 °C, deactivation occurs faster and results in distinctly lower final soot concentrations. The existence of maximum final soot yields as a function of temperature, as reported from previous work, was not confirmed by the present investigations.
- Soot growth phenomenology could be described by the reaction equation of an autocatalytic reaction, super-imposed by a deactivation process which is dominated by the reaction temperature. The soot formation model deduced in this way has proven to be capable of predicting soot concentrations in turbulent diffusion flames. The comparison of predictions with extensive soot concentration field measurement data from free turbulent axial-jet type diffusion flames of propane and methane showed good agreement; this could be improved provided that an adequate calculation model of soot oxidation will be added to the soot growth model.

References

1. Haynes, B. S., Wagner, H. G. (1981): Soot Formation, *Progr. Energy Combust. Sci.* **7**, 229–273.
2. Frenklach, M., Wang, H. (1994): Detailed Mechanism and Modelling of Soot Particle Formation, Soot Formation in Combustion (H. Bockhorn, Ed.), Springer-Verlag, Berlin-Heidelberg, 165–192.
3. Appel, J., Bockhorn, H., Frenklach, M. (2000): Kinetic Modelling of Soot Formation with Detailed Chemistry and Physics: Laminar Premixed Flames of C_2-Hydrocarbons, *Combust. Flame* **121**, 122–136.
4. Jander, H., Wagner, H. G., Petereit, N. (1989): Rußbildung in Diffusionsflammen höherer Kohlenwasserstoffe, *VDI-Berichte*, **765**, 109–120.
5. Bockhorn, H., Schäfer, T. (1994): Growth of Soot Particles in Premixed Flames by Surface Reactions, Soot Formation in Combustion (H. Bockhorn, Ed.), Springer-Verlag, Berlin-Heidelberg, 253–274.
6. Bockhorn, H., Fetting, F. et.al. (1984): Investigation of the Surface Growth of Soot in Flat Low Pressure Hydrocarbon Oxygen Flames, 20th Sympos. (Int.) on Comb., The Combustion Institute, Pittsburgh, 979–988.
7. Garo, A., Said, R., Borghi, R. (1994): Model of Soot Formation – Coupling of Turbulence and Soot Chemistry, Soot Formation in Combustion (H. Bockhorn, Ed.), Springer-Verlag, Berlin-Heidelberg, 527–550.

8. Kronenburg, A., Bilger, R.W., Kent, J.H. (2000): Modelling Soot Formation in Turbulent Methane-Air Jet Diffusion Flames, *Combust. Flame* **121**, 24–40.
9. Sivathanu, K.R., Faeth, G.M. (1990): Soot Volume Fractions in the Overfire Region of Turbulent Diffusion Flames, *Combust. Flame* **81**, 133–149.
10. Desjardin, P.E., Frankel, S.H. (1999): Two-dimensional Large Eddy Simulation of Soot Formation in the Near-Field of a Strongly Radiating Nonpremixed Acetylene-Air Turbulent Jet Flame, *Combust. Flame* **119**, 121–132.
11. Gore, J.P., Faeth, G.M. (1986): Structure and Spectral Radiation Properties of Turbulent Ethylene/Air Diffusion Flames, 21th Sympos. (Int.) on Combustion, The Combustion Institute, Pittsburgh, 1115–1124.
12. Bartenbach, B., Huth, M., Leuckel, W.: Untersuchungen zum Rußwachstum unter den Bedingungen turbulenter Diffusionsflammen, SFB 167, Forschungsbericht 1990–1992, 153–179.
13. Bartenbach, B., Leuckel, W.: Untersuchungen zum Rußwachstum und Rußabbrand unter den Bedingungen industrieller Diffusionsflammen, SFB 167, Forschungsbericht 1993–1995, 87–109.
14. Heilos, A.: Spektrale Analyse der thermischen Strahlungswechselwirkung in Kohlenwasserstoffflammen, PhD Thesis, University of Karlsruhe (submitted).
15. Brookes, S.J., Moss, J.B. (1999): Measurements of Soot Production and Thermal Radiation from Confined Turbulent Jet Diffusion Flames of Methane, *Combust. Flame* **116**, 49–61.
16. Huth, M. (1992): Untersuchung der Rußbildung bei partieller Brennstoffoxidation und Pyrolyse, PhD Thesis, University of Karlsruhe.
17. Bartenbach, B. (1998): Untersuchungen zur Rußbildung und Rußoxidation unter den Bedingungen turbulenter Diffusionsflammen, PhD Thesis, University of Karlsruhe.
18. Huth, M., Leuckel, W. (1990): Experiments on Soot Formation from Propane under Partial Oxidation Conditions in a Turbulent Plug-Flow Reactor, 23rd Symp. (Int.) on Combustion, The Combustion Institute, Pittsburgh, 1493–1499.
19. Bartenbach, B., Huth, M., Leuckel, W. (1993): Investigation on Soot Mass Growth in a Plug Flow Isothermal Reactor Combined with Soot Concentration Field Measurements in Turbulent Diffusion Flames, Proc. Anglo-German Combustion Sympos., The British Section of the Combustion Institute, Cambridge, 979–985.
20. Lege, R., Bartenbach, B., Mueller, A., Leuckel, W. (1993): Application of a Multiple Wavelength Extinction Technique for Soot Particle Determination in Turbulent Diffusion Flames, Proc. Anglo-German Combustion Sympos., The Brit. Section of the Combustion Inst., Cambridge, 483–486.
21. Boehm, H., Hesse, D., Jander, H. et al. (1988): The Influence of Pressure and Temperature on Soot Formation in Premixed Flames, 22nd Symposium on (Int.) on Combustion, The Combustion Institute, Pittsburgh, 403–411.
22. Kent, J.H., Honnery, D.R. (1994): Soot Mass Growth in Laminar Diffusion Flames – Parametric Modelling, Soot Formaton in Combustion (H. Bockhorn, Ed.), Springer-Verlag, Berlin-Heidelberg, 199–220.
23. Baumgaertner, L., Jander, H., Wagner, H.G. (1983): Rußbildung in verschiedenen Brennstoff-Luft-Flammen, *Berichte Bunsengesellschaft f. Phys. Chemie* **87**, 1077–1080.
24. Bartenbach, B., Hirsch, C., Huth, M., Leuckel, W. (1994): Modelling und Validation of Soot Concentration Patterns of Turbulent Diffusion Flames Based on Data from Plug Flow Reactor Experiments, *Chem. Engng. Progr.* **33**, 401–408.
25. Huth, M., Leuckel, W. (1994): Soot formation from Hydrocarbon in a Plug Flow Reactor – Influence of Temperature, Soot Formation in Combustion (H. Bockhorn, Ed.), Springer-Verlag, Berlin-Heidelberg, 371–381.

4 Heat Transfer and Radiation

Convective and Radiative Heat Transfer in Combustors

Achmed Schulz*

Efficiency and power density of gas turbines have increased drastically during recent years. Today's gas turbines for power generation reach thermal efficiencies close to 40 % at net power outputs beyond 270 MW. Using gas turbines for aero engines thermal efficiencies clear beyond 50 % can be realized. The improvements are primarily a result of increased thermodynamic parameters like pressure ratio and turbine inlet temperature (combustor exit temperature). Both parameters have direct impact on the heat load and hence cooling of the hot gas ducting gas turbine components. The highest combustor exit temperatures of 2000 K and above are approximately 800 K beyond the highest allowable material temperatures, while compressor exit temperatures (coolant temperatures up to 950 K) continually approach the material temperatures due to increased pressure ratios. Besides increased temperature differences on the hot gas side and the decreased temperature differences available on the coolant side, the introduction of new low pollutant combustion concepts with their higher need of primary air represents a new challenge for efficient combustor liner cooling. Combustor liner cooling, therefore, is in conflict with the parameters:

- hot gas temperature
- flame temperature
- radiation
- coolant temperature
- available coolant mass flux
- maximum allowable material temperature

Besides efficiency and power, especially the reduction of pollutant emissions is in focus of ground based gas turbine development. During the last ten years the emissions of nitrogen oxide could be reduced by one order of magnitude. This was primarily achieved by modifying the combustion process. By

* Institut für Thermische Strömungsmaschinen, Universität Karlsruhe, Kaiserstr. 12, 76128 Karlsruhe, Germany

using only one or a few burners, very long and almost stoichiometric flames with high flame temperatures were produced in old silo combustion chambers. Short lean pre-mixed flames with considerably lower temperatures are generated in today's combustion chambers. Together with the combustion process, the geometry of the combustor changed from big silo combustors with low power density to small highly loaded annular combustors with many burners along the circumference. The thermal load to the combustor walls in case of older combustion chambers was determined by the different combustion zones. In the primary zone the wall heat load is caused by a very hot flame with high radiation and by convection of the combustion gases. The radiative heat exchange with the walls, however, is reduced due to the long distance between flame and wall and the absorption by the colder combustion gases in between. Nevertheless, these flame tube areas had to be cooled intensively by establishing thick cooling films parallel to the wall or by the use of heat resistant mineral bricks forming the liner. The high temperatures of the primary zone were reduced to turbine entry level by adding cold compressor air. In the mixing zone and in the transition zone to the turbine the flame tube temperatures could be kept at an acceptable level by applying pure convection cooling to the outer surface.

Introducing pre-mixed flames in silo combustors and higher turbine inlet temperatures caused an strongly increased demand of primary air. Simultaneously the amount of coolant for properly cooling the transition area and the turbine increased. Due to the excess air, resulting in low flame temperatures, mixing air injection could be avoided. In addition, the burner technology, which uses a higher number of burners, allows for shorter flames and hence compact combustors. The cooled flame tube surface could be reduced considerably. Although the flame temperatures are low, the flame tube has to be film cooled in the flame zone since flames are close to the wall with accordingly high radiative heat transfer.

The most recent versions of ground based gas turbines are equipped with annular combustors which have three times higher power density compared to silo combustors operated in the pre-mixed mode. Such annular combustors penetrate into power density areas typical for jet engine combustors. The pre-mixed flame takes up almost the entire flame tube. The flame tube cooling has to be adjusted to the high thermal load. To reduce the expenditure of cooling, in many cases an impingement cooling is applied to the outer surface of the flame tube. Before blown through slots between flame tube tiles or ejected through holes in the wall, the coolant absorbs the heat transferred through the wall. On the inner surface the coolant forms a protective film.

Whereas combustion chambers of ground based gas turbines can be operated under very lean conditions – the difference between flame temperature and turbine inlet temperature is not much higher than 100 K – the combustion chambers of aero engines have to be run closer to stoichiometric conditions. The reasons on the one hand are the considerably higher turbine inlet temperatures despite simultaneously increased combustor inlet temperatures and the higher requirements on flame stability. On the other hand it is necessary to add compressor air to the hot gases for establishing an adequate radial temperature profile at the exit of the combustor. Since the turbine inlet temperatures of aero engines increased much faster than the maximum allowable material temperatures during recent years the cooling methods had to be improved considerably. The major goal is an adequate cooling of the flame tube at the highest allowable temperature level. This requires the application and combination of different cooling methods at the various locations of the flame tube. Most common is a combination of convective cooling on the outer surface of the flame tube and film cooling on the hot surface. Latest designs use full coverage film cooling for the combustor liner. In areas where film cooling is poor, the convective heat transfer on the outer liner wall can be locally intensified by use of impingement cooling.

For the adjustment of the various cooling schemes to the local requirements an in-depth understanding of the thermal boundary conditions is imperative. Remarkable contributions towards a better understanding of the heat load of combustor liners through fundamental experimental and theoretical studies were accomplished by the Collaborative Research Centre 167 "High Intensity Combustors – Steady Isobaric Combustion" with its part B. An excerpt of the comprehensive work is given in the following three sections.

Section 4.1 concentrates on high efficient cooling concepts for low emission combustors. Starting from film cooling configurations with heat transfer enhancing measures on the cold surface and their influence on the total cooling efficiency the paper describes the interaction of a wall normal mixing air jet with an existing cooling film in great detail. Adiabatic film cooling effectiveness as well as heat transfer coefficients are given in terms of surface contour plots. The flow is visualized by Schlieren photography. A further focus of the paper is full coverage film cooling. Since heat fluxes are highly three-dimensional in full coverage film cooled surfaces due to the interaction of various cooling mechanisms, it is important to know their share on the total cooling effectiveness. The paper attempts to separate the various cooling contributions in utilizing wall materials with different thermal conductivities. The cooling effectivenesses are shown as contour plots and as laterally averaged distributions over surface length.

A numerical procedure for predicting conjugate heat transfer in generalized coordinates is demonstrated in Section 4.2. The discretized equations at the fluid-solid interface are obtained using energy conservation principles and computational nodes are placed exactly on that interface, thus yielding the corresponding temperature directly, without the need for inter- or extrapolating from neighboring nodes. The temperature field over both, the fluid and the solid domains is computed implicitly, i. e. without iterating between the two. The computer code used for the computations was developed within the scope of the centre of excellence and is based on the finite volume method and the SIMPLE algorithm along with various turbulence models.

The successful treatment of complex wall cooling schemes is demonstrated in the light of a full coverage film cooled combustor wall. Results are compared to measured data obtained from the study described in Section 4.1.

Section 4.3 comprises experimental investigations and theoretical simulations of radiative heat transfer in combustors. Numerical methods suitable for predicting radiative heat transfer in gas turbine combustors will be discussed and appropriate experimental validation techniques will be presented. Moreover, experimental techniques for characterizing the radiative properties of typical gas turbine combustor liner materials, including thermal barrier coatings will be elucidated. Concluding, the application of the experimental techniques as well as the numerical methods to the analysis of the radiative transfer of different model combustors will be illustrated.

4.1 High Efficient Cooling Concepts for Low Emission Combustors

Moritz Martiny, Ralf Schiele, Michael Gritsch, Achmed Schulz, and Soksik Kim*

4.1.1 Introduction

Improvements in thermal efficiency and performance of modern gas turbines require higher turbine inlet temperatures and higher compressor pressure ratios. In combustors, rising temperature levels in combination with higher coolant temperatures result in a significantly increased heat load on the walls. Furthermore, the reduction of pollutant emissions asks for new combustor concepts such as **L**ean-**P**remixed-**P**revaporised- or **R**ich-**Q**uench-**L**ean-combustion. Both methods need a larger amount of combustion air than conventional concepts. As a consequence, less air is available for cooling purposes. A reduction of pollutant emissions is therefore directly dependent on the development of high-end cooling technologies which provide very high cooling effectiveness with a minimum amount of coolant. The performance of cooling techniques, however, is strongly influenced by the complex flow field in the combustor and the effects of radiative and convective heat transfer (Fig. 4.1.1). Hence, these cooling techniques must provide high performance at any location on the combustor liner under strongly varying operating conditions. These boundary conditions make high efficient cooling techniques one of the key demands in modern combustor designs.

Various cooling concepts are currently being used to prevent the combustor liners from damage. In general, these techniques can be subdivided into convective cooling and film cooling. Convective cooling of the rear surface of the flame tube is the simplest method to reduce the wall temperature. For good effectiveness it requires intensive heat transfer on the cold wall. Two examples of convective cooling are sketched in Fig. 4.1.2. A simple method is to guide the coolant through a thin slot on the rear side of the flame tube to induce large air velocities resulting in high heat transfer coefficients (Fig. 4.1.2, left).

---

* Institut für Thermische Strömungsmaschinen, Universität Karlsruhe, Kaiserstr. 12, 76128 Karlsruhe, Germany

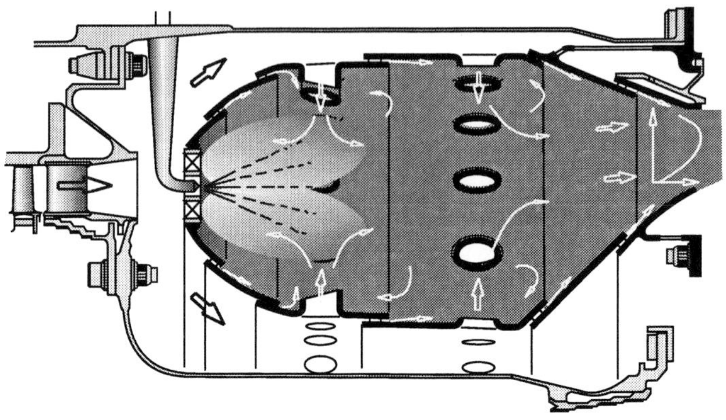

Figure 4.1.1: Primary and secondary flow in a combustor.

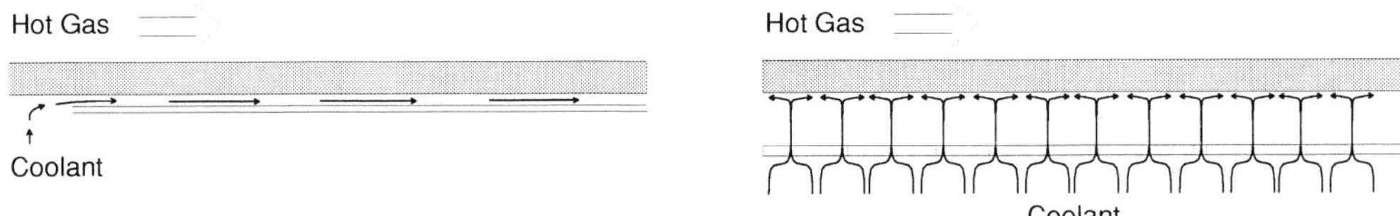

Figure 4.1.2: Convective cooling using a slot geometry (left) and impingement (right).

Additionally the slot surfaces may be equipped with an array of pins or ribs to increase the flow turbulence and the heat exchanging area. Impingement cooling (Fig. 4.1.2, right) is another concept to obtain high heat transfer rates on the outer liner. The high momentum jets are deflected parallel to the wall and very thin boundary layers in the impingement area induce high heat transfer coefficients. The concepts described can be of varying effectiveness depending on design and flow conditions in the coolant passage. Since the thermal conductivity of the wall material is comparatively low and the heat transfer on the inner combustor wall surface is both, finite and greater than zero, the maximum effectiveness of convective cooling will be limited to a certain value.

A more effective method to lower the wall temperature is to establish a coolant film on the inner combustor surface (Fig. 4.1.3), which considerably reduces the heat transfer between the combustion gas and the liner. Further downstream the cooling film will be diluted by mixing with the main flow. Accordingly, new films must be generated after a certain surface distance in order to guarantee an effective protection.

The cooling film may be created by blowing the coolant through a parallel or inclined slot, through a large number of inclined holes, or through a porous wall material (transpiration cooling). The temperature reduction in using film cooling can be considerably higher than by convective cooling.

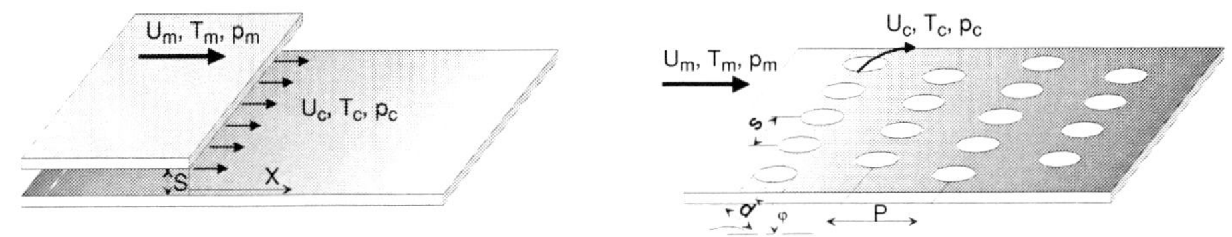

Figure 4.1.3: Film cooling on the inner combustor wall; slot (left), full coverage (right).

Furthermore, the heat flux transferred to the wall is largely reduced so that the net heat load on the combustor wall is smaller when applying film cooling. On the other hand, the performance of film cooling is highly dependent on the flow conditions in the combustor.

Therefore, the combination of convective cooling on the outer liner wall and effective film cooling on the inner surface appears to be the most promising concept to reduce the required amount of coolant.

From the comprehensive studies performed within the Collaborative Research Centre 167 in terms of heat transfer and cooling techniques, three topics which describe the key problems in combustor wall cooling will be discussed in detail. First, the convective cooling performance of different geometric arrangements will be compared by showing their relative performance in terms of cooling effectiveness. The second topic will present the behaviour of a cooling film interacting with a mixing jet blown perpendicular to the wall into the combustor. The experimental study reveals that the interaction strongly affects the performance of film cooling. Finally, as the state of the art in combustor cooling, the last subject will present an experimental analysis of the cooling performance of effusion cooling in combination with impingement on the back side. Tests under actual combustor conditions revealed a reduction in cooling air consumption of at least 30 % as compared to a conventional slot cooling geometry.

4.1.2 Combining Film-Cooling with Convective Cooling Schemes

Due to recent efforts on effectively controlling the combustion process in combustion chambers, only little air is left for cooling purposes. Although strong efforts have been made to improve film cooling schemes [1, 2], purely film cooling the combustor liner is not an efficient use of coolant. It is rather necessary to use the cooling potential of the coolant air before it is injected into the hot gas flow. These cooling concepts include effective cooling schemes for the back side of the combustor liner such as pin fin arrays, e. g. [3], ribbed channels, e. g. [4], or impingement cooling, e. g. [5].

As a part of the study on improving cooling techniques for combustor liners, the effect of combining film-cooling with additional convective cooling of the back side of the liner was investigated. Considerable efforts have been undertaken to provide an empirical data base for developing and/or evaluating the predictive capability of combustor liner models.

In the first set of experiments, coolant air was directly injected through a tangential slot into the hot gas boundary layer without convectively cooling the back side of the test plate. These data serve as a base case. In a second set of experiments, coolant air passes the backside of the test plate before being injected into the hot gas. This measure leads to an enhanced heat transfer on the back side of the plate and therefore reduces the temperatures on the inner surface. Several geometric configurations of the backside of the test plate have been tested to evaluate the effect of additional convective cooling on the overall performance of the cooling system. To give a good impression of the relative performance of the different cooling configurations, the data are presented as a function of the relative coolant mass flow $\dot{m}_C/\dot{m}_m$. ('c' for coolant and 'm' for main flow). Within the system used, this ratio could be varied up to values of approximately 45 % so that all conditions of practical importance were covered.

First, a simple smooth channel was used to provide a cross flow on the back side of the liner model. Figure 4.1.4 (left) nicely demonstrates the increase of the averaged film-cooling effectiveness η on the inner surface of the test model with growing coolant mass flow ratio. Decreasing the channel height leads to improved cooling effectiveness due to higher channel flow velocities augmenting the heat transfer on the back side.

Second, longitudinal ribs were introduced into the coolant channel. While the rib height was kept constant at $z = 2$ mm, two rib widths of $s = 4$ mm and 6 mm were tested (Fig. 4.1.4, right). Compared to the smooth channel results, the ribs further increase the overall cooling performance due to an increase of the heat transferring area. The differences between the two rib geometries were found to be rather small, particularly at small blowing rates.

Third, arrays of pin fins were used to increase the heat transferring area and to enhance the turbulent mixing in the coolant channel. Two configurations with narrow and wide spacing of the pin fins were investigated (Fig. 4.1.5, left). The height of the fins was kept constant at $z = 2$ mm. The increase of the overall performance was about comparable to that of the ribbed channel geometry. The spacing of the pin fins had only a minor effect on the overall cooling performance.

Last, the effect of increasing the heat transfer on the plate's back side by means of an impingement cooling scheme was investigated (Fig. 4.1.5, right). Different impingement cooling hole sizes were tested ($d = 1.7$ mm, 2.5 mm, 5 mm, 10 mm). The performance of the large holes was rather poor since the temperature distribution on the inner liner surface was inhomogeneous and temperatures were at a high level. These holes could not compete with

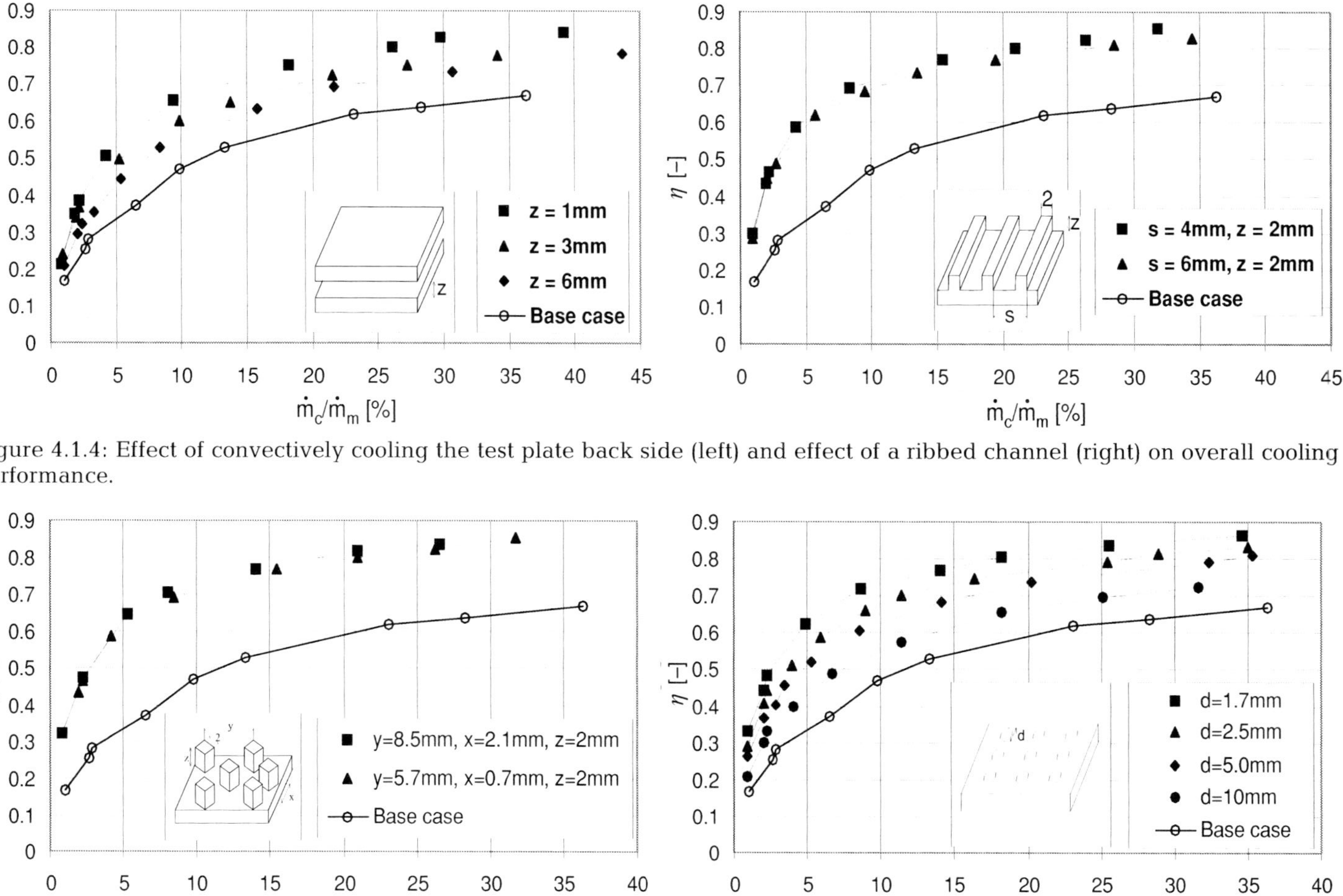

Figure 4.1.4: Effect of convectively cooling the test plate back side (left) and effect of a ribbed channel (right) on overall cooling performance.

Figure 4.1.5: Effect of pin fins in the coolant channel (left) and impingement cooling of the test plate's back side (right).

the ribs or pin fins. For the small holes, however, improved cooling performance was found.

Summarizing the results of the investigation, the ribbed coolant channel, the impingement cooling with small holes, and the wide spaced pin fin array were able to provide the most significant temperature reduction on the inner surface of the combustor liner. For a given temperature reduction, i. e. a constant cooling effectiveness, on the other hand, cooling air mass flows may be largely reduced. An η of 0.6 may for instance be achieved with one fourth of the amount of cooling air required for simple film cooling when the above described techniques are applied (see Figs. 4.1.4 and 4.1.5).

The cooling schemes presented, therefore, offer the potential to save large amounts of cooling air as compared to a purely film cooled combustor liner. This would result in an increase of gas turbine performance and a reduction of emissions because of better combustion.

4.1.3 Interaction of a Cooling Film with a Mixing Jet

As previously mentioned, flow conditions in combustors exhibit a highly variable three-dimensional structure. To induce strong mixing, secondary air is blown into the flame tube by forming high momentum jets directed perpendicular to the wall surfaces (see e. g. [6, 7]). By interacting with the cooling film and by influencing the overall flow field within the combustor, the mixing jets strongly affect the performance of the film (Fig. 4.1.1). Under engine conditions large areas of high surface temperatures extending downstream of the mixing jet ejection have been observed showing a deterioration of film protection.

A previous study within the Collaborative Research Centre 167 had already demonstrated the increase in heat transfer behind a two-dimensional jet entering a cross flow [8–10]. No cooling film was present in the flow situation considered at that time. While adding a cooling film on the surface of the combustor liner the aim of the study presented was to determine the reasons for the weak cooling protection downstream of a cylindrical jet ejection and to describe the main cooling parameters in the area of concern. To enable detailed investigations, a scaled-up test model consisting of a flat wall with a slot injected cooling film and a mixing jet was built. Using a laser light sheet the overall flow field was then visualized and qualitatively described. In a second step the flow was quantitatively investigated by applying Laser Doppler Velocimetry. For design purposes it is most important to know the distributions of the adiabatic effectiveness and the heat transfer coefficient on the wall. These distributions were determined in a third step by using heater foils and infrared thermometry. The data obtained complement one another and draw a comprehensive picture of the flow and heat transfer in the area under consideration.

4.1.3.1 Experimental Facility

The tests were conducted in an atmospheric wind tunnel. A detailed description of the test rig is given in [11]. The test section itself is shown in Fig. 4.1.6. The channel had a height of 105 mm and was 220 mm wide. The side walls were equipped with large windows for easy optical access. The channel top wall included a sapphire glass window for infrared thermography and laser light sheet investigations. The test plate was 12 mm thick and 550 mm long. The slot lip for tangential cooling film injection had a thickness of 2 mm and the slot measured 5 mm in height. The mixing hole diameter was fixed at 10 mm to avoid jet interaction with the test section ceiling. The jet injection was located on the channel centreline 50 mm downstream of the slot exit.

Based on experience gained in a previous study within the Collaborative Research Centre 167 (see [12, 13]) an infrared camera was employed for wall temperature measurements using a new highly sophisticated calibration technique reported in [14]. For the flow velocity measurements a two-component fibre-optic LDV-system was employed. The heat transfer analysis was carried out using the heater foil technique. The heater foil material was rubber with low thermal conductivity and temperature resistance up to 150 °C.

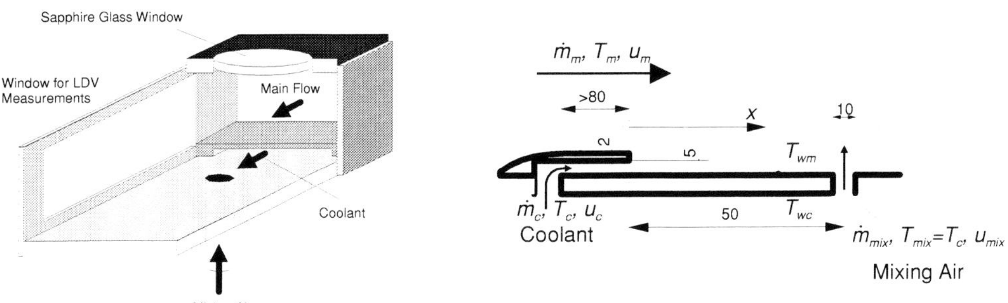

Figure 4.1.6: Test section and geometry of the test plate.

4.1.3.2 Results

The experiments were performed for three different operating conditions. The main flow Reynolds number (based on slot height) was 5800 for all test conditions which is representative for combustor flows. The different combinations of film cooling blowing ratio and jet momentum flux ratio are summarized in Tab. 4.1.1.

Table 4.1.1: Operating conditions.

	Re	$M = \dfrac{\rho_c \cdot u_c}{\rho_m \cdot u_m}$	$I = \dfrac{\rho_{mix} \cdot u_{mix}^2}{\rho_m \cdot u_m^2}$
A	5800	1	10
B	5800	1	7
C	5800	2	7

4.1.3.3 Flow Visualization

Figure 4.1.7 shows in its left picture a schematic view of the experimental set-up used for flow visualization. A laser optic was fixed above the channel top window and slightly inclined in streamwise direction to form a light sheet along the centreline of the channel starting from a location 10 mm upstream of the mixing jet entrance. Leaving the illuminated areas open, the hatching in the right pictures of Fig. 4.1.7 gives an indication of the extention of the laser light sheet as it was observed through the side walls of the test section. To visualize the flow either the coolant or the jet flow was seeded using oil droplets which generate pronounced light scattering as indicated through the dark regions in the right pictures of Fig. 4.1.7. To give an impression of the geometries involved the cooling slot is shown at its right vertical position in the pictures. The horizontal distance between slot and jet, however, is *not* correct to scale.

For the pictures in the upper left ($M = 1$, $I = 10$), upper right ($M = 1$, $I = 7$) and lower left ($M = 2$, $I = 7$) of Fig. 4.1.7 the seeding was introduced into the jet flow. It can be seen that the deflection of the jet is smallest for $M = 1$ combined with $I = 10$ and highest for $M = 2$ with $I = 7$. For the latter the momentum of the cooling film is higher than that of the main flow, leading to a strong deflection of the jet right after its injection. In the fourth picture (lower right, $M = 1$, $I = 10$) the coolant flow was seeded; the main as well as the jet flow remained unseeded. The unseeded jet causes a bright area above the exit of the hole where the seeded coolant flow is disrupted. Downstream of the mixing jet, pronounced light scattering can be detected. In this region the mixing jet interacts with the cooling film which is sucked off the wall.

LDV measurements performed in the symmetry plane of the test section nicely confirm the observations from the flow visualizations.

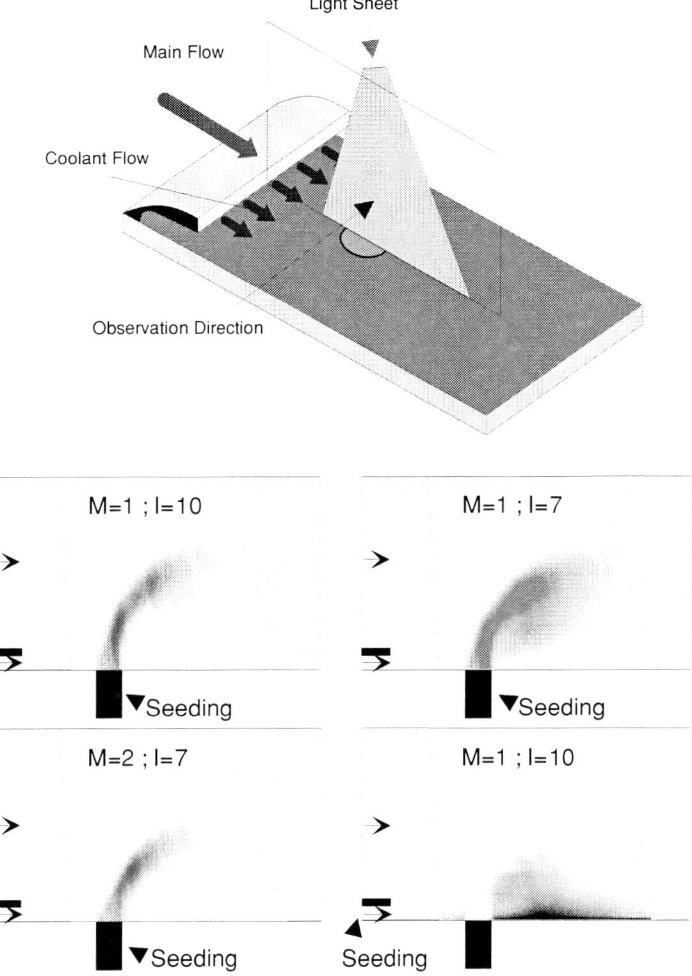

Figure 4.1.7: Light sheet set-up and scattered light intensities.

4.1.3.4 Heat Transfer and Adiabatic Effectiveness

The main purpose of this work was to provide data for heat transfer and adiabatic effectiveness within the area of cooling film and mixing jet interaction. The determination of these quantities was based on the superposition approach of film cooling proposed in [15]. The complete description of the evaluation procedure is reported in [11].

The superposition approach leads to the conclusion that the heat transfer coefficient h is linearly dependent on the dimensionless cooling temperature

$$\Theta = \frac{T_m - T_c}{T_m - T_w} \tag{1}$$

For adiabatic walls h becomes zero. Due to the fact that no convective effects are present in adiabatic walls any change in surface temperature is solely determined by the cooling film. The adiabatic wall temperature T_{aw}, which is usually presented in the dimensionless form of the adiabatic film cooling effectiveness

$$\eta_{ad} = \frac{T_m - T_{aw}}{T_m - T_c} \tag{2}$$

is, therefore, a good measure for the performance of the cooling film. Following the superposition theory the heat flux from the hot gas to the walls can be calculated either from

$$\dot{q}'' = h_0 \cdot (T_{aw} - T_w) \tag{3}$$

or

$$\dot{q}'' = h_\Theta \cdot (T_m - T_w) \tag{4}$$

with T_m and T_w being the temperatures of the main flow and the wall surface and h_0 and h_Θ being the heat transfer coefficients at isothermal ($T_m = T_c$) and non-isothermal ($T_m \neq T_c$) conditions, respectively.

4.1.3.4.1 Heat Transfer Coefficients for Isothermal Conditions

Figure 4.1.8 presents the distributions of the local heat transfer coefficient at $\theta = 0$. In the lateral direction the influence of the mixing air injection on the heat transfer coefficient can be observed in a band which extends from -3 to $+3$ diameters of the mixing hole in the y direction. The flow at the outer edges of this area shows no influence of the mixing jet and therefore leads to a heat transfer behaviour typical for film cooling without mixing jet injection. In the immediate vicinity of the mixing hole the influence of the mixing air supply results in an additional heat sink which leads to unrealistic heat transfer coefficients.

An important rise of the heat transfer coefficient can be detected downstream of the mixing air injection for all conditions under investigation. For a blowing ratio of $M = 1$ and a momentum ratio of $I = 10$, one diameter downstream of the injection, the heat transfer coefficient is almost twice as high as in the undisturbed region. These higher values decrease further downstream, approaching undisturbed conditions at a distance of 15 diameters downstream of the injection. The plot for $M = 1$ combined with $I = 7$ shows similar

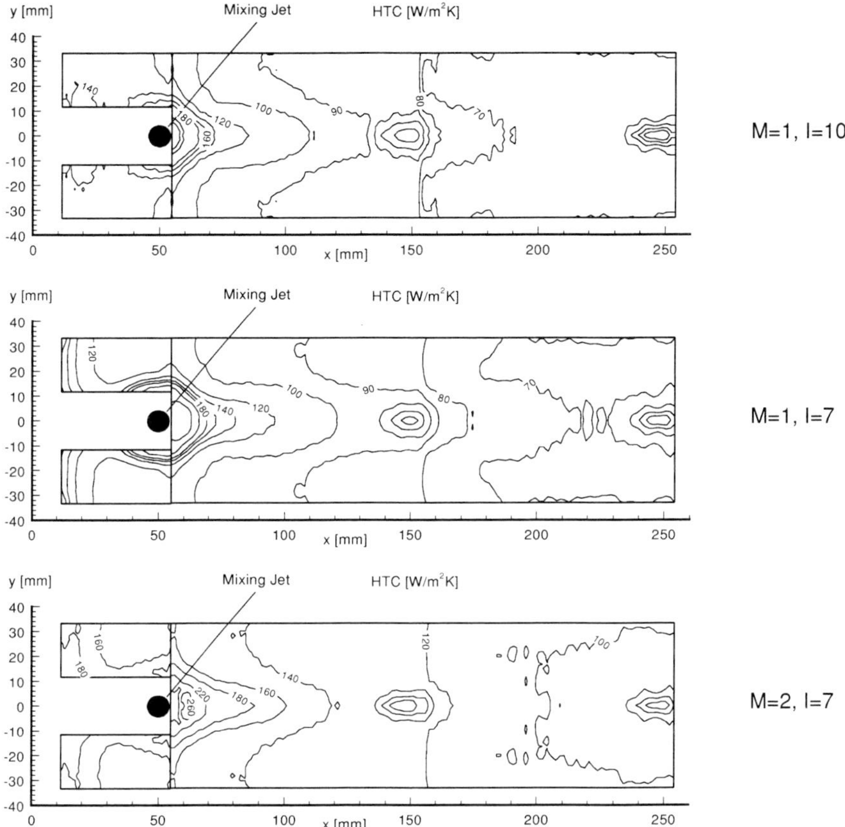

Figure 4.1.8: Heat transfer coefficients at isothermal conditions.

results. Due to larger cooling film velocities, the heat transfer level rises when the blowing ratio is increased to a blowing ratio of $M = 2$ (Fig. 4.1.8, bottom). At the same time the area of interaction between cooling film and jet becomes considerably shorter disappearing at a distance of approximately 10 mixing jet diameters downstream from the injection.

4.1.3.4.2 Adiabatic Effectiveness

Figure 4.1.9 shows the distributions of the adiabatic effectiveness, in which the influence of the mixing air injection again becomes evident.

For a blowing ratio $M = 1$ and a jet momentum ratio $I = 7$ the influence of the jet can be detected down to the end of the area observed. For the higher momentum ratio ($I = 10$) the effectiveness distribution seems to

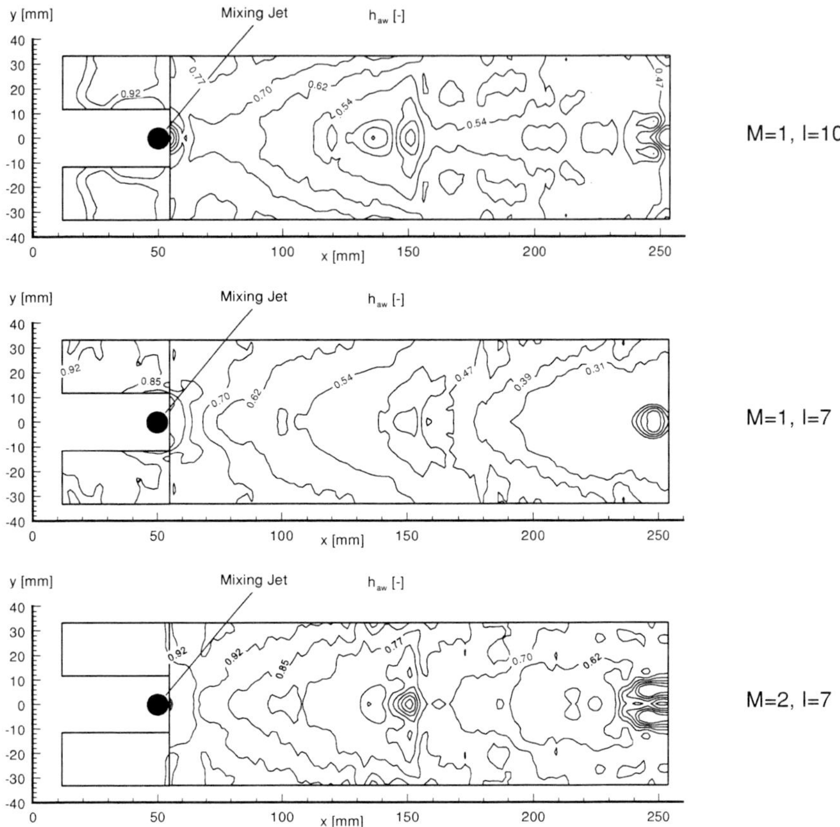

Figure 4.1.9: Adiabatic effectiveness distribution.

become undisturbed 12 diameters downstream of the injection location. The reason for this differing behaviour might be due to the deflection of the jet which is less for $I = 10$ than for $I = 7$ so that the jet is closer to the wall with $I = 7$ than at $I = 10$. As in the case of heat transfer, the influence of the jet is smallest for the blowing ratio $M = 2$, where the adiabatic effectiveness decreases by about 15 % in the region downstream of the injection and the cooling film seems to be recovered 12 diameters downstream of the injection. For the lower blowing ratio the decrease in effectiveness downstream of the injection is more than 25 %.

Knowing the heat transfer coefficient h_0 and the adiabatic wall temperature T_{aw} from the measurements, the heat load on the combustor wall model can now be calculated using Eq. (3). Since both values are much larger in the region downstream of the mixing air injection than in the undisturbed regions, the heat flux through the wall as well as the wall temperatures are increased considerably through the mixing jet.

Looking at the investigations presented, the interaction between a cooling film and a mixing air injection has been analysed for three different operating conditions. The laser light sheet pictures clearly demonstrate the suction effect of the counter-rotating jet vortices downstream of the injection. Moreover, this suction effect has a strong impact on the thermal behaviour of the film cooled wall as could be shown by detailed heat transfer analyses. Using electrical heater foils and applying the superposition approach of film cooling, both, adiabatic effectiveness and isothermal heat transfer coefficient could be determined with the same test plate. Depending on the combination of blowing ratio and momentum ratio of the mixing jet, the heat transfer is augmented by up to 100 % and the adiabatic cooling effectiveness decreases between 15 and 25 % in the interaction area downstream of the mixing air injection. It could be demonstrated that the influence of the mixing air injection is smaller for high blowing ratios combined with lower jet momentum ratios than for lower blowing ratios combined with high jet momentum ratios.

4.1.4 Effusion Cooling

It has been shown that neither convective cooling nor film cooling alone is suitable to provide sufficient thermal protection for the liners in modern combustors. A combination of both cooling methods, however, is capable to meet the cooling requirements in these highly loaded combustors.

As a consequence of the convection and film cooling investigations, the continuing activities concentrate on effusion cooling. Using this cooling technique on a turbine blade one of the pioneering investigations in this area was performed at the "Institut für Thermische Strömungsmaschinen" during the late 1970s [16–18]. For the experimental analysis presented a test surface equipped with an array of cooling holes was built (Fig. 4.1.10) and fixed in the test section previously described (Fig. 4.1.6). An extremely shallow hole angle (17°) together with small hole pitches were chosen for the test plate to provide a large area for convective cooling inside the holes. Furthermore, a very stable cooling film can be formed due to superposition of multiple cooling films. With respect to the results of the convective cooling study, this configuration was completed with impingement cooling on the back side.

The behaviour and performance of effusion cooling is strongly dependent on the blowing ratio. In the following section a survey of aerodynamic and aerothermal investigations will be presented. Schlieren optic was used to assess the jet flow as the coolant is deflected by the mainstream. In addition, the infrared temperature measuring technique described before was applied for measuring highly resolved surface temperature distributions, enabling the determination of the local cooling effectivenesses.

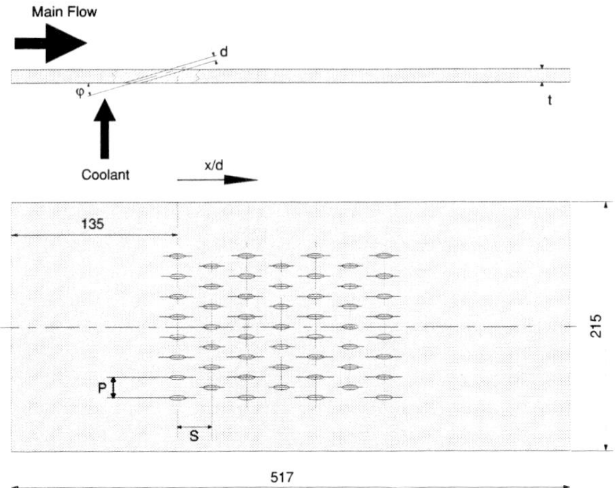

Figure 4.1.10: Geometry of the effusion cooling test plate.

4.1.4.1 Flow Visualization

Schlieren photography renders flow visualization at high temperature ratios possible. This technique was therefore applied to study the behaviour of the cooling jets under realistic temperature ratios (see [19] for a description of the optical set-up).

The results are presented in Fig. 4.1.11. The area observed includes the second and third row of holes. The main flow direction is from left to right. At the lowest blowing rate of $M = 0.5$ ($I = 0.15$) the momentum of the jets is very low and the coolant is deflected to the wall right after exiting the cooling hole. The film spreads downstream of the rows and remains attached to the wall. Due to the small coolant mass flux the film is rapidly absorbed by jet-main flow mixing. A blowing ratio of $M = 1.2$ ($I = 0.85$) leads to a localized jet detachment at the hole exit. Further downstream the film reattaches to the wall as the mixing proceeds. Due to the increased mass flux the cooling film itself is more resistant to the mixing with the main flow than for the smaller blowing rate of $M = 0.5$. At the highest blowing ratio investigated ($M = 4.0$; $I = 9.41$), the jets completely penetrate into the main flow and no film cooling effect is to be expected. In addition, the amount of air is so large that in the right picture of Fig. 4.1.11 the cooling air coming from the first row of effusion holes is still visible as a distinct jet.

From these results it becomes obvious that low blowing ratios are favourable, especially in the starting region of cooling films. For higher blowing ratios the jet detachment increases and no further augmentation in film cooling effectiveness can be expected for the first rows. Further downstream, the mixing region spreads in the normal direction thereby leading to a reattachment of the cooling film even for higher blowing ratios.

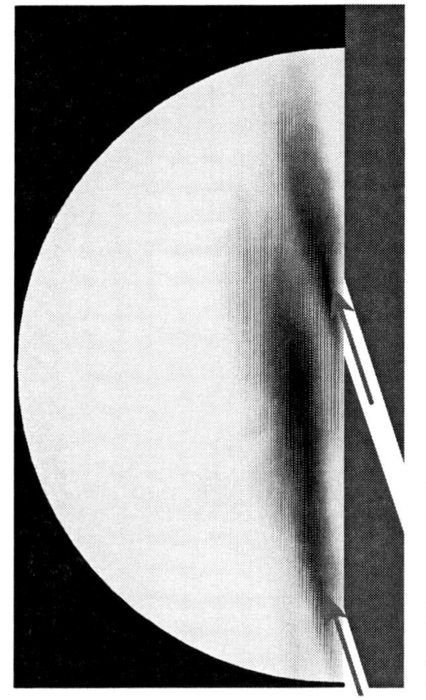

$M = 1.2 \quad I = 0.85$

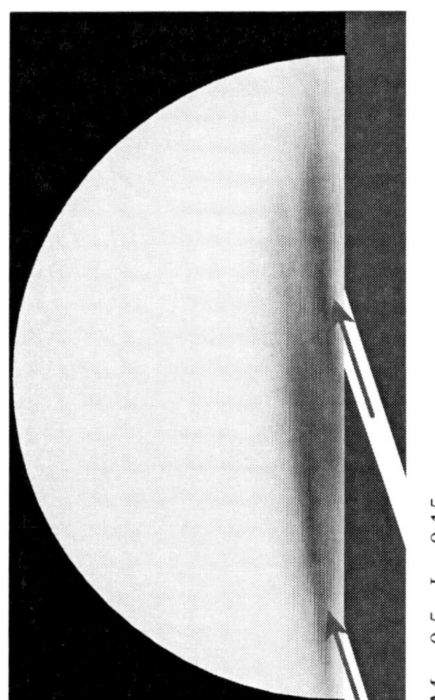

$M = 0.5 \quad I = 0.15$

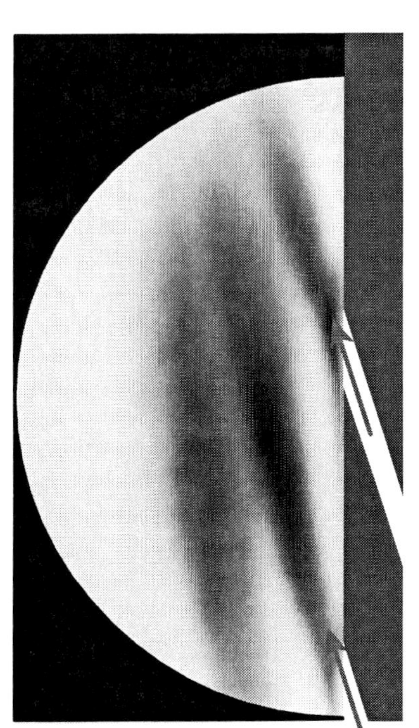

$M = 4.0 \quad I = 9.41$

Figure 4.1.11: Schlieren pictures from the cooling jets at varying blowing ratios M.

4.1.4.2 Near Adiabatic Wall Temperatures

The influence of the coolant jet flow on the adiabatic effectiveness was analysed by using the infrared measuring technique described previously. The high geometric and thermal resolution enables a very detailed study of the near adiabatic wall temperature distribution. The geometrical resolution is about 4 pixels per cooling hole diameter. Figure 4.1.12 presents the results obtained for different blowing ratios at a temperature ratio of $T_m/T_c = 1.85$. The main flow direction is from bottom to top. The gap in the infrared image at a location $x/D \approx 40$ is due to the limited optical access in the top wall of the flow channel, i. e. measurements were performed through two round sapphire glass windows.

In good agreement with the Schlieren pictures the low blowing rate ($M = 0.5$) leads to an attached cooling film. The superposition of the coolant streaks from row to row results in an almost homogeneous cooling film downstream of the last row of holes. Further downstream, the effectiveness decreases rapidly due to film-mainstream mixing. The temperature distribution downstream of the first hole at $M = 0.8$ shows a rather narrow streak of coolant which widens up further downstream, indicating the limit of the attached flow regime. The effectiveness after the last row is considerably higher than for $M = 0.5$. For blowing ratios beyond $M = 1.2$ a detached cooling regime can be observed. The increase of effectiveness within the first rows of holes is clearly slower compared to the attached regime. The

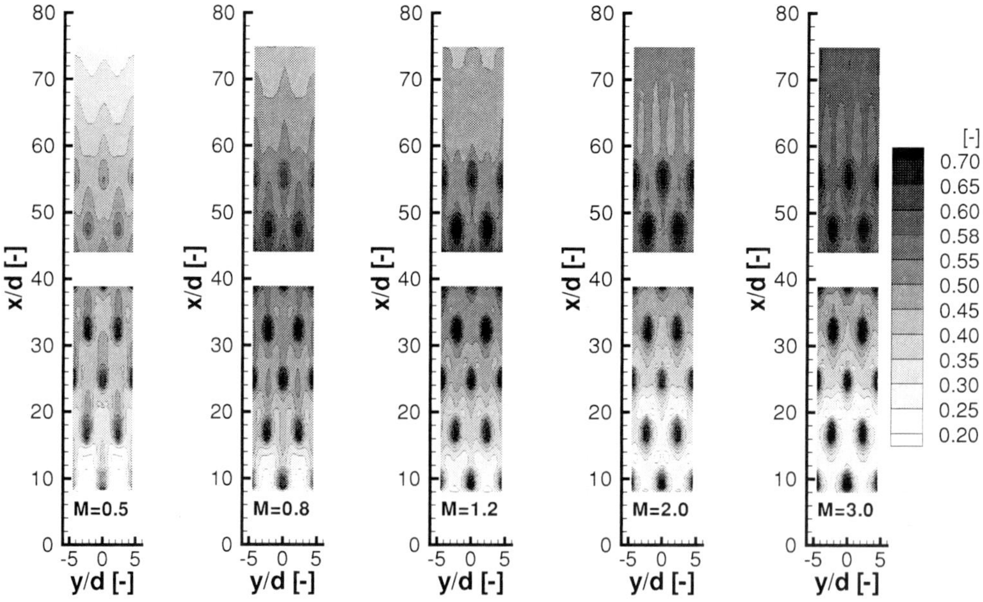

Figure 4.1.12: Near-adiabatic effectiveness distribution for plastic material.

reattachment at some distance downstream leads to a delayed increase in effectiveness. After the last row of holes the effectiveness attains the same level as for $M = 0.8$. At very high blowing ratios ($M = 2.0$ and $M = 3.0$) almost no film cooling occurs in the first rows. The slight temperature reduction observed has to be attributed to heat conduction in the test plate. According to the Schlieren pictures, the jets completely lift off and the hot main flow penetrates underneath the jets. Regarding film cooling only, it can be stated that even for very shallow hole angles attached flow regimes can only be established for blowing rates up to $M \approx 1.0$. This means that high blowing rates can not be converted sufficiently into effectiveness and should be avoided for a low coolant consumption design. The best performance within the effusion hole array is achieved for $M = 0.8$.

4.1.4.3 Overall Effectiveness for a Metallic Test Plate

In real applications wall temperatures are influenced by film cooling and convection cooling on the back side and in the holes. To study the overall effectiveness at realistic Biot numbers, a metallic test plate was manufactured. The results obtained are shown in Fig. 4.1.13.

The temperature distribution is very homogeneous for all blowing ratios. Furthermore it can be seen that the effectiveness increases with blowing ratio

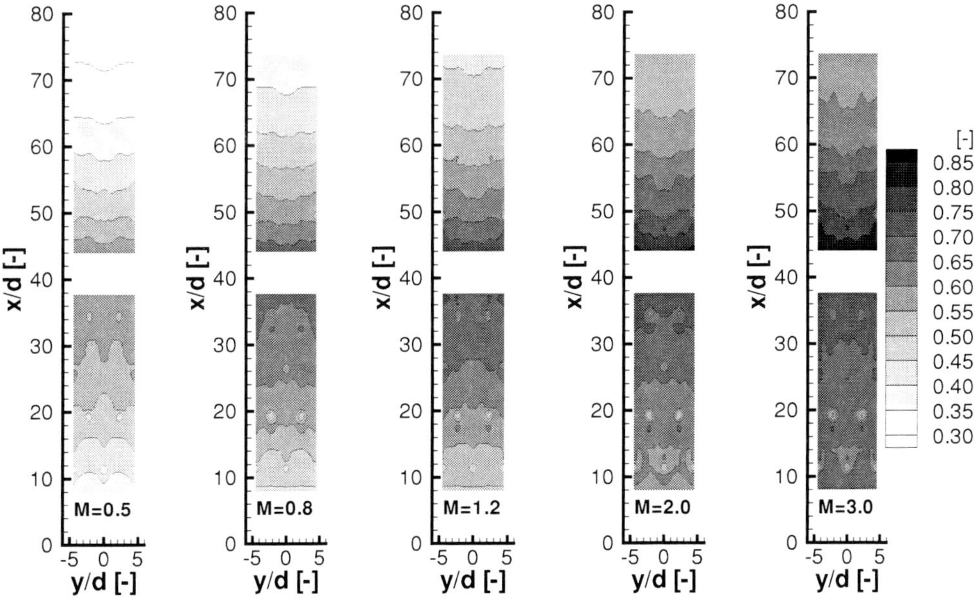

Figure 4.1.13: Overall effectiveness distribution for metallic test plate.

at any location on the test plate. This is a consequence of the pronounced convective cooling. In contrast to adiabatic effectiveness, convective heat transfer increases monotonously with blowing ratio. Therefore, the low effectiveness which may occur at high blowing ratios in the area of the first rows of holes is compensated by the intense convection within the holes. Downstream of the sixth row a rapid decrease in effectiveness occurs even at high blowing ratios. This is due to the missing convection in the holes. Taking these results into account, it can be concluded that a combination of film cooling and convective cooling may help to smooth the temperature distribution of the wall (see also [20]).

4.1.4.4 Effusion Cooling with Additional Impingement Cooling on the Back Side

Although effusion cooling already combines film cooling with convective cooling, a further increase in effectiveness can be achieved by using impingement cooling on the backside. For this, the previously discussed effusion test plate was equipped with an impingement configuration on the back side in the effusion hole area. The geometry was designed to generate 75 % of the total pressure drop in the impingement plate and 25 % of the pressure drop in the effusion holes.

The lateral averaged overall effectiveness for effusion cooling with and without impingement is shown in Fig. 4.1.14. It becomes obvious that the

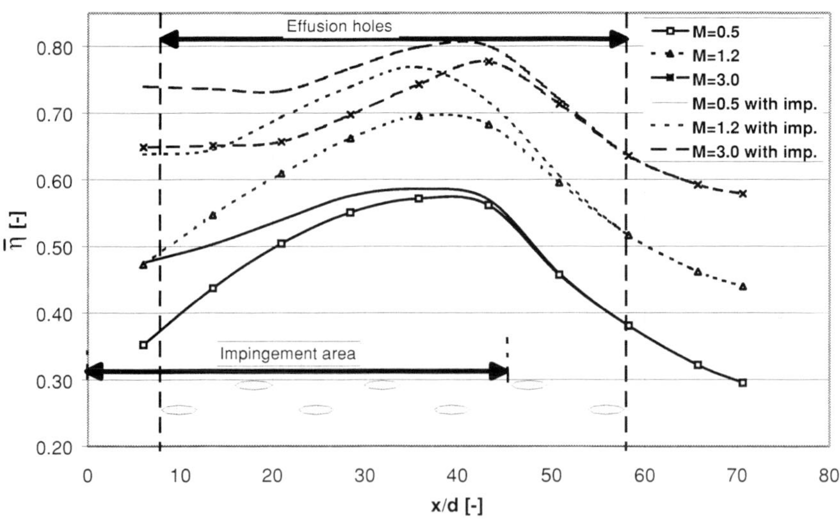

Figure 4.1.14: Comparison of lateral averaged overall effectiveness of effusion cooling with and without impingement on the back side.

enhancement is highest in the area where the cooling film is not very effective ($x/d < 25$). There, the large temperature difference between wall and coolant enables an intense convective cooling on the backside. At positions downstream of $x/d = 35$, the increase in effectiveness is less pronounced since the cooling film already offers good protection. The impingement area ends at $x/d = 45$. Further downstream the effectiveness approaches the same level of effusion cooling without impingement. Again these results demonstrate that film cooling and convective cooling support each other in a kind of "team play" and thus a combination of both is very promising to meet the current requirements of combustor liner cooling. Regions where only poor film cooling can be achieved may still be well protected if convective cooling is established on the back side of the liner.

4.1.5 Conclusions

Cooling is one of the key-technologies for any future improvement in combustor performance. Rising temperature levels in combination with higher coolant temperatures lead to significantly increased heat loads to the walls. At the same time, new concepts for low emission combustors ask for an increasing amount of combustion air. As a consequence, the amount of air available for cooling purposes is diminishing while the operating conditions become more demanding.

In this context the present paper gives an overview of new cooling concepts for combustors. Results presented lead to the conclusion that methods combining film and convective cooling have the highest potential to meet future requirements. If used separately, film cooling is more effective than convective cooling. On the other hand, film cooling is strongly dependent on the flow conditions in the combustor and its effect may be very small under certain conditions, e. g. in the vicinity of a mixing jet ejection. In contrast to film cooling, convective cooling schemes can be controlled more precisely. Enhancing the heat transfer by introducing means like pin fins, ribs, or impingement devices on the back side of the liners, convective cooling can contribute considerably to a reduction of material's temperatures. This is true even when effusion cooling is considered.

The comprehensive studies performed within the Collaborative Research Centre 167 have brought about valuable information on the performance of the two cooling techniques and their combined application. Flow field, heat transfer, and effectiveness measurements have strongly improved the understanding of the physical processes involved and represent an excellent data base for future developments. In a first application of the combined use of effusion and impingement cooling a considerable amount of coolant could be saved. Tests under actual engine conditions revealed a reduction in cooling air consumption of at least 30 % as compared to simple slot cooling geometries being widely used when the Collaborative Research Centre 167 was initiated.

References

1. Bittlinger, G., Schulz, A., Wittig, S. (1994): Film cooling effectiveness and heat transfer coefficients for slot injection at high blowing ratios. ASME Paper 94-GT-182.
2. Bittlinger, G. (1995): Filmkühlungen mit tangentialer Spaltausblasung: Experimentelle und numerische Ansätze zur Optimierung des Brennkammer-Kühlluftbedarfs. Dissertation, Lehrstuhl und Institut für Thermische Strömungsmaschinen, Universität Karlsruhe (TH).
3. Armstrong, J., Winstanley, D. (1988): A review of staggered array pin fin heat transfer for turbine cooling applications. *ASME – J. of Turbomachinery* **110**, 94–103.
4. Chandra, P. R., Fontenot, M. L., Han J.-C. (1998): Effect of rib profiles on turbulent channel flow heat transfer. *J. of Thermophysics and Heat Transfer* **12**, 116–117.
5. Metzger, D. E., Florschuetz, L. W., Takeuchi, D. I., Behee, R. D., Berry, R. A. (1979): Heat transfer characteristics for inline and staggered arrays of circular jets with crossflow of spent air. *ASME – J. of Engineering for Power* **101**, 526–531.
6. Wittig, S., Noll, B., Elbahar, O. M. F., Willibald, U. (1983): Einfluß von Mischluftstrahlen auf die Geschwindigkeits- und Temperaturverteilung in einer Querströmung. VDI-Bericht Nr. 487.
7. Wittig, S., Elbahar, O. M. F., Noll, B. (1984): Temperature profile development in turbulent mixing of coolant jets with a confined hot cross flow. *ASME – J. of Engineering for Gas Turbines and Power* **106**, 193–197.
8. Wittig, S., Scherer, V. (1987): Heat transfer measurements downstream of a two-dimensional jet entering a cross-flow. *ASME – J. of Turbomachinery* **109**, 572–578.
9. Scherer, V., Wittig, S. (1989): The influence of the recirculation region: A comparison of the convective heat transfer downstream of a backward-facing step and behind a jet in a cross flow. ASME Paper 89-GT-59.
10. Scherer, V., Wittig, S., Morad, K., Mikhael, N. (1991): Jets in a crossflow: effects of hole spacing to diameter ratio on the spatial distribution of heat transfer. ASME Paper 91-GT-356.
11. Martiny, M., Schulz, A., Wittig, S., Dilzer, M. (1997): Influence of a mixing-jet on film cooling. ASME Paper 97-GT-247.
12. Scherer, V., Pfeiffer, A., Wittig, S. (1988): Bestimmung der Wärmeübergangszahlen in abgelösten Strömungen mit Hilfe einer Infrarotkamera. Tagungsband DGLR-Workshop 2D-Meßtechnik, Markdorf, 245–253.
13. Scherer, V., Wittig, S., Bittlinger, G., Pfeiffer, A. (1993): Thermographic heat transfer measurements in separated flows. *Exp. Fluids* **14**, 17–24.
14. Martiny, M., Schiele, R., Gritsch, M., Schulz, A., Wittig, S. (1996): In situ calibration for quantitative infrared thermography. Quantitative Infrared Thermography, Eurotherm Seminar No. 50, 3–8.
15. Cho, H., Kays, W. M., Moffat, R. J. (1974): The superposition approach to film-cooling. ASME Paper 74-WA/GT-27.
16. Hempel, H., Friedrich, R. (1977): Untersuchungen zur Effusionskühlung von Gasturbinenschaufeln. *Forschung in der Kraftwerkstechnik*, 67–73.

17. Hempel, H. (1978): Untersuchungen zur Entwicklung einer effusionsgekühlten Gasturbinenschaufel. Dissertation, Lehrstuhl und Institut für Thermische Strömungsmaschinen, Universität Karlsruhe (TH).
18. Hempel, H., Friedrich, R., Wittig, S. (1980): Full coverage film-cooled blading in high temperature gas turbines: Cooling effectiveness, profile loss and thermal efficiency. *ASME – J. for Engineering for Power* **102**, 957–963.
19. Martiny, M., Schulz, A., Wittig, S. (1995): Full coverage film cooling investigations: Adiabatic wall temperatures and flow visualization. ASME Paper 95-WA/HAT-4.
20. Martiny, M., Schulz, A., Wittig, S. (1997): Mathematical model describing the coupled heat transfer in effusion cooled combustor walls. ASME Paper 97-GT-329.

4.2 Numerical Modelling of Combustor Liner Heat Transfer

Dietmar Giebert, Elias Papanicolaou, Carl-Henning Rexroth, Michael Scheuerlen, Achmed Schulz, and Rainer Koch*

Abstract

This paper discusses numerical methods for predicting conjugate heat transfer in generalized co-ordinates including the application to the heat transfer analysis of effusion cooled combustor liners. Comparisons of the streamwise velocity and the film-cooling effectiveness to detailed flow- and temperature field measurements, conducted at the Institut für Thermische Strömungsmaschinen at the University of Karsruhe, are presented for different blowing ratios and density ratios. Also, the influence of the thermal conductivity of the liner material on the film-cooling characteristic has been investigated.

The results reveal that all dominating flow structures and heat transfer phenomena of this jet in crossflow problems could be captured with good accuracy. For a low thermal conductivity of the liner material very good agreement for the laterally averaged film-cooling effectiveness is achieved, whereas for high thermal conductivity discrepancies are found, which are mostly caused by the setting of the boundary condition at the downstream cross-section of the test plate.

4.2.1 Introduction

Modern gas turbine combustors featuring high pressure and temperatures are subjected to very high thermal loads and can only be operated safely with intense cooling to prevent early distress. However, advanced low-emission combustor technology requires most of the air to be fed through the fuel

* Institut für Thermische Strömungsmaschinen, Universität Karlsruhe, Kaiserstr. 12, 76128 Karlsruhe, Germany

preparation system and reduction of the amount of liner cooling air is an important issue of modern design. Therefore, efficient cooling techniques like effusion cooling, together with a detailed knowledge of the turbulent flow structure and heat transfer characteristic are of major importance in the design process. Besides the radiative heat transfer, convective heat transfer as well as heat conduction inside the combustor structure play an important role. It is well known, that conjugate heat transfer has a significant impact on the cooling characteristics of combustor liners. The accuracy and effectiveness of the prediction of conjugate heat transfer are closely related to both, the modelling of the turbulent flow and the efficiency of the numerics.

In the present study, a numerical technique with implicit coupling of convection in the fluid and conduction in the adjacent solid surfaces is being presented. Special emphasis is put on the application to effusion-cooled combustor liners.

4.2.2 Numerical Method

4.2.2.1 Governing Equations and Turbulence Modelling of the Flow

The equations used to model the flow are the steady Reynolds averaged conservation equations of mass, momentum, and energy together with the equation of state. In this study, the gas is assumed as thermodynamic ideal. Turbulence is taken into account by the standard k,ε turbulence model [1] which allows the calculation of the eddy viscosity. The wall function approach [1, 2] is used to model turbulence effects in the near wall region. Throughout this study, a constant turbulent Prandtl number ($Pr = 0.86$) is employed limiting the closure problem on the specification of the turbulent viscosity.

4.2.2.2 Discretization and Solution Technique

The three-dimensional Navier-Stokes code has been developed at the Institut für Thermische Strömungsmaschinen at the University of Karlsruhe in the framework of the Collaborative Research Centre 167 [3, 4] and has been applied to a variety of technical flow problems [5–8]. The governing equations are formulated in a body-fitted non-orthogonal curvilinear co-ordinates system. It is based on a fully conservative, structured finite volume discretization method. The transport equations for the Cartesian velocity components and other scalars are solved using a non-staggered grid. All flow variables are stored at the same nodes ("cell-centered"). The solution procedure is iterative and the pressure-based SIMPLE algorithm [9] is used to derive a pressure-correction equation. An extension of this procedure [10] is applied to enable the computation of flows over a wide Mach number range from incompressible to highly compressible flows. To avoid checkerboard pressure-velocity oscillations, an interpolation scheme similar to that of Rhie and Chow [11] is applied. For the discretization of the diffusive terms, a second order central difference scheme is used, whereas the convective terms of all transport equations are discretized by the second order accurate Monotonized-Linear-Upwind scheme (MLU) [12]. This high resolution scheme is bounded and therefore, physically unrealistic over- and undershoots are avoided. Because of its stability, the scheme can be used also for the convective terms of the turbulence equations. For solving the system of the algebraic equations arising from discretization, a generalized conjugate gradient iterative procedure with an incomplete lower-upper decomposition (ILU-CG) [13] is used. Irrespective of the variety of iterative solvers implemented in the code (ILU-CG, SIP, BiCGSTAB) [13–15], this solver is used here due to its performance in terms of effectiveness, robustness, and computing time.

4.2.2.3 Solid-Fluid Coupling

To take conjugate heat transfer into account the fluid and solid regions have to be coupled via the energy equation. The general conservative form of this equation for a curvilinear, non-orthogonal co-ordinates system ξ_i can be written as:

$$\frac{\partial}{\partial \xi_i}(\rho U_i h) = \frac{\partial}{\partial \xi_i}\left[\frac{1}{J}\left(\frac{\mu}{\mathrm{Pr}} + \frac{\mu_t}{\mathrm{Pr}_t}\right) \cdot \left(q_{jm}\frac{\partial h}{\partial \xi_m}\right)\right] + J \cdot s(\xi_i) \tag{1}$$

Here: $q_{jm} = a_{1j}a_{1m} + a_{2j}a_{2m} + a_{3j}a_{3m}$, where the a are the co-factors of the Jacobian matrix J of the co-ordinates transformation [2] and U_i are the co-variant velocity components.

By integrating the enthalpy equation over the control volumes ΔV, the balance equation for each control volume takes the following form:

$$\int_{\Delta V} \frac{\partial I_\xi}{\partial \xi} \mathrm{d}V + \int_{\Delta V} \frac{\partial I_\eta}{\partial \eta} \mathrm{d}V + \int_{\Delta V} \frac{\partial I_\zeta}{\partial \zeta} \mathrm{d}V = \int_{\Delta V} J \cdot s(\xi,\eta,\zeta) \mathrm{d}V \tag{2}$$

where the symbol I denotes the total flux composed of the convective and normal diffusive flux. The source term s on the right-hand side consists of the cross-diffusion contribution, the pressure work and dissipation.

For control volumes located entirely inside the fluid or the solid domains, the integration of Eq. (2) is straightforward [2]. Cells located at the interface between the fluid and solid domain provide the coupling between the heat transfer and a modified discretization scheme is adopted to be described subsequently.

The main idea has already been outlined in a previous study by [16], where the discretized equations were derived for Cartesian co-ordinates. Nodes are placed on the solid-fluid interface. Their location coincides with the center of control volumes, as indicated by the shaded control volume in Fig. 4.2.1. By a balance of the energy fluxes, the temperature is directly determined at the interface. On the other hand, these nodes are treated as boundary nodes for the computation of the flow field.

The balance of the energy fluxes at the interface is performed separately for different sub-control volumes ΔV_{fluid} and ΔV_{solid} as indicated by the hatched areas in Fig. 4.2.1, each pertaining to either the fluid or solid region. This leads to the following expression for the unknown flux at face P on either side of the interface:

$$I_{\mathrm{P,f}} = A_{\mathrm{w}}(h_{\mathrm{P}} - h_{\mathrm{W}}) - A_{\mathrm{n,f}}(h_{\mathrm{N}} - h_{\mathrm{P}}) + A_{\mathrm{s,f}}(h_{\mathrm{P}} - h_{\mathrm{S}}) - A_{\mathrm{t,f}}(h_{\mathrm{T}} - h_{\mathrm{P}}) + A_{\mathrm{b,f}}(h_{\mathrm{P}} - h_{\mathrm{B}}) + S_{\mathrm{h,f}} \quad (3)$$

$$I_{\mathrm{P,s}} = A_{\mathrm{e}}(h_{\mathrm{E}}^{*} - h_{\mathrm{P}}^{*}) + A_{\mathrm{n,s}}(h_{\mathrm{N}}^{*} - h_{\mathrm{P}}^{*}) - A_{\mathrm{s,s}}(h_{\mathrm{P}}^{*} - h_{\mathrm{S}}^{*}) + A_{\mathrm{t,s}}(h_{\mathrm{T}}^{*} - h_{\mathrm{P}}^{*}) - A_{\mathrm{b,s}}(h_{\mathrm{P}}^{*} - h_{\mathrm{B}}^{*}) - S_{\mathrm{h,s}}^{*} \quad (4)$$

where the A's are the discretization coefficients [2]. For constant values of c_{p} in both, the fluid and solid, the enthalpy h^* in the solid can be linearly related to the enthalpy h of the fluid (for the general case of a non-zero enthalpy of formation Δh_{f}):

$$h^* = r_{\mathrm{C}}(h - \Delta h_{\mathrm{f}}) \quad (5)$$

where $r_{\mathrm{C}} = c_{\mathrm{p,solid}}/c_{\mathrm{p,fluid}}$.

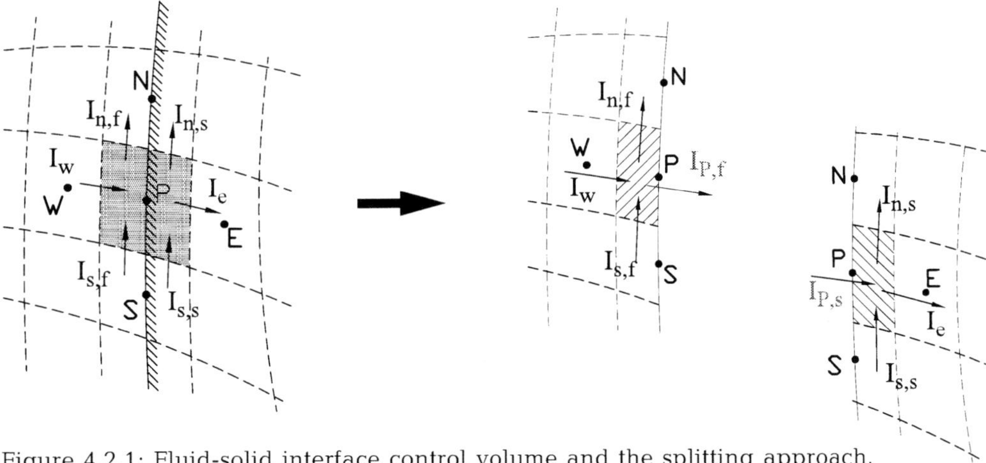

Figure 4.2.1: Fluid-solid interface control volume and the splitting approach.

By requiring continuity of the flux at the interface:

$$I_{P,f} = I_{P,s} \tag{6}$$

and insertion of Eqs. (3) and (4), the discretized equation is obtained for the interface nodes:

$$\begin{aligned} &A_e^*(h_E - h_P) - A_w(h_P - h_W) + A_n^*(h_N - h_P) - A_s^*(h_P - h_S) + A_t^*(h_T - h_P) \\ &- A_b^*(h_P - h_B) = S_{h,f} + r_C S_{h,s} \end{aligned} \tag{7}$$

where the discretization coefficients take the form

$$\begin{aligned} &A_e^* = r_C A_e,\ A_n^* = r_C A_{n,s} + A_{n,f},\ A_s^* = r_C A_{s,s} + A_{s,f},\ A_t^* = r_C A_{t,s} + A_{t,f},\ A_b^* \\ &= r_C A_{b,s} + A_{b,f} \end{aligned} \tag{8}$$

With this approach, coupling between the solid and fluid regions is achieved not only through points in normal direction to the interface, but through all adjacent points, i. e. lateral diffusion effects are also taken into account.

4.2.3 Full-Coverage Film-Cooling of Combustor Walls

4.2.3.1 Geometry and Flow Conditions

As representative test case for the computational study of combustor liner heat transfer, an effusion-cooled plate was selected which was experimentally investigated by Martiny et al. [17]. Details of both, the experimental setup and the measurement techniques are given in [17, 18]. The geometry is shown in Fig. 4.2.2. A flat plate with staggered rows of cylindrical holes, inclined at an angle of 17° in the flow direction, is considered.

The main supply of coolant comes from a plenum at the bottom, which is bounded vertically by adiabatic walls. The extremely small angle together with a small hole pitch provides a large area for convective cooling inside the holes. Due to the repetitive hole pattern in the lateral direction, it is sufficient to consider a computational domain between the axis of two adjacent longitudinal rows of holes (Fig. 4.2.3). This domain includes a total of seven half holes.

In order to identify the conjugate heat transfer effects, the standard k,ε turbulence model with wall functions was used for the first computations.

A H-type grid was used (Fig. 4.2.4). It includes mainstream channel, test plate, cooling holes, and plenum. The grid consists of a total of 351 000 grid

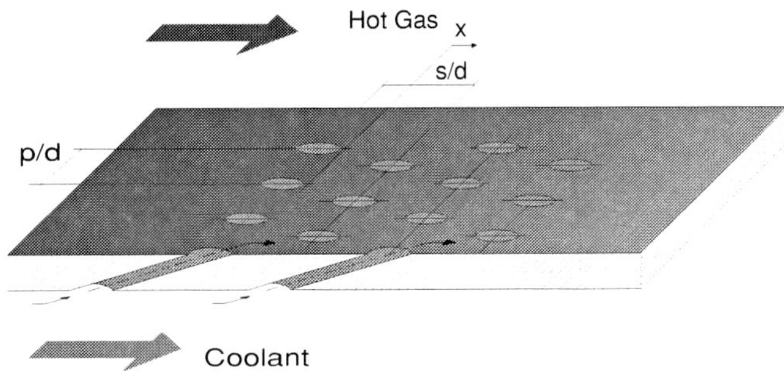

Figure 4.2.2: Schematic of the effusion-cooled plate.

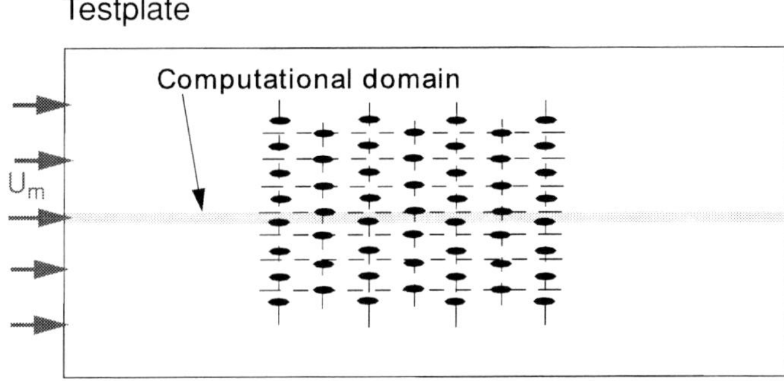

Figure 4.2.3: Test plate and computational domain.

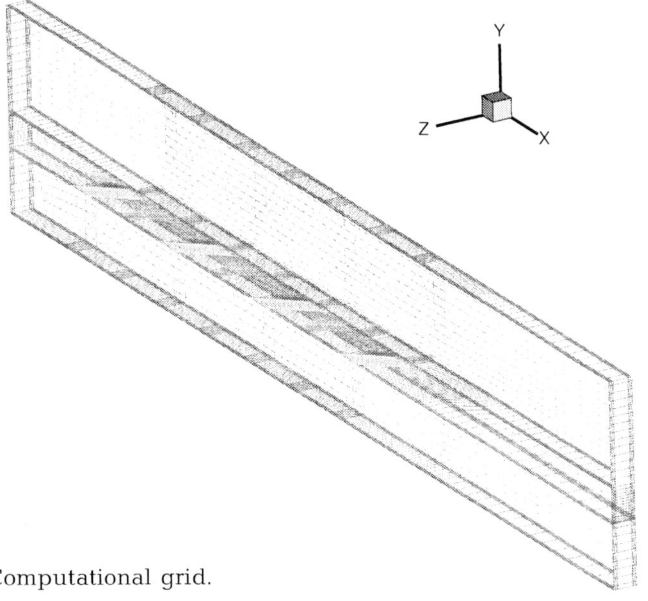

Figure 4.2.4: Computational grid.

points (mainstream channel $\approx$ 132 000, test plate with cooling holes $\approx$ 130 000, and plenum $\approx$ 89 000 grid points). It should be noted, that it is necessary to use a single block grid, as the interior of the plate is also of importance for the heat transfer. For the whole computational domain, the first grid nodes at the walls were placed in the fully turbulent region at an average dimensionless distance of $y^+ \approx 40 - 90$. It was found that by adding some curvature to the grid at the inlet of the holes, and thus avoiding acute angles in this region, better convergence behaviour is obtained. Curvature was found to be not very critical at the exit of the holes but for the sake of a more symmetrical appearance the same pattern was used. In the hole itself, the gridlines in the crossflow direction are gradually rotated to become perpendicular to the walls of the hole.

As inlet conditions for the main gas flow, the velocity profile is prescribed from the experimental condition. The velocity attains a constant value of $u_{\mathrm{m}} = 28$ m/s at a distance of 12 mm from the test plate bottom wall. In the plenum, vertical velocities are specified such that the resulting velocities in the injection holes yield the desired blowing ratio M. For the numerical calculations, both, isothermal and non-isothermal conditions for the main flow and coolant temperature were specified in order to compare the predicted results with the measured velocity and temperature fields. For the heat transfer investigations, the inlet temperature of the main flow is prescribed from the experimental condition and increases in a power-law fashion from 475 K on the plate surface to a constant value of 550 K far from the plate. The coolant inlet temperature was set to 315 K which leads then to a prescribed density ratio D of $D = \rho_{\mathrm{c}}/\rho_{\mathrm{m}} = 1.8$.

The turbulence level was set to 4 % at the inlet of the mainstream channel and 5 % at the plenum entrance. The turbulent length scales are set to 0.03 of the inlet heights for both inflow boundaries.

4.2.3.2 Results and Discussion

Detailed comparisons of numerical results with flowfield measurements of mean velocity are subsequently presented for two different blowing ratios and a density ratio of unity. Additionally, experimental data of film-cooling effectiveness and temperature profiles are compared for blowing ratios of $M = 0.5$, $M = 1.2$, and $M = 3.0$ and a density ratio of 1.8 for a plastic and metallic test plate. For all results, the origin of the x co-ordinate is located at the trailing edge of the first row of the cooling holes (Fig. 4.2.5). The y co-ordinate is oriented in normal direction to the test plate.

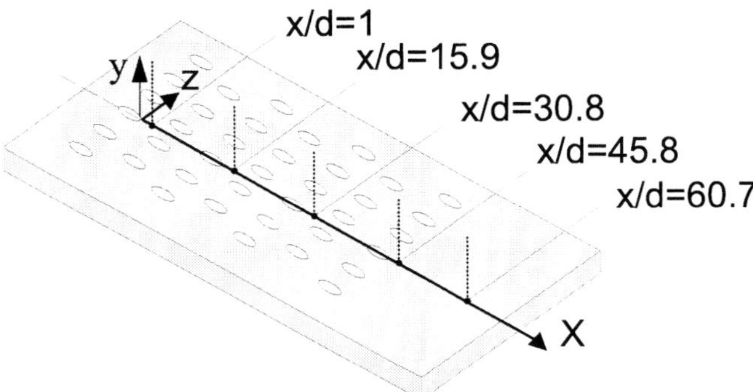

Figure 4.2.5: Sketch of the test plate with measurement locations for the streamwise velocity.

4.2.3.2.1 Flowfield Comparison

A comparison of the streamwise velocity profiles at the centerline at different axial locations x/d (Fig. 4.2.5) for blowing ratios of 0.5 and 3.0 are plotted in Figs. 4.2.6 and 4.2.7.

For the lower blowing ratio of $M = 0.5$ (Fig. 4.2.6) good agreement is obtained for all axial locations $x/d \geq 15.9$. Slight differences between prediction and measurement are found at $x/d = 1$, downstream of the first row of holes. The experimental data shows a significant momentum deficit at the wall which is not adequately captured by the prediction. The reason may be rooted to differences in the shape and contour of the jet in experiment and prediction. While the calculation reveals a jet mainly concentrated along the centerline plane, the experiment indicates a jet out-flow over a large portion of the hole outlet plane with lower velocities. A similar case was also analyzed by Giebert et al. [7] where film-cooling from scaled-up holes was investigated experimentally and numerically. It was found there, that the location and shape of the jet is strongly affected by the way the coolant enters the film cooling hole as well as by the correct prediction of the turbulence intensity. Since the calculation of this study shows a too low turbulence production at the inlet of the holes in the first row, it seems very likely that this effect together with discrepancies of the jet at the hole inlet may cause the deviations in the velocity profiles.

Figure 4.2.7 shows the same comparison for the blowing ratio of 3.0. For wall distances $y/d \geq 1.5$ the prediction agrees very well with the experimental data. However, for smaller wall distances ($y/d < 1.5$) significant discrepancies are found which are due to deficits of the numerical calculation. The use of the wall function approach prohibits a fine resolution of the near wall boundary layer. Therefore, the jet lift-off which is observed in the experiment at that high blowing ratio, cannot be resolved adequately in the calculation. Additionally, the standard k,ε turbulence model is well

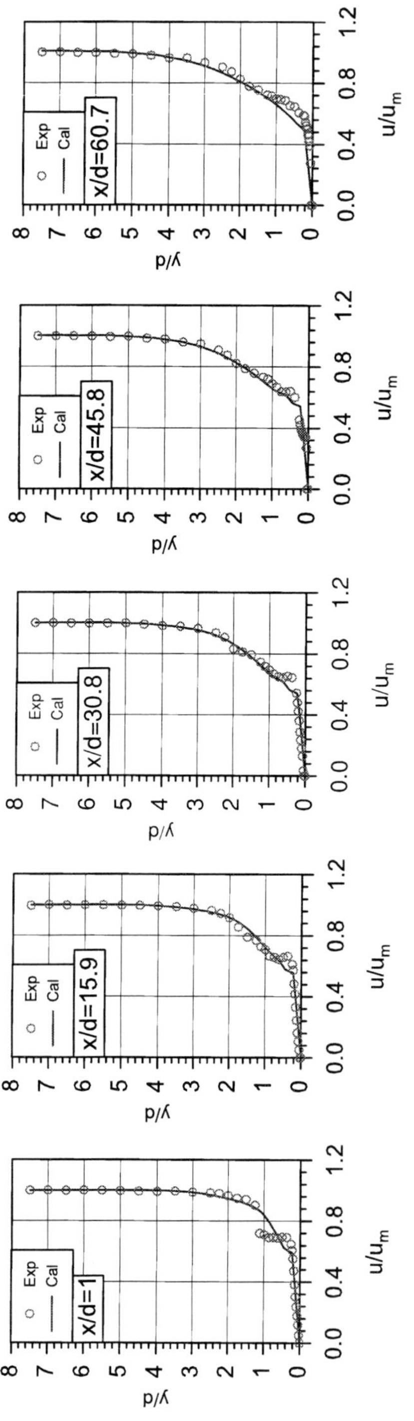

Figure 4.2.6: Streamwise velocity profiles for blowing ratio of $M = 0.5$.

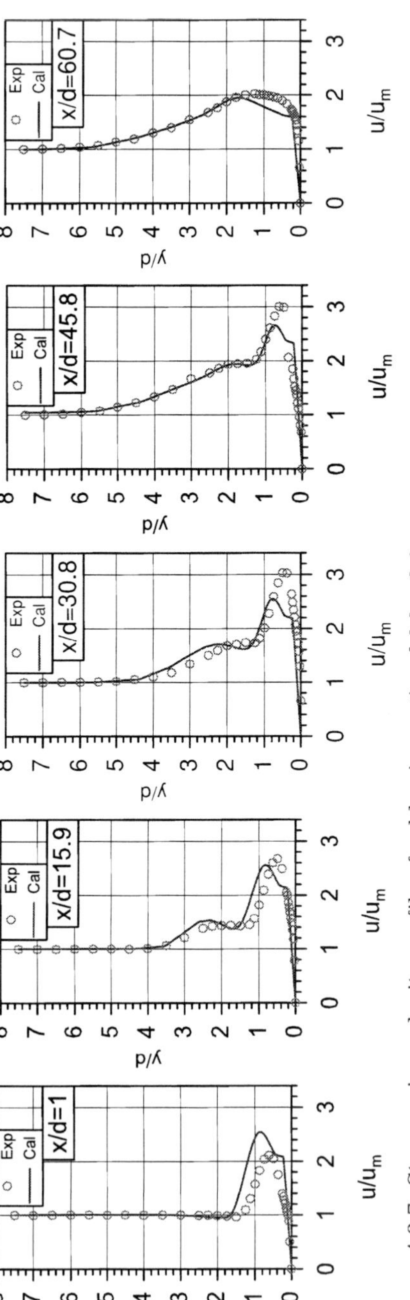

Figure 4.2.7: Streamwise velocity profiles for blowing ratio of $M = 3.0$.

known [7, 19] to underpredict the rate of the lateral spreading of the jets downstream of the hole. As a consequence, the interaction between adjacent jets is underestimated resulting in lower velocities near the wall.

4.2.3.2.2 Comparison of the Heat Transfer

This section focuses on the heat transfer to the film-cooling plate in terms of the film cooling effectiveness

$$\eta = \frac{T_m - T_w}{T_m - T_c} \tag{9}$$

and the normalized temperature Θ

$$\Theta = \frac{T(y) - T_c}{T(y)_0 - T_c} \tag{10}$$

T_w corresponds to the temperature at the hot gas channel wall, T_m is the temperature at the channel half-height, and T_c is the temperature of the coolant at the inlet. The temperature Θ is normalized by the temperature T_0 at the inlet of the hot gas channel in order to eliminate its influence on the film cooling effect.

Plastic Plate

Figure 4.2.8 shows the calculated overall film-cooling effectiveness for different blowing ratios for the film-cooling plate made of Tekapeek, a plastic material with low thermal conductivity (λ = 0.33 W/mK).

The calculated thermal fields reveal the same trend in the temperature distributions with increasing blowing ratio as observed in the experimental investigation [18]: The predictions show a non-homogenous temperature field where the flow pattern of the cooling jets can be clearly identified. Further it can be seen that an increase in the blowing ratio from $M = 0.5$ to $M = 1.2$ leads to higher values of the effectiveness downstream of the first three rows of holes, whereas a further enhancement of the blowing ratio up to $M = 3.0$ is accompanied by a reduction of the effectiveness due to lift-off of the cooling jets. In the upstream region of the first row of holes, an increase of the effectiveness due to heat conduction inside the test plate is revealed for higher blowing ratios. However, since heat conduction is the sole mechanism of heat transfer, there is no optimum of the blowing ratio for which a maximum in the effectiveness can be obtained in that region.

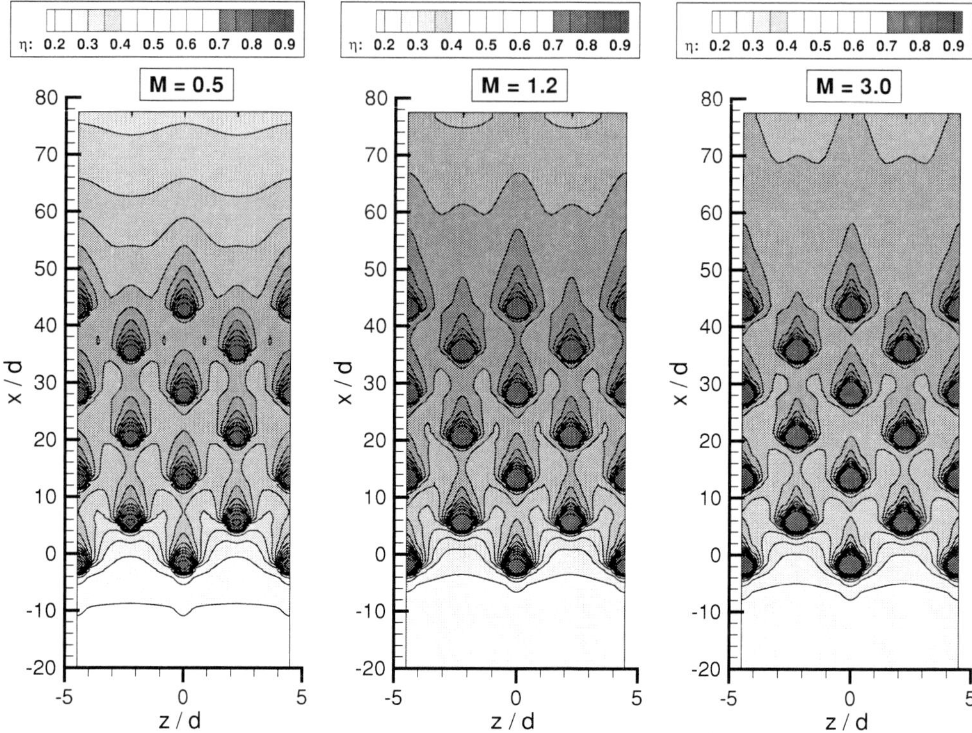

Figure 4.2.8: Calculated film cooling effectiveness distribution for plastic material.

A more quantitative analysis is shown in Fig. 4.2.9 where the laterally averaged effectiveness at different streamwise positions is compared.

For blowing ratios of $M = 0.5$ as well as $M = 1.2$ the agreement is quite good. Very good predictions are obtained for the maximum value of the effectiveness. However, its location slightly differs from the experimental data. For the highest blowing ratio ($M = 3.0$), the measured effectiveness shows a dip with a following increase in the streamwise direction. This effect is attributed to the detachment and subsequent reattachment of the jets. As mentioned previously, the lift-off of the jets cannot be adequately captured by the calculation because of the use of the wall functions. However, compared to the predicted results for the blowing ratio of $M = 1.2$, smaller values for the effectiveness are obtained indicating stronger penetration of hot gas flow underneath the jets.

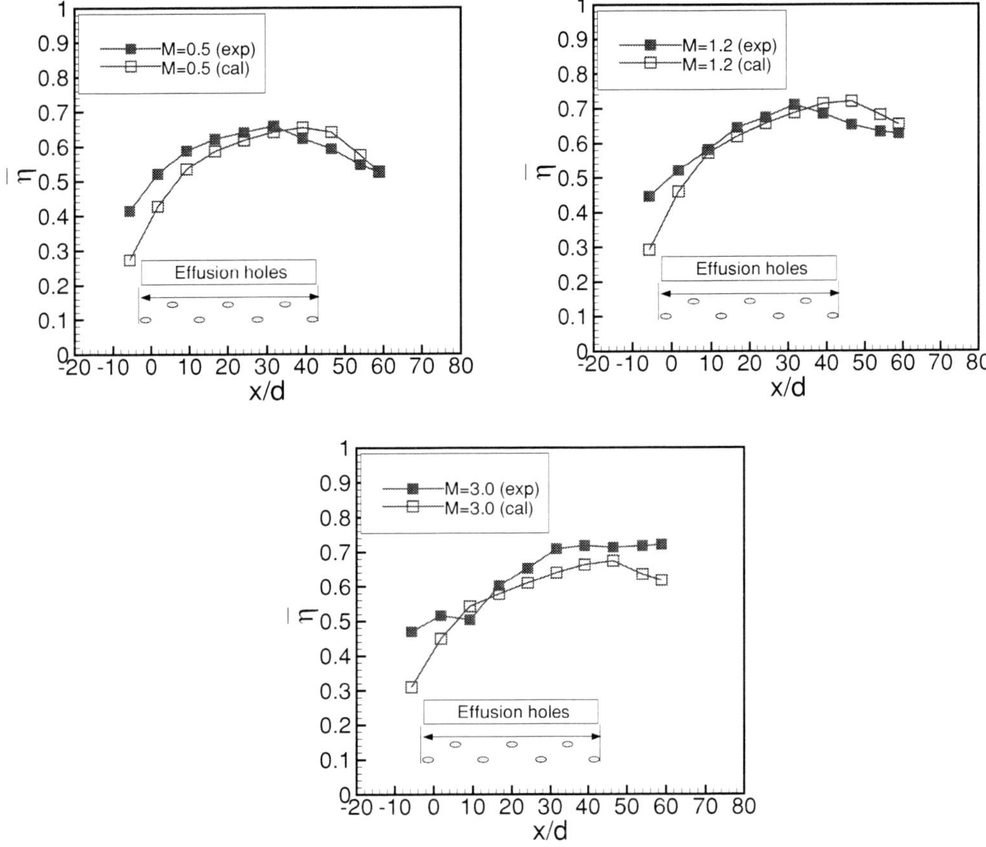

Figure 4.2.9: Laterally averaged effectiveness for the plastic test plate.

Metallic Test Plate

The calculated overall film-cooling effectiveness for the same three blowing ratios of the metallic test plate (Incoloy (800 H), $\lambda = 12.2$ W/mK) is shown in Fig. 4.2.10. In contrast to the plastic material, the predicted temperature distribution was found to be very homogenous at all blowing ratios, which is in good agreement to the experiment [18]. The experimental study also revealed that heat conduction inside the plate strongly affects cooling, in particular at higher blowing ratios.

This effect is very well reflected by the prediction: Due to the high thermal conductivity of the metallic plate, the flow pattern of the cooling jets cannot be identified. Moreover, an increase in the blowing ratio is accompanied by a monotonous enhancement of the film cooling effectiveness η even at the highest blowing ratio. Unlike the situation for the plastic material, lower

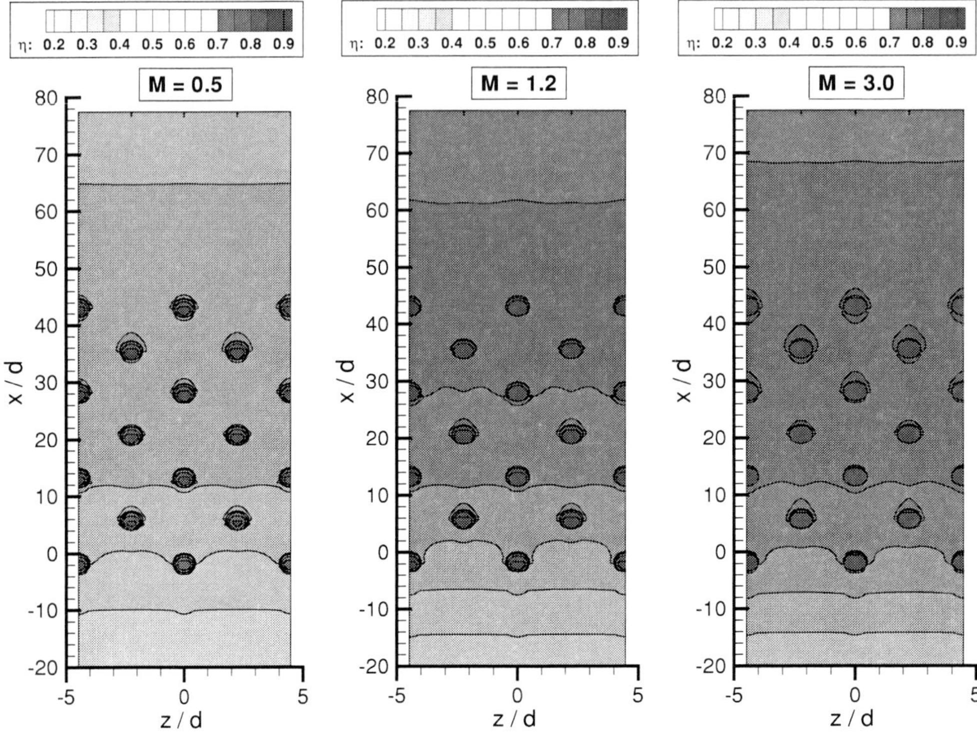

Figure 4.2.10: Calculated film cooling effectiveness distribution for the metallic test plate.

values of η are not found in the area of the first row of holes for $M = 3.0$ due to jet lift-off downstream the jet outlet. This phenomenon is completely compensated by the strong heat conduction inside the plate. Another less obvious effect is the intense cooling of the plate section upstream of the holes with increasing blowing ratio. Due to the higher thermal conductivity, this effect is more pronounced for the metallic test plate compared to the plastic material (Fig. 4.2.8).

The quantitative comparison of the laterally averaged effectiveness for the metallic test plate is shown in Fig. 4.2.11. Quite satisfactory results are obtained for η for downstream distances $x/d \leq 35$ at all blowing ratios even in the upstream portion of the first row of holes ($x/d < 0$). Further downstream ($x/d > 35$), significant discrepancies between calculation and experiment are found. The experimental data show a sharp decrease in η due to the missing conductive influence of the film cooling holes, whereas a moderate increase is found in the prediction with its maximum at $x/d \approx 60$. Those discrepancies are caused by an inadequate specification of the thermal wall boundary condition in the calculation. At the downstream end plane of the test plate a zero wall heat flux has been specified. However, in the experi-

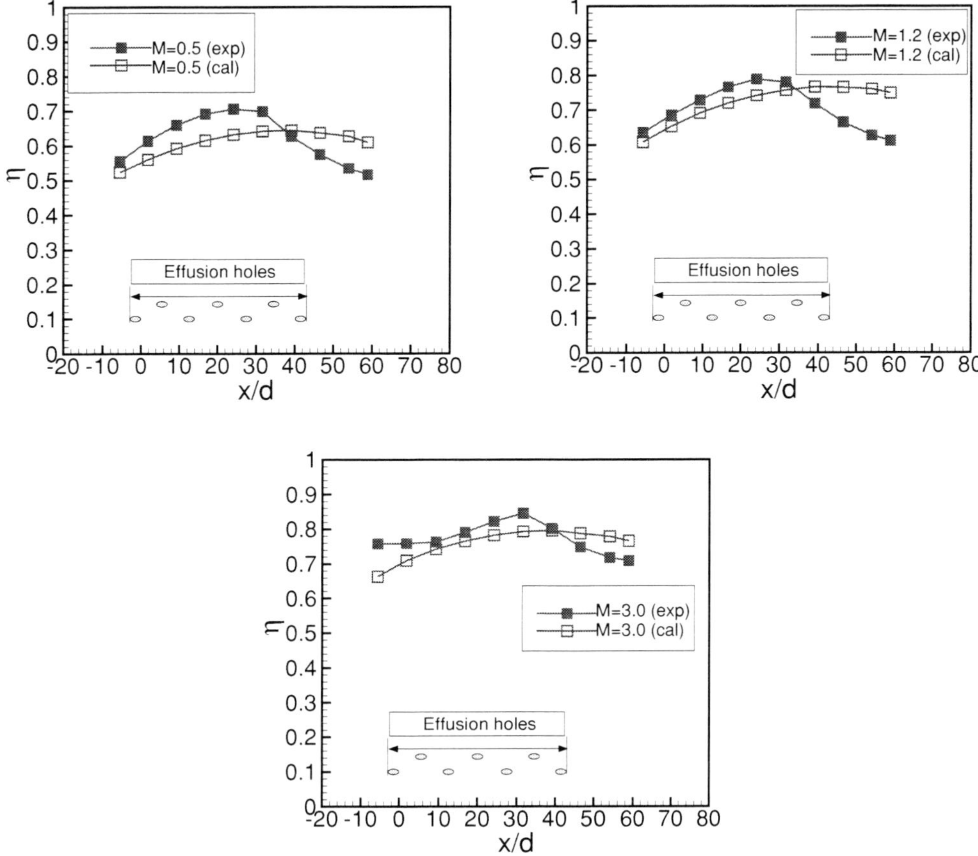

Figure 4.2.11: Laterally averaged effectiveness for the metallic material.

ment, there was a non-negligible heat flux through the seal material between the end plane of the test plate and the surrounding containment. It should be noted, that for the plastic plate (Fig. 4.2.9) the specification of the zero flux thermal wall condition at the end plane has a minor effect due to the low thermal conductivity of the material.

Temperature Profiles

Figure 4.2.12 shows, as a typical result, the comparison of the normalized temperature profiles for a blowing ratio of $M = 1.2$ and the metallic test plate. The profiles correspond to the centerline locations along the streamwise direction (see Fig. 4.2.13).

Remarkable good agreement between prediction and experiment was found at all locations x/d. Even the temperature profile in the near wall

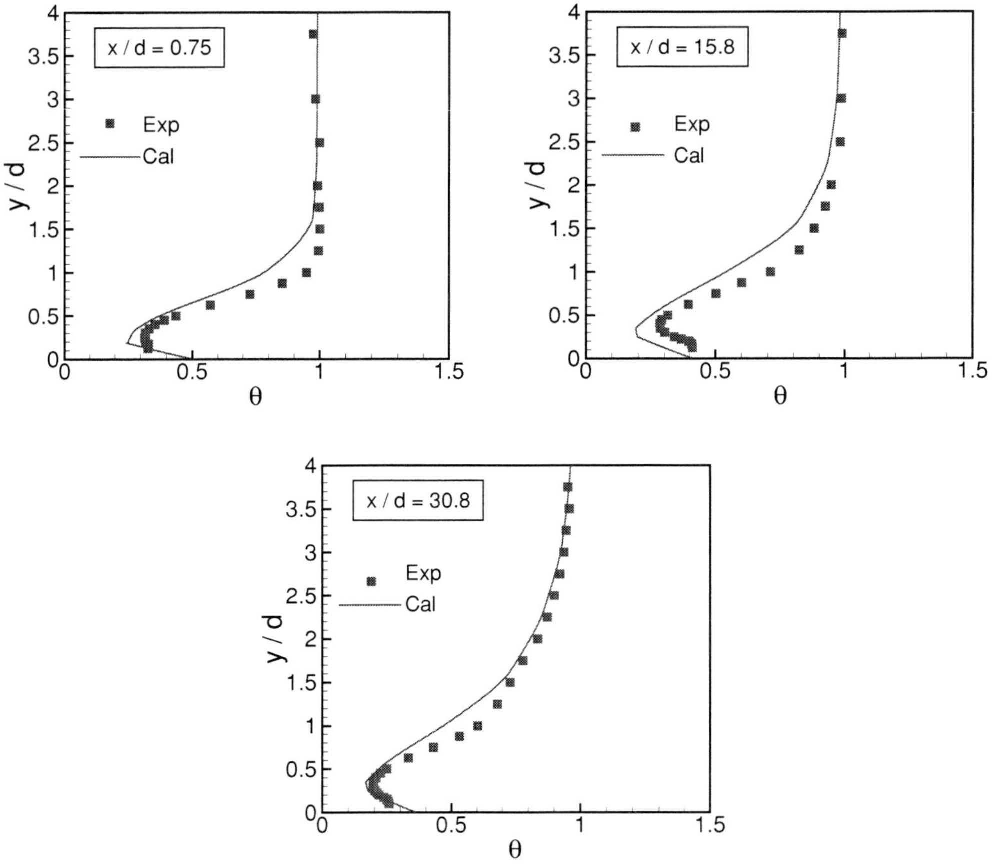

Figure 4.2.12: Calculated and measured temperature profiles for $M = 1.2$ and metallic test plate.

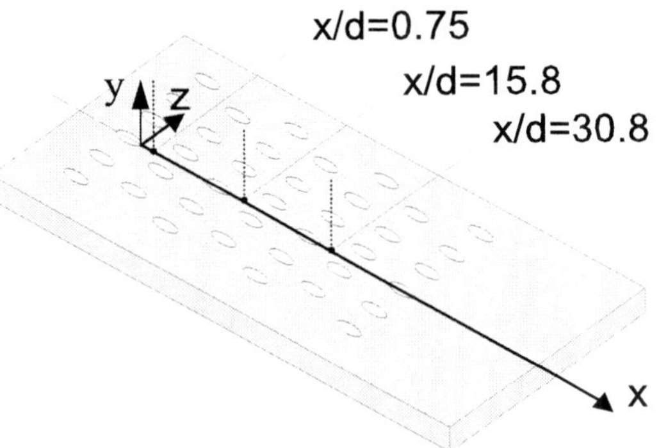

Figure 4.2.13: Sketch of the test plate with measurement locations for the temperature profiles.

region $y/d < 0.5$, which indicates cooling of the test plate, is very well captured by the calculation. However, a closer inspection reveals that the measured temperature profiles show a strong variation whereas the calculation gives a linear distribution near the wall because of the use of the wall functions. It is well known [20], that in order to capture this variation of the temperature profiles near the wall, a low Reynolds number (LRN) extension of the turbulence model must be used enabling a fine resolution of the near wall region. In an accompanying 2D study of highly complex flow [21], it was demonstrated that quite promising results can be obtained, when conjugate heat transfer phenomena are predicted by a LRN turbulence model.

4.2.4 Summary and Conclusions

A numerical procedure based on the finite volume method for predicting conjugate heat transfer in combustor liners has been developed and tested. The method has been shown to capture the heat transfer phenomena of an effusion cooled combustor liner where heat conduction inside the liner is a decisive factor. However, the application of the wall function approach was found to impose restrictions, particularly in capturing of both, the lift-off of the cooling jets at high blowing ratios and the temperature variation in the near wall region. Switching to a low Reynolds number extension of the turbulence model may overcome these deficiencies.

References

1. Launder, B. E., Spalding, D. (1974): The Numerical Computation of Turbulent Flows. *Computer Methods in Applied Mechanics and Engineering* **3**, 269–289.
2. Ferziger, J. H., Peric, M. (1996): Computational Methods for Fluid Dynamics, Springer-Verlag, Berlin-Heidelberg.
3. Giebert, D., Kurreck, M., Bauer, H.-J., Wittig, S. (1995): Dreidimensionale Strömungsvorgänge in Brennkammern mit komplexer Geometrie, Sonderforschungsbereich 167, Hochbelastete Brennräume – stationäre Gleichdruckverbrennung, Teilprojekt B1, 267–293.
4. Benz, E., Wittig, S., Noll, B. (1992): Dreidimensionale Strömungsvorgänge in Brennkammern mit komplexer Geometrie, Sonderforschungsbereich 167, Hochbelastete Brennräume – stationäre Gleichdruckverbrennung, Teilprojekt B1, 283–304.
5. Giebert, D., Koch, R., Schulz, A., Wittig, S. (1998): Evaluation of Advanced Low-Reynolds Number k-ε Turbulence Models for Predicting Convective Heat Transfer, 2nd EF Conference in Turbulent Heat Transfer, Manchester, UK.

6. Benz, E., Wittig, S. (1992): Prediction of the Interaction of Coolant Ejection with the Main Stream at the Leading Edge of a Turbine Blade: Attached Grid Application, Int. Symp. on Heat Transfer in Turbomachinery, Athen, Greece.
7. Giebert, D., Gritsch, M., Schulz, A., Wittig, S. (1997): Film-Cooling from Holes with Expanded Exits: A Comparison of Computational Results with Experiments, ASME Paper 97-GT-163.
8. Kurreck, M., Wittig, S. (1994): Prediction of Turbulent Three-Dimensional Combustor Flows on Parallel Computing Systems. Technical Report, Paderborn Center for Parallel Computing/TR-009-94, 74–76.
9. Patankar, S. V., Spalding, D. (1972): A Calculation Procedure for Heat, Mass and Momentum Transfer in Three-Dimensional Parabolic Flows. *Int. J. Heat Mass Transfer* **15**, 1787–1806.
10. Karki, K. Patankar, S. V. (1989): Pressure Based Calculation Procedure for Viscous Flows at All Speeds in Arbitrary Configurations. *AIAA Journal* **27**, No. 9, 1167–1174.
11. Rhie, C., Chow, W. (1983): Numerical Study of the Turbulent Flow Past an Airfoil with Trailing Edge Separation. *AIAA Journal* **21**, No. 11, 1525–1532.
12. Noll, B. (1992): Evaluation of a Bounded High-Resolution Scheme for Combustor Flow Computations, *AIAA Journal* **30**, No. 1, 64–69.
13. Noll, B., Wittig, S. (1991): Generalized Conjugate Gradient Method for the Efficient Solution of Three-Dimensional Fluid Flow Problems, *Numerical Heat Transfer*, Part B, **20**, 207–221.
14. Stone, H. (1968): Iterative Solution of Implicit Approximations of Multidimensional Partial Differential Equations, *SIAM Journal of Numerical Analysis* **5**, No. 3, 530–558.
15. Van der Vorst, H. A. (1992): Bi-CGSTAB: A Fast and Smoothly Converging Variant of Bi-CG for the Solution of Nonsymmetric Linear Systems, *SIAM Journal on Scientific and Statistical Computing* **13**, No. 2, 631–644.
16. Papanikolaou, E., Jaluria, Y. (1993): Mixed Convection from a Localized Heat Source in a Cavity with Conducting Walls, *Numerical Heat Transfer*, Part A, **23**, 463–484.
17. Martiny, M., Schulz, A., Wittig, S. (1995): Full-coverage Film Cooling Investigations: Adiabatic Wall Temperatures and Flow Visualization, ASME Paper 95-WA/GT-4.
18. Martiny, M., Schiele, R., Gritsch, M., Schulz, A., Kim, S. (1998): High Efficient Cooling Concepts for Low Emission Combustors, Collaborative Research Centre 167, High Intensity Combustors – Steady Isobaric Combustion.
19. Leylek, J., Zerkle, R. (1994): Discrete-Jet Film Cooling: A Comparison of Computational Results with Experiments, *ASME Journal of Turbomachinery* **116**, 358–368.
20. Giebert, D., Koch, R., Schulz, A., Wittig, S. (1998): Evaluation of Advanced Low-Reynolds Number k-ε Turbulence Models For Predicting Convective Heat Transfer, 2nd International Conference on Turbulent Heat Transfer II, Manchester, May 31–June 4.
21. Papanicolaou, E. L., Giebert, D., Koch, R., Schulz, A. (1998): A Discretization Approach for Conjugate Heat Transfer and Application to Turbomachinery Flows, Proceedings of the ASME, Heat Transfer Division – 1998 (HTD-Vol. 361 / PID-Vol. 3), edited by R. A. Nelson, Jr., L. W. Swanson, M. V. A. Bianchi, C., 349–362.

4.3 Experimental Investigation and Numerical Prediction of Radiative Heat Transfer

Rainer Koch*, Benedikt Ganz, Werner Krebs, Berthold Noll, and Sigmar Wittig

4.3.1 Introduction

Radiative heat transfer is an important mode of heat transfer in many combustion applications. In some cases, e. g. power plant boilers, it may even be dominating convective heat transfer. In gas turbine combustors, the contribution of radiation to the thermal heat load of the liner is approximately of the same order of magnitude as convective heat transfer.

Radiation also affects the temperature field of the reacting flow and thus many other processes, like droplet evaporation, chemical reactions, and pollution formation. Modern low-emission gas turbine combustors are characterized by high bulk flow velocities and short residence times. Therefore, radiation has only a minor impact on the temperature inside the combustor, but it is still sufficient to significantly affect the NO_x emissions. Figure 4.3.1 illustrates the effect of radiation on the temperature field of the diffusion flame to be described in Section 4.3.7.2

Subsequently, numerical methods suitable for predicting radiative heat transfers in gas turbine combustors will be discussed and appropriate experimental validation techniques will be presented. Also, experimental techniques for characterizing the radiative properties of typical gas turbine combustor liner materials, including Thermal Barrier Coatings will be elucidated. Concluding, the application of the experimental techniques as well as the numerical methods for the analysis of the radiative transfer of different model combustors will be illustrated.

* Institut für Thermische Strömungsmaschinen, Universität Karlsruhe, Kaiserstr. 12, 76128 Karlsruhe, Germany

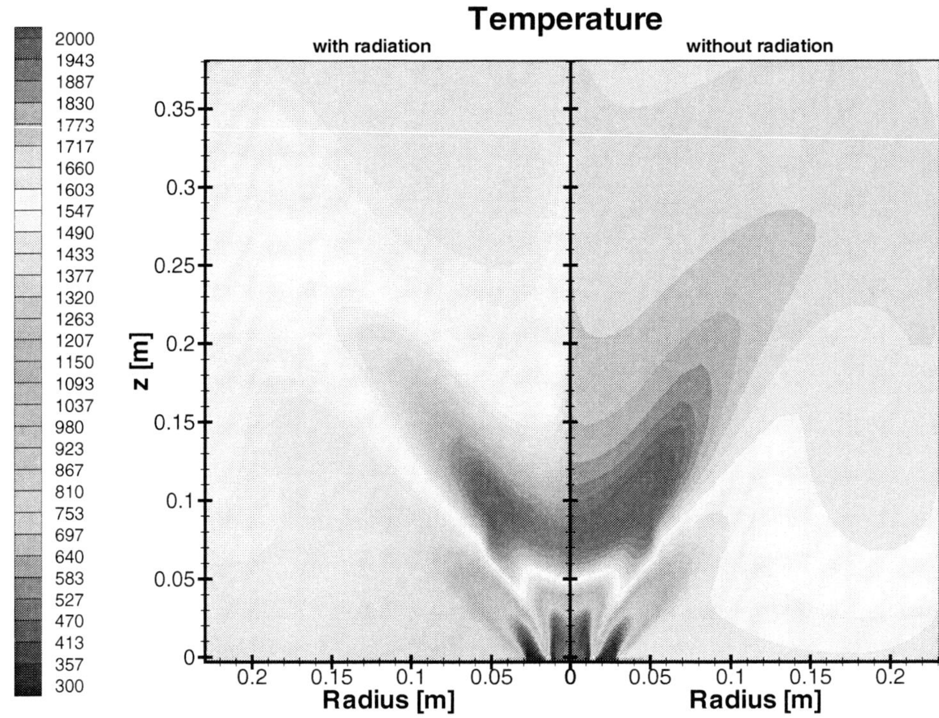

Figure 4.3.1: Effect of radiation on the temperature field of a diffusion flame.

4.3.2 Fundamentals of Radiative Transfer

Radiative heat transfer in combustion systems is primarily determined by the temperature, the concentration of participating species (CO, CO_2, H_2O, Soot), and the radiative properties of the combustor walls. Those quantities determine the absorption coefficient a, the scattering coefficient s and the emissive power I_b within a control volume as well as the emissivity ε of the walls. Radiative transfer itself is essentially a 5-dimensional problem, characterized by three spatial co-ordinates $\vec{r} = (x, y, z)$ and two directional co-ordinates $\vec{\Omega} = (\Theta, \Phi)$, as illustrated in Fig. 4.3.2.

The radiative transfer equation (RTE)

$$\vec{\Omega} \cdot \vec{\nabla} I(\vec{r},\vec{\Omega}) = -(a+s)\, I(\vec{r},\vec{\Omega}) + a I_b(\vec{r}) + \frac{s}{4\pi} \int\limits_{\Omega'} I(\vec{r},\vec{\Omega}')\, P(\vec{\Omega} \to \vec{\Omega}')\, \mathrm{d}\Omega' \quad (1)$$

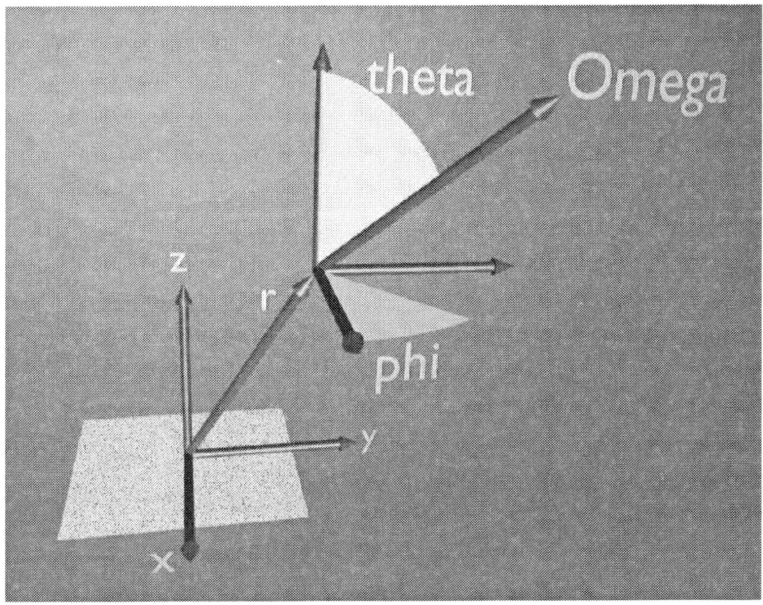

Figure 4.3.2: The spatial and directional co-ordinates of radiative transfer.

can readily be derived from a balance along a ray traversing a control volume containing an absorbing, emitting, and scattering medium. The corresponding boundary condition at the wall of the enclosure

$$I(\vec{r_b},\vec{\Omega}^+) = \varepsilon I_b(\vec{r_b}) + \int\limits_{\vec{n}\cdot\vec{\Omega}^-<0} I(\vec{r_b},\vec{\Omega}^-)\, R(\vec{\Omega}^- \rightarrow \vec{\Omega}^+)\, \mathrm{d}\Omega^- \tag{2}$$

is obtained from a balance of incident (direction $\vec{\Omega}^-$) and emerging (direction $\vec{\Omega}^+$) radiation at a wall element.

For problems of practical relevance, i. e. multi-dimensional radiative transfer and non-homogeneous distribution of the temperature and absorption coefficient, the RTE can be solved only numerically. Numerical methods appropriate for gas turbine combustors will be discussed in Section 4.3.3.

However, is has to be emphasized that the temperature and concentrations inside the combustor and the temperature of the liner must be known before starting with radiative transfer calculations. They may either be predicted by a combustion code or determined from experiment. Within a complete numerical prediction, an iterative procedure (cf. Fig. 4.3.3) is required because the reacting flow and radiative heat transfer are mutually affecting each other, in particular if soot is present.

4.3.3 Numerical Prediction of Radiative Transfer

The numerical methods required for predicting radiative transfer in combusting systems can be subdivided into two groups:

- methods for determining the radiative properties
- methods for solving the RTE

Additionally, also the interaction between radiation and reacting flow has to be accounted for, requiring an iterative coupling procedure.

In Fig. 4.3.3 it is illustrated schematically, which information has to be provided and exchanged during the different steps of a complete modeling of a combusting system. The first step is to determine the radiative properties of the different participating media, i. e. gases, particles and solid materials. These properties as well as the mesh are passed to the RTE solver. Once the solution of the RTE is obtained, the enthalpy source terms due to radiation are fed into the solution procedure of the reacting flow. The converged solution of the reacting flow in turn provides the temperatures and concentrations required for determining the radiative properties.

Subsequently, the methods for predicting the radiative properties and the methods for solving for the RTE will be discussed in detail.

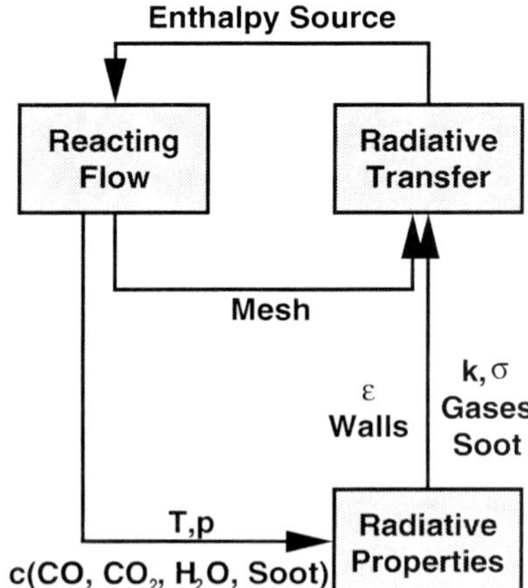

Figure 4.3.3: Schematical flow chart for predicting reacting flow and radiative transfer.

4.3.3.1 Radiative Properties

When considering gas turbine combustors, the radiative properties of the participating gases CO, CO_2, H_2O, and of the soot particles have to be determined. Additionally, also the reflectivity/emissivity of the liner has to be known.

Various methods with different degrees of sophistication have been suggested for the representation of the radiative properties [1]. The methods differ mainly with respect to the achievable accuracy, and it is still a matter of discussion which one is most appropriate for a certain application. The major issue is whether a spectral representation is required or whether wavelength averaging, i. e. the gray medium approximation, is sufficient. In general, a spectral representation provides significantly higher accuracy, however, at the expense of much higher computational effort.

With respect to gas turbine combustors, the selection is mostly a matter of the flame characteristics. In modern premixing, low-nox combustors no soot is present, and the radiative intensity varies strongly with wavelength, particularly in the range from 1.5 to 5.0 μm where most of the radiative energy concentrated, as illustrated in Fig. 4.3.4. Those combustors require a detailed spectral representation of the radiative properties.

In contrast, conventional combustors with diffusion flame exhibit significant amount of soot in the flame, and the radiation spectra are reasonably smooth. For those cases a spectrally averaged representation of the radiative properties might be sufficient.

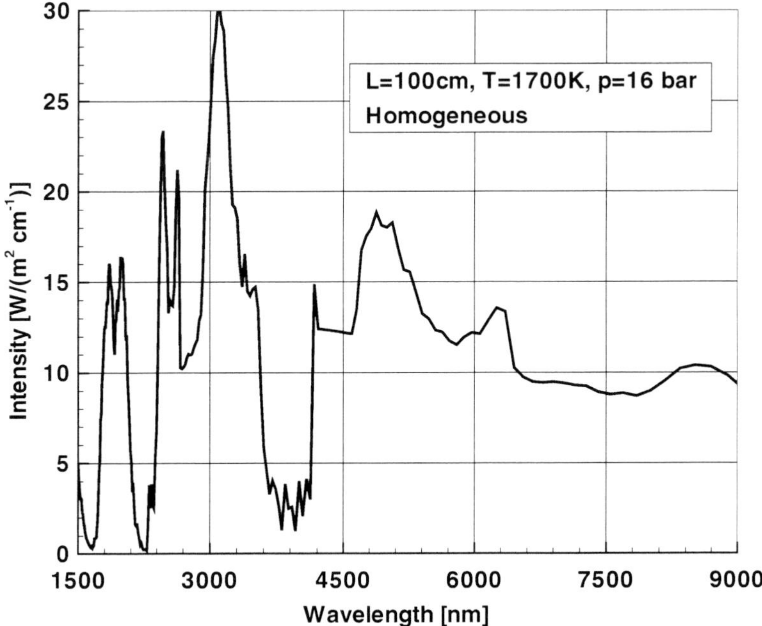

Figure 4.3.4: Intensity spectrum of a premixing gas turbine combustor.

4.3.3.1.1 Gases

For the representation of the radiative properties of gases, a narrow band representation based on the work of Soufiani and Taine [2] has proven to be an adequate choice for gas turbine combustors. Narrow band models do not take into account the details of the fine structure of the gaseous radiation bands. The fine-structure is implemented by means of the statistical model of Goody [3].

Comparisons of different gray gas approximations [4–6] to spectral narrow band representation revealed that gray gas models may lead to errors in the radiative flux of about 60 % [7], in particular when steep temperature gradients are present inside the flame [8]. Therefore, gray gas approximations have to be used with great care when gaseous radiation is predominant.

4.3.3.1.2 Soot

The radiative properties of soot particles can be calculated from Rayleigh theory, as the soot particle size is in the regime of 10–50 nm, and thus much smaller than the wavelength of radiation (1–10 μm).

According to Rayleigh theory, the extinction coefficient is given by

$$k_{\text{ext,Rayleigh}} = \frac{\pi}{\lambda} \cdot f_V \cdot \frac{36nk}{(n^2 + k^2)^2 + 4(n^2 - k^2) + 4} \tag{3}$$

The index of refraction $m = n - ik$ depends on the composition of the soot and is affected by the fuel and the flame characteristics [9, 10]. Equation (3) reveals that the extinction coefficient is proportional to the soot volume fraction f_V and proportional to $1/\lambda$. Also, the soot particle size has no effect on the extinction coefficient.

A closer examination [11] reveals that scattering from soot particles is isotropic, i. e.

$$P(\vec{\Omega} \rightarrow \vec{\Omega}') \tag{4}$$

and can be neglected in comparison to absorption because of the small size of the soot particles.

4.3.3.1.3 Liner Materials

Besides nickel based metallic alloys, pure ceramics, or thermal barrier coatings (TBCs) are utilized for thermally high loaded combustor liners. The radiative properties of the liner cannot be calculated from first principles. They have to be determined by extensive experimental investigations.

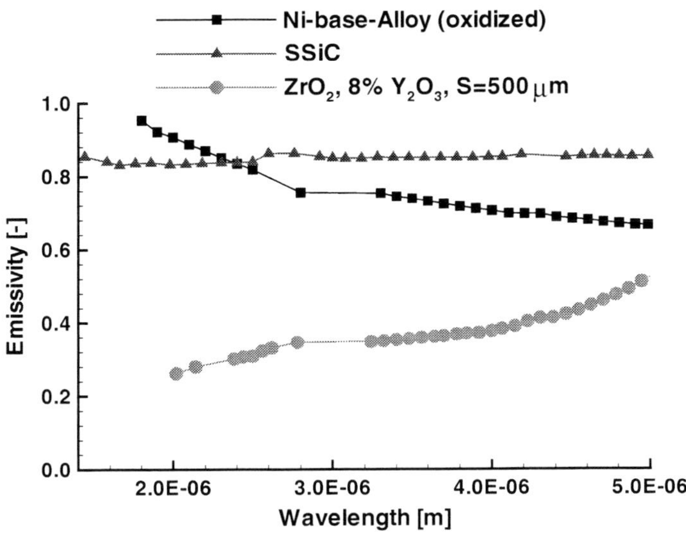

Figure 4.3.5: Normal spectral emissivity of liner materials.

Some of the appropriate experimental techniques will be discussed in Section 4.3.5.

In order to illustrate the radiative properties of typical gas turbine liner materials, the emissivity of oxidized nickel base alloy (Inconel 617), sintered SiC (SSiC) and plasma-sprayed ZrO_2 TBC are compared in Fig. 4.3.5 [12].

As to be discussed also in Section 4.3.5, most liner materials can generally be assumed to be diffusely reflecting without significant loss of accuracy. Thus, the reflection term within the boundary condition (Eq. (2)) can be written as

$$\int_{\vec{n}\cdot\vec{\Omega}^-<0} I(\vec{r_b},\vec{\Omega}^-)\, R(\vec{\Omega}^- \rightarrow \vec{\Omega}^+)\, \mathrm{d}\Omega^- = \frac{(1-\varepsilon)}{\pi} \int_{\vec{n}\cdot\vec{\Omega}^-<0} I(\vec{r_b},\vec{\Omega}^-)|\vec{n}\cdot\vec{\Omega}^-|\mathrm{d}\Omega^- \quad (5)$$

4.3.3.2 Radiative Transfer

Among the various methods for predicting radiative transfer in participating media [13], the Discrete Ordinates Method (DOM) [14, 15] is presently judged as the best compromise between accuracy and computational effort.

4.3.3.2.1 The Discrete Ordinates Method

The DOM is capable to resolve the non-isotropic directional characteristics of radiative heat transfer by subdividing the directional space into discrete solid angles. A typical widely used angular discretization scheme, the S-4 quadrature [16] is illustrated in Fig. 4.3.6.

However, the discretization into a finite number of solid angles gives raise to the major deficiency of the Discrete Ordinates Method, the so-called ray effects [17, 18]. These effects are particularly pronounced in cases where hot spots (e. g. flames) are enclosed by a colder medium. Figure 4.3.7 illustrates the ray effects by considering a cubical enclosure. The wall of the cube are assumed to be non-reflecting and to be at low temperature, i. e. a emitted flux (I_b/π) of unity. The absorbing, emitting medium inside the cube is set to the same temperature with the exception of the central control volume and set to a non-dimensional emitted flux (I_b/π) of hundred.

Because of the geometry of the test case, a smooth, symmetrical distribution of fluxes at the surface of the cube is expected with the highest fluxes at the center of the faces and the lowest at the vertices. Due to pronounced ray effects the actual flux distribution is highly non-homogeneous and resembles the arrangement of the discrete ordinates.

Ray effects are inherent to the DOM, and thus can't be avoided completely. However, they can be reduced significantly by using higher order quadrature schemes with a finer discretization of the solid angle [19, 20].

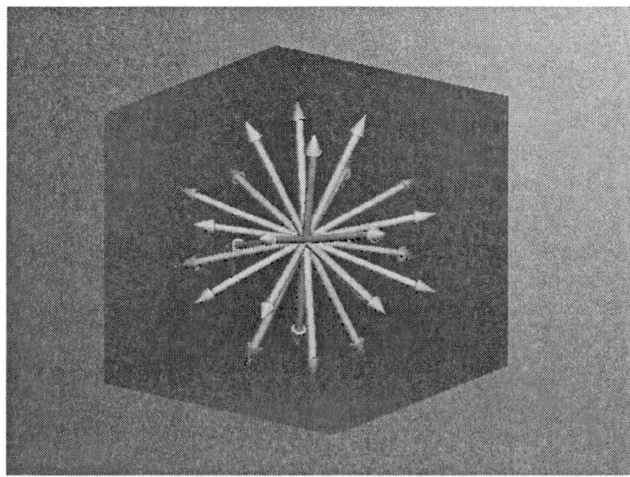

Figure 4.3.6: Solid angle discretization of the S-4 quadrature.

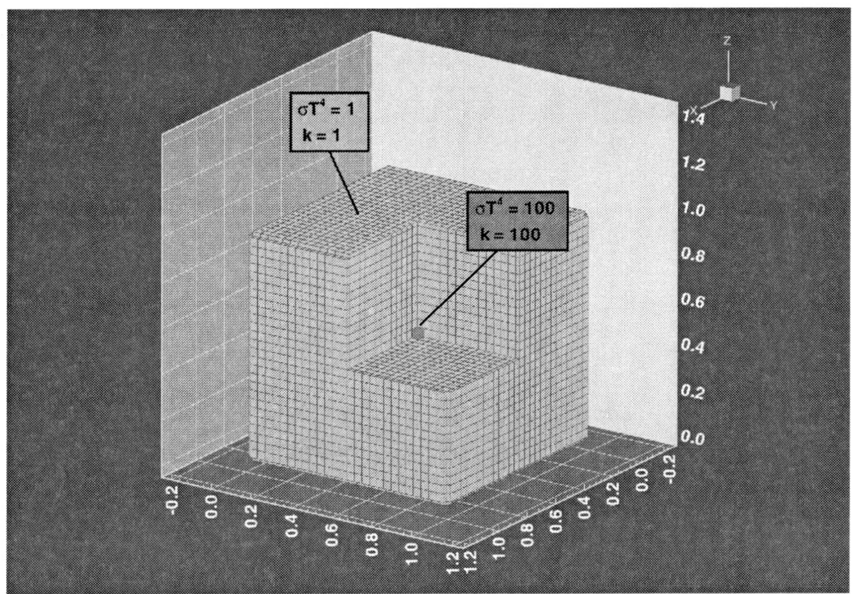

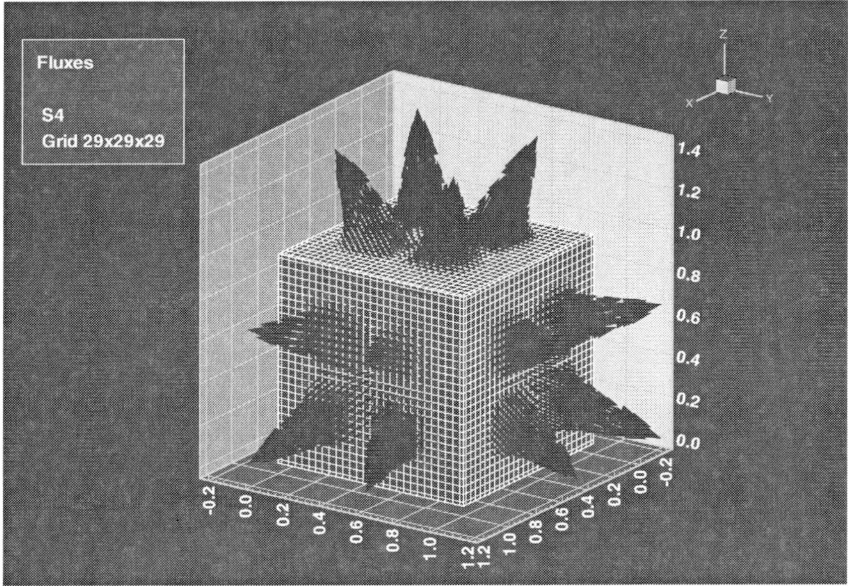

Figure 4.3.7: Ray effects of a cubical enclosure with hot center element.

4.3.3.2.2 Numerical Solution of the Discrete Ordinates Equations

After preselecting a set of M representative discrete directions $\vec{\Omega}_m$, $m = 1,\ldots,M$ together with the corresponding weights w_m from a quadrature scheme [19–21], the radiative transfer equation as well as the corresponding boundary conditions can be written as a set of equations for each direction.

$$\vec{\Omega}_m \cdot \vec{\nabla} I(\vec{r},\vec{\Omega}_m) = -(a+s)I(\vec{r},\vec{\Omega}_m) + aI_b(\vec{r}) + \frac{s}{4\pi}\sum_{m'=1}^{M} w_{m'} I(\vec{r},\vec{\Omega}_{m'}) \tag{6}$$

$$I(\vec{r_b},\vec{\Omega}_m) = \varepsilon I_b(\vec{r_b}) + \frac{(1-\varepsilon)}{\pi}\sum_{m'}^{\vec{n}\cdot\vec{\Omega}_{m'}<0} w_{m'} |\vec{n}\cdot\vec{\Omega}_{m'}| I(\vec{r_b},\vec{\Omega}_{m'}) \tag{7}$$

In Eqs. (6) and (7) the simplifications of isotropic scattering and diffuse reflection have been incorporated.

Currently two different numerical approaches for solving the DOM equations have been established. The most obvious one is to solve Eqs. (6) and (7) directly. However, because of the hyperbolic nature of Eq. (6), artificial numerical diffusion has to be added to ensure stability and convergence of the numerical solution. This is generally achieved by using discretization schemes of the upwind type [22].

The other possibility is to use the even-parity formulation [23] of the RTE. The adoption of this method to radiative transfer is described comprehensively by Song et al. [24]. The basic idea of the even-parity formulation is to transform the set of first order differential equations (Eq. (6)) into an equivalent set of second order differential equations by introducing the variables

$$F(\vec{r},\vec{\Omega}_m) = I(\vec{r}, +\vec{\Omega}_m) + I(\vec{r}, -\vec{\Omega}_m) \tag{8}$$

$$G(\vec{r},\vec{\Omega}_m) = I(\vec{r}, +\vec{\Omega}_m) - I(\vec{r}, -\vec{\Omega}_m) \tag{9}$$

By adding and subtracting both, the RTE and the boundary conditions for each pair of opposite directions $\vec{\Omega}_m, -\vec{\Omega}_m$, the even-parity formulation in terms of the variable F is derived:

$$\begin{aligned}\vec{\Omega}_m \cdot \vec{\nabla}\left[\frac{1}{(a+s)}\vec{\Omega}_m \cdot \vec{\nabla} F(\vec{r},\vec{\Omega}_m)\right] &= (a+s)F(\vec{r},\vec{\Omega}_m) - 2aI_b(\vec{r}) \\ &\quad - \frac{s}{4\pi}\sum_{m'=1}^{M/2} 2w_{m'} F(\vec{r},\vec{\Omega}_{m'})\end{aligned} \tag{10}$$

$$F(\vec{r_b},\vec{\Omega}_m) - \frac{1}{(a+s)} sign(\vec{n}\cdot\vec{\Omega}_m)\vec{\Omega}_m \cdot \vec{\nabla} F(\vec{r_b},\vec{\Omega}_m) = \varepsilon I_b(\vec{r_b}) \tag{11}$$

$$+\frac{(1-\varepsilon)}{\pi}\sum_{m'=1}^{M/2} w_{m'}|\vec{n}\cdot\vec{\Omega}_{m'}|\left[F(\vec{r_b},\vec{\Omega}_{m'}) + sign(\vec{n}\cdot\vec{\Omega}_{m'})\frac{1}{(a+s)}\vec{\Omega}_{m'} \cdot \vec{\nabla} F(\vec{r_b},\vec{\Omega}_{m'})\right]$$

In contrast to the conventional discrete ordinates method, there are only $M/2$ unknowns $F(\vec{r},\vec{\Omega}_1)\ldots F(\vec{r},\vec{\Omega}_{M/2})$ to be determined directly within the even-parity method. The remaining quantities $F(\vec{r},\vec{\Omega}_{M/2+1})\ldots F(\vec{r},\vec{\Omega}_M)$ are then derived by back-substitution using Eqs. (8) and (9).

Equation (10) is of second order and of the parabolic type and can be discretized without the need to add artificial diffusion.

4.3.4 Experimental Techniques

For both, the radiation measurements of flames and combustors as well as the investigation of the radiative properties of combustor liner materials an IR-spectroscope has been developed. A schematic of the spectroscope is shown in Fig. 4.3.8.

The IR-spectroscope is capable to perform spectral measurements from 1.4 to 5.4 µm, with a resolution of 20 nm at $\lambda < 2.8$ µm and of 30 nm at $\lambda > 2.8$ µm. The time response of the spectroscope covers the range up to 1 kHz, enabling time resolved studies of turbulence induced radiation fluctuations.

In the case of combustor measurements, the radiation incident normally onto the combustor walls is collected by the spectroscope. For liner material investigations, the directional, spectral intensity emitted by a heated specimen is recorded.

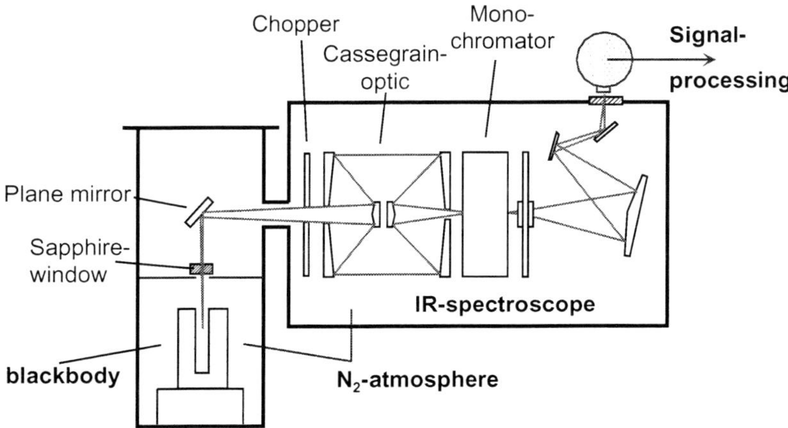

Figure 4.3.8: Infrared spectroscope.

4.3.5 Investigation of Gas Turbine Liner Materials

Extensive experimental investigations have been carried out at the Institut für Thermische Strömungsmaschinen [12, 25] in order to characterize the radiative properties of gas turbine liner materials. Typical results have been discussed in Section 4.3.3.1 in order to illustrate the characteristics of the different liner materials (nickel based alloys, TBC coatings, Si based ceramics). The emissivities have been recorded from heated specimen at a surface temperature of 973 K and are in good agreement with other data [26].

Recent experimental investigations at the Institut für Thermische Strömungsmaschinen have been focused on emissivity measurements of TBCs and the determination of the bi-directional reflectance distribution function of all relevant liner materials. The emphasis of this section will be mainly on these two topics.

4.3.5.1 Spectral Emissivity of Thermal Barrier Coatings

The application of TBCs can significantly reduce the thermal heat load of the combustor liner due to lower thermal conductivity and lower radiative emissivity and absorptivity. As illustrated by Fig. 4.3.5, the emissivity can be lowered by zirconia barrier coatings by about 50 % compared to pure metallic or ceramic materials leading to significant reduction of radiative heat transfer to the liner.

For gas turbine applications, TBCs are applied in thin layers on nickel- or cobalt based super-alloy substrates by plasma-spraying. The radiative emissivity of a TBC coated liner depends not only on the composition of the coating but also on its thickness.

Figure 4.3.9 illustrates for typical plasma-sprayed zirconia TBCs, how the spectral normal emissivity is affected by the thickness of the coating. Zirconia TBCs can be regarded as dense highly scattering media and, thus, the coating is semi-transparent at lower thicknesses. It was found from the experimental investigations [27] that zirconia TBCs can be considered as opaque above a thickness of 0.6 mm. At lower thickness, the resulting emissivity is given by a superposition of the emissivity of the TBC and the bond-coating which is located between the metallic substrate and the TBC in order to prevent hot gas corrosion.

Generally, stabilizing agents have to be added to zirconia TBCs in order to prevent martensitic conversion at higher temperatures. The composition of the TBC due to different stabilizing agents affects the thermal conductivity as well as the emissivity of the TBC, and a compromise in the composition has to be found in order to reduce the total heat transfer to the liner. Figure 4.3.10 illustrates the influence of different stabilizing agents on the emissivity of zirconia TBCs.

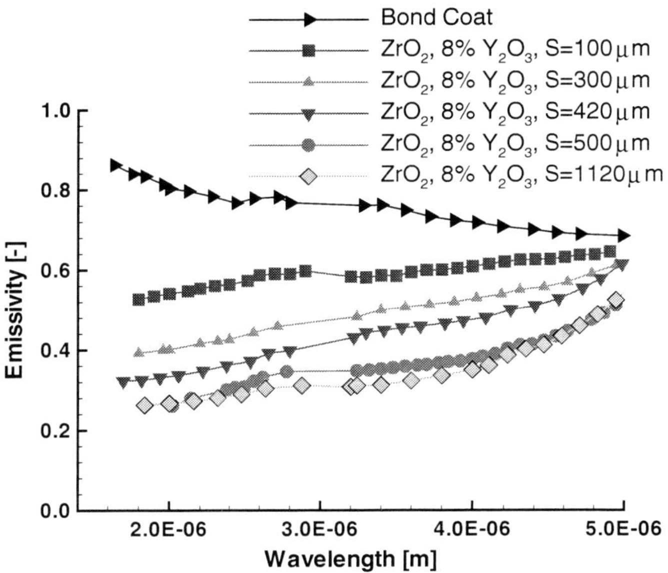

Figure 4.3.9: Effect of the thickness of zirconia TBCs.

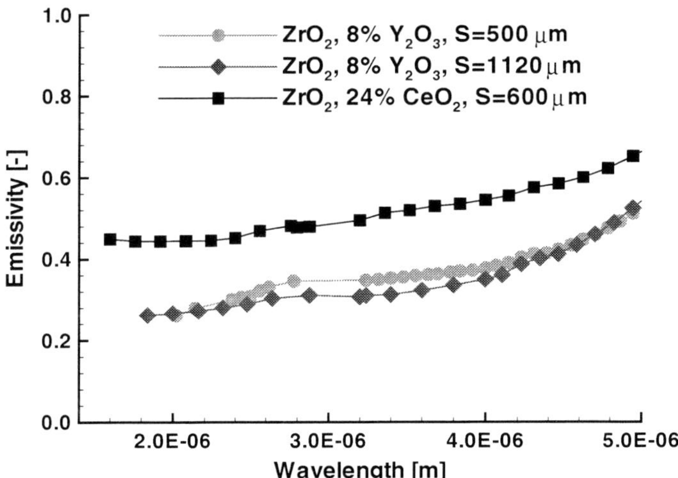

Figure 4.3.10: Effect of TBC stabilizing agent.

4.3.5.2 Reflectivity of Liner Materials

It is well known that plane metallic materials exhibit rather specular than diffuse reflection characteristics. Various experimental and numerical investigations have been performed at the Institut für Thermische Strömungsmaschinen to study the effect of non-diffuse reflection characteristics with particular emphasis on the assessment of the errors induced by approximating specular or partly specular reflecting surfaces by diffuse reflecting surfaces. All these investigations were performed in the wavelength regime of visible light in order to simplify the experimental set-up.

In a first step, the reflection characteristics of typical gas turbine liner materials have been recorded in terms of the bidirectional reflectance distribution function $R(\vec{\Omega}^- \rightarrow \vec{\Omega}^+)$ (c.f. Eq. (1)). The definition of the bidirectional reflectance distribution function is illustrated in Fig. 4.3.11.

Figure 4.3.12 shows the reflection characteristics of a plasma sprayed zirconia TBC. The plot at the left side gives the directional reflectance for a single angle of incidence (30°), the plot at the right side the total hemispherical reflectivity as function of the angle of incidence.

As to be expected, it was found that the reflectivity of zirconia TBCs is nearly diffuse due to the porous and rough surface. The hemispherical reflectivity decreases with increasing angle of incidence and corresponds well to the Lambertian law of reflection, $\rho = \rho_n \cos\Theta_i$, with a normal hemispherical reflectivity of $\rho_n = 0.626$. The value of ρ_n is in good agreement with emissivity measurements of similar samples at infra-red wavelengths [25, 12].

It is well known that the reflection characteristics of surfaces are shifted towards specular reflection at higher ratios of wavelength to surface roughness [28]. Therefore, as the measurements were performed with visible light, specular or even mirror like reflectance is to be expected in the infra-red. In order to study the influence of the ratio of wavelength to surface roughness, aluminum sheet metals with different surface roughness were investigated. The surface roughness was varied by exposing the sample to different surface treatments including blasting with sand or glass spheres.

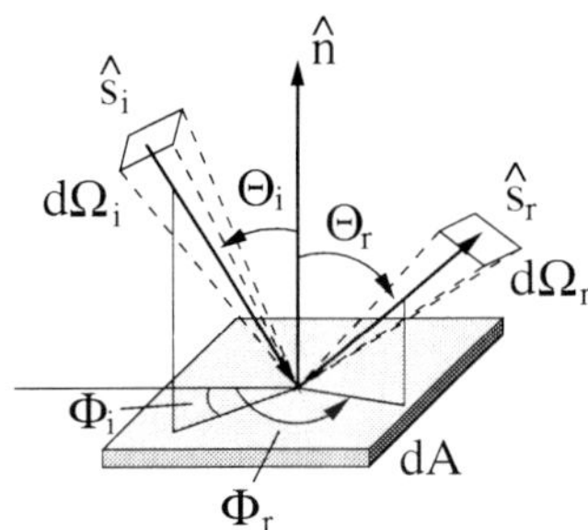

Figure 4.3.11: Definition of the quantities determining the surface reflection.

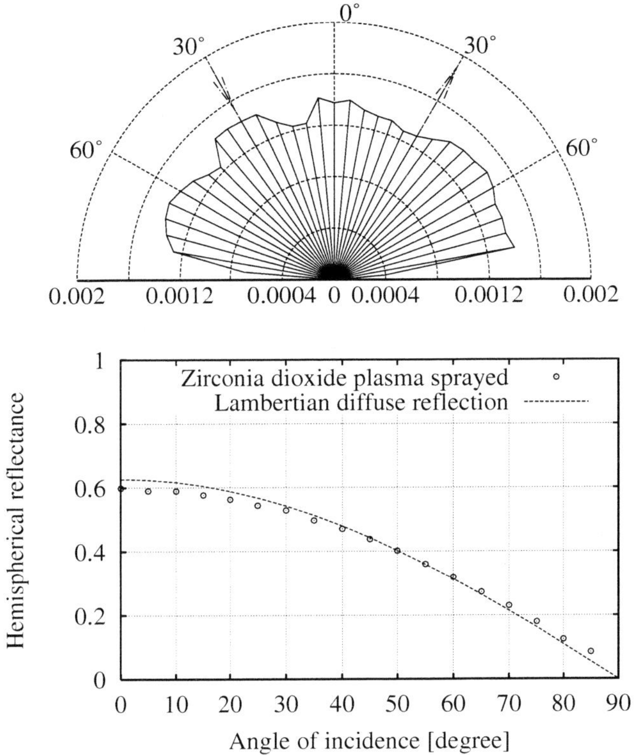

Figure 4.3.12: Reflectivity of zirconia TBC.

The typical surface roughness of sand blasted aluminum was in the range of 1–6 µm. Figure 4.3.13 reveals that the sand blasted aluminum surface, which exhibits a lower ratio of surface roughness to wavelength compared to the zirconia TBC, shows a more specular reflectance characteristic which becomes more pronounced at greater angles of incidence. However, it is remarkable that the maximum of reflected intensity is not found at the mirror like direction, but at an angle closer to the surface. Nevertheless, the hemispherical reflectivity of sand blasted aluminum can also be well represented by the Lambertian law.

Shining aluminum shows almost mirror like reflectance (Fig. 4.3.14), with most of the reflected radiation confined to an angle of approximately 10 degrees around the mirror like direction. As the surface wasn't polished or otherwise machined, the hemispherical reflection is considerably lower compared to the sand blasted sample due to the oxidized surface. Hence, the normal reflectance of $\rho_n = 0.28$ for an zero angle of incidence was found to be much lower than values of $\rho_n = 0.4$ to 0.9 given in other publications [26].

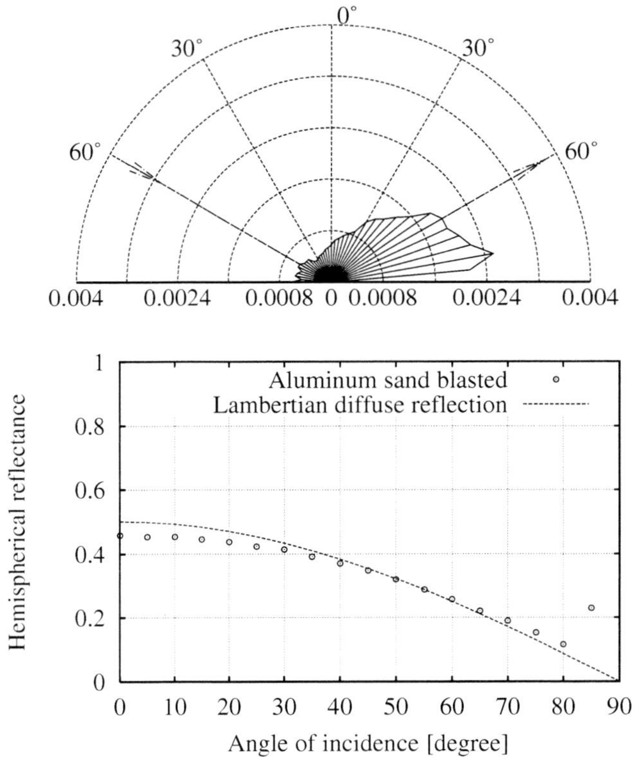

Figure 4.3.13: Reflectivity of sand blasted aluminum.

The hemispherical reflectance of the shining aluminum shows an increase with towards greater angles of incidence. As consequence, the reflection characteristics of shining surface can't be represented by Lambert's law.

4.3.6 Radiation in Enclosures with Non-Diffuse Reflecting Surfaces

In most radiative heat transfer predictions the combustors walls are assumed to emit and reflect diffusely, mainly because specular reflection can't be implemented accurately into most radiative transfer models, like the Discrete Ordinates method, Flux method, Discrete Transfer method, and P-1 method. Therefore, it is of interest to quantify the errors induced by representing specular or partly specular reflecting surfaces by diffuse emitting surfaces.

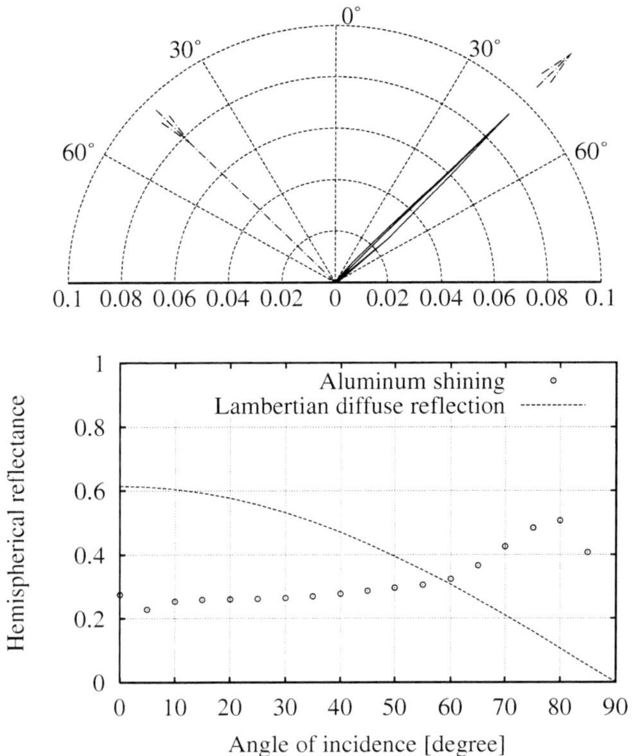

Figure 4.3.14: Reflectivity of shining aluminum.

A simple test facility resembling an enclosure with an extended radiation source and free inter-reflectance between the walls was build. The measured angular intensity distribution was compared to numerical predictions from different radiative heat transfer models in order to analyze the accuracy of the methods.

4.3.6.1 Test Section and Experimental Techniques

The test section is essentially a cubic enclosure with inner dimensions of 190 mm (right part in Fig. 4.3.15). The extended radiation source is provided by one of the walls of the enclosure. This wall consists of scattering glass illuminated by an extended light source which is set up by an array of fluorescent light tubes placed in front of a mirror channel (left part in Fig. 4.3.15). The other walls of the enclosure consist of solid material of well known reflection characteristic.

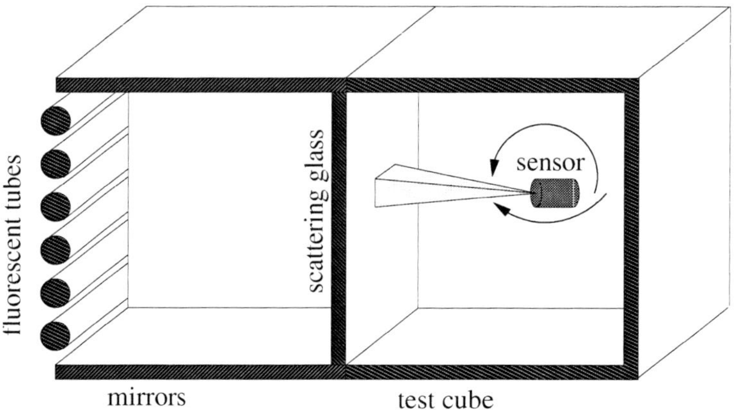

Figure 4.3.15: Schematic of the test section.

A photodiode sensor was installed 65 mm away from the right wall and 65 mm from the back wall at half height of the cubical enclosure. For the measurements, the sensor was rotated around its aperture with a full range of 360 degrees in azimuthal and 90 degrees in polar direction in order to record the angular intensity distribution within a complete hemisphere.

4.3.6.2 Comparison with Radiative Transfer Models

Using the Ray Tracing method, the Zone method [29] and the Discrete Ordinates method [30], the radiative intensity at the location of the probe was calculated. Within the Ray Tracing method, the specular or partly specular reflection characteristics of the walls have been accounted for. In case of the Zone method and Discrete Ordinates method, diffuse reflecting walls were assumed, with the normal reflectivity set to the best fit according to the Lambertian law of reflection. For the Zone method, the walls have been subdivided into 20 x 20 elements. The discrete ordinates predictions are based on a S-8 quadrature [19].

Figure 4.3.16 shows the comparison of the predicted angular intensity distributions for the walls of the enclosure consisting of sand blasted and shining aluminum. All intensities are normalized by the normal emission of the scattering glass. Additionally, also the cross-section of the enclosure is displayed in Fig. 4.3.16. The illuminated wall is located at the left side.

For sand blasted aluminum, the predictions of all radiative transfer models are in good agreement with the experimental data. In the case of shining aluminum, the problems associated with non-diffuse reflecting surfaces become apparent: For the directions not directly facing the illuminated wall, i. e. where reflection is predominant, the intensity is underestimated

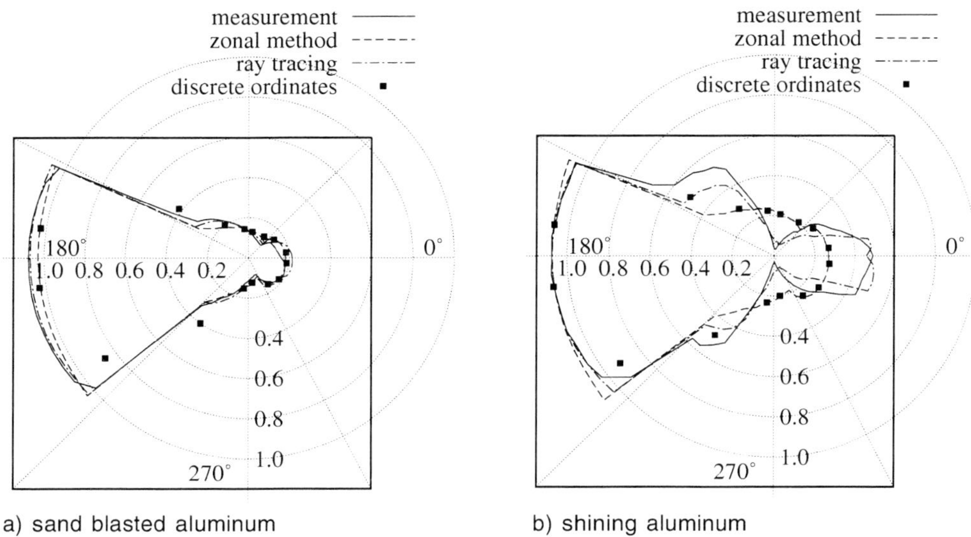

Figure 4.3.16: Comparison of the angular intensity distributions.

by over 50 % by the Zone and Discrete Ordinates method, because within both methods specular reflection is approximated by diffuse boundary conditions. The Ray Tracing method is capable to predict the intensity distribution quite correctly, because specular reflection is accounted for. However, for all methods the error in the radiosity $I = \int_{4\pi} I \mathrm{d}\Omega$ is below 10 %. Thus, for predicting integral quantities like volumetric radiative energy source terms and the wall heat flux, as required for engineering applications, it seems justified to neglect the specular reflection characteristics of liner materials. Therefore, most combustor liner materials may be assumed to be diffusely reflecting.

4.3.7 Radiation in Combustors

In this section the prediction of the reacting flow including radiative heat transfer will be compared to experimental data. Additionally, the impact of radiation on the reacting flow at realistic combustion conditions will be discussed.

4.3.7.1 Numerical Techniques

For predicting the reacting flow, a 3-D finite-volume method based code [31–33] was used. The standard k,ε-model [34] as well as an extended k,ε-model [35], which is capable to account for the non-isotropic turbulence of swirling flows, have been employed. The chemical reactions are modeled by an eddy-dissipation-model.

The 3-D radiative heat transfer code is based on the Discrete Ordinates approach. It utilizes also a finite-volume discretization. For the present calculations, the same mesh was used for predicting the reacting flow as well as the radiative transfer. The Discrete Ordinates method is based on the hyperbolic equations (Eqs. (6), (7)). The spectral radiative properties of the combustion gases have been calculated by the methods described in Section 4.3.3.1. The walls of the combustor were assumed to be black and diffusely emitting.

The final solution is achieved iteratively by calculating alternatively the reacting flow and the radiative heat transfer and by taking into account the radiative energy source term within the energy balance of the reacting flow (cf. Section 4.3.3).

4.3.7.2 Model Combustor

The model combustor (Fig. 4.3.17) has been developed at the Engler-Bunte Institut. The combustor is fired by natural gas. The thermal power is 150 kW at an air fuel ratio of 0.83 and a swirl number of 0.9. The walls of the combustion chamber are cooled by water. The geometry (D = 0.5 m), the flame configuration (type-II swirling diffusion flame) and the highly turbulent flow conditions are characteristic for industrial combustors.

The burner nozzle is located at the bottom of the cylindrical combustion chamber. It consists of a swirl generator with movable block design [36] which enables to adjust the swirl number of the combustion air over a wide range. The combustion air passes the swirl generator and is then fed through a concentric orifice (D = 60 mm) into the combustion chamber. Natural gas is injected centrally by an axial jet (outer diameter 26 mm, inner diameter 20 mm).

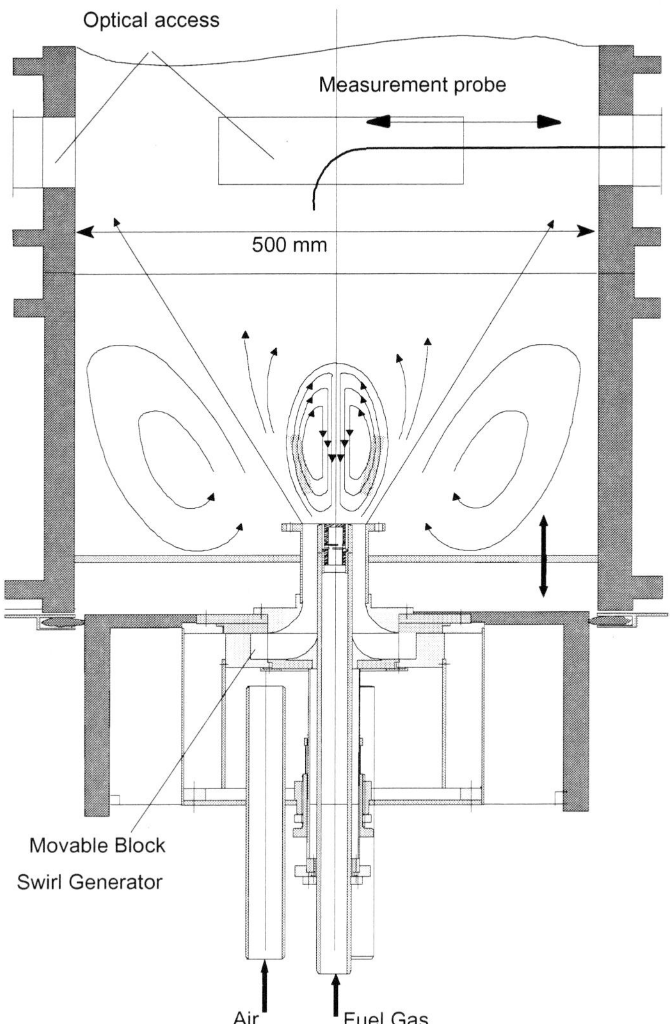

Figure 4.3.17: The model combustor.

4.3.7.3 Temperature Field

Induced by the swirling flow, a ring vortex structure is formed with an inner and outer recirculation zone. The flame is located at the shear layer between the outer and the inner recirculation zone. This is the main reaction zone with nearly stoichiometric conditions. Because of the injection of cold combustion air, the temperature decreases in radial direction and finally increases again in the outer recirculation zone.

Figure 4.3.18 shows the comparison of temperature field as determined experimentally and as predicted by the reacting flow code. The displayed predictions have been attained by using the extended k,ε-model and by taking into account radiative transfer.

Major discrepancies are found at the center-axis directly above the burner orifice and in the reaction zone. Both are attributable to a small recirculation bubble just above the burner orifice that was present in the predictions but could not be confirmed by the measurements. This recirculation bubble prevents hot reaction gases from flowing back to the burner orifice and delays ignition of the mixture. The recirculation bubble decreases the mixing of the recirculating hot gases with the cold air stream emerging from the burner. This effect causes lower temperatures within the reaction zone.

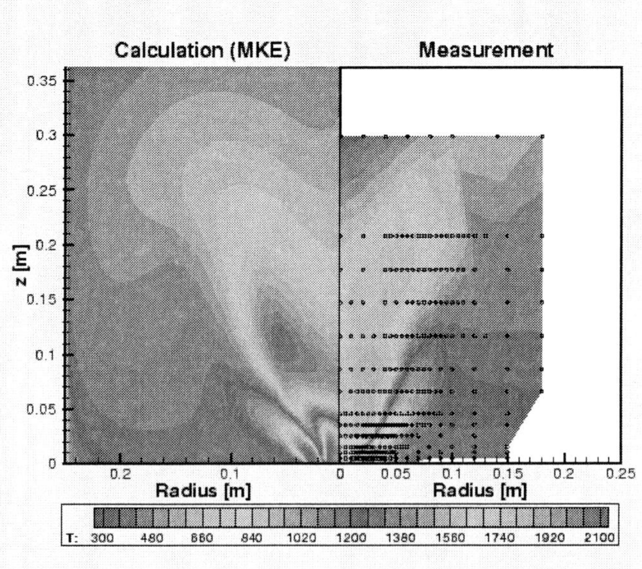

Figure 4.3.18: Comparison of measured and predicted temperature field.

4.3.7.4 Effect of Radiation on the Temperature Field

In order to illustrate the effect of radiation on the temperature field, the difference of the predicted temperature field when taking into account and when neglecting radiation is shown in Fig. 4.3.19.

Due to the large dimensions of the flame, the temperature in the outer recirculation zone and particularly in the main reaction zone is decreased by up to 400 K by radiative losses. On the other hand, the temperature of the cold inlet air stream and the natural gas stream is increased by radiative heating. These temperature differences accentuate the importance of radiative heat transfer when calculating reacting flows. Because of the strong temperature dependence of the NO_x formation rate, considering radiative transfer is crucial for predicting the NO_x emission of combustion systems.

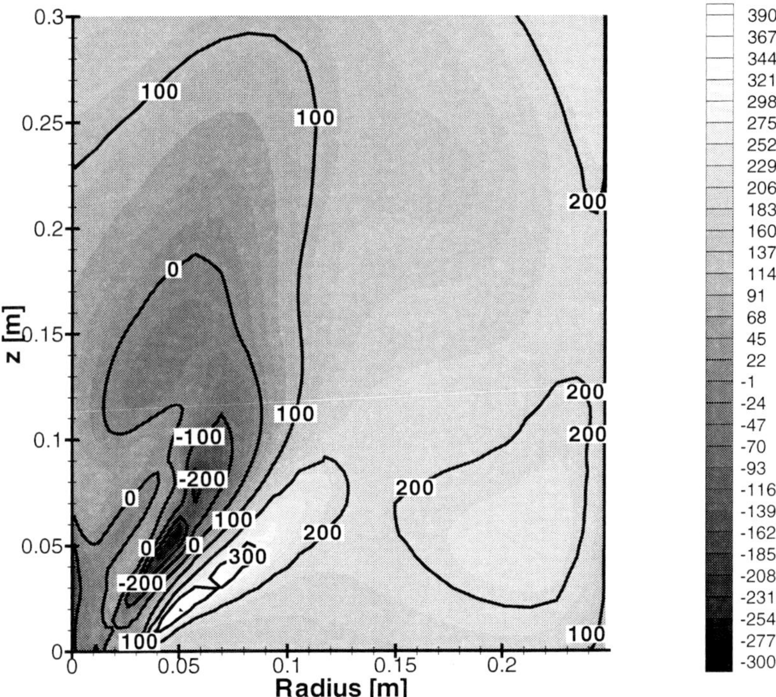

Figure 4.3.19: Predicted temperature change due to radiation.

4.3.7.5 Radiation Spectra

The radiation spectra and the corresponding time fluctuations have been recorded through openings of the combustor walls at eleven planes downstream the burner. Subsequently, the recorded and predicted radiation spectra will be compared for two typical positions: One at the reaction zone 30 mm downstream the nozzle and one 300 mm downstream the nozzle where the combustion reactions are almost completed.

The recorded and predicted radiation spectrum for the plane 30 mm downstream the burner nozzle is displayed in Fig. 4.3.20. Due to lean combustion conditions no soot radiation was detected and the flame emission is confined to the well-known radiation bands of water vapor and carbon dioxide.

The calculated radiation spectrum agrees reasonably well with the experimental data. Deviations in the CO_2-band at 4.3 μm are due to reabsorption by the cold air that was required for flushing the sapphire window that separates the combustion chamber from the IR-spectroscope.

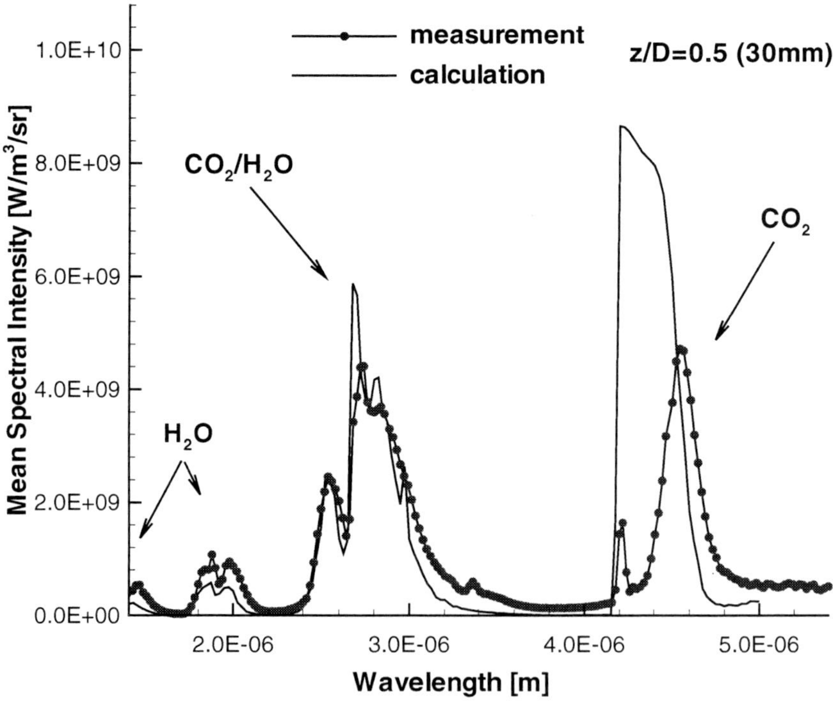

Figure 4.3.20: Spectral intensity 30 mm downstream of the nozzle.

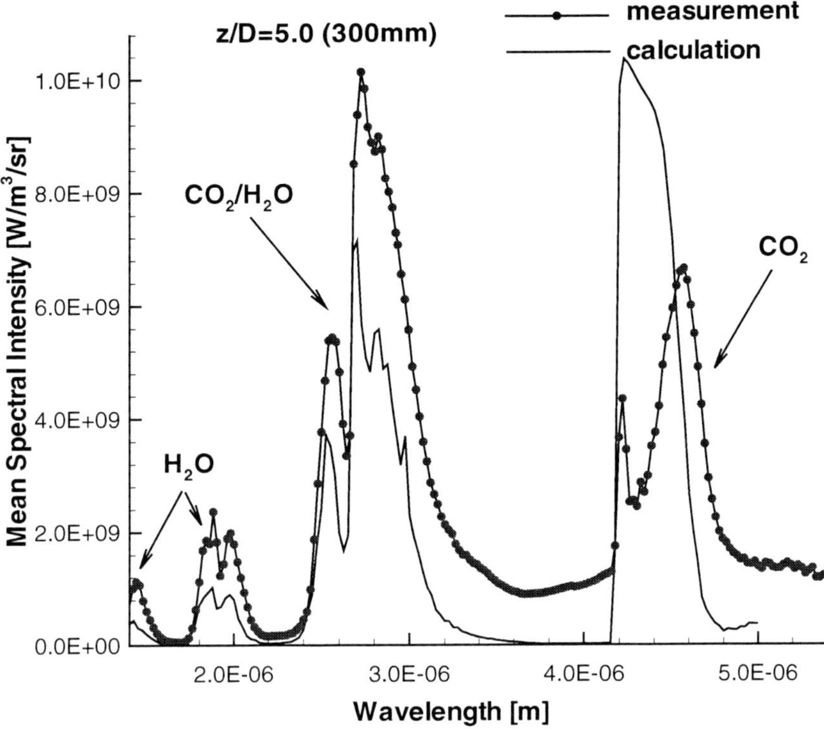

Figure 4.3.21: Spectral intensity 300 mm downstream of the nozzle.

Figure 4.3.21 displays the comparison of the spectra 300 mm downstream the nozzle. The agreement between experiment and prediction is also quite good. However, the measured spectrum shows an offset in the wavelength range above 2.5 μm. This background radiation is caused by the emission of another window located at the opposite side of the opening of the combustor, which was not considered in the predictions.

References

1. R. Viskanta and M.P. Mengüç (1987): Radiation heat transfer in combustion systems. *Progr. Energy Combust. Sci.*, **13**, 97–160.
2. A. Soufiani and J. Taine (1997): High temperature gas radiative property parameters of statistical narrow-band model for H_2O, CO_2 and CO, and correlated k model for H_2O and CO_2. *Int. J. Heat Mass Transfer*, **40(4)**, 987.
3. R.M. Goody (1952): A statistical model for water vapor absorption. *Quart. Journal of the Royal Meteorological Society*, **78**, 165–169.
4. P.B. Taylor and P.J. Forster (1974): Some gray gas weighting coefficients for CO_2-H_2O-soot mixtures. *Int. J. Heat Mass Transfer*, **18**, 1331–1342.
5. M.F. Modest (1991): The weighted-sum-of-gray-gases model for arbitraty solution methods in radiatve transfer. *J. Heat Transfer*, **113**, 650–656.
6. M.K. Denisson and B.W. Webb (1993): A spectral line-based weighted-sum-of-gray-gases model for arbitraty RTE solvers. *J. of Heat Transfer*, **115**, 1004–1012.
7. A. Soufiani and E. Djavdan (1994): A comparison between weighted sum of gray gases and statistical narrow band models for combustion applications. *Combustion and Flame*, **97**, 240–250.
8. M. Cano-Wolff (1996): Bewertung von Graugasmodellen zur Berechnung des Strahlungsaustauschs in Heißgasströmungen. Diplomarbeit, Institut für Thermische Strömungsmaschinen, Universität Karlsruhe.
9. S.C. Lee and C.L. Tien (1981): Optical constants of soot in hydrocarbon flames. In: *18th Symposium (Int.) on Combustion*, Pittsburgh. The Combustion Institute.
10. W.H. Dazell and A.F. Sarofim (1969): Optical constants of soot and their application to heat-flux calculations. *J. Heat Transfer*, **91**.
11. M. Kerker (1969): The Scattering of Light and other Electromagnetic Radiation. Academic Press, New York.
12. B. Ganz, W. Krebs, R. Koch and S. Wittig (1998): Spectral radiation measurements of thermal barrier coatings. *7th AIAA/ASME Joint Int. Thermophysics and Heat Transfer Conference*.
13. J.R. Howell (1988): Thermal radiation in participating media: The past, the present, and some possible futures. *J. Heat Transfer, 110*, 1220–1229.
14. S. Chandrasekhar (1960): Radiative Transfer. Dover Publications Inc., New York.
15. B.G. Carlson and K.D. Lathrop (1968): Transport theory – the method of discrete ordinates. In Greenspan, Kelber, and Okrent, editors, Computing Methods in Reactor Physics. Gordon & Breach, New York.
16. K.D. Lathrop and B.G. Carlson (1965): Discrete angular quadrature of the neutron transport equation. Report LA-3186, Los Alamos Scientific Laboratory.
17. K.D. Lathrop (1968): Ray effects in the discrete ordinates equations. *Nucl. Sci. Eng.*, **32**, 357–368.
18. R. Koch, W. Krebs, S. Wittig and R. Viskanta (1995): A parabolic formulation of the discretes ordinates method for the treatment of complex geometries. In M.P. Mengüç, editor, *1. Symposium on Radiative Transfer, Kusadasi*, Turkey, 1995. ICHMT, Begel House.
19. B.G. Carlson (1971): Tables of equal weight quadrature eq_n over the unit sphere. Report LA-4737, Los Alamos Scientific Laboratory.
20. R. Koch, W. Krebs, S. Wittig, and R. Viskanta (1995): Discrete ordinates quadrature schemes for multidimensional radiative transfer. *J. Quant. Spect. Rad. Transfer*, **53**, 353–372.

21. W. A. Fiveland (1991): Three-dimensional radiative heat transfer solutions by the discrete-ordinates method. In *HDT*, volume 160, pages 89–96, New York, ASME.
22. J. C. Chai and S. V. Patankar (1994): Evaluation of spatial differencing practices for the discrete-ordinates method. *J. Thermophysics and Heat Transfer*, **8**, 140–144.
23. E. E. Lewis and W. F. Miller (1960): Computational Methods of Neutron Transport. John Wiley & Sons, New York.
24. T. H. Song and C. W. Park (1992): Formulation and application of the second order discrete ordinate method. In Bu-Xuan Wang, editor, Transport Phenomena Science and Technology, Beijing, China, Higher Education Press.
25. W. Krebs (1995): Mehrdimensionaler Strahlungswärmeübergang an Gasturbinen-Brennkammerwände: Entwicklung und Überprüfung von Berechnungsverfahren. PhD thesis, Institut für Thermische Strömungsmaschinen, Universität Karlsruhe.
26. Y. S. Touloukian and D. P. DeWitt (1970): *Thermal Radiative Properties*. IFI, New York.
27. B. Ganz, R. Koch, S. Wittig, P. Schmittel and B. Lenze (1997): Spectral radiation measurements of a confined turbulent gas diffusion flame. In M. P. Mengüç, editor, *2. Symposium on Radiative Transfer*, Kusadasi, Turkey, ICHMT.
28. K. E. Torrance and E. M. Sparrow (1966): Off-specular peaks in the directional distribution of reflected thermal radiation. *J. Heat Transfer*, **88(2)**, 223–230.
29. M. F. Modest (1993): *Radiative Heat Transfer*. McGraw-Hill, New York.
30. B. Ganz, P. Schmittel, R. Koch and S. Wittig (1998): Validation of numerical methods at a confined turbulent natural gas diffusion flame considering detailed radiative transfer. *ASME Int. Gas Turbine Conference*, 98-GT-228.
31. B. Noll and S. Wittig (1991): A generalized conjugate gradient method for the efficient solution of three-dimensional fluid flow problems. *Numerical Heat Transfer, Part B*, **20**, 207–221.
32. M. Kurreck and S. Wittig (1994): Numerical simulation of combustor flows on parallel computers – potential, limitations and practical experience. *ASME Int. Gas Turbine Conference*, 94-GT-404.
33. M. Kurreck, R. Koch and S. Wittig (1995): Numerical simulation of turbulent three-dimensional flow problems on parallel computing systems. In F.-K. Hebeker, R. Rannacher, and G. Wittum, editors, Numerical Methods for the Navier-Stokes Equations, Notes on Numerical Fluid Mechanics. Vieweg-Verlag.
34. B. E. Launder and D. B. Spalding (1972): Lectures in Mathematical Models of Turbulence. Academic Press, London.
35. C. Hirsch (1996): A curvature correction for the k-ε model in engineering applications. In W. Rodi and G. Bergeles, editors, Engineering Turbulence Modelling and Experiments, volume 3. Elsevier.
36. W. Leuckel (1967): Swirl intensities, swirl types and energy losses of different swirl generating devices. Technical Report G02/a/16, International Flame Research Foundation (IFRF).

5 High Temperature Materials

Deformation and Damage Behaviour of Structural Materials

Detlef Löhe and Otmar Vöhringer*

Components, which are crucial for the lifetime of power generation, industrial combustion systems or aircraft propulsion are exposed to complex thermal-mechanical loads. For example, the blade of a gas turbine experiences high centrifugal forces, which result in creep loading at rather high temperatures. Furthermore, there is isothermal high cycle fatigue caused by vibrations of the blade and thermally induced low cycle fatigue due to start-up and shut-down procedures as well as oxidative and/or corrosive attack.

On the other hand, materials in combustion chambers experience much lower mechanically induced loads, but severe thermally induced stresses. Again, there is high cycle fatigue caused by the vibrations of the structure and low cycle thermal-mechanical fatigue as well as oxidation and corrosion. As far as metallic structural materials for combustion chambers are concerned, they must meet the following demands: 1.) availability as thin sheet material with good formability and weldability to ensure an economic fabrication of complex combustion chamber structures, 2.) high creep and relaxation strength at very high temperatures, 3.) high resistance against mechanically induced high cycle fatigue and thermally induced low cycle fatigue and 4.) sufficient resistance against oxidation and corrosion. However, the metallic materials used in combustion chambers do not have adequate resistance against the high thermal loading and the accompanying oxidation and corrosion. Therefore, coatings are used to prevent the structural material against damage.

In the frame of the Collaborative Research Centre 167, systems were considered in which the structural materials are Ni- or Co-based superalloys. Contrarily to the blade materials treated above, little or no γ'-hardening is used because of its reduced effectiveness at very high temperatures. The main strengthening mechanisms are solid solution hardening and carbide precipitation hardening. Typical materials are NiCr22Co12Mo9 (Inconel 617) and CoCr22Ni22W14 (Haynes 188). These structural materials are protected by a bond coat (BC) against oxidation and corrosion. Important BC's

* Institut für Werkstoffkunde I, Universität Karlsruhe, Kaiserstr. 12, 76128 Karlsruhe, Germany

are of the MCrAlY-type, in which the amounts of Ni, Co, Fe (= M) and the other elements are adjusted to the substrate material. An important function of the BC is to provide an optimal adhesion of the thermal barrier coating (TBC), which is mostly based on ZrO_2 and which reduces the temperature of the metallic materials in the coating system.

From the above, the following setting of tasks in the framework of the Collaborative Research Centre 167 was evident:

- Analyses of the deformation and damage behaviour of nickel- and cobalt-superalloys as metallic structural materials (creep: *[→ U. T. Schmidt et al.]*; relaxation behaviour, isothermal fatigue behaviour, thermal-mechanical fatigue behaviour: *[→ M. Moalla et al.]*)
- Analyses of the alteration of the deformation and damage behaviour of metallic structural materials by technological treatments (influence of pre-strain, influence of coating processes: *[→ U. T. Schmidt et al.]*, influence of repair and refurbishment processes)
- Modelling of the material behaviour and the behaviour of metallic structures in combustion chambers (physically based modelling of the creep, relaxation: *[→ U. Martin et al.]* and fatigue behaviour, use of constitutive models for the analyses of the deformation behaviour, and the development of damage of structures *[→ J. Aktaa et al.]*)
- Analyses of the damage behaviour of material composites (structural metallic material + BC + TBC) during mechanical and thermal-mechanical loading (creep *[→ U. T. Schmidt et al.]*, isothermal fatigue, thermal-mechanical fatigue)

The contributions in this chapter deal with some results of these investigations, as indicated by the hints printed in italics. Further experimental and theoretical results of investigations of the behaviour of these structural materials during tensile deformation, creep, and relaxation [1–8] as well as isothermal fatigue and thermal-mechanical fatigue [9–18] were published as indicated in different periodicals and proceedings.

References

1. D. Viereck, G. Merckling, D. Löhe, O. Vöhringer, E. Macherauch (1988): Plastic Deformation Behaviour of Thin Sheet Nickel-Base Materials at Temperatures up to 1473 K. *Proc. Int. Conf. Strength Materials* **8** (ICSMA 8), Aug. 1988, Tampere/ Finland, Pergamon Press, 929–934.
2. G. Merckling, K.-H. Lang, D. Eifler, O. Vöhringer, E. Macherauch (1991): Creep-Fatigue Behaviour of Solid Solution Hardened Superalloys at Temperatures up to 1473 K. (ICSMA9), (Eds. D.G. Brandon, R. Chaim, A. Rosen), Freund Publ. Co., London, 443–450.
3. D. Viereck, D. Löhe, O. Vöhringer, E. Macherauch (1991): Stress Relaxation Behaviour of NiCr22Co12Mo9 at Temperatures up to 1473 K. (ICSMA9), (Eds. D.G. Brandon, R. Chaim, A. Rosen), Freund Publ. Co., London, 699–705.
4. G. Merckling, D. Viereck, D. Löhe, O. Vöhringer, E. Macherauch (1992): Kurzzeitkriechverhalten der Nickelbasis-Superlegierung NiCr22Cu12Mo9 im Temperaturbereich 873 K $\leq T \leq$ 1473 K, *Z. Metallkunde* **83**, 441–448.
5. C. Antes, U. Schmidt, O. Vöhringer, D. Löhe, U. Mühle, U. Martin, H. Oettel (1997): Deformation Behaviour and Development of Microstructure During Creep of the Superalloy CoCr22Ni22W14. In: Creep and Fracture of Engineering Materials and Structures (Ed. by J.C. Earthman, F.A. Mohamed). The Minerals, Metals & Materials Society, Warrendale, 96–108.
6. D. Breuer, W. Pantleon, U. Martin, U. Mühle, H. Oettel, O. Vöhringer (1995): Modelling of the Creep Behaviour of a Cobalt-Base Alloy by Applying the Kocks-Mecking-Model. *phys. stat. sol.* (a) **150**, 281–295.
7. U. Mühle, U. Martin, H. Oettel, O. Vöhringer (1997): Microstructure and Modelling of the Deformation Behaviour of CoCr22Ni22W14 in Hot-Tensile- and Creep-Tests. *Mater. Sci. Eng.* **A 230**, 81–87.
8. M. Jerenz, U. Martin, F. Schurack, U. Schmidt, O. Vöhringer, H. Oettel (1999): Mikrostruktur und Ausscheidungsverhalten von Superlegierungen mit Wärmedämmschichtsystemen. *Prakt. Metallogr.* **36,** 13–21.
9. J. Schwertel, G. Merckling, K. Hornberger, B. Schinke, D. Munz (1991): Experimental investigations on Ni-base superalloy IN617 and their theoretical description. In: High-temperature constitutive modeling: theory and application, Eds: Freed, A. D., Walker, A. K., MD-Vol. 26, AMD-Vol. 121, ASME, New York, 285–295.
10. B. Kleinpaß, K.-H. Lang, D. Löhe, E. Macherauch (1996): Thermal-Mechanical Fatigue Behaviour of NiCr22Co12Mo9. In: J. Bressers, L. Remy (Eds.), Fatigue under Thermal and Mechanical Loading, Proc. Int. Symp. 22–24 May 1995, Petten, NL, Kluwer, Academic Publishers, Dordrecht/Boston/London, 327–337.
11. B. Kleinpaß, K.-H. Lang, D. Löhe, E. Macherauch (1996): Entwicklung der Riß- und Korngrenzenschädigung bei in-phase thermisch-mechanischer Beanspruchung von NiCr22Co12Mo19. In: Brocks (Hrsg.) 28. Vortragsveranstaltung des DVM-Arbeitskreises Bruchvorgänge in Bremen, DVM, Berlin, 25–34.
12. A. Möndel, K.-H. Lang, D. Löhe, E. Macherauch (1996): Two Step Fatigue of CoCr22Ni12W14 at 850 °C. In: G. Lütjering, H. Nowack (Eds.), Fatigue '96, Proc. 6th Int. Fatigue Congress, 6–10 May 1996, Berlin, Elsevier Science Ltd., Oxford, UK, 843–848.
13. A. Möndel, K.-H. Lang, D. Löhe, E. Macherauch (1997): Creep-fatigue behaviour of CoCr22Ni22W14 in the temperature range 850 °C $\leq T \leq$ 1200 °C., *Mater. Sci. Eng.* **A234–236**, 715–718.

14. B. Kleinpaß, K.-H. Lang, D. Löhe, E. Macherauch (1998): Influence of the Minimal Cycle Temperature on the Thermal-Mechanical Fatigue Behaviour of NiCr22-Co12Mo9. In: F. Lecomte-Beckers, P. Schubert, J. Ennis (Eds.) Materials for Advanced Power Engineering, Proc. 6th Liège Conf. on Materials for Advanced Power Engineering, 5.–7. Oktober 1998, Liège, Belgien, Schriften des Forschungszentrums Jülich, Reihe Energietechnik/Energy Technology; Vol. **5**, Part III, ISBN 3-89336-228-2, 1369–1377.
15. M. Moalla, K.-H. Lang, D. Löhe, E. Macherauch (1998): Isothermal High Temperature Fatigue Behaviour of NiCr22Co12Mo9 under superimposed LCF and HCF Loading. In: K. T. Rie, P. D. Portella (Eds.) Proc. 4th Int. Conf. on Low Cycle Fatigue and Elasto-Plastic Behaviour of Materials, 7.–11. September 1998, Garmisch-Partenkirchen, Elsevier Science Ldt., Oxford, UK, ISBN 0-08-043326-X, 27–32.
16. B. Kleinpaß, K.-H. Lang, D. Löhe, E. Macherauch (2000): Influence of the Mechanical Strain Amplitude on the In-Phase and Out-of-Phase Thermal-Mechanical Fatigue Behaviour of NiCr22Co12Mo9. In: H. Sehitoglu and H. J. Maier (Eds.): Thermo-Mechanical Fatigue Behaviour of Materials, Third Volume, ASTM-STP 1371, American Society for Testing and Materials, West Consohocken, PA, USA, 36–50.
17. R. Kühner, J. Aktaa, L. Angarita, K.-H. Lang (2000): The Role of Temperature-Rate Terms in Viskoplastic Modelling – Theory and Experiments. In: H. Sehitoglu and H. J. Maier (Eds.): Thermo-Mechanical Fatigue Behaviour of Materials, Third Volume, ASTM-STP 1371, American Society for Testing and Materials, West Consohocken, PA, USA, 103–115.
18. M. Moalla, K.-H. Lang, D. Löhe (1999): Hochtemperatur-Rißausbreitungsverhalten von NiCr22Co12Mo9. DVM-Bericht 231 „Bruchmechanische Bewertungskonzepte im Leichtbau", DVM, Berlin, 163–172.

5.1 Systematic Investigation of the High-Temperature Deformation Behaviour of Selected Materials for Combustion Chambers in Different Component Conditions

Uli T. Schmidt, Otmar Vöhringer, Detlef Löhe, and Eckard Macherauch*

Abstract

Tensile tests and creep tests were performed to determine the microstructural features governing the macroscopic behaviour of combustion chamber alloys (NiCr22Co12Mo9 and CoCr22Ni22W14) and to understand the influences of technological treatments that are performed in manufacturing and refurbishment processes. Depending on temperature, different modes of dislocation/solute atom interaction lead to stress-strain interdependencies with work hardening, serrated flow or no work hardening, respectively. Plastic deformation behaviour at a given temperature is governed by precipitation structure which can be severely influenced by mechanical and thermal treatments.

5.1.1 Introduction

The manufacturing process of hot section gas turbine parts, e. g. combustion chambers, is a series of manifold processing steps including non-cutting shaping, heat treatment, welding and – as a state of the art processing – the deposition of a thermal barrier coating system (TBC). The benefits of the application of a TBC is an increased service life of the component and/ or an increased efficiency and reduced fuel consumption by increased turbine inlet temperature. These benefits are based on the insulating and

* Institut für Werkstoffkunde I, Universität Karlsruhe, Kaiserstr. 12, 76128 Karlsruhe, Germany

reflecting properties of the top ceramic layer and on the oxidation and hot corrosion resistance of the metallic bond coat (BC) beneath it. In addition to these deliberately utilized features side effects e. g. the influence of the coating and the coating process as well as all other processing steps on the behaviour of the substrate have to be known. Thus, the causes of these influences and therefore the microstructural features governing the macroscopic behaviour of the substrate have to be understood.

5.1.2 Experimental Set-up and Specimen Parameters

To understand the governing principles of plastic deformation of high temperature alloys for combustion chamber applications tensile and creep tests were performed using the Ni-base alloy NiCr22Co12Mo9 (Nicrofer 5520 Co/Inconel 617) and the Co-base alloy CoCr22Ni22W14 (Haynes 188). Both alloys are solid solution and carbide precipitation strengthened materials having a face centred cubic (fcc) lattice. The Co-base alloy is characterized by a stacking fault energy γ_S of about 20–30 erg/cm^2, which is a factor of 2 lower as the γ_S-value of the investigated alloy NiCr22Co12Mo9 [1].

As a reference, the behaviour of the as received state, being hot and cold rolled, solution annealed, quenched in air and water, and cleaned in a brushing process was examined in the range between −196 °C and 1200 °C in tensile tests and in the range between 850 °C and 1200 °C in creep tests [1–6].

The influence of technological treatments that are part of the manufacturing process of combustion chamber components on creep behaviour was studied after prestrain, after application of a TBC and after performing a heat treatment to simulate the temperature/time path of the coating process. Prestrain was performed at room temperature at a strain rate $\dot{\varepsilon} = 3 \cdot 10^{-4}\ s^{-1}$ to plastic strains of 1 %, 5 %, 10 %, and 20 %. The coating process consisted of grit blasting, plasma spraying of the bond coat, diffusion annealing at solution annealing temperature with slow furnace cooling, and plasma spraying of the thermal barrier coating. No further processing of the thermal barrier coating system was performed. To make a distinction between the influence of the heat treatment accompanying the coating process and the influence of the coating itself (through e. g. interdiffusion processes) the solution annealing with slow furnace cooling was carried out as a stand alone treatment.

5.1.3 Results and Discussion

Figure 5.1.1 shows a compilation of the plastic deformation behaviour of CoCr22Ni22W14 (Fig. 5.1.1a) and NiCr22Co12Mo9 (Fig. 5.1.1b) in tensile tests at a strain rate of $\dot{\varepsilon} = 3 \cdot 10^{-4}\ s^{-1}$. For both materials the plastic deformation behaviour in tensile tests can be classified into four temperature ranges. The first temperature range ($-196\,°C \leq T \leq 20\,°C$) is related to smooth stress-strain curves. Yield strength, tensile strength, and strain to rupture decrease with increasing temperature. Second, serrated flow due to the Portevin-LeChatelier effect is found at temperatures between 200 °C and 700 °C (NiCr22Co12Mo9) and 750 °C (CoCr22Ni22W14), respectively. Above these temperatures only small amounts of work hardening in the third deformation regime (NiCr22Co12Mo9, $750\,°C \leq T \leq 800\,°C$) or no work hardening (fourth temperature range, $T > 800\,°C$) can be observed. In general both, yield strength and tensile strength decrease with increasing temperature, whereas strains to rupture exhibit minimum values in the temperature range of serrated flow.

To illustrate the plastic deformation regimes yielding the above described macroscopic behaviour Fig. 5.1.2 shows transmission electron micrographs of NiCr22Co12Mo9 after 10 % plastic strain at different temperatures. In the first, low temperature range plasticity is governed by planar slip as can be seen in Fig. 5.1.2a showing a great number of activated slip bands. In the second, serrated flow range and third, restricted

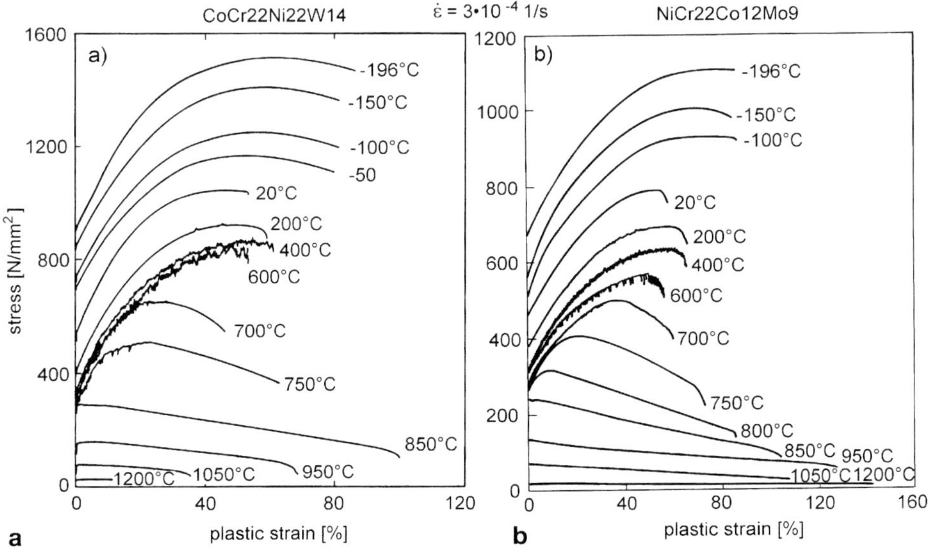

Figure 5.1.1: Stress strain dependencies of CoCr22Ni22W14 (a) and NiCr22Co12Mo9 (b) at different temperatures ($\dot{\varepsilon} = 3 \cdot 10^{-4}\ s^{-1}$).

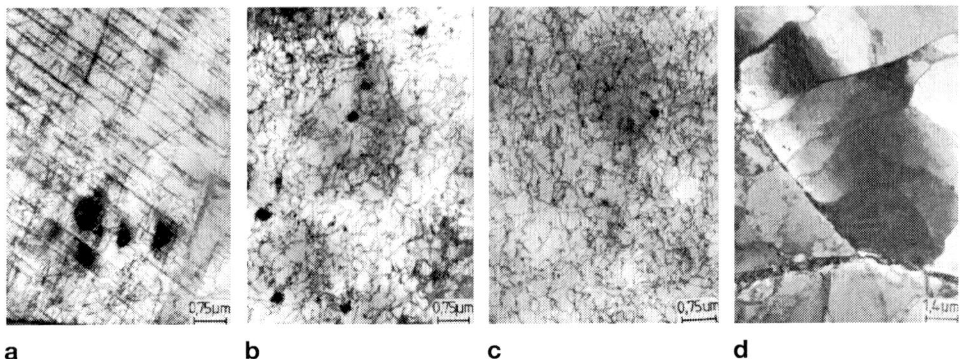

Figure 5.1.2: Dislocation structures of NiCr22Co12Mo9 after 10 % strain at different temperatures; a) T = 20 °C, b) T = 600 °C, c) T = 850 °C, and d) T = 1050 °C.

work hardening range plasticity is governed by the elastic interaction of dislocations and diffusing foreign atoms yielding a diffuse dislocation structure as shown in Figs. 5.1.2b and 5.1.2c. Recovery processes at higher temperatures cause the formation of a subgrain structure (Fig. 5.1.2d) and inhibit work hardening.

Thus, it can be stated that temperature influences plasticity, as always, by contributing an increasing amount of energy to overcome the thermal flow stress component with increasing temperature as well as by changing the governing flow regime from planar slip, to serrated flow, to viscous flow, and to recovery controlled deformation. This influence can mainly be understood by a change in the kind of elastic interaction between foreign atoms and dislocations.

The interaction between precipitations and dislocations and thus precipitation hardening can severely be influenced by technological treatments e. g. prestrain at room temperature [7]. In Fig. 5.1.3 the influence of prestrain on creep curves is shown in a creep strain-time-plot and a creep rate-creep strain-plot at a given temperature (T = 850 °C) and a given nominal stress (σ_n = 150 N/mm^2). An increase in prestrain to a maximum value $\varepsilon_{p,pre}$ = 20 % leads to an increase in creep time and to a decrease in strain to rupture and minimum creep rate.

A compilation of transmission electron micrographs in Fig. 5.1.4 shows the development of the dislocation structure in CoCr22Ni22W14 at room temperature with increasing prestrain. The resulting dislocation and precipitation structure after 5 h temperature compensation at creep temperature, just before applying creep load is shown in Fig. 5.1.5a to 5.1.5d. Increased dislocation densities after prestrain represent increased supply of nuclei for the precipitation of carbides and accelerated diffusion (pipe diffusion) so that inside grains carbides precipitate in increased numbers and volume fractions and decreased size after prestrain.

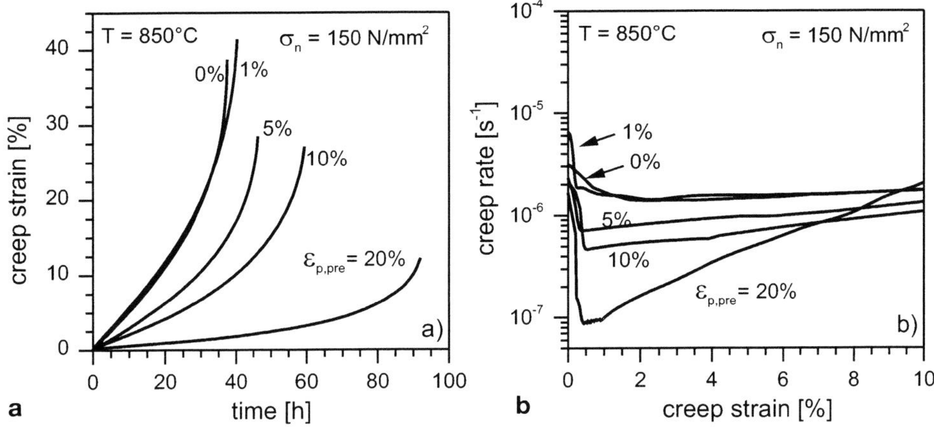

Figure 5.1.3: Influence of prestrain $\varepsilon_{p,pre}$ on creep curves of CoCr22Ni22W14; a) creep strain vs. time, b) creep rate vs. creep strain at $T = 850\,°C$ and nominal stress $\sigma_n = 150$ N/mm^2.

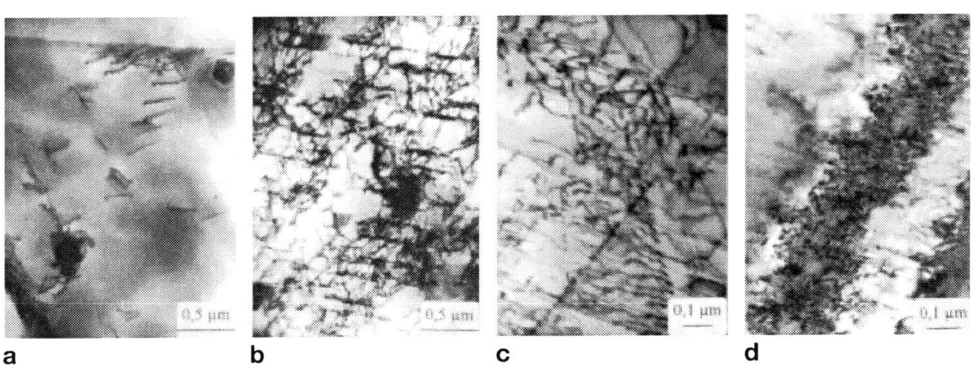

Figure 5.1.4: Dislocation structures of CoCr22Ni22W14 after a) 0, b) 5 %, c) 10 %, and d) 20 % prestrain at room temperature.

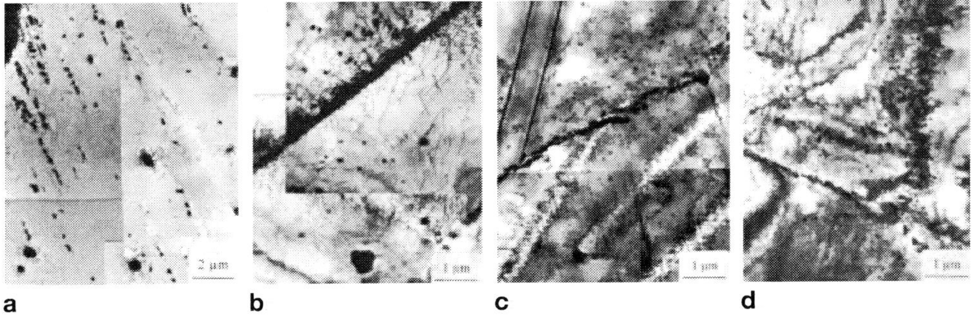

Figure 5.1.5: Dislocation and precipitation structure of CoCr22Ni22W14 after a) 0, b) 5 %, c) 10 % and d) 20 % prestrain at 20 °C and 5 h temperature balance at 850 °C.

Table 5.1.1: Norton exponents of CoCr22Ni12W14 after different prestrains and at different creep temperatures.

Temperature	Norton exponent n				
	$\varepsilon_{\mathrm{p,pre}} = 0$	$\varepsilon_{\mathrm{p,pre}} = 1\,\%$	$\varepsilon_{\mathrm{p,pre}} = 5\,\%$	$\varepsilon_{\mathrm{p,pre}} = 10\,\%$	$\varepsilon_{\mathrm{p,pre}} = 20\,\%$
850 °C	6.6	7.1	7.9	8.6	9.1
950 °C	5.8	6.2	6.6	6.8	8.3

Under creep load both precipitation hardening mechanisms and work hardening mechanisms in terms of high dislocation densities and structures stabilized by precipitates are effective. These mechanisms cause the afore described drop in creep rate $\dot{\varepsilon}_{min}$ and an increase in the Norton exponent n ($\dot{\varepsilon}_{min} \sim \sigma^n$) from $n = 6.6$ (no prestrain) to $n = 9.1$ ($\varepsilon_{\mathrm{p,pre}} = 20\,\%$) and from $n = 5.8$ (no prestrain) to $n = 8.3$ ($\varepsilon_{\mathrm{p,pre}} = 20\,\%$) at $T = 850\,°\mathrm{C}$ and $T = 950\,°\mathrm{C}$, respectively (see Tab. 5.1.1).

The effects of prestrain on the precipitation microstructure show such a large extend because plastic deformation was performed on a supersaturated solid solution obtained in a solution annealing and quenching process.

The diffusion annealing process carried out in the course of a thermal barrier coating process consists of a heat treatment at the same temperature as the solution annealing but the cooling rate is significantly lower. Thus the nucleation and the growth of precipitates proceed during cooling which leads to a coarse distribution of carbides both, on grain boundaries as well as within grains [8]. There is no significant difference in carbide distribution, whether the heat treatment with furnace cooling is carried out in the course of a coating process or as a stand alone treatment, which can be seen in Fig. 5.1.6.

Besides the coarsening of carbide precipitations there is an additional loss of precipitation hardening due to decarburisation by diffusion of C-atoms into the bond coat or into the annealing atmosphere (vacuum) when no coating is applied. Evidence can be found in Fig. 5.1.7, showing a carbide free zone, some 50 µm underneath the bond coat of a coated specimen (Fig. 5.1.7a) and a carbide free surface hem in Fig. 5.1.7b (no coating, only annealing and furnace cooling).

Bearing this in mind it can easily be understood that the coating procedure leads to a decrease in creep rupture time and increase in creep rate as can be seen in Fig. 5.1.8 for creep tests at $T = 950\,°\mathrm{C}$ at different applied creep loads. This accelerating effect is significant at low creep loads whereas no influence can be found at high creep loads. Thus there is a change in the stress dependence of creep rate (Fig. 5.1.9) that can be expressed in a drop in Norton exponent as shown in Tab. 5.1.2.

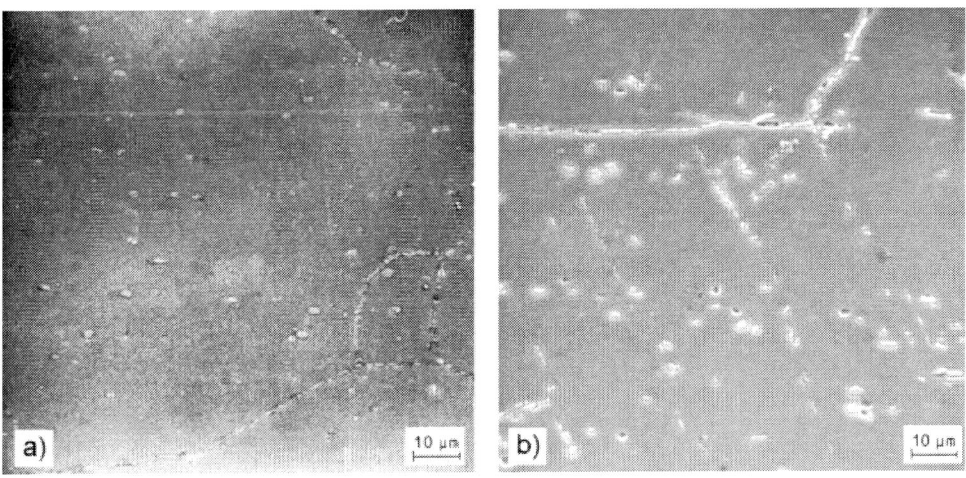

Figure 5.1.6: Precipitation structure of NiCr22Co12Mo9; a) in coated condition, b) after solution annealing and furnace cooling.

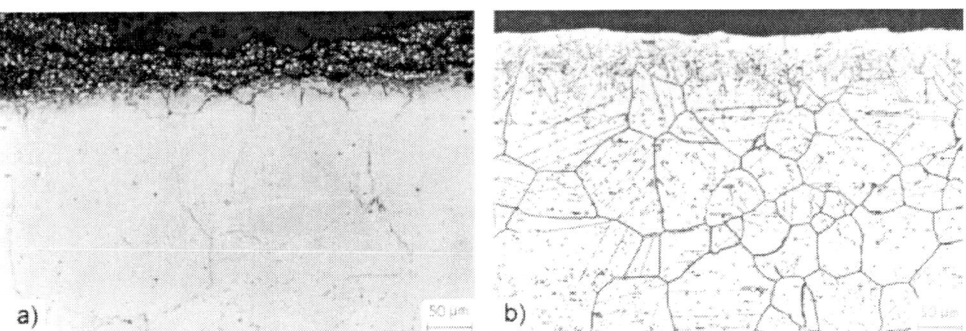

Figure 5.1.7: Decarburized region of NiCr22Co12Mo9 a) in coated condition, b) after solution annealing and furnace cooling.

Table 5.1.2: Norton exponents of NiCr22Co12Mo9 at different temperatures in different component conditions.

Temperature	Norton exponent n		
	solution annealed and quenched	solution annealed and furnace cooled	coated
850 °C	7.4	5.9	5.5
950 °C	6.6	5.3	5.1
1050 °C	5.0	5.4	5.0

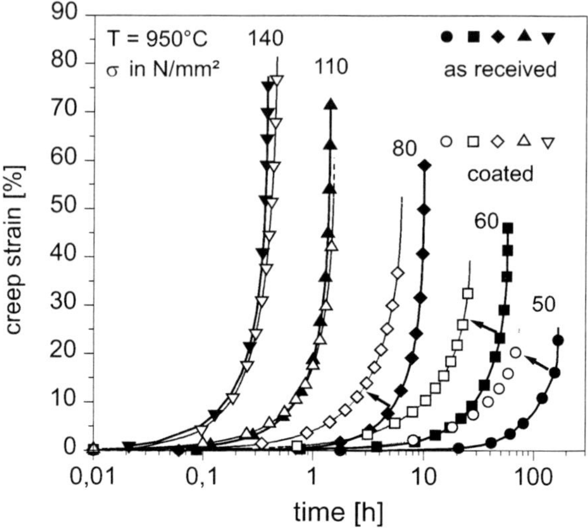

Figure 5.1.8: Influence of coating process on creep strain vs. time curves of NiCr22-Co12Mo9 at 950 °C and different creep stresses.

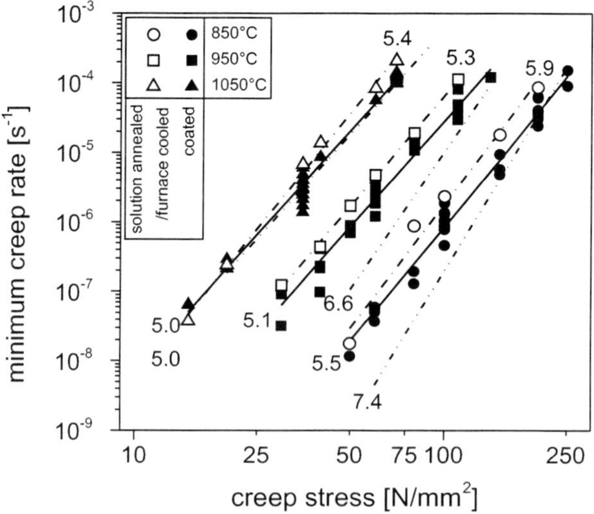

Figure 5.1.9: Influence of coating processes on minimum creep rate vs. creep stress of NiCr22Co12Mo9 at different temperatures.

No significant influence of the coating or the annealing with furnace cooling is observed at $T = 1050\,°C$. At this temperature coarsening mechanisms are active to such an extend that any fine carbide distribution would undergo severe Ostwald ripening which would result in a comparable microstructure as after solution annealing and furnace cooling.

Acknowledgements

The support of the investigations by the DFG is gratefully acknowledged.

References

1. Viereck, D. (1990): Das Zug- und Relaxationsverformungsverhalten von Hochtemperaturblechwerkstoffen im Temperaturbereich 78 K $\leq T \leq$ 1473 K, Dissertation Universität Karlsruhe (TH), *VDI Fortschritt-Berichte*, Reihe 5, Nr. **202**, VDI Verlag.
2. Merckling, G. (1989): Kriech- und Ermüdungsverhalten ausgewählter metallischer Werkstoffe bei höheren Temperaturen, Dissertation Universität Karlsruhe (TH).
3. Antes, C. (1993): Statisches und quasistatisches Hochtemperaturverformungsverhalten der Brennkammerlegierung CoCr22Ni22W14, Dissertation Universität Karlsruhe (TH).
4. Viereck, D., Merckling, G., Lang, K.-H., Eifler, D., Löhe, D. (1989): Verformungsverhalten der Legierung NiCr22Co12Mo9 unter statischer, quasistatischer und zyklischer Beanspruchung bei Temperaturen bis zu 1473 K, In: K. Schneider: Berichte DGM. Symposium Festigkeit und Verformung bei hoher Temperatur, Bad Nauheim, DGM Oberursel, 201–218.
5. Viereck, D., Löhe, D., Vöhringer, O., Macherauch, E. (1991): Stress Relaxation Behaviour of the Cobalt-Base Superalloy CoCr22Ni22W14, In: M. Jono, T. Inoue: Proc. of the 6th Int. Conf. on the Mechanical Behaviour of Materials, ICM 6, Kyoto, Japan, 607–612.
6. Merckling, G., Viereck, D., Löhe, D., Vöhringer, O., Macherauch, E. (1992): Kurzzeitkriechverhalten der Nickelbasis-Superlegierung NiCr22Co12Mo9 im Temperaturbereich 873 K $\leq T \leq$ 1473 K, *Z. Metallkunde* **83**, 441–448.
7. Schmidt, U., Löhe, D., Vöhringer, O. (1996): Zum Einfluß plastischer Vorverformungen auf das Kriech- und Zeitstandverhalten der Hochtemperaturlegierung CoCr22-Ni22W14, Werkstoffwoche '96, Symposium 3, H. W. Grünling (Ed.), DGM-Informationsgesellschaft Verlag, 163–168.
8. Schmidt, U. (1998): Das Kriechverformungs-, Zeitstand- und Versagensverhalten plasmagespritzter Wärmedämmschichtverbunde, Dissertation Universität Karlsruhe (TH).

5.2 Fatigue Behaviour of NiCr22Co12Mo9 under Isothermal and Thermal-Mechanical Fatigue Loadings

Mourad Moalla, Karl-Heinz Lang, and Detlef Löhe*

5.2.1 Introduction

Repeated start up and shut down processes of gas turbines cause complex thermal-mechanical loadings, which lead to low cycle fatigue (LCF) of the wall of the combustion chamber. Additionally, a high cycle fatigue loading (HCF) is superimposed as a result of mechanical vibrations and unsteady combustion processes. Furthermore, creep and relaxation processes occur during stationary operation at maximum temperature. In the present study, the cyclic deformation behaviour and the fatigue life of a typical material for combustion chambers was investigated under different loading conditions. To determine the effect of different deformation rates and the influence of creep on the isothermal cyclic behaviour, total strain controlled isothermal high temperature fatigue tests at different frequencies were performed. Based on a study [1] about the high temperature superalloy CoCr22Ni22W14, the validity of the lifetime models according to the frequency-modified relationships of Coffin-Manson and Basquin was analysed. The influence of thermal fatigue due to the repeated start-stop cycles of gas turbines was investigated in out-of-phase thermal-mechanical fatigue experiments (TMF) at different maximum temperatures. According to previous investigations on NiCr22Co12Mo9 under combined isothermal LCF and HCF loading, a superimposed HCF loading may change the cyclic deformation behaviour significantly and may reduce the fatigue lifetime by more than 90 % of the value which is measured without superimposed HCF loadings [2]. Other studies on cast irons showed a considerable reduction of lower frequent isothermal as well as of thermal-mechanical fatigue lifetimes due to a superposition by HCF loadings [3, 4]. Therefore, the influence of different superimposed HCF loadings on the cyclic deformation behaviour and the fatigue life during out-of-phase TMF experiments has also been investigated.

* Institut für Werkstoffkunde I, Universität Karlsruhe, Kaiserstr. 12, 76128 Karlsruhe, Germany

5.2.2 Material

The material investigated is the solid solution and carbide precipitation hardened nickel base superalloy NiCr22Co12Mo9 (Nicrofer 2250 Co, Inconel 617). It was supplied as round bars which were annealed at 1200 °C and water quenched. The chemical composition of this material is given in Tab. 5.2.1.

In the as received state, the microstructure of NiCr22Co12Mo9 shows grains with a high density of twin boundaries and a nearly homogeneous distribution of primary precipitates of the type M_6C and Ti(C,N) (Fig. 5.2.1). From the supplied bars, round solid specimens were manufactured with a cylindrical gauge length of 10 mm, a diameter of 7 mm within the gauge length and conical gripping heads.

Table 5.2.1: Chemical composition of NiCr22Co12Mo9 (all quantities in wt.-%).

Ni	Cr	Co	Mo	Al	Fe	Ti	Si	C
bal.	22.25	11.45	8.88	1.28	0.56	0.4	0.11	0.06

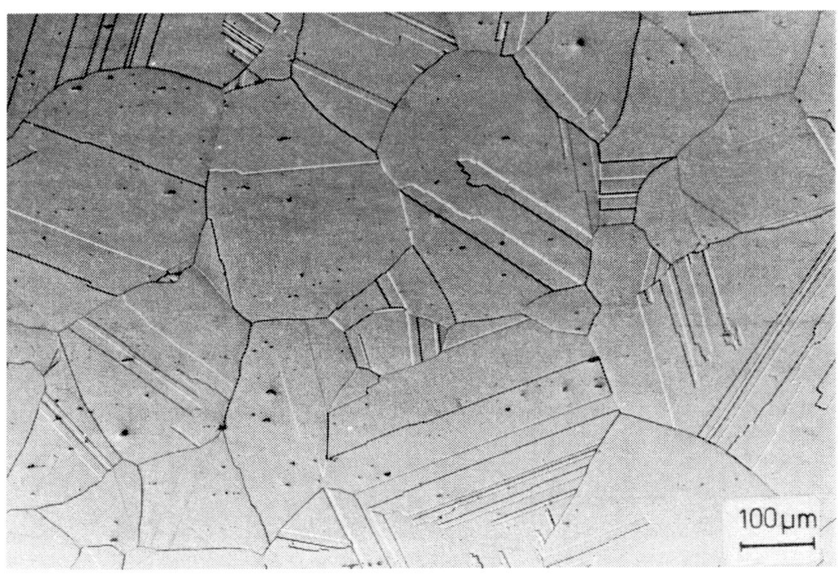

Figure 5.2.1: Microstructure of NiCr22Co12Mo9 in the as received state.

5.2.3 Experimental Details

All experiments were carried out on a servo hydraulic fatigue testing machine with a maximum loading capacity of 100 kN. For the strain measurement, a high temperature capacitive extensometer was used. The specimens were heated up by an inductive furnace with closed loop control and the temperature was measured with a Ni-CrNi thermocouple, which was spot welded close to the gauge length of the specimens. During TMF tests, the specimens were cooled by thermal conduction to the water cooled grips and, if necessary, additionally by a proportionally controlled air jet, that was blown on to the specimens surface.

The isothermal fatigue tests were performed at $T = 850\,°C$ and $1000\,°C$ under total strain control with zero mean strain, triangular waveforms and frequencies between 0.005 Hz and 5 Hz.

The TMF experiments without and with superimposed HCF loading were also carried out under total strain control. For all TMF tests, the minimum temperature was $T_{min} = 200\,°C$ and the maximum temperature was varied between $T_{max} = 750\,°C$ and $1200\,°C$. The heating and cooling rate was 14 K/s resulting in cycle periods ranging from 79 s to 143 s.

At the beginning of a TMF test, the specimen is first heated up to the mean temperature T_m. Then it is subjected to three triangle-shaped temperature cycles illustrated in Fig. 5.2.2 without any mechanical loading. To determine the thermal expansion and contraction, the total strain of each specimen is measured during these cycles. Afterwards, the testing machine is switched to total strain control and the TMF loading is started.

During the TMF experiments without superimposed HCF loading the total strain, $\varepsilon_t = \varepsilon_t^{me} + \varepsilon^{th}$, is kept constant throughout the test. Therefore, a mechanical strain amplitude $\varepsilon_{a,t}^{me}$ equal to the thermal strain amplitude ε_a^{th} is induced (Fig. 5.2.2). The phase shift between the temperature and the mechanical strain is 180°. Thus, tensile stresses are acting at low temperatures and compressive stresses at high temperatures. Such experiments are well known as out-of-phase TMF tests.

During TMF tests with superimposed HCF loading, the out-of-phase loading described above was combined with a sinusoidal HCF loading with a frequency of 5 Hz and an amplitude $\varepsilon_{a,t}^{HCF}$ between 0.05 % and 0.2 %.

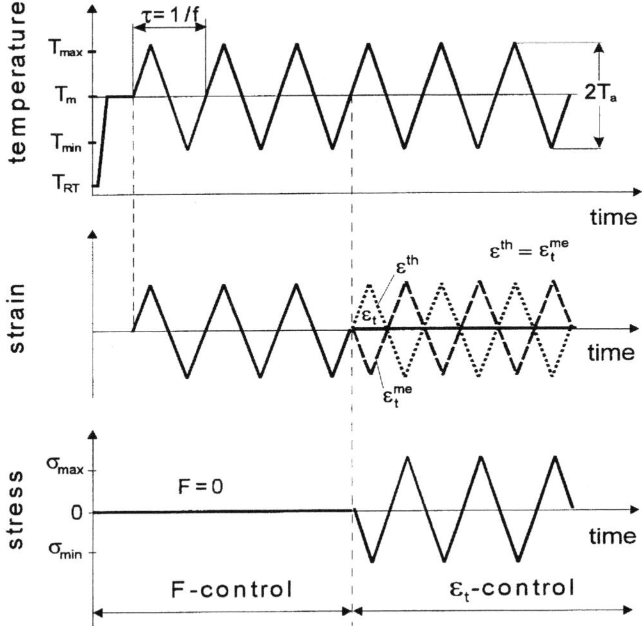

Figure 5.2.2: Experimental procedure for out-of-phase TMF tests without superimposed HCF loading.

5.2.4 Results and Discussion

5.2.4.1 Isothermal Fatigue Tests

Figure 5.2.3 shows the development of the plastic strain amplitude $\varepsilon_{a,p}$ (top) as well as the stress amplitude σ_a (bottom) during isothermal fatigue tests at $\varepsilon_{a,t} = 0.25\,\%$, $T = 850\,°C$ and different frequencies. Independent of the frequency, the same values of $\varepsilon_{a,p}$ and σ_a were measured in the first cycle. At $f = 0.005$ Hz, $\varepsilon_{a,p}$ as well as σ_a remain approximately constant until macro crack initiation and propagation finally lead to a steep decrease of $\varepsilon_{a,p}$ and σ_a. At $f \geq 0.05$, the decrease of $\varepsilon_{a,p}$ and the increase of σ_a indicate the cyclic hardening behaviour of the material investigated. After cyclic hardening, a transition to neutral cyclic deformation behaviour can be seen that appears the later the higher the frequency is. With decreasing frequency and/or with increasing temperature a change occurs in the mechanisms, which controls the cyclic deformation behaviour. At high frequencies, e. g. $f = 5$ Hz, the cyclic deformation is mainly determined by thermally activated dislocation motion and dynamic strain ageing processes [5]. With decreasing frequency,

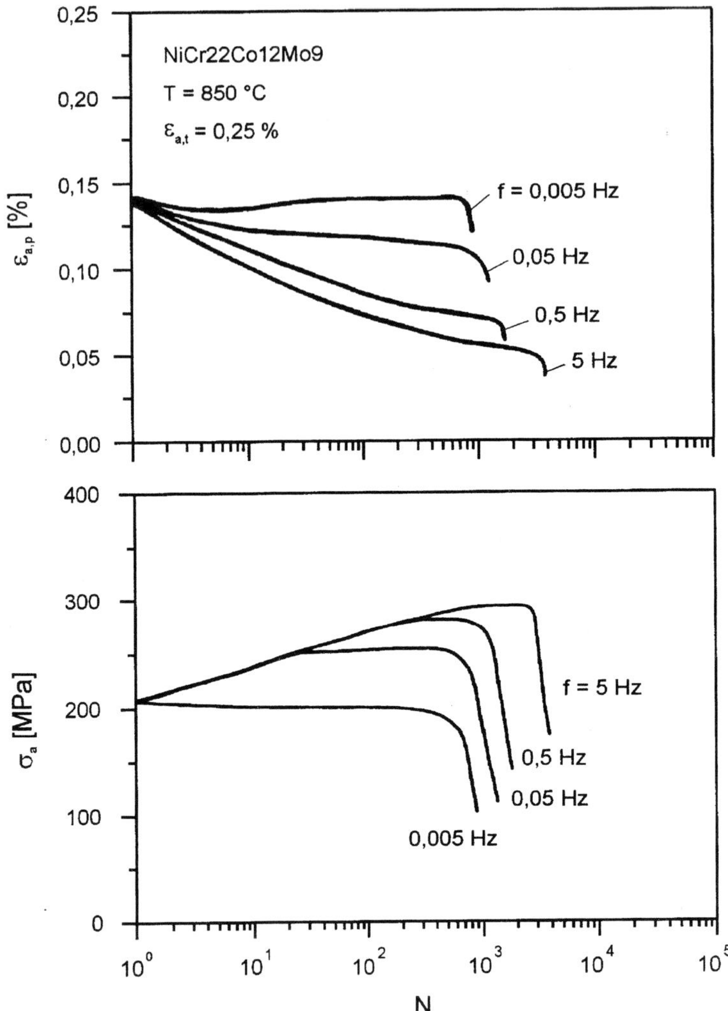

Figure 5.2.3: Plastic strain amplitude (top) as well as the stress amplitude (bottom) vs. the number of cycles at $\varepsilon_{a,t} = 0.25\,\%$, $T = 850$ °C and different frequencies.

diffusion controlled creep processes more and more influence the cyclic deformation behaviour until at $f = 0.005$ Hz a neutral cyclic deformation behaviour is found (Fig. 5.2.3).

In Fig. 5.2.4, the number of cycles to failure N_f is plotted vs. the frequency in double logarithmic scaling for fatigue tests at $\varepsilon_{a,t} = 0.25\,\%$ and $T = 850\,°C$ as well as $1000\,°C$. At both temperatures, N_f increases linearly with increasing frequency. Additionally, Fig. 5.2.4 contains the isochrones for constant times to failure of 0.1, 1, 10, and 100 h (dotted lines). At both temperatures the time to failure decreases with increasing frequency.

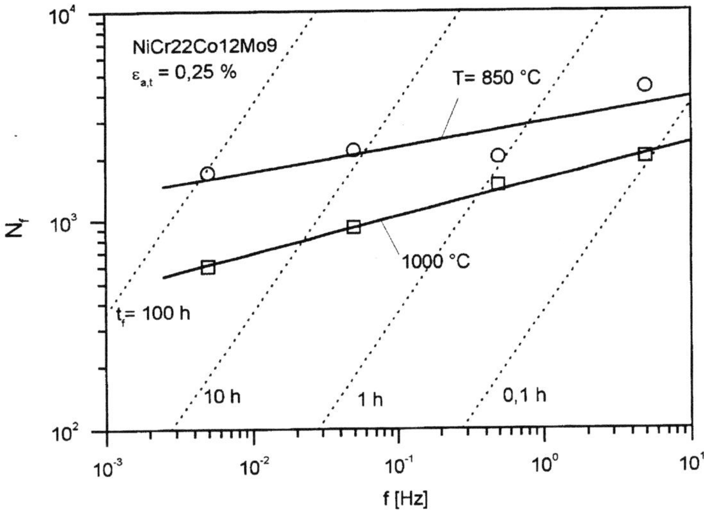

Figure 5.2.4: Number of cycles to failure vs. frequency at $\varepsilon_{a,t} = 0{,}25\,\%$, $T = 850$ °C and 1000 °C.

The study of Coffin [6] has shown that, for plastic strain controlled fatigue tests, the relationship between the frequency f and the time to failure t_f is given by the relationship

$$f^k\, t_f = C_1 \tag{1}$$

with k and C_1 being constants. C_1 depends on the plastic strain amplitude according to the relationship

$$C_1^a\, \varepsilon_{a,p} = C_2 \tag{2}$$

where a and C_2 are constants. With $t_f = N_f/f$ and the combination of Eqs. (1) and (2), the frequency-modified Coffin-Manson relationship can be described as

$$\varepsilon_{a,p}\, f^{a(k-1)}\, N_f^a = C_2 \tag{3}$$

As shown in the following, this equation is also valid for results of total strain controlled fatigue tests on the material investigated, if the plastic strain amplitude at half of the number of cycles to failure $N_f/2$ is used.

The fatigue ductility exponent a was determined from total strain controlled fatigue tests at 850 °C with $f = 5$ Hz. For the material investigated, $a = 0.97$ was found out. To estimate the constants k and C_1, an additional fatigue test with $\varepsilon_{a,t} = 0.3\,\%$ at $f = 5$ Hz was performed, which yielded $\varepsilon_{a,p} = 0.096\,\%$ at $N_f/2$. The corresponding data point is plotted together with the results from the tests at $\varepsilon_{a,t} = 0.25\,\%$, in the $\varepsilon_{a,p}$,f-diagram shown in Fig. 5.2.5. Then, the data point at $\varepsilon_{a,t} = 0.25\,\%$ with the same $\varepsilon_{a,p}$ was determined and transferred in the lg t_f-lg$f$-diagram as illustrated in Fig. 5.2.6. Thus, a curve with constant $\varepsilon_{a,p}$ at different frequencies was constructed. According to the Eq. (1), k is given by the slope of this curve and C_1 by

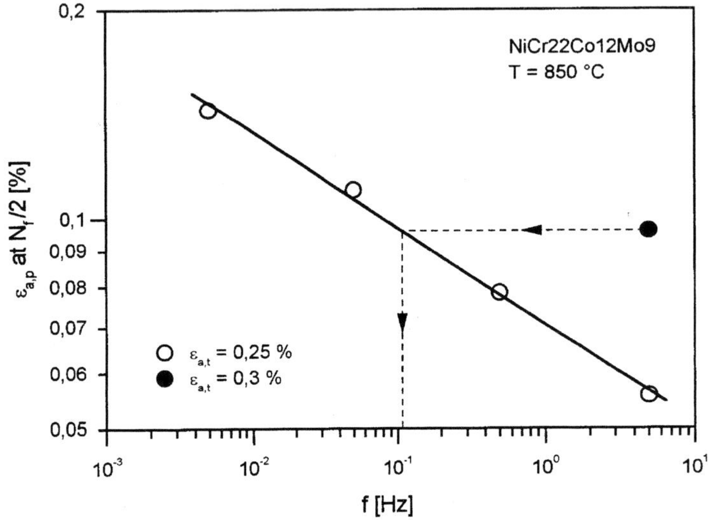

Figure 5.2.5: Plastic strain amplitude at $N_f/2$ vs. frequency at $\varepsilon_{a,t} = 0.25\,\%$ as well as $\varepsilon_{a,t} = 0.3\,\%$ and $T = 850\,°C$.

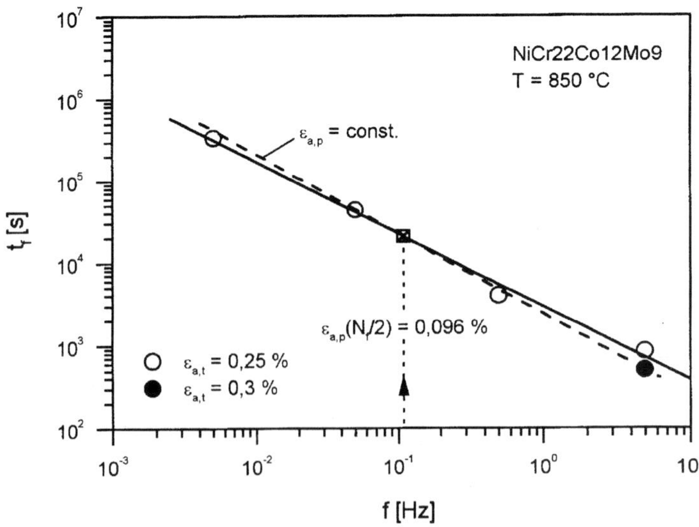

Figure 2.5.6: Time to failure vs. frequency at $\varepsilon_{a,t} = 0.25\,\%$ as well as $\varepsilon_{a,p} = 0.096\,\%$ and $T = 850\,°C$.

the intersection with the t_f axis. Using Eq. (2), the constant C_2 was calculated from C_1 and a.

For $T = 850\,°C$, the frequency-modified plastic strain amplitudes $\varepsilon_{a,p} f^{a(k-1)}$ at $N_f/2$ are plotted as a function of N_f in Fig. 5.2.7 in a double logarithmic scaling. As can be seen, all data points are described well by the frequency-modified Coffin-Manson relationship, independent of the frequency.

In a similar way, the Basquin relationship which describes the dependence between the elastic strain amplitude $\varepsilon_{a,e}$ and the number of cycles to failure, was modified

$$\varepsilon_{a,e}\, f^{\beta(k_1-1)}\, N_f^{\beta} = C_3 \tag{4}$$

The fatigue strength exponent $\beta = 0.091$ and the constants k_1 and C_3 of the material investigated are specified in the same way as it has been described for the frequency-modified Coffin-Manson relationship. In Fig. 5.2.8, the frequency-modified elastic strain amplitude $\varepsilon_{a,e}\, f^{\beta(k_1-1)}$ at $N_f/2$ is plotted as a function of N_f. The validity of the frequency-modified Basquin relationship at different frequencies in the range between 0.5 and 0.005 Hz is clearly demonstrated for the material investigated at $T = 850\,°C$.

Using the combination of the frequency-modified Basquin relationship Eq. (3) and the frequency-modified Coffin-Manson relationship Eq. (4), the total strain amplitude ($\varepsilon_{a,1} = \varepsilon_{a,e} + \varepsilon_{a,p}$) can be described as a function of the number of cycles to failure and the frequency

$$\varepsilon_{a,t} = C_3\, \varepsilon_{a,e}\, f^{-\beta(k_1-1)}\, N_f^{-\beta} + C_2\, \varepsilon_{a,p}\, f^{-a\,(k-1)}\, N_f^{-a} \tag{5}$$

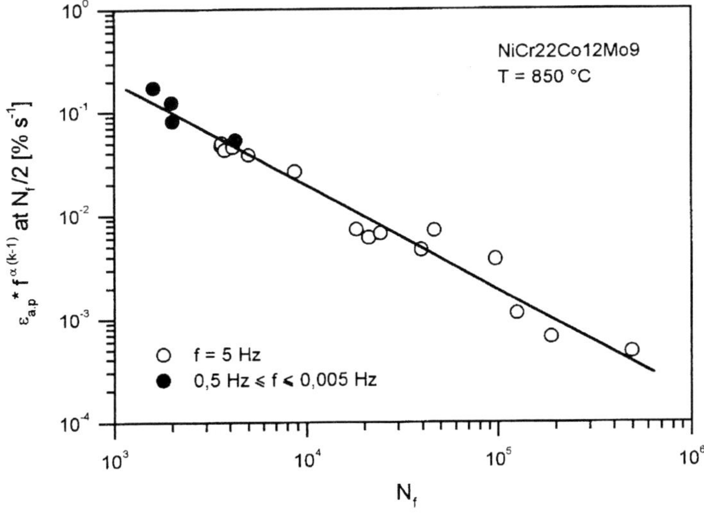

Figure 5.2.7: Frequency-modified plastic strain amplitude vs. the number of cycles to failure at different frequencies and $T = 850\,°C$.

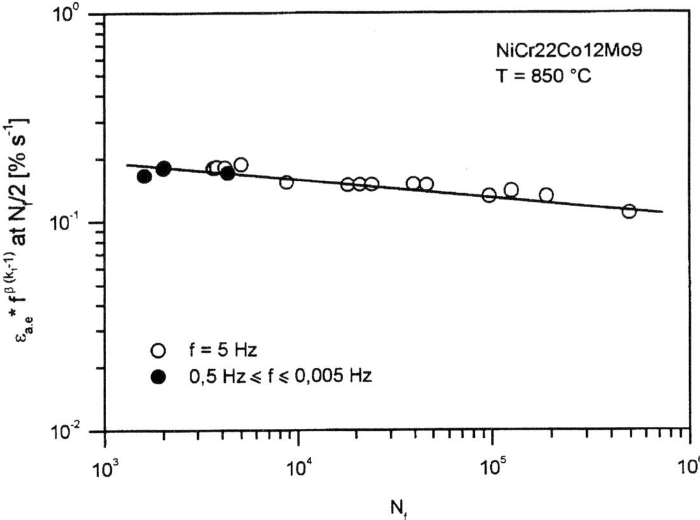

Figure 5.2.8: Frequency-modified elastic strain amplitude vs. the number of cycles to failure at different frequencies and $T = 850\,°\text{C}$.

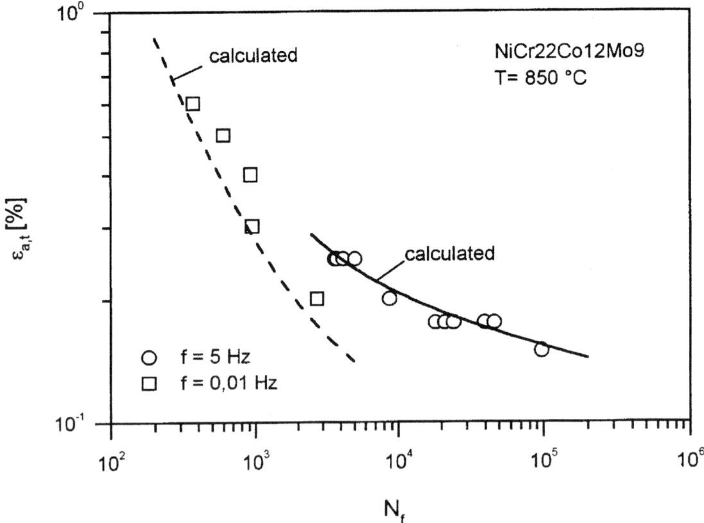

Figure 5.2.9: Comparison between the calculated fatigue life and the experimental results from fatigue tests at $T = 850\,°\text{C}$ and $f = 5$ Hz as well as $f = 0.01$ Hz.

With this equation, the prediction of the number of cycles to failure for total strain controlled fatigue tests with a given total strain amplitude and frequency should be possible. As can be seen in Fig. 5.2.9, the measured data points (open circles) at $f = 5$ Hz, which are used for the determination of the parameters of the frequency modified Coffin-Manson and Basquin relationships, can be fitted with the Wöhler curve (solid line), which is calculated with Eq. (5). To verify the validity of this relationship, additional fatigue tests at $f = 0.01$ Hz were performed. The comparison between the experimental data points (open squares) resulting from these experiments and the predicted fatigue life (broken line) calculated with the combination of the frequency-modified relationships of Basquin and Coffin-Manson shows a somewhat conservative estimation of the number of cycles to failure.

5.2.4.2 Thermal-Mechanical Fatigue Tests

In Fig. 5.2.10, the stress-total mechanical strain hysteresis loops evaluated at $N_f/2$ during out-of-phase TMF experiments at different T_{max} without superimposed HCF loading are plotted. With increasing T_{max} the stress amplitudes σ_a which are induced by the TMF loading decrease, whereas the $\varepsilon_{a,p}$ values increase as a result of thermally activated dislocation motion. The fluctuations of the nominal stress, that are rather weak during TMF at $T_{max} = 1000\,°C$ and more pronounced during TMF at $T_{max} = 1200\,°C$, are the consequence of dynamic strain ageing processes resulting from the interaction between moving dislocations and diffusing solution atoms [7]. Moreover, all hysteresis loops, particularly those from TMF tests at $T_{max} \geq 850\,°C$, show a significant dynamic relaxation of the induced compressive stresses during heating up to T_{max} after exceeding a certain combination of stress and tem-

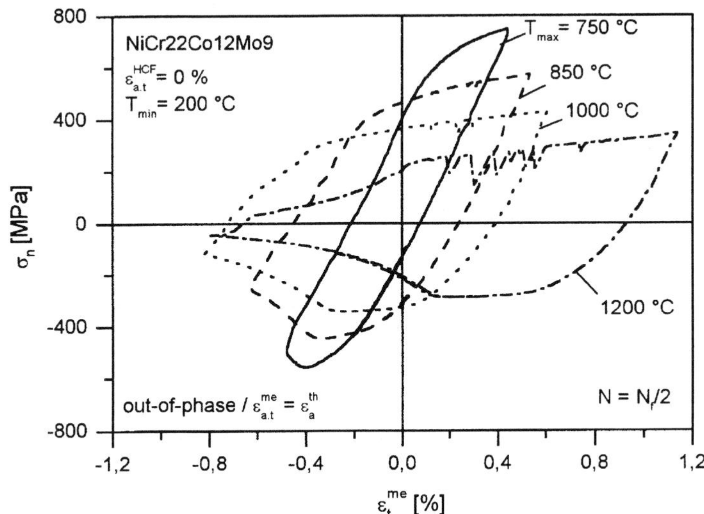

Figure 5.2.10: Stress-total mechanical strain hysteresis loops at $N_f/2$ evaluated from out-of-phase TMF tests at $\varepsilon_{a,t}^{HCF} = 0\,\%$ and different T_{max}.

perature. For this reason, the hysteresis loops are shifted to tensile mean stresses σ_m during out-of-phase TMF tests at all maximum temperatures investigated.

Figure 5.2.11 shows the development of $\varepsilon_{a,p}$ (top) as well as of σ_a and σ_m (bottom) during TMF tests at different maximum temperatures. At all T_{max}, cyclic hardening takes place which is shown by the decrease of $\varepsilon_{a,p}$ and the increase of σ_a during the TMF test. The cyclic hardening is the less pronounced the higher T_{max} is, as with increasing maximum temperature the cyclic deformation is more and more influenced by diffusion controlled

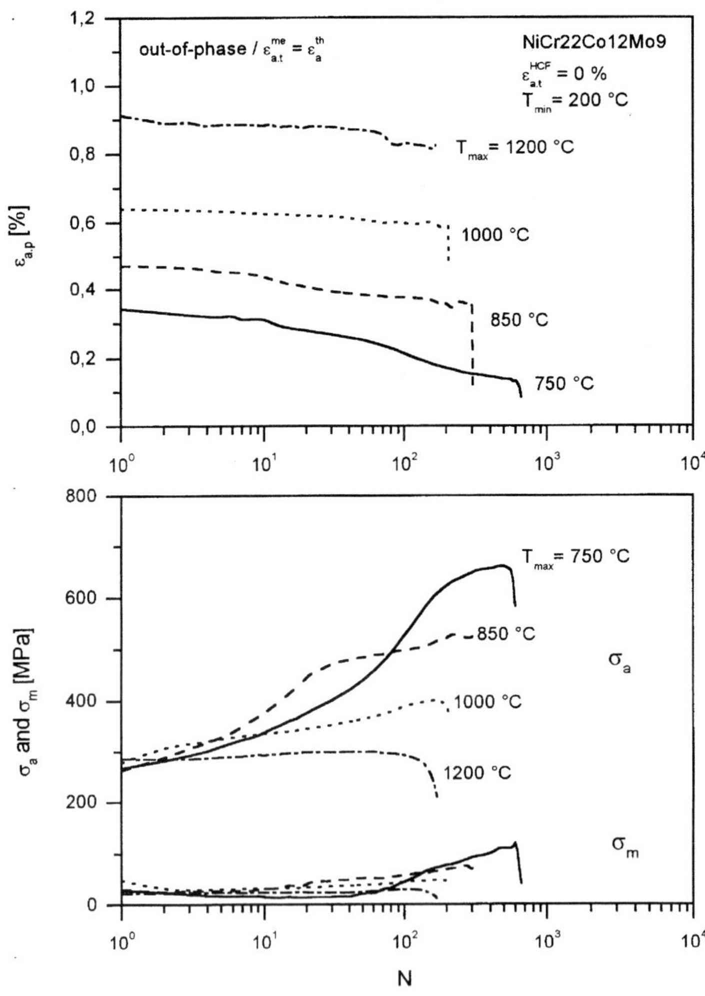

Figure 5.2.11: Development of the plastic strain amplitude (top) as well as the stress amplitude and the mean stress (bottom) during out-of-phase TMF tests at $\varepsilon_{a,t}^{HCF} = 0\,\%$ and different T_{max}.

creep and relaxation processes. Due to this fact, cyclic hardening vanishes, when T_{max} is raised up to 1200 °C. During TMF at $T_{max} = 750$ °C, there is a continuous cyclic hardening resulting from strong interactions between dislocations and small carbides which precipitate during the TMF test. Cyclic hardening is even stronger during the first 30 cycles at $T_{max} = 850$ °C, but weakens with a higher number of cycles. This finding is mainly due to the higher rates of carbide precipitation and carbide coarsening during TMF at $T_{max} = 850$ °C. As mentioned above, tensile mean stresses develop during all TMF tests performed due to dynamic relaxation processes at higher temperatures. In all cases, almost the same σ_m values are present after the first cycle. At $T_{max} = 750$ °C, σ_m remains approximately constant within the first 100 cycles and then increases continuously until macro crack initiation and propagation. At T_{max} 850 °C, the increase of σ_m during the test is less pronounced then at $T_{max} = 750$ °C.

In Fig. 5.2.12, the hysteresis loops evaluated at $N_f/2$ from TMF tests with superimposed HCF loadings at $T_{max} = 850$ °C show the influence of different HCF amplitudes $\varepsilon_{a,t}^{HCF}$ on the out-of-phase TMF deformation behaviour. Whereas serrations due to dynamic strain ageing processes during TMF tests without superimposed HCF loading at $N_f/2$ are only observed at T_{max} higher than 850 °C, these effects already appear well pronounced during TMF at $T_{max} = 850$ °C, while the specimen is cooled down to the minimum temperature (Fig. 5.2.12). Particularly at $\varepsilon_{a,t}^{HCF} = 0.05$ %, the effects of the

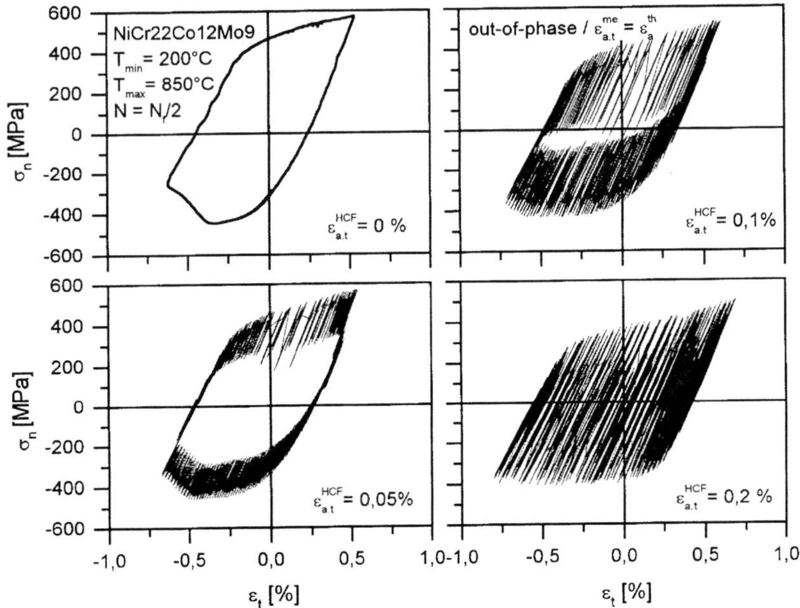

Figure 5.2.12: Stress strain hysteresis loops at $N_f/2$, $T = 850$ °C and different superimposed $\varepsilon_{a,t}^{HCF}$.

dynamic strain ageing are clearly discernible. Figure 5.2.12 also indicates that the plastic strain amplitude which is evaluated as half of the width of the hysteresis loop at mean stress, enhances with increasing superimposed HCF loading. The σ_a values measured at $N_f/2$ show a weak diminution with increasing $\varepsilon_{a,t}^{HCF}$, because the lifetime is reduced by the superimposed HCF loadings. Hence, cyclic hardening is terminated after a smaller number of cycles.

Figure 5.2.13 exemplarily shows the effects of a superimposed HCF loading on the cyclic deformation behaviour during out-of-phase TMF at

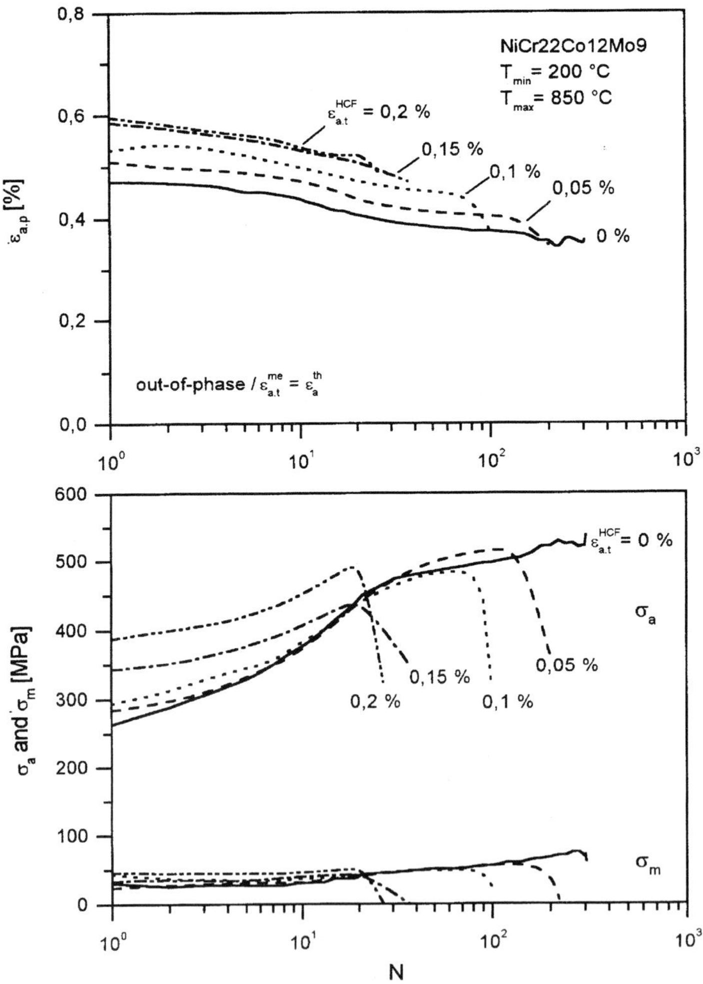

Figure 5.2.13: Development of the plastic strain amplitude (top) as well as the stress amplitude and the mean stress (bottom) during out-of-phase TMF tests at $T_{max} = 850\,°C$ and different superimposed $\varepsilon_{a,t}^{HCF}$.

T_{max} = 850 °C. For all $\varepsilon_{a,t}^{HCF}$, a cyclic hardening is indicated by a decrease of $\varepsilon_{a,p}$ (top) and an increase of σ_a (bottom) during the TMF tests. Generally, the measured values of σ_a and $\varepsilon_{a,p}$ increase with increasing $\varepsilon_{a,t}^{HCF}$. An augmentation of $\varepsilon_{a,t}^{HCF}$ from 0.15 % to 0.2 %, however, hardly enhances $\varepsilon_{a,p}$, because at higher values of $\varepsilon_{a,t}^{HCF}$, the cyclic deformation behaviour is more and more determinate by the superimposed HCF loading.

During all TMF experiments with superimposed HCF loadings, tensile mean stresses are induced. They slightly increase with growing amplitudes of the superimposed HCF loadings and with increasing number of cycles.

To determine the effect of the superimposed HCF loadings on the out-of-phase TMF lifetime at different T_{max}, the total strain amplitude $\varepsilon_{a,t} = \varepsilon_{a,t}^{me} + \varepsilon_{a,t}^{HCF}$ is plotted as a function of N_f in a double logarithmic scaling in Fig. 5.2.14. In order to compare the lifetime evaluated from pure out-of-phase TMF tests ($\varepsilon_{a,t}^{HCF}$ = 0 %, broken line, solid symbols) with N_f from isothermal fatigue tests, additionally total strain Wöhler curves at T = 850 °C and 1000 °C (dotted lines) are graphed in the figure. As can be seen, it is possible to conservatively estimate N_f of pure out-of-phase TMF test at T_{max} = 1000 °C from isothermal fatigue test at T = 1000 °C, whereas the TMF life at T_{max} = 850 °C is distinctly less than N_f obtained from isothermal fatigue test at 850 °C.

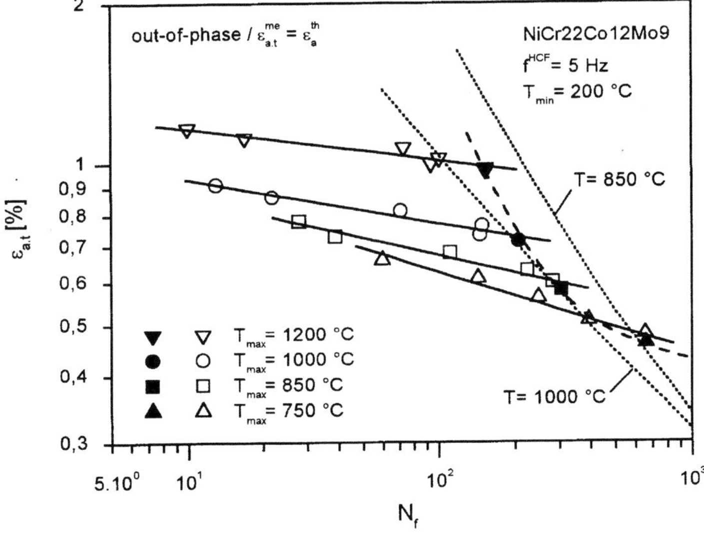

Figure 5.2.14: Total strain amplitude $\varepsilon_{a,t} = \varepsilon_{a,t}^{me} + \varepsilon_{a,t}^{HCF}$ vs. the number of cycles to failure from out-of-phase TMF tests at different $\varepsilon_{a,t}^{HCF}$ and T_{max} as well as isothermal total strain Wöhler curves at T = 850 °C and 1000 °C.

Superimposed HCF loadings cause a pronounced reduction of the TMF lifetimes. This is particularly shown in Fig. 5.2.15, where the lifetime reduction and the lifetime ratio (N_f divided by N_f at $\varepsilon_{a,t}^{HCF} = 0\,\%$), respectively, are plotted as a function of the superimposed HCF amplitude. The amount of the lifetime reduction increases with increasing $\varepsilon_{a,t}^{HCF}$ and may approach 93 % of the fatigue lifetimes determined by pure TMF tests.

The straight lines valid for TMF tests with superimposed HCF loadings in Fig. 5.2.14 can be described with a potential function for each maximum temperature

$$\varepsilon_{a,t} = A N_f^{-b} \tag{6}$$

where A and b are constants. The exponent b which is given by the slope of the straight line, decreases with increasing maximum temperature. Obviously, there is a dependence between this exponent and the material properties at different temperatures. This is proved in Fig. 5.2.16, where the exponent b is plotted as a function of the tensile strength at maximum temperature. By this empirical linear relationship and Eq. (6), it is possible to estimate the amount of the out-of-phase TMF lifetime reduction caused by a superposition of a higher frequent HCF loading.

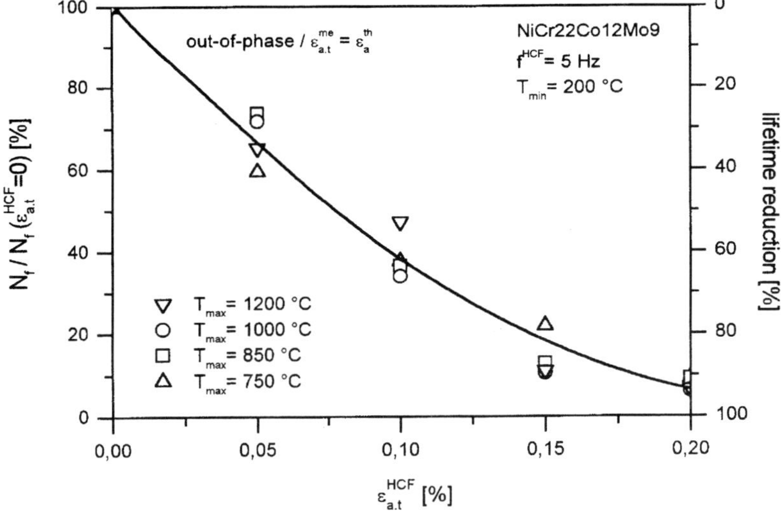

Figure 5.2.15: Lifetime ratio, N_f divided by N_f at $\varepsilon_{a,t}^{HCF} = 0\,\%$ and the lifetime reduction, respectively, as a function of $\varepsilon_{a,t}^{HCF}$ for the TMF tests at different T_{max}.

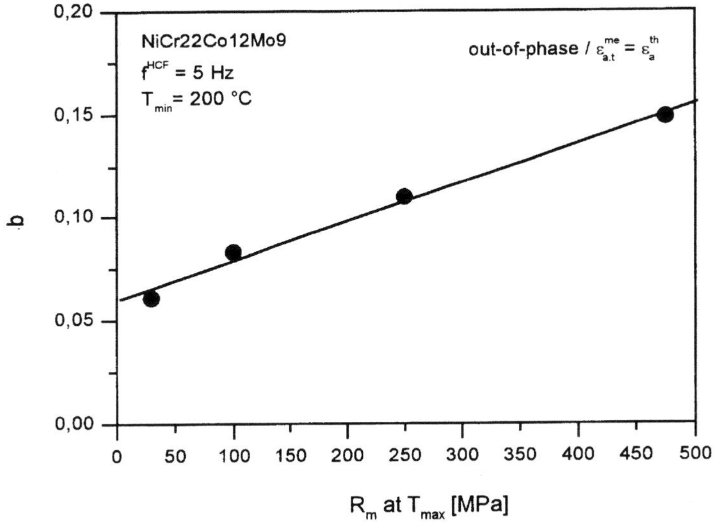

Figure 5.2.16: Exponent *b* vs. the tensile strength at the maximum temperature.

5.2.5 Summary

Isothermal, total strain controlled fatigue tests at different deformation rates and temperatures show that the cyclic deformation at high frequencies and/or low temperatures is mainly determined by thermally activated dislocation motion and dynamic strain ageing processes. With decreasing frequency and/or with increasing temperature, the cyclic deformation is more and more influenced by diffusion controlled creep processes. As a consequence, a change from cyclic hardening to cyclic neutral behaviour is observed. The fatigue life as a function of the frequency in double logarithmic scaling indicates a linear dependence of the number of cycles to failure N_f as well as of the time to failure t_f on the frequency. With decreasing frequency N_f decreases and t_f increases. Furthermore, the validity of the frequency-modified Coffin-Manson and Basquin relationships for these total strain controlled fatigue tests using the plastic and the elastic strain amplitude at half of the number of cycles to failure was proved. With the combination of both relationships the fatigue life was estimated conservatively under different loading conditions.

During total strain controlled out-of-phase TMF experiments without and with superimposed HCF loadings cyclic hardening is observed, which is the less pronounced the higher the maximum temperature is. With increasing superimposed HCF amplitudes the cyclic deformation behaviour is obviously more and more determined by the superimposed HCF loadings. Thus, the plastic strain amplitude remains approximately constant after an augmentation of $\varepsilon_{a,t}^{HCF}$ from 0.15 % to 0.2 %. Due to the dynamic relaxation processes at higher temperatures, tensile mean stresses develop during all TMF tests performed. As a result of superimposed HCF loadings the out-of-phase TMF lifetimes decrease significantly. This lifetime reduction increases with growing superimposed HCF amplitudes and may approach 93 % of the fatigue lifetime obtained from pure TMF experiments. Furthermore, for each maximum temperature the dependence between the total strain amplitude ($\varepsilon_{a,t} = \varepsilon_{a,t}^{me} + \varepsilon_{a,t}^{HCF}$) and the number of cycles to failure can be described as a potential function. The exponent of this relationship decreases with increasing maximum temperature and is obviously dependent on the material properties at different temperatures. A linear dependence between the exponent and the tensile strength at T_{max} was proved. Using these empirically determined relationships it is possible to estimate the out-of-phase TMF lifetime reduction caused by a superposition of a higher frequent HCF loading.

References

1. Möndel, A., Lang, K.-H., Löhe, D., Macherauch, E. (1997): Creep-Fatigue Behaviour of CoCr22Ni22W14 in the Temperature Range 850 °C ≤ T ≤ 1200 °C, 11th Int. Conf. on the Strength of Materials, *Mat. Sci. and Eng.* **A234–236**, Elsevier Sci., 715–718.
2. Moalla, M., Lang, K.-H., Löhe, D., Macherauch, E. (1998): Isothermal High Temperature Fatigue Behaviour of NiCr22Co12Mo9 under Superimposed LCF and HCF Loading, Fourth Int. Conf. on Low Cycle Fatigue and Elasto-Plastic Behaviour of Materials (LCF4), Elsevier Sci., 27–32.
3. Lang, K.-H. (1991): Einfluß betriebsnaher Beanspruchungen auf die Lebensdauer der Zylinderkopfwerkstoffe GG-30 und GGV30. Werkstoffkunde „Beiträge zu den Grundlagen und zur interdisziplinären Anwendung". DGM-Verlag, 79–88.
4. Hallstein, R. (1991): Das Verhalten von Gußeisenwerkstoffen unter isothermer und thermisch-mechanischer Wechselbeanspruchung, Dr.-Ing. Thesis, University of Karlsruhe, Germany.
5. Merckling, G. (1989): Kriech- und Ermüdungsverhalten ausgewählter metallischer Werkstoffe bei erhöhter Temperaturen; Dr.-Ing. Thesis, University of Karlsruhe, Germany.
6. Coffin Jr., L. F. (1969): Predictive Parameters and their Application to High Temperature, Low Cycle Fatigue, Int. Conf. on Fracture, Vol. **2**, Brighton, 643–654.
7. Kleinpaß, B., Lang, K.-H., Löhe, D., Macherauch, E. (1995): Thermal-Mechanical Fatigue Behaviour of NiCr22Co12Mo9, Symposium on Fatigue under Thermal and Mechanical Loading, Kluwer Academic Publishers, 327–337.

5.3 Microstructure and Deformation Behaviour of Carbide-Hardened Superalloys

Ulrich Martin, Heinrich Oettel*, Uwe Mühle, and Otmar Vöhringer**

Abstract

The present study deals with the application of hot deformation experiments and microstructure investigations in order to predict the creep, the tensile, and the relaxation behaviour of the highly loaded superalloys NiCr22-Co12Mo9 and CoCr22Ni22W14. The results of mechanical tests and transmission electron microscopy (TEM) investigations have been used as input data in two models which describe the high temperature plastic deformation. It is the aim of the microstructure investigations to estimate parameters like dislocation density and carbide spacing for a microstructure related modelling using an effective stress and a constitutive model. A common log $\dot{\varepsilon}$-log σ plot of tensile and creep data shows a systematic shift of about 10 % towards higher stresses in the case of tensile loading. In creep tests the dislocation densities obtain their steady-state values at the points of minimum strain rates, whereas in tensile tests the dislocation densities increase at any given time until the tensile strength is reached at significantly higher strains. The dislocation densities measured by TEM, agree quite well with those, determined by the evolution equation of the applied two models, introducing a time dependent term. The simulation by the effective and the constitutive model yield qualitatively correct predictions of the creep, tensile, and the relaxation behaviour of the carbide hardened superalloys.

5.3.1 Introduction

In the field of high temperature materials carbide precipitation hardened superalloys have been designed on the base of nickel (e. g. NiCr22Co12Mo9) or cobalt (e. g. CoCr22Ni22W14) for enhanced high temperature applications

* Institut für Metallkunde, TU Bergakademie Freiberg, Gustav-Zeuner-Str. 5, 09596 Freiberg, Germany

** Institut für Werkstoffkunde I, Universität Karlsruhe, Kaiserstr. 12, 76128 Karlsruhe, Germany like sheet material in combustion chambers of gas turbines. Both superalloys exhibit excellent high temperature mechanical properties accompanied by high resistance against oxidation and hot gas corrosion. In practical applications these superalloys are coated with a duplex thermal barrier system (TBC) of a plasma-sprayed ZrO_2 thermal barrier deposited on top of an NiCoCrAlY bond coat.

Referring to the unidirectional deformation behaviour of high temperature sheet materials [1–7], especially concerning hot tension experiments and creep of NiCr22Co12Mo9 [3, 4, 7, 14, 15] and also CoCr22Ni22W14 [2, 5, 6, 16] a large number of data are available from former investigations. In principle, similar microstructure processes were found in hot tensile, creep, or relaxation experiments under tensile loading.

The motivation of modelling using these data has been the description of the deformation behaviour of high temperature material under thermal and mechanical load on the base of physical equations. The present study deals with the application of modelling hot tensile, creep, and relaxation tests of the superalloys NiCr22Co12Mo9 and CoCr22Ni22W14 to predict the high temperature deformation behaviour. Estimating parameters like dislocation density and carbide spacing by TEM the hot deformation behaviour has been connected with the evolution of microstructure. Therefore, the results of the mechanical tests and of microstructural TEM observations are transferred to a constitutive model (by Kocks-Mecking-Estrin [12, 18, 19]) and a modified effective stress model (by Haasen-Alexander [13]) of high temperature plastic deformation. The predictions obtained from these models have been compared with the experiments. The investigations have been focussed on the cobalt base superalloy CoCr22Ni22W14 [5, 6, 9, 10].

5.3.2 Material Characterization

The used superalloys are solid solution and carbide precipitation hardened and do not contain any γ'-particles because high volume fractions of these particles reduce the ductility at the operating temperature. The fcc matrix is particle strengthened by carbides of the types M_6C und $M_{23}C_6$. Carbides of M_6C-type being rich of tungsten (CoCr22Ni22W14) or molybdenum (NiCr22-Co12Mo9) were already found in the as received state. Usually their diameter lies between 1 and 2 μm, the mean spacing is about 6 μm. These parameters do not change during loading. Moreover, $M_{23}C_6$ carbides with a high chromium content precipitate after an exposure time of a few hours occur (see Fig. 5.3.1) often already during heating and balancing the temperature before load is applied. The carbide particle strengthening of the finely distributed $M_{23}C_6$ carbides is essential for the creep and also for the stress-strain tensile curves. Their crystallographic orientation is identical to the one of the matrix, but a

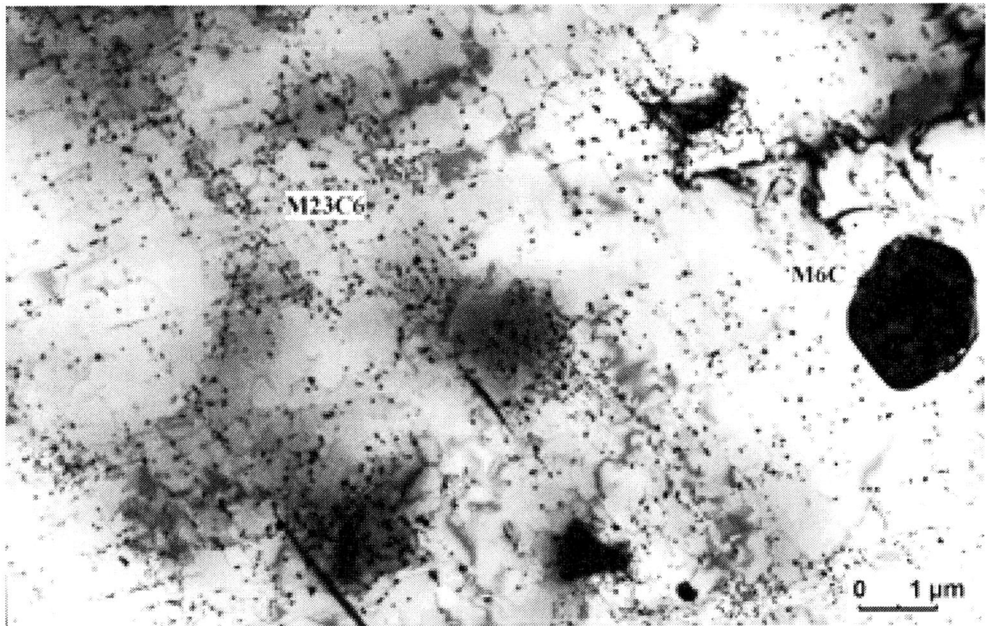

Figure 5.3.1: Various kinds of carbides in the uncoated superalloy NiCr22Co12Mo9 (M_6C and $M_{23}C_6$) after isothermal treatment 850 °C/33 h.

slight difference in lattice parameters is balanced with geometrically necessary dislocations. In the case of NiCr22Co12Mo9 at 850 °C, the mean diameter of these carbides is about 40 nm; their average distance amounted to 100 nm [14]. In the case of CoCr22Ni22W14 at 850 °C the mean diameter of those carbides is about 80 nm; their average spacing amounted to 350 nm defined by the first maximum of the pair correlation function [6, 8]. The precipitation of $M_{23}C_6$ carbides appear in rows, which represent traces of {111} planes, combined with obstacle free channels.

5.3.3 Hot Deformation Tests and Experimental Details

The uniaxial creep tests (Fig. 5.3.6) were performed in cooperation within the Collaborative Research Centre 167 of the Karlsruhe University [5, 7] under constant nominal stresses σ_0. Hot tensile experiments (Fig. 5.3.7) with strain rates, which are in correlation with the rates adjusted in the creep experiments, were carried out with constant displacement rate v of the traverse within the temperature range of 750 °C to 1050 °C [6]. Before application of load temperature was held constant for temperature balance. In order to investigate the evolution of the microstructure during loading, the deformation of some specimens has been interrupted at distinct plastic strains. The true stresses σ during creep and the true strain rates $\dot{\varepsilon}$ in tensile tests were determined under the assumption of a constant specimen volume.

The evaluation of the $\dot{\varepsilon}$-values as a function of the true stress in both tests allows to plot tensile and creep data in one log $\dot{\varepsilon}$-log σ-diagram. Using the values of the maximum true stresses $\sigma_{max} = R_m[1 + \varepsilon_{max}]$ (R_m = ultimate tensile strength) at different strain rates, an Norton equivalent relationship has been obtained. Both ways of testing lead to the same Norton exponent $n \approx 6.6$. The regression curve of the data points of the tensile tests in the log $\dot{\varepsilon}$-log σ-diagram (Fig. 5.3.2) differs by about 10 % from that of the creep tests towards higher stresses.

This shift to higher tensile stresses appears systematically in the temperature range between 750 °C and 950 °C (Fig. 5.3.2). At higher stresses the secondary part of the creep curves follows the slope of Norton's law. However, the creep curve at 100 MPa is significantly steeper corresponding to an exponent of about 12 in the secondary region. This behaviour results from the influence of a changing microstructure. Whereas the experiments with nominal stresses of 200 up to 315 MPa took only a few hours, the experiment at 100 MPa, however, lasted more than 300 hours. As mentioned in Section 5.3.5, there are changes of the dislocation and particle structure with increasing annealing time, which lead to a change of Norton's exponent.

At the applied relaxation technique [6] the crosshead motion of a preceded tensile test with constant displacement rate v of the cross head was

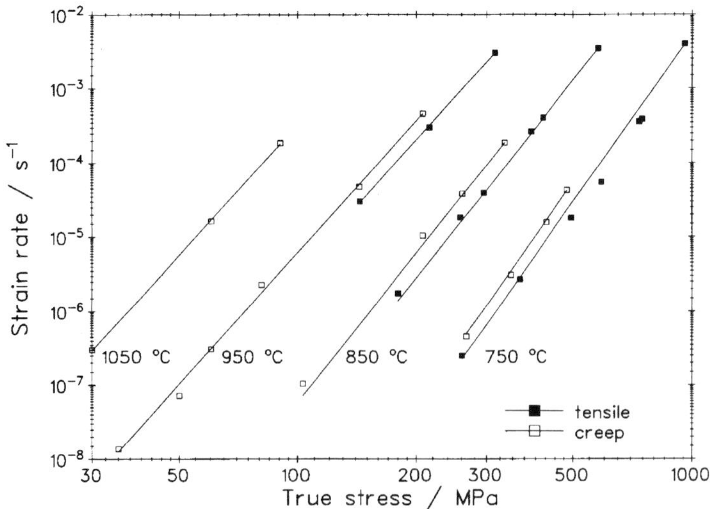

Figure 5.3.2: Norton law for the uncoated superalloy CoCr22Ni22W14 for different temperatures (□: from creep curves; after [5], ■: from tensile stress-strain curves).

stopped, after that, in the relaxation phase the course of the stress has been recorded as function of time. The stop of the cross head motion corresponds to the constant total strain ε_{tot},

$$\varepsilon_{tot} = \varepsilon_{pl} + \varepsilon_{el} = \varepsilon_{pl} + \frac{\sigma}{E} = \text{const.} \tag{1}$$

referring to the different contributions plastic ε_{pl} and elastic strain ε_{el}. The elastic contribution can be determined from the measured stress σ and the Youngs' modulus E [6]. Due to increased strain rates during the preceding tensile phase the applied true stresses increased also at the beginning of a new relaxation cycle. The elevated initial relaxation stresses were reduced by a steep slope in the primary relaxation phase followed by a flatter slope of the stress curve [6].

Thin foils for transmission electron microscopy were prepared from the fractured specimens as well as from specimens of the interrupted experiments. For a computer aided estimation of the dislocation density ϱ [11] according to the method of sectioned lines by Ham [11, 17], TEM-images were used. The distribution of carbides was estimated by particle spacing (determination of the pair correlation function) using image analysis of TEM or SEM images (back scattered electron contrast). To perform the numerical calculations necessary for modelling, a commercial table calculation (MS Excel) was used.

5.3.4 Modelling of the High Temperature Deformation

To describe the deformation behaviour quantitatively, the constitutive model purposed by Kocks and Mecking [12] and the effective stress model published by Alexander and Haasen [13] were used.

5.3.4.1 Constitutive Model

Similar to the Norton relationship the general assumption of the constitutive model is that the stress is correlated to a structural parameter $\hat{\sigma}$

$$\dot{\varepsilon} = \dot{\varepsilon}_0 \left(\frac{\sigma}{\hat{\sigma}}\right)^m \tag{2}$$

where $\dot{\varepsilon}_0$ is a comparable reference strain rate, which includes thermally activated dislocation movement [12]. The structural parameter $\hat{\sigma}$ is related to the dislocation density ϱ analogous to the Taylor relationship $\hat{\sigma} = \alpha \text{MGb} \sqrt{\varrho}$ (M: Taylor factor, G: shear modulus, b: Burgers vector).

The change of the dislocation density is described by the evolution equation

$$\frac{d\varrho}{d\varepsilon} = \frac{d\varrho^+}{d\varepsilon} + \frac{d\varrho^-}{d\varepsilon} \tag{3}$$

where the generation term $d\varrho^+/d\varepsilon$ represents the strengthening processes of dislocations and the annihilation term $d\varrho^-/d\varepsilon$ constitutes the reduction of the dislocation density by annihilation. The dislocation generation term of

$$\frac{d\varrho^+}{d\varepsilon} = \frac{M}{b\, L_P} \tag{4}$$

is correlated with the density of obstacles to the dislocation motion in regard to the mean free path L_P for a moving dislocation between two obstacles, for example carbide particles. In the range of steady-state-deformation the microstructure does not change according to the definition of deformation at constant structure. This leads with Eq. (3) to equation

$$\frac{d\varrho^+}{d\varepsilon_{SS}} = -\frac{d\varrho^-}{d\varepsilon_{SS}} \tag{5}$$

The recovery processes are described by the relationship [19]:

$$\frac{d\varrho^-}{d\varepsilon} = -k_2\varrho \tag{6}$$

The recovery term in the evolution equation (3) is derived from the knowledge of the mechanical and microstructural parameters in the steady-state-condition. Consequently, for the evolution of the dislocation density an equation was developed, which includes different kinds of obstacles for dislocation multiplication. Exclusive dislocation hardening, exclusive particle hardening, and combined dislocation and particle hardening are distinguished. In the present case of complex carbide-hardened superalloys, the third variant (hybrid model) has been applied. The efficiency of the respective strengthening mechanisms in the "hybrid model" is related with a weighting parameter p [19] and leads to

$$\frac{d\varrho}{d\varepsilon} = M\left(k_0(1-p) + k_1\, p\, \sqrt{\varrho} - k_2\, \varrho\right) \tag{7}$$

for the dislocation evolution. The weighting parameter p depends on the nominal stress and can be determined by stress change tests during creep as a function of the nominal stress. In the case of CoCr22Ni22W14 the internal back stress σ_i was estimated by [5] to a constant part of 37.1 MPa, which is attributed to the particle back stress σ_p. Then, the difference between σ_i

and σ_p is due to the strengthening effect of the forest dislocations. The relation

$$k_0 = \frac{1}{bL_P} \tag{8}$$

expresses the influence of particle hardening, using the mean particle distance L_P. The second hardening term $k_1\sqrt{\varrho}$ describes the dislocation hardening. Recently performed tensile tests [3], carried out without preceded temperature balancing time did not indicate any sign of secondary $M_{23}C_6$ carbides. The extrapolation of the differentiated tensile curves ($d\sigma/d\varepsilon$-σ-curves in regime III with $d\sigma/d\varepsilon = \Theta$) to $\sigma = 0$ leads to the work hardening property k_1 using the relation

$$k_1 = \frac{\Theta_{III,0} \;/\; G}{M^2 \; \alpha \; b} \tag{9}$$

and with the steady-state condition to k_2

$$k_2 = \frac{k_0\;(1-p) \;+\; k_1\; p\; \sqrt{\varrho_{SS}}}{\varrho_{SS}} \tag{10}$$

To model the creep curve, the structural parameter $\hat{\sigma}$ is estimated from the actual value of $\varrho(\varepsilon)$, using the Taylor relationship. Then, the Norton law provides the actual strain rate. $\dot{\varepsilon}_0$ results from the fitting of the strain rate towards the steady-state value $\dot{\varepsilon}_{SS}$. The value m, the so-called strain rate sensitivity, is related to the exponent of the Norton equation [19] by

$$\frac{1}{n} = \frac{1}{n'} + \frac{1}{m} \tag{11}$$

For superalloys it has been experienced that $m << n'$, so that the approximation $m \approx n$ can be used.

5.3.4.2 Effective Stress Model

The Orowan-equation is the starting point of the effective stress model consideration by Alexander and Haasen [13].

$$\dot{\varepsilon} = \varrho_{Mob} \cdot \frac{b}{M} \cdot v_{\varrho} \tag{12}$$

The (plastic) creep rate $\dot{\varepsilon}$ of materials at high temperatures is combined with $\varrho_{Mob,}$ the density of the free mobile dislocations, and their average dislocation velocity v_ϱ. To describe the dislocation velocity v_ϱ a power law has been chosen

$$v_\varrho = v_0 \cdot \left(\frac{\sigma_{eff}}{G}\right)^{n^*} \tag{13}$$

The stress exponent n^* is connected with the Norton exponent n from creep tests according to the relation $n^* = n - 2$. By introducing an effective stress σ_{eff}, describing the difference between the applied stress σ and the internal back stress σ_i, these ideas were completed.

$$\sigma_{eff} = \sigma - \sigma_i \tag{14}$$

Thereby σ_i may contain different contributions

$$\sigma_i = \sigma_\varrho + \sigma_P + \sigma_{SB} + \sigma_{GB} \tag{15}$$

Since, the investigated superalloys exhibit a large grain size diameter of 70 µm, the contribution of grain boundary strengthening σ_{GB} can be neglected, because of a constant grain structure before and after the creep tests. In the case of CoCr22Ni22W12 complete subgrains were found only at high strains, thus subgrain hardening σ_{SB} is negligible either. Concerning the particle hardening term σ_P of these superalloys by second-phase particles like carbides a lot of experimental and microstructural data exists [2–10]. The back stress σ_ϱ caused by dislocation interactions is given by the Taylor relationship. The plot of the average dislocation distance $1/\sqrt{\varrho}$ versus MGb/σ leads to the elastic interaction constant α, which characterizes the strengthening interaction of dislocations.

The most important microstructural parameter is the dislocation density as carrier of plastic deformation in Eq. (12) as well as contribution to internal back stress in Eq. (14). For a further modelling of the relaxation behaviour after tensile deformation the time dependent decrease of dislocation density is described by a cubical relationship, whereas the softening constant k_3 has a constant value. Therefore a further constant in the dislocation evolution (Eq. (7)) was introduced, which is time dependent. Assuming, that the diffusion of vacancies through dislocation pipes is the process, which determines the velocity of time dependent recovery, [6] finds:

$$\frac{d\varrho}{dt} = -k_3 \cdot \varrho^3 \tag{16}$$

Since the measured dislocation densities are all within an order of magnitude, the description of softening should be possible with a unique constant for the actual test temperature. By integration of Eq. (16) the relation for k_3 is, therefore,

$$k_3 = \frac{\frac{1}{\varrho_1^2} - \frac{1}{\varrho_2^2}}{2 \cdot \Delta t} \tag{17}$$

where (ϱ_1) is the dislocation density before, (ϱ_2) the value after the relaxation phase and Δt the time of relaxation (see [6]). According to the method of the Kocks-Mecking-model [12] the final equation used for simulation of the dislocation density evolution within the effective stress and also the constitutive model is:

$$\frac{d\varrho}{d\varepsilon} = M\left(k_0 \cdot (1-p) + k_1 \cdot p \cdot \sqrt{\varrho} - k_2 \cdot \varrho - k_3 \cdot \frac{\varrho^3}{M \cdot \varepsilon}\right) \quad (18)$$

5.3.5 Results and Discussion

5.3.5.1 Dislocation and Carbide Structure

The most important strengthening contribution of precipitations in NiCr22-Co12Mo9 and CoCr22Ni22W14 results from chromium rich $M_{23}C_6$ carbides (Fig. 5.3.1). Figure 5.3.3 shows the evolution of dislocation density of CoCr22-Ni22W14 for the nominal stress of 100 MPa at 850 °C, combined with the dislocation structure at defined values of strain. At first, the dislocation density increases very strongly and then reaches its steady-state value at a strain of 3.5 %. In creep tests the dislocation density reaches its steady-state value at the point of minimum deformation rate, whereas in tensile tests, the dislocation density increases until tensile strength is reached at significantly higher values of strain [6].

The further, very slight increase should be correlated with the increase of true stress with increasing strain. Subgrains were only found at strains much larger than the strain corresponding to the minimum strain rate. The existence of subgrains does not significantly influence strain rate. The average diameter of subgrains has a constant value of 2.2 ± 0.2 μm at any strain.

At higher stresses the distribution of dislocations remains homogeneous until fracture. According to the Taylor relationship (Section 5.3.4.2) the dislocation interaction constant α was determined. The nearly constant value of the dislocation density after reaching the minimum strain rate justifies a similar treatment of the dislocation densities obtained by fractured and by distinctly deformed specimens using the true stress as a reference value. In CoCr22Ni22W14 there is a significant decrease of α at intermediate dislocation distances of about 80 nm down to the relatively low value of 0.18 [8]. Such low constants of interaction are correlated with channelling structures of dislocations. If the dislocation distances at a high forest dislocation density are smaller than the spacings between the precipitates, the higher value of α is dominant, correlated with a more homogeneous dislocation distribution.

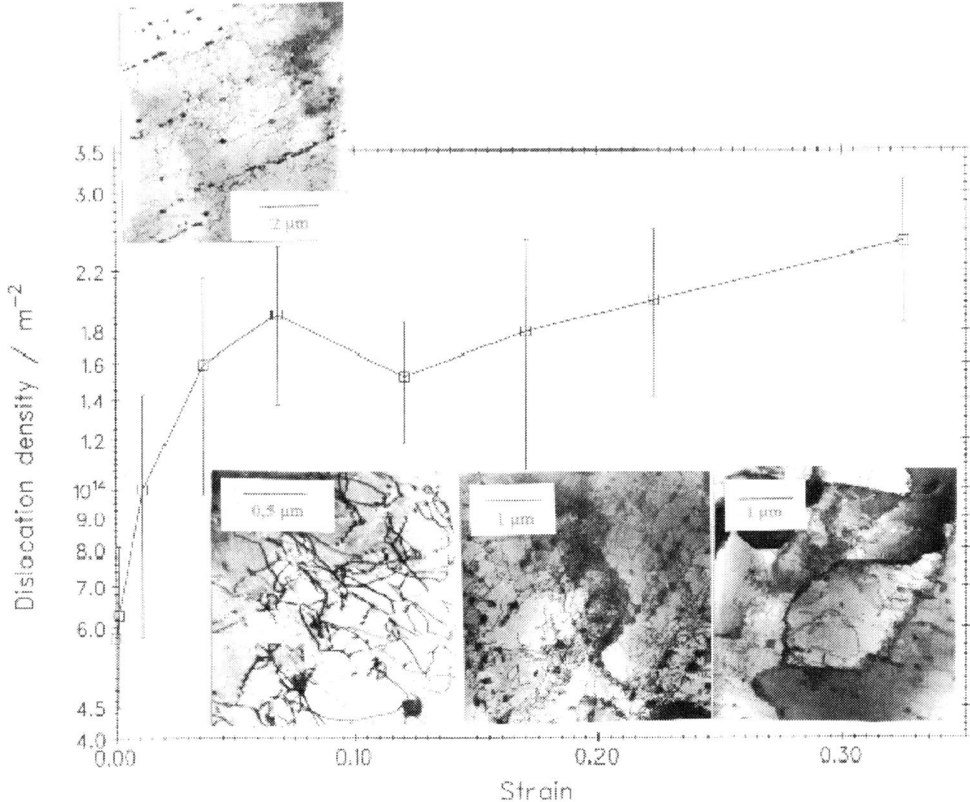

Figure 5.3.3: Microstructure and dislocation density evolution for creep of CoCr22-Ni22W14 at 850 °C and 100 MPa.

During the tensile tests the evolution of the dislocation density up to its maximum value is rather slow in comparison to creep tests (Fig. 5.3.5). As implied by the tensile deformation curves in Fig. 5.3.7, a steady-state regime of deformation is not detectable. Simultaneously, there is a significantly stronger formation of subgrain boundaries (Fig. 5.3.4). The constant of dislocation interaction α is significantly higher than in creep tests. At higher tensile strengths and dislocation spacings smaller than 80 nm, α reaches a value of about 0.5. With decreasing stress, the α-value then decreases to 0.32 at a mean dislocation distance of about 80 nm. Assuming that the subgrain structure contributes also to the strengthening, this transition can be understood [6, 9]. An estimation of the creep resistance which is based only on tensile experiments leads to a prediction of higher rupture times of the materials as truly observed. In practice, a higher creep rate than predicted by the hot tensile tests is measured.

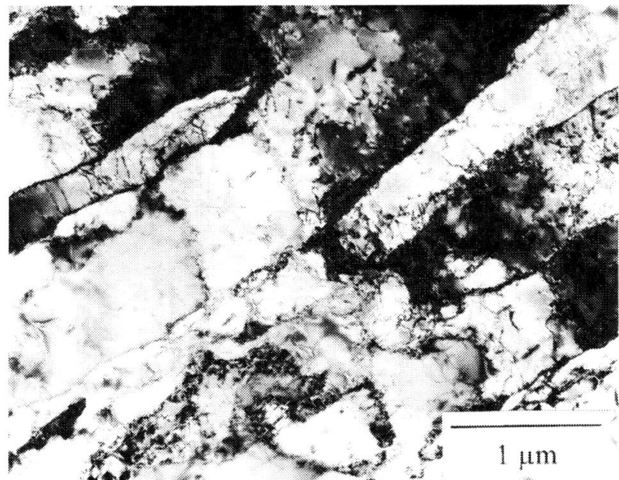

Figure 5.3.4: Subgrains after tensile deformation of CoCr22Ni22W14 at 850 °C and a strain rate $3 \cdot 10^{-5}\ s^{-1}$.

After relaxation tests the microstructure of NiCr22Co12Mo9 exhibits a local dehomogenization of the dislocation structure, which leads to a rearrangement of dislocations into subgrains. The estimated dislocation density decreases slightly during the holding time of the relaxation phase by time dependent annihilation processes like climbing.

5.3.5.2 Modelling of the Deformation Behaviour of the Superalloys

In order to predict the deformation behaviour of the superalloys, the estimated microstructure parameters have been used as input data for an effective stress model and a modified Kocks-Mecking model, which describe the high temperature deformation. The estimated dislocation densities, agree quite well with those, determined by the evolution equation of the used two models, introducing a time dependent term. Using the values of the dislocation density the effective stress model and also the constitutive model yield qualitatively correct predictions for both dislocation evolution and creep behaviour (Figs. 5.3.5 and 5.3.6). Especially the change from normal transient behaviour at low stresses to sigmoidal primary creep behaviour at higher stresses has been described in a correct manner by the two models (Fig. 5.3.6) The microstructure related modelling of primary creep reflects the type of the curves and the starting point of the steady-state region rather

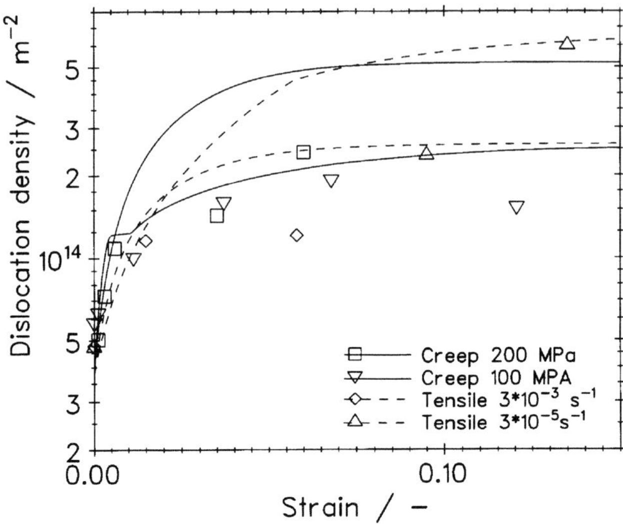

Figure 5.3.5: Modelled dislocation densities (dashed and full lines) for creep and tensile tests at 850 °C in comparison to the measured values (unfilled symbols); CoCr22-Ni22W14.

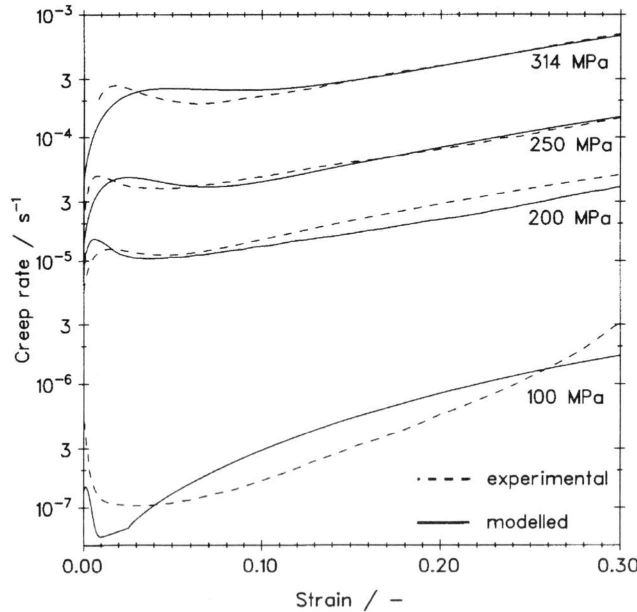

Figure 5.3.6: Constitutive model simulation of the $\dot{\varepsilon} - \varepsilon$ in creep curves (solid lines) in comparison to the experimental creep curves at 850 °C; CoCr22Ni22W14 (different loading conditions; dashed lines).

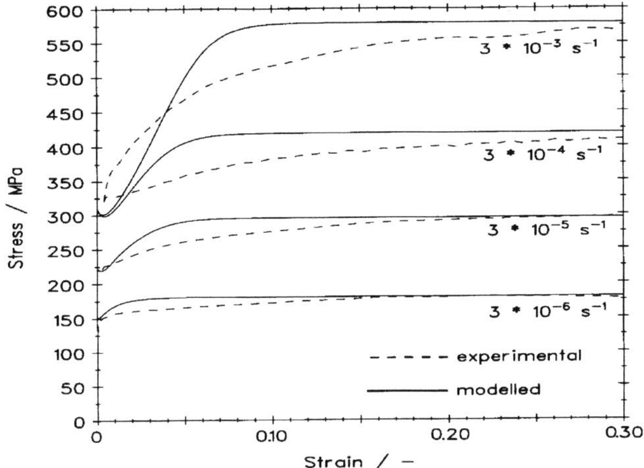

Figure 5.3.7: Effective stress model simulation of the tensile curve (solid lines) in comparison to the experimental tensile curves at 850 °C (different loading conditions; dashed lines); CoCr22Ni22W14.

well. The exclusive use of microstructurally determined parameters without further fit parameters for the creep curves leads to relatively small deviations from the experimental findings. On the other hand, Fig. 5.3.7 shows the results of the modelled tensile curves in comparison with the experiment using Eq. (2) with a value of 2.78 for n^*. The modelling of the tensile curves (Fig. 5.3.7) was carried out, using only one fit parameter v_0. For modelling the dislocation evolution and also the simulation of tensile curves following values of the constants (after [6]) have been used: α = 0.14 (tensile)/0.10 (creep); M = 3.06; G = 63.5 GPa; $p = p(\sigma)$; σ_P = 37.1 MPa; $\varrho = \varrho(\varepsilon)$; L_P = 350 nm; b = 0.254 nm; $k_0 = 1.12 \cdot 10^{12}$ cm^{-2}; $k_1 = 5.85 \cdot 10^6$ cm^{-1}; $k_2 = k_2\ (\dot{\varepsilon}_{ss}, \varrho_{ss})$; $k_3 = 8.0 \cdot 10^{-27}$ cm^{-2}s^{-1}. The variation of v_0 represents the influence of internal friction by solute atoms on the movement of dislocations in the first part of the σ–ε tensile curves. The exclusive use of microstructurally determined parameters without further fit parameters for the creep curves leads to relatively small deviations from the experimental findings.

Using the estimated values of the dislocation density the effective stress model according to Alexander and Haasen [13] yields qualitatively correct predictions for creep, tensile, and also relaxation behaviour. Especially the change from normal creep transient behaviour at low stresses to sigmoidal primary behaviour at higher stresses has been described in right manner by the model. An interesting result for practical use is the correct prediction of the time for reaching the failure strain in creep tests.

Figure 5.3.8 shows the evolution of true stress during a one hour relaxation phase at 850 °C for CoCr22Ni22W14. The strain rate of preceding tensile tests was $3 \cdot 10^{-5}$ s^{-1}. Due to increased strain rate or decreased temperature during the preceding tensile phase the applied true stresses increased also at the beginning of a new relaxation cycle. After the stop of the traverse motion the stress drops very considerably at first, after that a pass over in a flatter part of stress plot takes place. This steep slope is clearly markable after prestrain at high strain rates. In the further course of the relaxation curves the plots from both tests arranged parallel with a difference of nearly 20 MPa. Taking into account that the different strain rates during the tensile phase lead to different dislocation densities, consequently to different internal back stresses, which do not approximate to each other during the relaxation phase, this parallel course is understandable. The steeper slope in the primary region may be explained by the higher effective stress connected with increased strain rate. During the one-hour relaxation phase the microstructure of the superalloys exhibit a local dehomogenization of the dislocation structure, which leads to a rearrangement of dislocations into subgrains.

Caused by alterations of the dislocation density, the dislocation configuration and also the particle back stress, the modelling of the relaxation behaviour of both superalloys was carried out successfully (Fig. 5.3.8) by application of the two mentioned models with a time depending recovery constant [6].

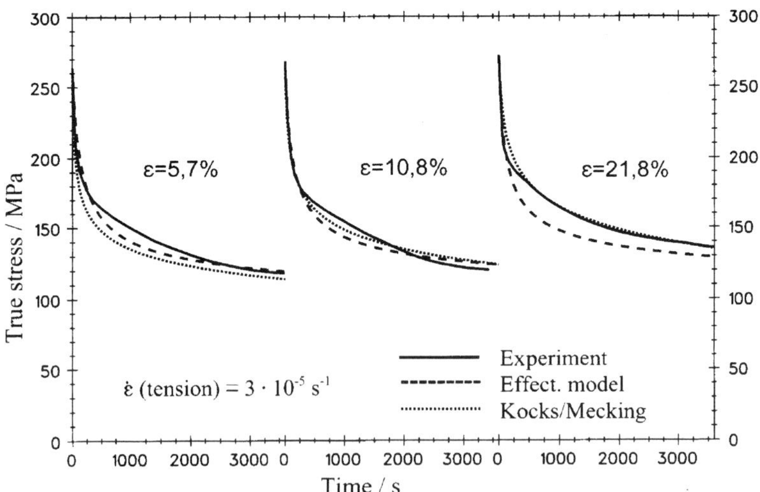

Figure 5.3.8: Modelling of the stress relaxation in CoCr22Ni22W14 at 850 °C for different values of strain and a preceding strain rate of $3 \cdot 10^{-5}$ s^{-1}.

5.3.6 Conclusions

The most important strengthening contribution of precipitations in CoCr22-Ni22W14 results from chromium rich $M_{23}C_6$ carbides. Comparing creep and hot tensile tests, there are identical Norton exponents, but a systematic difference in the absolute values of strength. A log $\dot{\varepsilon}$-log σ-diagram indicates a systematic deviation of about 10% of the straight line according to the Norton relationship towards higher stresses for tensile loading. Neglecting these deviations and evaluating the creep strength only on the basis of tensile tests would lead to an overestimation of the rupture times of the material. This deviation reflects a modified microstructure evolution during creep and tensile tests. In creep tests the dislocation density reaches its steady-state value at the point of minimum strain rate, whereas in tensile tests the dislocation density increases until the tensile strength is reached at significantly higher values of strain. Subgrain formation is more pronounced after tensile than after creep deformation, presuming equal values of true stress. The dislocation densities measured by TEM, agree quite well with those, determined by the evolution equation of the used two models, introducing a time dependent term. The simulation by the effective and the constitutive model yields to qualitatively correct predictions of the creep, tensile, and the relaxation behaviour of the carbide hardened superalloys. The effective stress model describes quite well the deformation behaviour in the primary creep range. The microstructure related modelling of primary creep reflects the type of the creep or tensile curves and also the starting point of deformation rather well. Stress relaxation in tensile loaded superalloys has been simulated successfully, if the lowered dislocation density and the particle spacing are taken into consideration.

Acknowledgements

The investigations were sponsored by the Deutsche Forschungsgemeinschaft within the frame of the Collaborative Research Centre 167, sub-project C 10, at the University of Karlsruhe. We convey our special thanks to Dr. C. Antes and Dr. U. Schmidt, University of Karlsruhe, for helpful discussions and making their unpublished results of mechanical tests available.

References

1. Schneider, W., Hammer, J., Mughrabi, H. (1992): Proc. 7th Int. Symp. on Superalloys, Champion, Pennsylvania, September 20–24, 1992, TMS Publication, Warrendale.
2. Mühle, U., Martin, U., Oettel, H., Forschungsbericht des SFB 167, (1993–1995), 531–563, University of Karlsruhe.
3. Viereck, D. (1990): Das Zug- und Relaxationsverhalten von Hochtemperaturblechwerkstoffen im Temperaturbereich 78 K $\leq T \leq$ 1473 K, Fortschrittsberichte VDI, Reihe 5, Grund- und Werkstoffe, Nr. 2, VDI Verlag Düsseldorf.
4. Merckling, G. (1989): Kriech- und Ermüdungsverhalten ausgewählter metallischer Werkstoffe bei höheren Temperaturen, Ph. D. Thesis; University of Karlsruhe.
5. Antes, C. (1993): Statisches und quasistatisches Hochtemperaturverformungsverhalten der Brennkammerlegierung CoCr22Ni22W14, Ph. D. Thesis; University of Karlsruhe.
6. Mühle, U. (1997): Mikrostruktur und Modellierung des einsinnigen plastischen Verformungsverhaltens der Superlegierung CoCr22Ni22W14, Ph. D. Thesis; Freiberg, University of Mining and Technology.
7. Schmidt, U. (1998): Über das Kriechverformungs-, Zeitstand- und Versagensverhalten des plasmagespritzten Wärmedämmschichtverbundes NiCr22Co12Mo9-NiCoCrAlY-ZrO_2/7Ma.-%Y_2O_3, Ph. D. Thesis; University of Karlsruhe.
8. Breuer, D., Pantleon, W., Martin, U., Mühle, U., Oettel, H., Vöhringer, O. (1995): Creep Behaviour Modelling of a Cobalt Based Alloy by Modifying the Kocks-Mecking-Model, *phys. stat. sol. (a)*, **150**, 281–295.
9. Mühle, U., Martin, U., Oettel, H. (1997): Microstructure and Modelling of the Deformation Behaviour of CoCr22Ni22W14 in Hot-Tensile- and Creep-Tests, *Mat. Sci. and Engng.* **A 230**, 81–87.
10. Martin, U., Mühle, U., Oettel, H. (1997): Microstructure and Modelling of the Deformation Behaviour of Superalloys at High Temperatures, *Comput. Mater. Sci.* **9**, 92–98.
11. Martin, U., Mühle, U., Oettel, H. (1995): The Quantitative Measurement of Dislocation Density in the TEM, *Prakt. Metallogr.* **32**, 467–477.
12. Kocks, U., Mecking, H. (1981): Kinetics of Flow and Strain-Hardening, *Acta metall.* **29**, 1865–1875.
13. Alexander, H., Haasen, P. (1968): Dislocations and Plastic Flow in the Diamond Structure, *Solid State Phys.* **22**, 27–158.
14. Schurack, F.(1997): Mikrostruktur und Interdiffusion in den Grenzflächenbereichen beschichteter Superlegierungen nach Hochtemperaturbeanspruchung, Diplomarbeit, Freiberg University of Mining and Technology.
15. Schmidt, U., Löhe, D., Vöhringer, O. (1997): Über das Kriechverformungs- und Versagensverhalten des Schichtverbundes NiCrCo12Mo9-NiCoCrAlY-ZrO_2/7 Ma.-%Y_2O_3, Proc. Workshop Hochwarmfeste Werkstoffe, 24.–25. 2. 1997, Aachen, 87–90.
16. Jerenz, M., Martin, U., Schurack, F., Oettel, H. (1997): Mikrostruktur und Kriechverhalten beschichteter Superlegierungen, Proc. Workshop Hochwarmfeste Werkstoffe, 24.–25. 2. 1997, Aachen, 81–84.
17. Ham, R.K., (1961): *Phil. Mag.* **6**, 1183.
18. Mecking, H., Estrin, Y. (1987): in: Andersson (Ed.) "Constitutive Relations and their Physical Basis", 8th Risø Int. Symposium on Metallurgy and Material Sciences, 1987, 123.
19. Mecking, H., Estrin, Y. (1984): A Unified Phenomenological Description of Work Hardening and Creep Based on One-Parameter-Model, *Acta Metall.* **32**, 57–70.

5.4 Advances in the Inelastic Failure Analysis of Combustor Structures

Holger Kiewel, Jarir Aktaa, and Dietrich Munz*

Abstract

To investigate the inelastic deformation behaviour of complex components several non-linear, unified material models have been implemented into commercial finite element codes. The calculations which have been performed require much computer resources. This especially is a decisive problem when simulating cyclic loadings of components having a high number of cycles to failure. Several strategies have been developed to overcome this problem. A well-known strategy consists in reducing the structure under study to a partial structure containing the most critical regions of the component. For cyclically loaded components such as combustor structures another, much more efficient method was developed. The main idea consists in extrapolating the complete set of internal variables. To demonstrate the capabilities of this new scheme, a complete failure analysis has been carried out for a ring combustor of a gas turbine. The material model used for this investigation is based on the well-known Chaboche model in combination with the damage model by Kachanov/ Rabotnov.

5.4.1 Introduction

During the runtime of the Collaborative Research Centre 167, a great variety of viscoplastic material models [1–4] have been implemented into commercial finite element codes. They all base on the concept of internal variables, the number of which is quite large to describe reliably the actual material behaviour. However, their evolution in time is mostly non-linear. Therefore, integration of the respective evolution equations requires a large number of time increments. In the case of cyclic loadings as for combustor structures under

* Institut für Zuverlässigkeit und Schadenskunde im Maschinenbau, Universität Karlsruhe, Kaiserstr. 12, 76128 Karlsruhe, Germany

non-stationary loading conditions, there are a lot of changes in the direction of loading. As a consequence, the maximum length of an increment has to be distinctly smaller than a single cycle. For this reason, the number of increments up to failure of the component increases enormously as compared to the simulation of a pure tensile test. An optimisation of the applied integration scheme to calculate the deformation of complex structures becomes essential [5]. Although these optimisations reduce the computing time significantly, simulations of loadings of more than 1000 cycles are not practicable, if the complete loading path is divided into increments only. To overcome these problems, an extrapolation method for internal variables has been developed.

5.4.2 Chaboche/Rabotnov Model

The viscoplastic model chosen for the final investigations of the present project was developed by Chaboche. For relatively high temperatures with low strain rates, fatigue is negligible even for cyclic loadings, and creep becomes the dominant damage process. The model by Kachanov/Rabotnov then describes reliably the damage behaviour of materials. The interaction of deformation and damage is obtained by coupling these two models within the hypothesis of strain equivalence [6]. Here, the version used shall be described briefly.

The total strain $\boldsymbol{\varepsilon}$ is additively composed of the elastic strain $\boldsymbol{\varepsilon}^{\mathrm{el}}$, the thermal strain $\boldsymbol{\varepsilon}^{\mathrm{th}}$ and a single inelastic strain $\boldsymbol{\varepsilon}^{\mathrm{in}}$:

$$\boldsymbol{\varepsilon} = \boldsymbol{\varepsilon}^{\mathrm{el}} + \boldsymbol{\varepsilon}^{\mathrm{th}} + \boldsymbol{\varepsilon}^{\mathrm{in}} \tag{1}$$

An additive subdivision of the inelastic term into a purely plastic and a pure creep term is not possible. This property characterises the class of unified models to which the Chaboche model belongs. The elastic part is given by Hooke's law:

$$\boldsymbol{\sigma} = (1 - D) \cdot \mathbf{E} : \boldsymbol{\varepsilon}^{\mathrm{el}} \tag{2}$$

where $\boldsymbol{\sigma}$ and $\mathbf{E}$ denote Cauchy's stress tensor and the elasticity tensor, respectively. Here, we have introduced a further internal variable D, which describes isotropic damage of the material under study. Virgin material is specified by $D = 0$. A value of $D = 1$ stands for maximum damage. As obvious, increasing damage leads to decreasing elastic stiffnesses. The thermal strain is determined by the linear expansion law

$$\boldsymbol{\varepsilon}^{\mathrm{th}} = \alpha(T)(T - T_0) \cdot \mathbf{1} \tag{3}$$

with the thermal expansion coefficient $\alpha(T)$ depending on the temperature T. For the inelastic strain rate the flow rule

$$\dot{\boldsymbol{\varepsilon}}^{\text{in}} = \frac{3}{2}\left\langle\frac{\overline{\Sigma} - K - k}{Z}\right\rangle^{n}\frac{\boldsymbol{\Sigma}}{\overline{\Sigma}} \tag{4}$$

is assumed to be valid. Here, we use the so-called McAuley brackets:

$$\langle x\rangle = \begin{cases} x & \text{for} \quad x \geq 0 \\ 0 & \phantom{\text{for}} \quad x < 0 \end{cases} \tag{5}$$

In Eq. (4) n, k, and Z denote material parameters. The overstress $\boldsymbol{\Sigma}$ is defined by

$$\boldsymbol{\Sigma} = \frac{\mathbf{s}}{1 - D} - \mathbf{a} \tag{6}$$

where $\mathbf{s}$ denotes the deviator of $\boldsymbol{\sigma}$ and $\mathbf{a}$ the back-stress tensor. The quantity $\overline{\Sigma}$ follows from $\boldsymbol{\Sigma}$:

$$\overline{\Sigma} = \sqrt{\frac{3}{2}\boldsymbol{\Sigma} : \boldsymbol{\Sigma}} \tag{7}$$

For the drag-stress K and the back-stress $\mathbf{a}$, the evolution equations read:

$$\dot{K} = c(K_s - K)\overline{\dot{\varepsilon}^{\text{in}}}, \tag{8}$$

$$\dot{\mathbf{a}}_i = \frac{2}{3}H_i\dot{\boldsymbol{\varepsilon}}^{\text{in}} - \left[D_i\overline{\dot{\varepsilon}^{\text{in}}} + R_i\overline{a}_i^{(m_i-1)}\right]\mathbf{a}_i \tag{9}$$

with the material parameters c, K_s, H_i, D_i, R_i, and m_i. In the last equation, the first term contains hardening effects, the latter two terms describe dynamic and static recovery, respectively. For a better fit to experimental results, two back-stresses were used for our calculations [7]:

$$\mathbf{a} = \sum_{i=1}^{2}\mathbf{a}_i \tag{10}$$

The invariants of the back-stress and inelastic strain are defined through:

$$\overline{a_i} = \sqrt{\frac{3}{2}\mathbf{a}_i : \mathbf{a}_i} \tag{11}$$

$$\overline{\dot{\varepsilon}^{\text{in}}} = \sqrt{\frac{2}{3}\dot{\boldsymbol{\varepsilon}}^{\text{in}} : \dot{\boldsymbol{\varepsilon}}^{\text{in}}} \tag{12}$$

To gain access to the effect of damage, we have chosen the simplest formulation of creep damage, which was developed by Kachanov [8] and Rabotnov [9]. The evolution equation reads:

$$\dot{D} = \left\langle \frac{\chi(\boldsymbol{\sigma})}{A(1-D)} \right\rangle^{r} (1-D)^{-\kappa} \tag{13}$$

where A, r, and κ denote material parameters. The quantity

$$\chi(\boldsymbol{\sigma}) = \alpha\sigma_1 + \beta\sigma_v + \gamma \operatorname{Trace}(\boldsymbol{\sigma}) \tag{14}$$

with

$$\alpha + \beta + \gamma = 1 \tag{15}$$

was first introduced by Hayhurst [10] and maps a multi-axial stress state to a uniaxial one. Here, σ_1 represents the maximum principal stress and

$$\sigma_v = \sqrt{\frac{3}{2}\mathbf{s} : \mathbf{s}} \tag{16}$$

For selected temperatures, material parameters were determined mostly from uniaxial experimental work (see Tab. 5.4.1). The respective parameters for any temperature are then calculated by linear interpolation except for the parameters of static recovery R_i, which are determined by exponential interpolation.

5.4.3 Extrapolation Method

Monotonic loadings like a tensile test are characterised by a monotonic evolution of all internal variables during loading history. Therefore, a numerical integration of the respective evolution equations requires a comparatively small number of increments. The computing expenditure is relatively small. In contrast, cyclic loadings can produce strong variations of internal variables. This behaviour becomes particularly distinct for variables whose evolution equation enables non-monotonic behaviour in time. Even a change of sign during a single cycle may occur. The evolution equation for kinematic hardening, Eq. (9), may serve as an example. When considering the values of these variables at equivalent points of time for different cycles, a monotonic behaviour similar to that of monotonic loading can be noticed in general. In principle, this behaviour of internal variables during the complete lifetime of a component can be divided into three periods (Fig. 5.4.1):

Table 5.4.1: Parameters of the applied material model determined for the combustor material NiCr22Co12Mo9 (trade name Nicrofer 5520 Co, similar to IN 617).

Temperature [K]	293	873	1123
E[MPa]	$2.100 \cdot 10^5$	$1.660 \cdot 10^5$	$1.467 \cdot 10^5$
ν	0.3	0.3	0.3
$\alpha[K^{-1}]$	$1.4 \cdot 10^{-5}$	$1.4 \cdot 10^{-5}$	$1.4 \cdot 10^{-5}$
k[MPa]	83.0	83.0	11.5
$Z[MPa \cdot s^{1/n}]$	100	100	199
n	46.7	46.7	15.8
H_1[MPa]	$1.62 \cdot 10^5$	$1.62 \cdot 10^5$	$3.10 \cdot 10^5$
D_1	$1.33 \cdot 10^3$	$1.33 \cdot 10^3$	$3.23 \cdot 10^3$
$R_1[MPa^{1-m_1}s^{-1}]$	10^{-60}	$5.32 \cdot 10^{-6}$	$4.55 \cdot 10^{-1}$
m_1	0	2.51	0.30
H_2[MPa]	$4.72 \cdot 10^3$	$4.72 \cdot 10^3$	$1.84 \cdot 10^3$
D_2	50.9	50.9	2.29
$R_2[MPa^{1-m_2}s^{-1}]$	10^{-30}	10^{-30}	$1.65 \cdot 10^{-3}$
m_2	0	0	1.30
c	1.00	1.00	7.79
K_S[MPa]	287	187	20.4
$A[MPa \cdot s^{1/r}]$	4247.6	4247.6	1293
r	7.38	7.38	6.50
κ	7.62	7.62	8.50
α	0.00	0.00	0.00
β	0.75	0.75	0.75
γ	0.25	0.25	0.25

- The starting period is characterised by enormous changes of nearly all variables due to large hardening effects at the beginning of loading.
- In the following period only small, nearly linear variations occur. Some variables show a quasi-stationary behaviour. This period takes up to 99 % of the entire lifetime.
- The last period leads to the failure of the component. Especially, the damage variable D shows an enormous increase, since Eq. (13) is strongly non-linear.

Obviously, the second of these three periods is predestined for employment of an extrapolation method. Since the second period covers nearly the entire lifetime, an enormous reduction of computing time is to be expected by using an efficient extrapolation scheme.

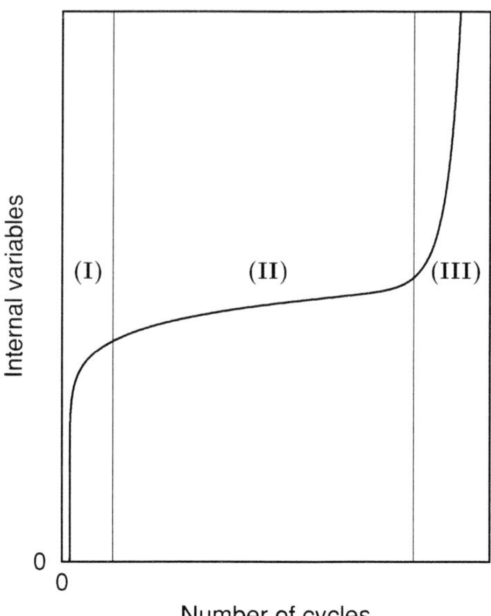

Figure 5.4.1: Evolution of internal variables during cyclic loading until failure: (I) starting region, (II) plateau region, (III) region of high damage.

Based on this principal behaviour of internal variables, we have developed the following extrapolation scheme. Our calculation starts with a regular integration of the evolution equations for the first loading cycles. For this purpose, the embedding formula of Runge-Kutta of 4th/5th order is applied. It is implemented into the commercial finite element code ABAQUS [11]. The results of this calculation are the starting point for the first extrapolation step. For each integration point of the structure and for each internal variable, a piece-wise polynomial or spline function is constructed:

$$p(t) = \sum_{j=1}^{k} c_{ji} \frac{(t - t_i)^{j-1}}{(j-1)!} \quad \text{for} \quad t_i \leq t < t_{i+1} \tag{17}$$

where t_i are the breakpoints and $k - 1$ is the degree of the spline. This degree has to be chosen as small as possible to avoid oscillations caused by high powers in Eq. (17). In practise, degrees 2 and 3 (cubic splines) have been chosen. The coefficients c_{ji} are determined under the constraint of the spline function passing through the values of the internal variables at the end of the respective cycles. Then, the spline is extrapolated from the end of the last cycle t_N for a step size h:

$$y_{\text{sp}} = p(t_N + h) \tag{18}$$

To obtain an error estimation, this extrapolated value is compared with those of a linear extrapolation y_{lin}. We define an error function for the complete structure and all internal variables:

$$E = \max_{n,i} \left\{ w_i \cdot \left| y_{\text{sp}}(n,i) - y_{\text{lin}}(n,i) \right| \right\} \tag{19}$$

Here, the index n runs over all integration points, i numbers the internal variables. We have introduced a weight function w_i which guarantees that the different kinds of internal variables make approximately the same contribution to E. By varying the step size, it is ensured that E is below a suitably chosen maximum error δ to guarantee stability of the procedure:

$$E \leq \delta \tag{20}$$

Obviously, the integration point with the most non-linear behaviour determines the step size.

The values of the internal variables after the first extrapolation step are the input values for the next finite element calculation consisting of one cycle only. It guarantees that the equilibrium conditions are fulfilled. The finite element results are then combined with the results of previous cycles to construct a spline function and extrapolate the internal variables again. In principle, this procedure can be repeated up to failure of the structure. Only the step size will be reduced drastically in the third region of Fig. 5.4.1. As a consequence, the share of the extrapolation method in the integration becomes negligible. The stiff system of evolution equations in the last period of lifetime is solely integrated with the help of the embedding formula of Runge-Kutta, which has been proven to integrate reliably stiff systems [11].

To test the reliability of the new extrapolation method and its implementation into a finite element code, the deformation behaviour of simple systems is investigated. For this purpose, convergence tests have been applied. Figure 5.4.2 and Tab. 5.4.2 present results for a displacement-controlled cyclic loading of a single element with an amplitude of $\varepsilon_{\text{max}} = 0.0008$ and a strain rate of $\dot{\varepsilon} = 8 \cdot 10^{-7}\text{s}^{-1}$. Since this rate is comparatively low, fatigue is negligible and the application of the pure creep damage law of Kachanov/ Rabotnov is justifiable [12]. The lifetime is nearly identical for values of $\delta = 5 \cdot 10^{-6}$ up to $\delta = 5 \cdot 10^{-7}$ and amounts to approximately 40 500 cycles. For $\delta = 5 \cdot 10^{-6}$ the extrapolation scheme has converged.

Nearly the complete computing time is consumed by the finite element calculation. The time for extrapolating the internal variables is mostly negligible, especially for complex structures. As a consequence, the ratio R which is defined as the lifetime obtained for the minimum value of δ divided by the number of extrapolation steps is a suitable measure to value the reduction of computer time. In the case of the latter investigation, the reduction for $\delta = 5 \cdot 10^{-6}$ amounts to 35.0, i.e. the computer time is reduced by a factor of 35.

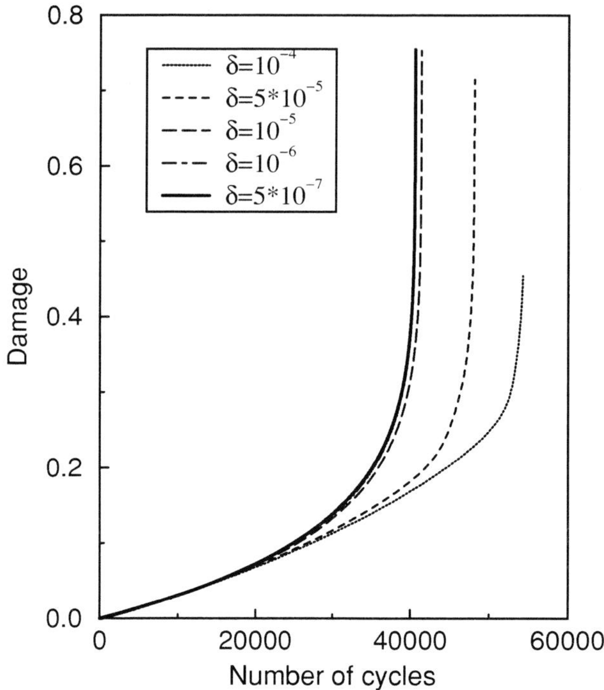

Figure 5.4.2: Damage vs. number of cycles for a displacement-controlled cyclic loading of a single element with $\varepsilon_{\text{max}} = 0.0008$ and $\dot{\varepsilon} = 8 \cdot 10^{-7}\text{s}^{-1}$. The curves are labelled by their different δ-values. This figure corresponds to Tab. 5.4.2.

Table 5.4.2: Results for a displacement-controlled cyclic loading of a single element with $\varepsilon_{\text{max}} = 0.0008$ and $\dot{\varepsilon} = 8 \cdot 10^{-7}\text{s}^{-1}$. To measure the reduction of computer time, we have introduced a ratio R which is defined as the lifetime obtained for a value of $\delta = 5 \cdot 10^{-7}$ divided by the number of extrapolation steps. This table corresponds to Fig. 5.4.2.

	Extrapolation method			
δ	lifetime [cycles]	extrapolation steps	error [percent]	R
10^{-3}	148076	69	265.5	587.2
$5 \cdot 10^{-4}$	78327	130	93.3	311.7
10^{-4}	54312	310	34.0	130.7
$5 \cdot 10^{-5}$	48044	437	18.6	92.7
10^{-5}	41312	873	2.0	46.4
$5 \cdot 10^{-6}$	40790	1158	0.7	35.0
10^{-6}	40535	2187	0.0	18.5
$5 \cdot 10^{-7}$	40518	2853	–	14.2

For structures with very high numbers of cycles to failure an additional numerical problem appears. Although the maximum degree of the spline function used is 3, oscillations of internal variables may occur and cause an upper limit for the step size. Basis of an approximation to avoid this problem is that above a certain number of cycles N_{st} all variables are quasi-stationary except for the damaging variable D. The value of N_{st} has to be determined very carefully. All integration points and their behaviour in time have to be investigated to ascertain a value for N_{st}. After finding this value, all internal variables are kept constant except for D. Therefore, the extrapolation is confined to the single variable D. Since the values for D are very small during most of the lifetime and due to its monotonic behaviour, the step size can be raised enormously. This approximation can be used successfully for a failure analysis of complex components as is demonstrated in the following section.

5.4.4 Failure Analysis for a Ring Combustor

In the following, a failure analysis for the ring combustor of a gas turbine shall be presented. Previous analyses of this component were restricted to the first cycle of flight [13]. In the absence of an extrapolation scheme, further investigations were impossible due to an enormous need in computing time.

Before starting the finite element calculations, the thermomechanical boundary conditions have to be determined. Therefore, the temperature distribution inside the combustor wall is determined using thermocolours. The development of temperature during a single cycle of flight for the hottest point of the structure is displayed in Fig. 5.4.3. Two parts of a cycle are of special interest. The first is the starting phase of flight. During this short period of approximately 40 seconds, nearly the whole increase in damage occurs (see below). During the flight phase of about 3 hours, the maximum temperature is 870 K which is too low to produce a significant additional damage. But it is sufficient by far for a decisive amount of static recovery which leads to stress relaxations. As a consequence, a relocation of the point of maximum damage occurs (see below). Other parts of the flight are of less importance.

Values for the most important quantities on the inner surface of the combustor during the first cycle of flight are displayed in Fig. 5.4.4. Maximum temperatures are found in the region of the secondary holes which appear as semicircles due to the utilisation of the combustor symmetry. To assess the loading, the stress distribution is necessary as well. As for the temperature, the maximum stresses are concentrated around the secondary holes. Far from the secondary holes, the stress is approximately zero. According

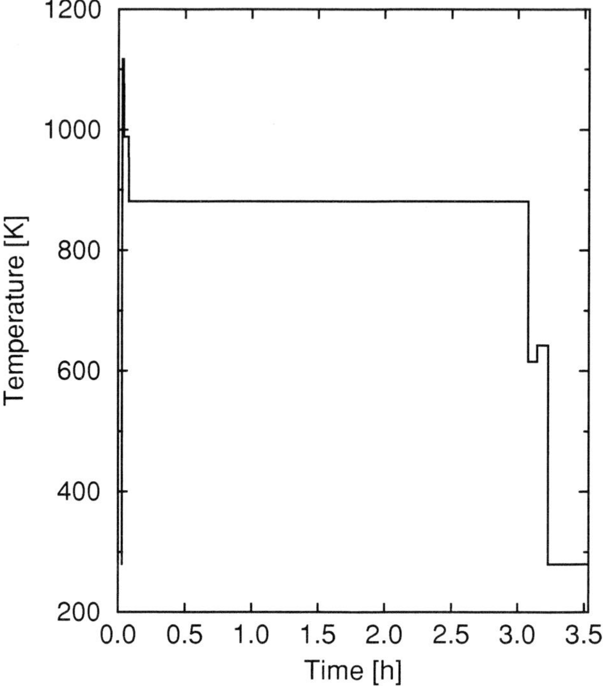

Figure 5.4.3: Development of temperature during a single cycle of flight for the hottest point of the structure.

to the loading of the individual regions, the deformation is purely elastic or includes inelastic parts. The latter characterises the regions close to the secondary holes. The damage shows the respective behaviour. For the following it is decisive that the damage around the upper secondary hole is significantly larger than for the lower one (see Fig. 5.4.4d).

Since nearly the complete inelastic deformation is concentrated around the secondary holes and the upper secondary hole possesses the maximum damage values, another strategy to reduce computer time can be applied. The structure containing 1/48 of the whole ring combustor is reduced to a partial structure containing the most deformed region only. Therefore, a nearly rectangular region centred around the upper secondary hole is cut out of the complete structure along the element boundaries. This region contains approximately the complete inelastic deformed region around this secondary hole. The new free surfaces are loaded in each cycle with those displacements calculated with the aid of the complete structure.

To determine the actual position of failure, we investigate the values for the damage variable D depending on the number of cycles. Figure 5.4.5 shows D after 1 and 100 cycles, respectively, on the inner surface of the

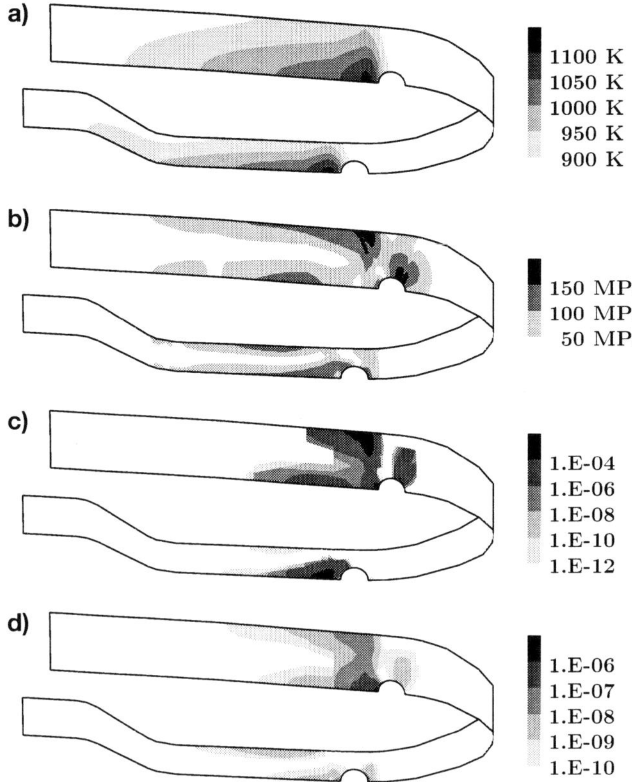

Figure 5.4.4: Distributions of system variables on the inner surface of the ring combustor. a) Temperature at the end of the starting phase of flight, b) stress (von Mises) at the end of the starting phase of flight, c) accumulated inelastic strain at the end of the first cycle, d) distribution of damage at the end of the first cycle.

wall. As obvious, there is a significant relocation of maximum damage from the groove root to an internal position. If the term for static recovery in Eq. (9) is omitted, this relocation does not appear. Therefore, this effect of maximum damage relocation can be explained by static recovery effects and the resulting stress relocation during the flight phase.

As shown by more detailed investigations, the relocation of maximum damage occurs after approximately 13 cycles of flight. After this relocation the position of maximum damage is constant until failure (Fig. 5.4.6). To find out, whether this relocation rather is of singular nature or whether it is a general behaviour, the outer surface of the ring combustor was studied additionally. For this surface the effect is even more distinct. At the beginning of the loading history up to 200 cycles, the maximum damage is right of the secondary hole. After this period, the position of the maximum changes to the left of the secondary hole. The effect of relocation is completed after

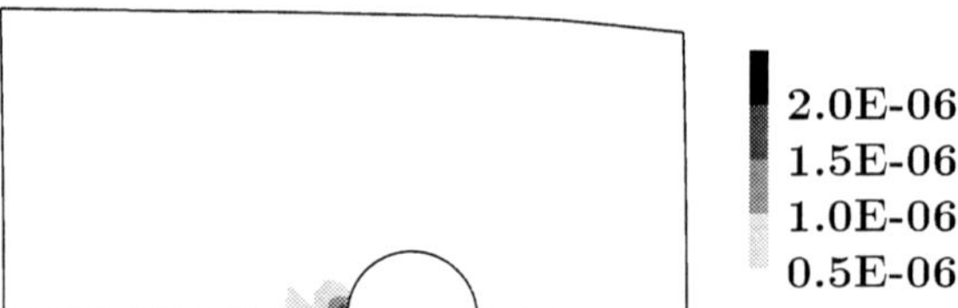

Figure 5.4.5a: Partial structure. Distribution of damage after the first cycle.

Figure 5.4.5b: Distribution of damage after 100 cycles.

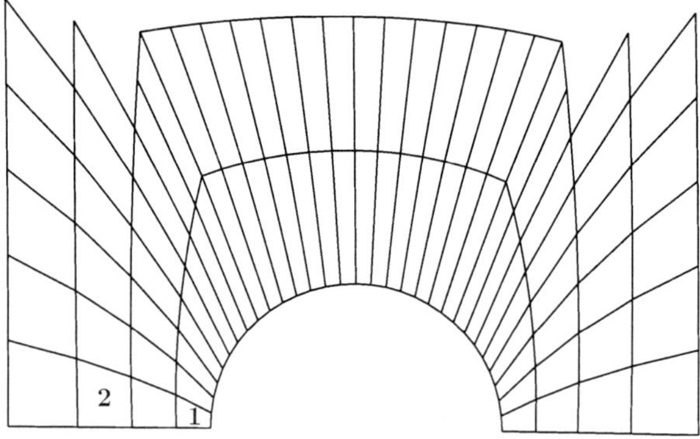

Figure 5.4.6a: Position of maximum damage on the inner surface.

more than 100 000 cycles, which is more than 10 % of the lifetime (see below). To determine reliably the most likely position for a crack initiation, it is therefore indispensable to investigate a large amount of cycles of at least 10 or even 20 % of lifetime. These different properties of the inner and the outer surface are due to the varying hardening behaviour.

After calculating the most likely position of a crack initiation, we now determine the lifetime of the ring combustor (Fig. 5.4.7). Due to numerical

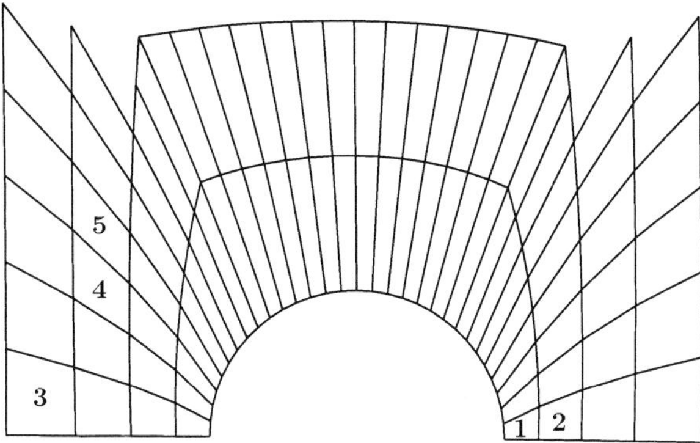

1 0 cycles - 21 cycles

2 22 cycles - 227 cycles

3 641 cycles

4 1090 cycles - 101256 cycles

5 137617 cycles - failure

Figure 5.4.6b: Position of maximum damage on the outer surface.

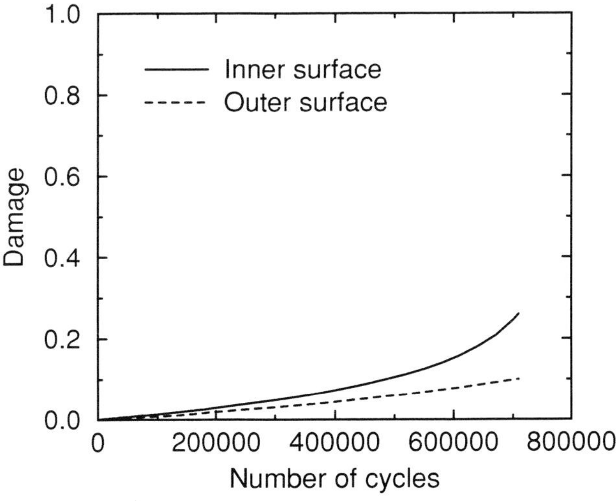

Figure 5.4.7: Maximum damage of the inner and outer surface of the ring combustor during lifetime.

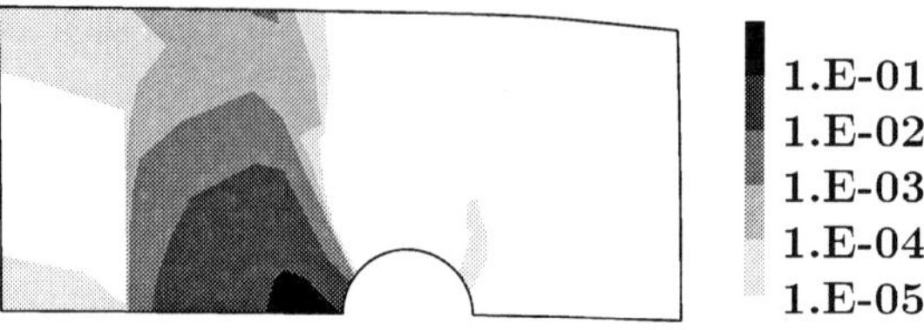

Figure 5.4.8a: Distribution of damage of the inner surface after 710 305 cycles.

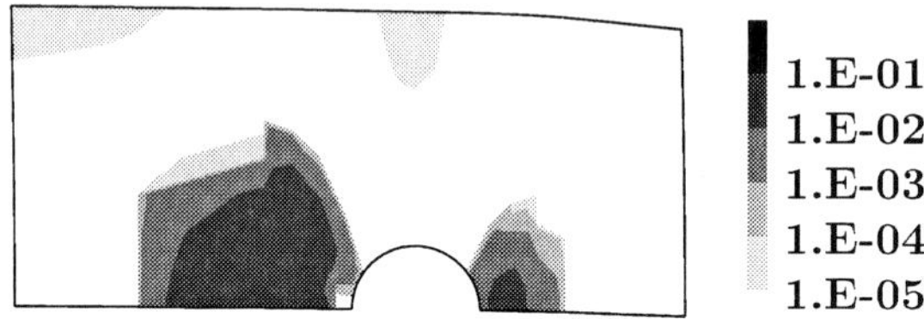

Figure 5.4.8b: Distribution of damage of the outer surface after 710 305 cycles.

reasons, the upper limit for the damage variable D is 0.25 in the present version of the extrapolation method. Since the evolution law for D in Eq. (13) shows a strongly non-linear behaviour above this value, the additional time up to a value of $D = 1$ is very small in comparison to the whole lifetime. Considering Fig. 5.4.7, a lifetime of approximately 800 000 cycles is obtained.

To understand the mechanism of failure, the distribution of damage close to the time of a crack initiation is of special interest (Fig. 5.4.8). On the inner surface there is a finite region with $D > 0.1$, which is surrounded by a relatively large region of $D > 0.01$. The respective region can be also found on the outer surface. Therefore, a mesh of small cracks will be generated and destroy the wall.

To evaluate the efficiency of our extrapolation method, a short comparison of computer times shall be given (Tab. 5.4.3). The introduction of the partial structure described above results in a factor of 8 for the computer

Table 5.4.3: Estimated computer times (CRAY J916-4096) for the failure analysis of a ring combustor with 800 000 cycles of lifetime. We employ the material model of Chaboche/Rabotnov with two back-stresses.

pure finite element method	1/48 ring combustor	680 000 h (78 years)
	partial structure	85 000 h (10 years)
extrapolation method	partial structure	210 h

time, while the application of the extrapolation scheme gives an additional factor of 400. Without this additional reduction of computer time, the above calculations would be impossible.

5.4.5 Conclusions

The customary way to carry out a failure analysis consists in employing a purely elastic material law. On the one hand, this proceeding saves computer time, on the other hand, the applicability of the results obtained is strongly limited. Only the maximum loaded regions inside a component during the first cycle can be determined.

The combination of the modern viscoplastic material law, damage model, and extrapolation scheme gives a powerful tool to calculate the distribution of any system variable for any cycle. Regions with maximum values for the damage variable D depending on the number of cycles can be found. Therefore, the most likely position for a crack initiation is given. Additionally, a value for the lifetime of a component is obtained. However, the reliability of this value depends on the used damage model. In the case of a ring combustor we calculate a lifetime of 800 000 cycles for the Kachanov/Rabotnov model.

Acknowledgements

The contributions of Dr. H. Stamm, Dr. H. Hornberger, Prof. Dr. B. Schinke, and Dr. Th. Fleig within their former activities in the Collaborative Research Centre 167 are acknowledged. The theoretical calculations have been carried out with the help of the staff and the equipment of the Forschungszentrum Karlsruhe.

References

1. Chaboche, J. L. (1977): Viscoplastic constitutive equations for the description of cyclic and anisotropic behavior of metals, Boulletin de l'academie polonaise de sciences, Série Science et Techniques Volume XXV, 33–42.
2. Chaboche, J. L. and Rousselier, G. (1983): On the plastic and viscoplastic constitutive equations, *J. Press. Vessel Techn.* **105**, 153–164.
3. Robinson, D. N. (1978): A unified creep-plasticity model for structural metals at high temperature, ORNL Report TM-5969.
4. Robinson, D. N. and Bartolotta, P. A. (1985): Viscoplastic constitutive relationships with dependence on thermomechanical history, NASA CR-174836.
5. Hornberger, K. (1988): Anwendung viskoplastischer Stoffgesetze in Finite-Elemente-Programmen, Dissertation, Universität Karlsruhe.
6. Lemaitre, J. and Chaboche, J. L. (1990): Mechanics of solid materials, Cambridge University Press, Cambridge.
7. Chaboche, J. L. (1993): Cyclic viscoplastic constitutive equations, Part I: A thermodynamically consistent formulation, *J. Appl. Mech.* **60**, 813–828.
8. Kachanov, L. M.(1958): Time of the rupture process under creep conditions, TVZ Akad. Nauk. S.S.R. *Otd. Tech. Nauk.*, **8**, 26–31.
9. Rabotnov, Y. N. (1968): Creep rupture, Proc. of the 12th Int. Congress of Applied Mechanics, Stanford.
10. Hayhurst, D. R. (1972): Creep rupture under multiaxial state of stress, *J. Mech. Phys. Solids* **20**, 381–390.
11. Fleig, Th., Aktaa, J., and Schinke, B. (1994): Implementation of a viscoplastic constitutive model coupled with damage in the FE-code ABAQUS, Proc. of the 5th Conf. Materials for Advanced Power Engineering 1994, Oct. 3–6, 1994, Liège, Belgium, 611–620.
12. Spera, D. A. (1969): The calculation of elevated temperature cyclic life considering low cycle fatigue and creep, NASA TN D-5317.
13. Schwertel, J., Merckling, G., Hornberger, K., Schinke, B., and Munz, D. (1991): Experimental investigations on Ni-base superalloy IN617 and their theoretical description, in: "High-temperature constitutive modeling: theory and application", eds: Freed, A. D., Walker, A. K., MD-Vol. 26, AMD-Vol. 121, ASME, New York, 285–295.

5.5 Modeling of the Non-linear Deformation and Damage Behaviour of Combustor Structure Materials

Jarir Aktaa and Dietrich Munz*

Abstract

The present project focused on the modeling of the deformation and damage of combustion chamber structures. To describe the inelastic deformations occurring in combustion chambers, several viscoplasticity models have been chosen. During past application periods, these models were studied with regard to their suitability for selected structural materials. The models included that by Chaboche, which was chosen for further verification calculations. These calculations covered both, the isothermal two-step as well as the non-isothermal cyclic loading tests that had been performed under the projects C5 and C7 of the Collaborative Research Centre 167 using the materials NiCr22Co12Mo9 and CoCr22Ni22W14.

The continuum mechanics concept chosen for damage determination and lifetime prediction is characterised by the combination of a damage model and a deformation model. This allows the interaction between deformation and damage to be taken into account. On the basis of the experience gained from the use of previously known models and lifetime prediction rules, a new model has been developed. It describes the time-dependent damage at high temperatures without making the classic distinction between creep and fatigue damage. For this purpose, the dependence of the damage rate on internal strain hardening and, hence, on the deformation history was introduced. The new damage model coupled with the Chaboche model was fitted to the behaviour of the material NiCr22Co12Mo9 and verified afterwards. For this, isothermal as well as non-isothermal uniaxial cyclic tests were considered. In general, qualitatively and sometimes even quantitatively satisfying predictions were obtained.

---

* Institut für Zuverlässigkeit und Schadenskunde im Maschinenbau, Universität Karlsruhe, Kaiserstr. 12, 76128 Karlsruhe, Germany

5.5.1 Introduction

Highly thermomechanically loaded components like combustors are always subjected to inelastic deformations. For a reliable dimensioning of such components, it becomes increasingly necessary to consider inelastic deformations and the resulting complex damage behaviour. The dimensioning task can be performed efficiently using the finite element method (FEM). It allows to take into account the inelastic material behaviour, provided that suitable material models are available for the mathematical modeling of major inelastic material behaviour phenomena, such as the Bauschinger effect, cyclic hardening or softening as well as creep and relaxation processes, with sufficient approximation [1].

In the past, the class of so-called unified models was developed for this purpose [2]. Models of this type allow the description of basic phenomena of the inelastic material behaviour. They are based on the internal variables concept of irreversible thermodynamics where the state of a material is described unambiguously by the external variables, i.a. stress and temperature, and a set of internal variables.

When calculating a component, its expected lifetime is of particular interest. For lifetime prediction, however, calculation of deformation is not sufficient and damage has to be determined in addition. For this purpose, damage models have been developed, which are formulated to be combined with the existing deformation models. Damage is introduced into the deformation models as an additional internal variable. Using this approach, the interaction of damage and deformation is taken into account by modeling.

Within the framework of the projects C2/C3 of the Collaborative Research Centre 167, deformation and damage models were studied in more detail, further developed and efficient algorithms for parameter determination were developed [3, 4]. Furthermore, they were applied to describe the deformation and damage behaviour of the combustor materials NiCr22Co12Mo9 (trade name Nicrofer 5520 Co, similar to IN 617) and CoCr22Ni22W14 (Haynes 188) under isothermal and non-isothermal loadings. The results obtained within these activities shall be summarised in the present report.

5.5.2 Modeling of the Deformation Behaviour

To describe the time-dependent deformation behaviour of metallic materials numerous viscoplastic models were developed in the last three decades. The activities were focused in particular on the so-called unified models. By these models, contrary to traditional models, the inelastic deformation is handled in a unified manner, i.e. it is no longer distinguished between a pure time-dependent (creep) part and a pure time-independent (plastic) part. A subset of these models was formulated within the framework of irreversible thermodynamics by introducing internal variables to characterise the change of the internal state due to inelastic deformation. Hence, the inelastic deformation rate results as a function of external as well as internal variables of state.

Over the runtime of the Collaborative Research Centre 167 the unified viscoplasticity models by Robinson, Walker, and Chaboche have been investigated extensively. Among them, the model by Chaboche is given particular attention. It is characterised by a relatively simple mathematical setup which facilitates the determination of necessary model parameters, the application of the model, and the model modifications possibly required.

5.5.2.1 Chaboche's Viscoplasticity Model

The viscoplasticity model by Chaboche has been presented in detail in [5]. Here, it shall therefore be dealt with briefly only. For infinite deformations, the total strain rate can be decomposed additively into an elastic, an inelastic, and thermal fraction:

$$\dot{\varepsilon} = \dot{\varepsilon}^{\mathrm{el}} + \dot{\varepsilon}^{\mathrm{in}} + \dot{\varepsilon}^{\mathrm{th}} \tag{1}$$

The elastic fraction is derived from the Hooke's law

$$\varepsilon^{\mathrm{el}} = \mathbf{E}^{-1} : \boldsymbol{\sigma} \tag{2}$$

where $\boldsymbol{\sigma}$ and $\mathbf{E}$ denote Cauchy's stress tensor and the elasticity tensor, respectively. The thermal fraction is determined by means of the linear expansion law

$$\varepsilon^{\mathrm{th}} = \alpha\,(T - T_0) \cdot \mathbf{1} \tag{3}$$

with α and T as the thermal expansion coefficient and the temperature, respectively. The inelastic strain rate is given by a system of coupled differential equations of first order. It consists of the law of flow

$$\dot{\varepsilon}^{\mathrm{in}} = \frac{3}{2}\,\dot{p}\,\frac{\boldsymbol{\sigma}' - \boldsymbol{\Omega}}{J(\boldsymbol{\sigma}' - \boldsymbol{\Omega})} \tag{4}$$

where

$$\dot{p} = \left\langle \frac{J(\boldsymbol{\sigma}' - \boldsymbol{\Omega}) - K - k}{Z} \right\rangle^{n} \tag{5}$$

and

$$J(\boldsymbol{\sigma}' - \boldsymbol{\Omega}) = \sqrt{\frac{3}{2}(\boldsymbol{\sigma}' - \boldsymbol{\Omega}) : (\boldsymbol{\sigma}' - \boldsymbol{\Omega})} \tag{6}$$

with

$$\langle x \rangle = \begin{cases} x & : \ x \geq 0 \\ 0 & : \ \text{else} \end{cases} \tag{7}$$

the evolution equation for the kinematic hardening variable

$$\boldsymbol{\Omega} = \sum_{i=1}^{2} \boldsymbol{\Omega}_i \tag{8}$$

$$\dot{\boldsymbol{\Omega}}_i = \frac{2}{3} H_i \dot{\boldsymbol{\varepsilon}}^{\text{in}} - D_i \boldsymbol{\Omega}_i \dot{p} - R_i [J(\boldsymbol{\Omega}_i)]^{m_i - 1} \boldsymbol{\Omega}_i \tag{9}$$

and the evolution equation for the isotropic hardening variable

$$\dot{K} = c(R_\infty - K)\dot{p} \tag{10}$$

$\boldsymbol{\sigma}'$ denotes the deviatoric part of the stress tensor. In addition to the elastic constants and the thermal expansion coefficient, k, Z, n, H_i, D_i, R_i, m_i, c, and R_∞ are material- and temperature-dependent parameters. They are applied for fitting the model to a given material. Their temperature dependence reflects the temperature dependence of the simulated material behaviour. An additional temperature dependence of the material behaviour can be simulated in the model by incorporating terms which are proportional to the temperature rate into the structure of the equations. Such terms are not encountered under isothermal conditions. This possibility has been mentioned in [6] already and good results were obtained by Schwertel [3] in a first calculation, where Eq. (9) was modified as

$$\dot{\boldsymbol{\Omega}}_i = \frac{2}{3} H_i \dot{\boldsymbol{\varepsilon}}^{\text{in}} - D_i \boldsymbol{\Omega}_i \dot{p} - R_i [J(\boldsymbol{\Omega}_i)]^{m - 1} \boldsymbol{\Omega}_i + \frac{1}{H_i} \frac{\delta H_i}{\delta T} \boldsymbol{\Omega}_i \dot{T} \tag{11}$$

The determination of the parameters for a given material was a partial activity in the Collaborative Research Centre 167. An approach was developed, which allows the parameters identification on the basis of data from standard tensile, creep, and cyclic tests. It is given in [3] and shall be presented briefly here.

Implementation of this model in a FE program allows the calculation of the deformation of a structure taking into account the inelastic material behaviour, provided that the values of material parameters are known. By the additional use of a damage model, material damage can also be considered during deformation analysis.

In the following sections, the Chaboche model shall be investigated in further detail. For this purpose, simulations of isothermal and non-isothermal loading paths are computed and compared with the experimental results. The experiments were performed at the Institute for Materials Science I of Karlsruhe University within the framework of the projects C4, C5, and C7 of the Collaborative Research Centre 167 using the materials CoCr22Ni22W14 and NiCr22Co12Mo9.

5.5.2.2 Application of Chaboche's Model

5.5.2.2.1 Determination of the Parameters

To determine the parameters, the model is fitted to the experimentally determined material behaviour by a least-square minimisation procedure

$$\chi^2 = \sum_{\substack{\text{Experiment}\\ \text{types}}} \sum_{i}^{ND} \left[\frac{\sigma_i^{\text{exp}} - \sigma_i^{\text{cal}}(\varepsilon, \dot{\varepsilon}, \vec{\theta})}{G_i} \right]^2 \stackrel{!}{=} \text{minimum} \tag{12}$$

$\vec{\theta} = (\theta_1, \theta_2, \ldots, \theta_{NP})$ denotes the set of parameters to be determined, while σ^{cal} and σ^{exp}, respectively, represent the calculated and measured stresses. ND is the number of data points of the experiment considered. The weighting of the experimental data G_i is chosen in such a way that each experiment should have the same weight. For the fitting procedure, the following data are used:

- strain-time creep curves
- stress-strain tensile curves
- saturated cyclic stress-strain curves
- cyclic hardening curves (stress amplitude vs. cycle number)

The minimisation algorithm used is a modification of the Davidon-Fletcher-Powell algorithm [7]. The gradients $\nabla\chi^2$ required are calculated by central finite differences. For further details see [3, 8].

For reasons of stability of the model equations as well as for the computing time, the following procedure has been applied for the fit:

1. First, a preliminary fit of the tension curves with a large strain rate is performed. That means, static recovery can be neglected ($R_i = 0$). Some manual calculations give reasonable starting values for the fit.

2. In a second step, all curves are fitted with the exception of the saturation loops. For this, the constants in the flow law as well as the parameters in the backstress evolution should be kept constant.

3. In a last step, all experiments and all parameters are fitted simultaneously. The fit results from steps 1 and 2 are used as starting values.

In Annex A, the resulting parameter sets for combustor materials CoCr22-Ni22W14 and NiCr22Co12Mo9 at different temperatures are listed. The quality of the fit is demonstrated exemplarily in Fig. 5.5.1.

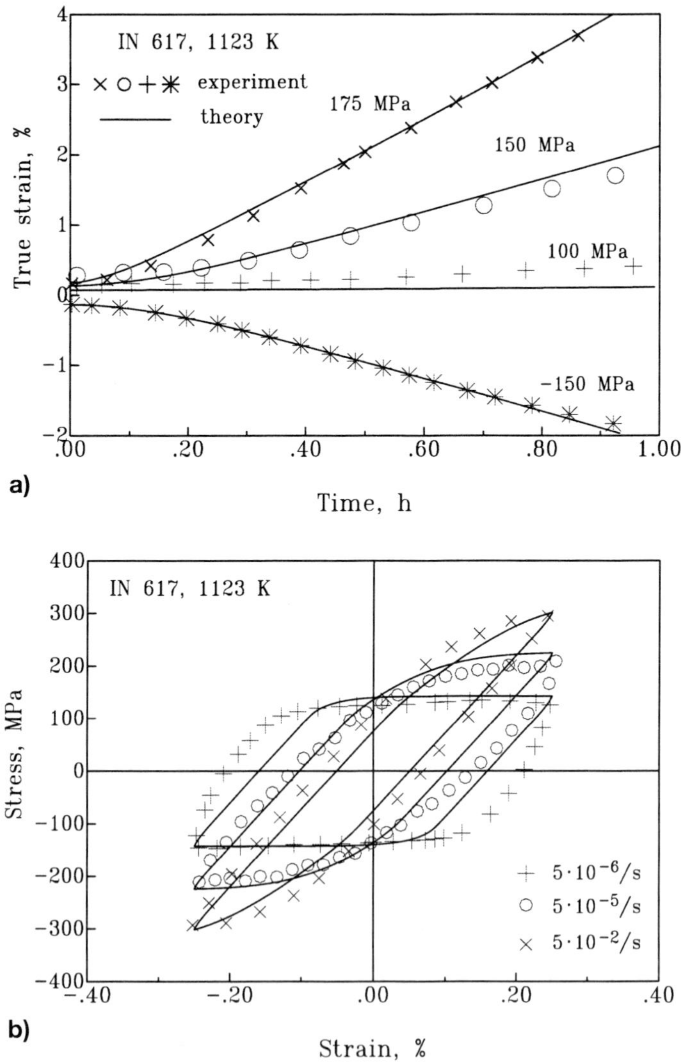

Figure 5.5.1: Comparison of measured and predicted creep curves (a) and saturation hysteresis loops (b) of NiCr22Co12Mo9 at 850 °C.

5.5.2.2.2 Verification of Fitted Model

Application to Isothermal Two-step Loading

For two-step loading, specimens of the material CoCr22Ni22W14 were subjected to strain-controlled cyclic loading at 850 °C. Two test series were performed. During the first series, the strain amplitude was increased from 0.35 % to 0.5 %. In the second series, it was decreased from 0.5 % to 0.35 % again. Transition from the first to the second load amplitude took place after 1 % and 10 % of the specimen lifetime determined in the single-step preliminary experiments, respectively.

The saturation cycles of the single-step experiments and the modeling results are shown in Fig. 5.5.2. The symbols denote the experimental data, whereas modeling is represented by full lines. The parameters of the model were obtained among others from these saturation hystereses of the single-step experiments.

Figures 5.5.3a and 5.5.3b show the predictions of the model for the two-step experiments as well as the data of the assigned experiments, which were not used for parameter determination. The decreasing stress amplitudes at the end of the lifetime in the experiment have to be attributed to ongoing damage processes and, in principle, cannot be described by a pure deformation model. For this, a damage model has to be used in addition to the deformation model. The modeling of damage proposed will be presented in Section 5.5.3.

The range before reaching the end of the lifetime is characterised by cyclic hardening processes which tend to a saturation value for each strain

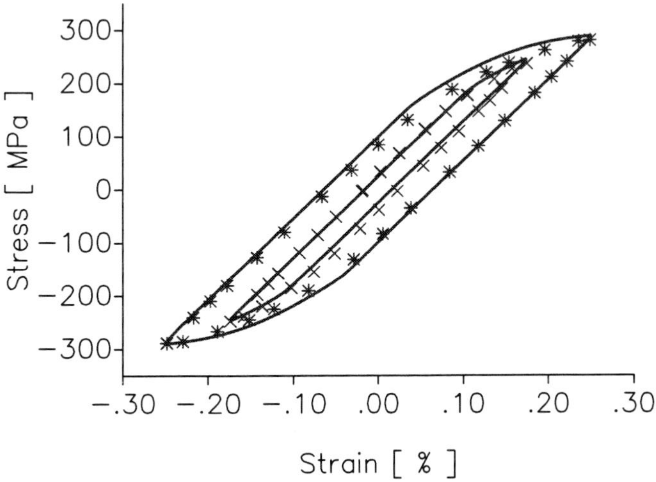

Figure 5.5.2: Comparison of measured (symbols) and prediced (solid lines) saturation hysteresis loops of CoCr22Ni22W14 at 850 °C (single-step experiments).

amplitude. This behaviour is also predicted by the deformation model. For experiments with an increase in the strain amplitude (cf. Fig. 5.5.3a) in particular, model prediction is in good agreement with the experiments. In case of a decrease in the strain amplitude, the specimens still exhibit increasing hardening, provided that the transition takes place at small cycle numbers. This is also predicted by the model. Here, however, the cycle numbers are smaller than those observed in the experiment (cf. Fig. 5.5.3b).

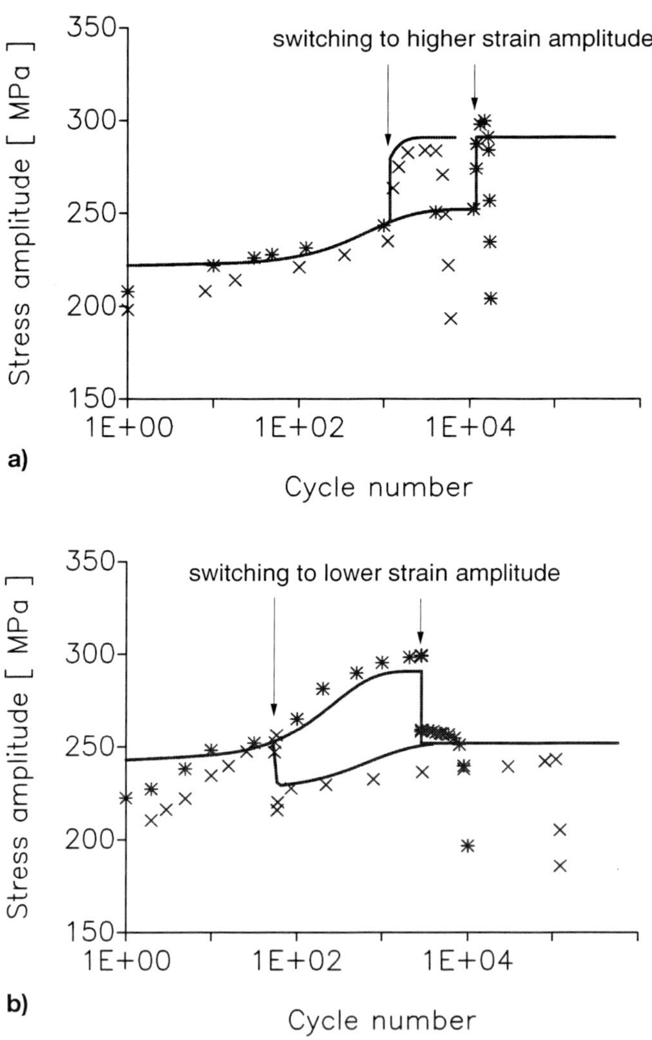

Figure 5.5.3: Comparison of measured (symbols) and predicted (solid lines) cyclic hardening curves with increasing (a) and decreasing (b) strain amplitude of CoCr22-Ni22W14 at 850 °C.

Application to Non-isothermal Single-step Loading

The non-isothermal load histories were performed using specimens of the material NiCr22Co12Mo9. In these experiments, mechanical strain was controlled independently of thermal strain. Two principally different load histories were imposed: In the in-phase tests, the temperature and the mechanical strain were increased simultaneously such that maximum strain was imposed at maximum temperature. In the out-of-phase tests, mechanical strain was decreased, while the temperature was increased. As a result, minimum strain was exerted at maximum temperature. The temperature rate amounted to 14 K/s in all experiments both, during the heating and the cooling phase. The mechanical strain rate was determined with the duration of a strain cycle corresponding to the duration of a temperature cycle. Both, the in-phase and the out-of-phase tests were performed at temperatures ranging from 200 °C to 850 °C and 1050 °C, respectively. In each temperature interval, the mechanical strain amplitudes of $\Delta\varepsilon_m = 0.5\,\%$ and $\Delta\varepsilon_m = 1.25\,\%$ were applied. A survey of the load histories used is given in Tab. 5.5.1.

For calculating the model prediction, the already existing parameter sets were used [3]. None of the experiments presented here was applied for parameter determination. The parameters were kept constant at the value of the last temperature node when the temperature was below 600 °C or exceeded 1000 °C. Between 600 °C and 1000 °C, all parameters except for n were interpolated linearly between the values of the adjacent temperature interpolation nodes. The exponent of the law of flow n was determined by linear interpolation over $1/n$, as linear interpolation over n would have caused larger deviations from the experimental results.

The results of the model predictions and the respective experiments are compared in Figs. 5.5.4 to 5.5.7. The first cycle and the cycle at half of the number of cycles to failure are indicated. The stress is plotted over the mechanical strain. The quality of the material behaviour prediction made by the model ranges from a nearly complete agreement with the experiment (cf. Fig. 5.5.4a) to deviations of 200 MPa for the maximum stress (cf. Fig. 5.5.5b). It cannot be noted that the quality of the prediction depends on whether in-phase tests or out-of-phase tests have been modeled. As a tendency, however, the prediction of the saturation cycles at half of the number of cycles to failure is worse than that of the first cycles (cf. Figs. 5.5.4 to 5.5.7). This may suggest that the effects of microstructure formation under thermo-mechanical loading can be predicted from the behaviour under isothermal

Table 5.5.1: Loading conditions of the non-isothermal tests considered.

	In-phase tests				Out-of-phase tests			
$T_0 =$	850 °C		1050 °C		850 °C		1050 °C	
$\Delta\varepsilon_M =$	0.5 %	1.25 %	0.5 %	1.25 %	0.5 %	1.25 %	0.5 %	1.25 %

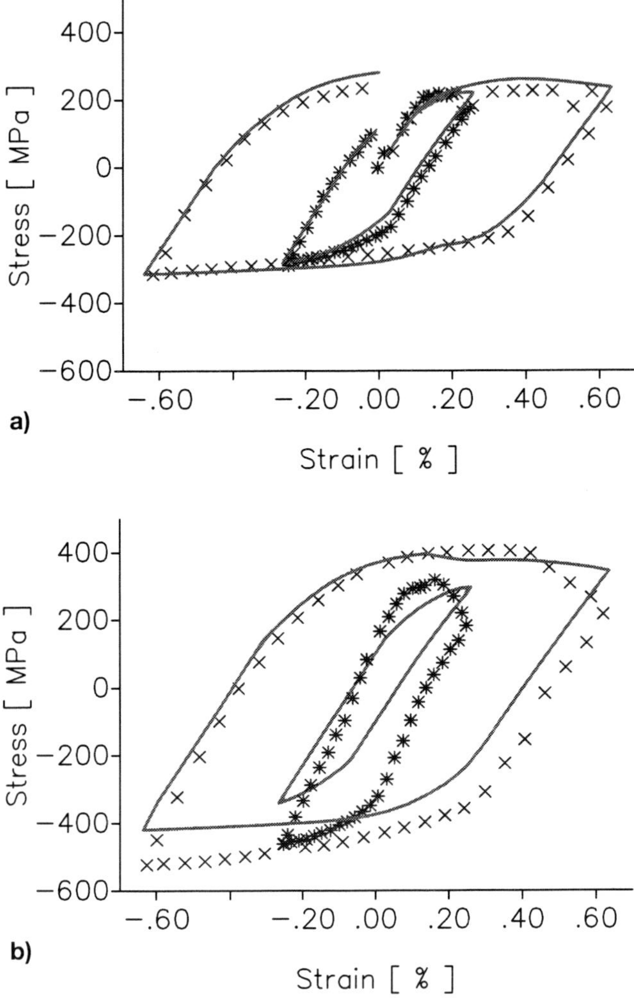

Figure 5.5.4: Comparison of measured (symbols) and predicted (solid lines) first cycles (a) and saturation cycles (b) of the in-phase tests with $T_0 = 850\,°C$.

loading with certain reservations only. It was not checked, whether these deviations might also be due to the small number of only three interpolation nodes for the parameter set. To settle this question, isothermal experiments will have to be performed at more temperatures for subsequent determination of the parameters of Chaboche's model.

To sum up, it must be stated that it may be possible to assess the material behaviour of NiCr22Co12Mo9 under non-isothermal conditions from the data obtained under isothermal loading, provided that the requirements made on the prediction quality are not too high.

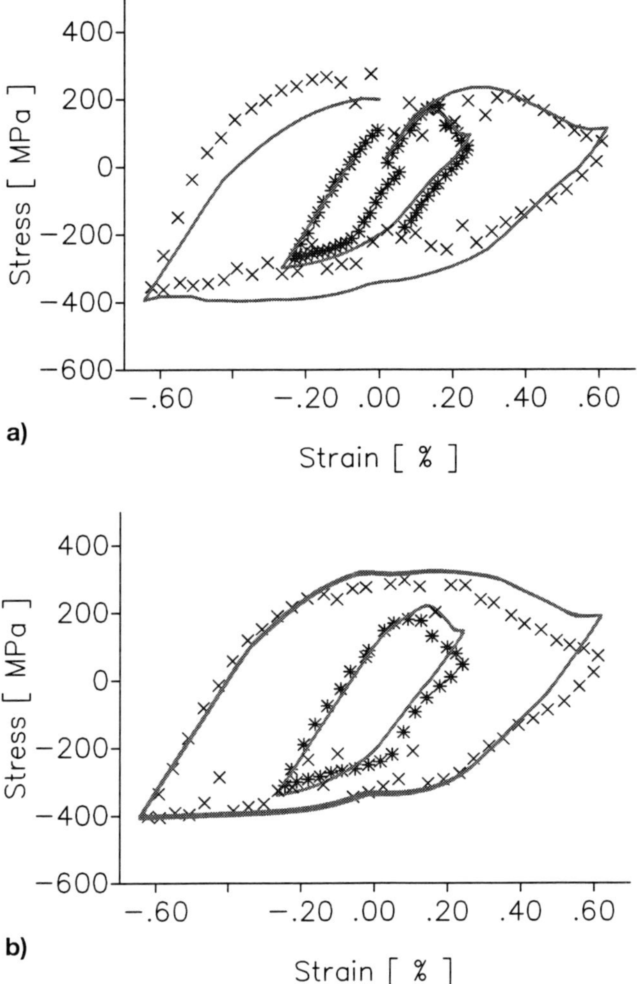

Figure 5.5.5: Comparison of measured (symbols) and predicted (solid lines) first cycles (a) and saturation cycles (b) of the in-phase tests with $T_0 = 1050\,°C$.

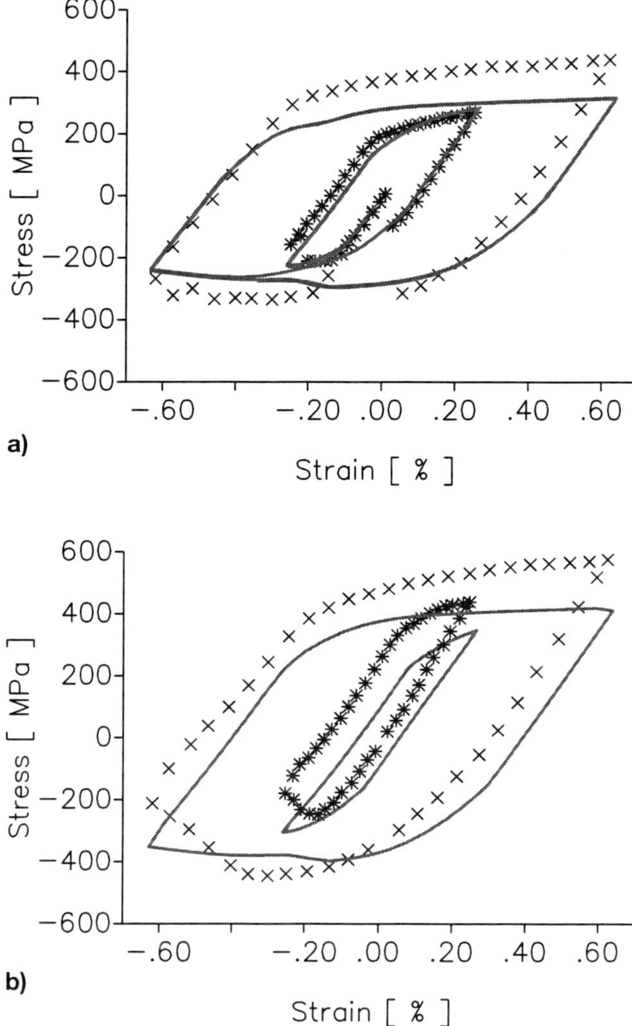

Figure 5.5.6: Comparison of measured and predicted first cycles (a) and saturation cycles (b) of the out-of-phase tests with $T_0 = 850\,°C$.

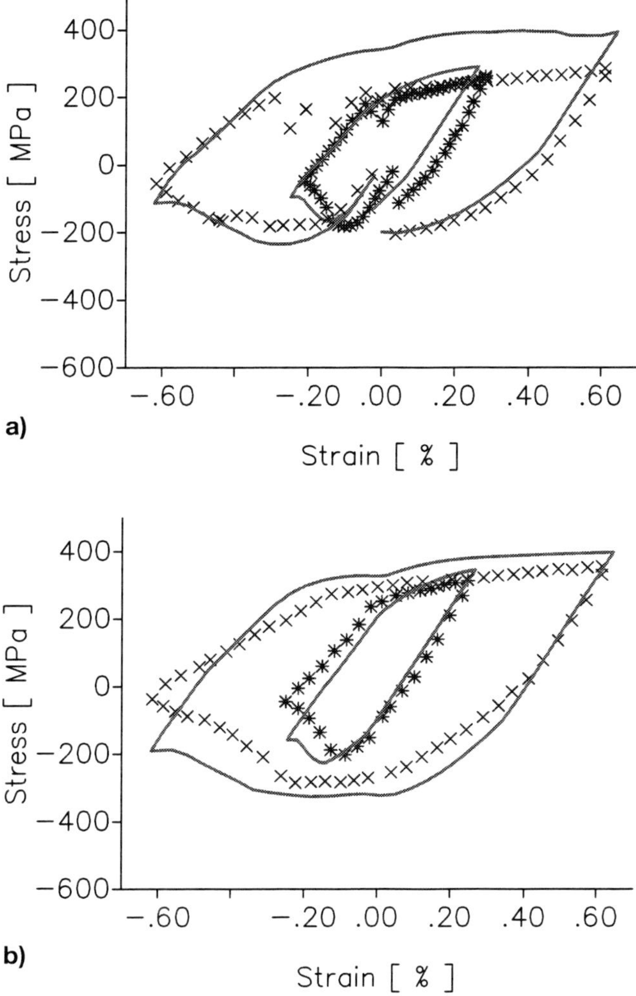

Figure 5.5.7: Comparison of measured (symbols) and predicted (solid lines) first cycles (a) and saturation cycles (b) of the out-of-phase test with $T_0 = 1050\,°C$.

5.5.3 Modeling of Damage

By means of a deformation model, the load is calculated in the form of stresses and strains. To make a lifetime prediction, however, the damage caused by these stresses and strains has to be determined. For this, a continuum-mechanics concept was chosen, which consists of both, a deformation model and a damage model coupled together. The damage model describes the evolution of a scalar variable of state introduced for the damage. Within the framework of the CDM-concept[1] adopted, the ISRM-model is developed, which allows the determination of damage under loads typically occurring in combustion chamber structures. In the sections below, the CDM-concept and the ISRM-model shall be described in further detail. Finally, application results shall be presented.

5.5.3.1 CDM-Concept

In the CDM-concept an internal state variable for the damage is introduced. The definition of this variable is based on a proposal made by Kachanov [9]. He defines the damage of a surface element of δa and orientation $\vec{n}$ at a material point as follows:

$$D_{\vec{n}} = \frac{\delta \tilde{a}_D}{\delta a} \tag{13}$$

$\delta \tilde{a}_D$ corresponds to the reduction of the effective load-carrying surface. It may be interpreted as the mapping of a partial surface δa_D which is covered by damage features, such as micropores and microcracks. By this mapping the changes in the microscopic load distribution induced by the form and distribution of the damage features are considered. Furthermore, D has a range with the following limits:

- $D_{\vec{n}} = 0$ for an undamaged material point
- $D_{\vec{n}} = 1$ *for the rupture of a volume element at the material point along the surface element considered*

If $D_{\vec{n}}$ is independent on the orientation, the damage of the material point is referred to be isotropic. The scalar variable $D = D_{\vec{n}}$ then becomes the damage variable. The definition of this variable is completed by the mapping rule of

[1] CDM stands for continuum damage mechanics.

δa_D to $\delta \tilde{a}_D$. It is given indirectly by the hypothesis of strain equivalence [10, 11][2]:

Under the influence of the stress $\boldsymbol{\sigma}$, the damaged material exhibits the same deformation behaviour than the undamaged material under effective stress $\tilde{\boldsymbol{\sigma}}$:

$$\tilde{\boldsymbol{\sigma}} = \boldsymbol{\sigma} \frac{\delta a}{\delta a - \delta \tilde{a}_D} = \frac{\boldsymbol{\sigma}}{1 - D} \tag{14}$$

Selection of this hypothesis allows indirect and simple measurement of the damage variable D [11]. This may be of particular importance when setting up its evolution equation [11–13]. Furthermore, the interaction between deformation and damage can be taken into account by using the effective stress instead of the stress in the deformation model.

In the case of anisotropic damage, a tensorial variable is required for characterisation [14]. Hereinafter, damage shall be regarded to be isotropic such that the variable defined above is sufficient for determination.

The load-induced variation of the damage variable D versus time is described by an evolution equation or damage model. Here, the damage rate is expressed as a function of the relevant influencing parameters. These parameters may include all variables that characterise the state of the material. These are the external variables of state, such as the stress $\boldsymbol{\sigma}$ and the temperature T, as well as internal variables of state, such as the damage variable D and the hardening variable q_i:

$$\dot{D} = f(\boldsymbol{\sigma}, T, D, q_i) \tag{15}$$

Depending on the loading, three types of damage, namely ductile damage, creep, and fatigue damage are distinguished, as in classical lifetime prediction. Hence, several models have been developed for D as a function of the type of damage [9, 15–17]. In case of a mixed load or various overlapping types of damage, the respective models are coupled. In order to avoid problems in particular when describing interaction effects, models for a unified treatment and modeling of damage have been developed [12, 13, 18–20].

Usually, non-isothermal cyclic loadings with long holding times are encountered in combustion chamber structures. To determine damage under these relatively complex loads, a model was developed within the framework of the present partial project. The model, called ISRM-model (**I**nelastic **S**train **R**ate **M**odified), covers particularly at high temperatures time-dependent damage without distinguishing between the creep and fatigue fractions. This is achieved by introducing the dependence of the damage rate on the inelastic strain rate. The influence of the loading history on the internal state and, hence, on the damage is also taken into consideration.

[2] This hypothesis is also referred to in literature as the "Concept of Effective Stress"

5.5.3.2 ISRM-Model

This model represents, in contrast to classical models another possibility of describing damage at high temperatures. It does not distinguish between creep and fatigue damage, as they hardly occur separately at high temperatures. A dependence of the damage rate on the internal state is introduced with the latter being determined by the damage and the deformation history. This is achieved implicitly by introducing a dependence on the inelastic strain rate which is also given by the internal state and can be measured indirectly. The experience described in literature, such as that gained by Franklin [21] and Danzer [22] using the Robinson rule for creep damage, and own indirect damage measurements were used as a basis [4, 12, 13]:

$$\dot{D} = \left\langle \frac{\chi(\tilde{\sigma})}{A^*} \right\rangle^r \left(\frac{\dot{p}}{\dot{\varepsilon}_s^{\text{in}}} \psi(\boldsymbol{\sigma}, \dot{\boldsymbol{\varepsilon}}^{in}) \right) \frac{\left[1 - (1-D)^{1+\kappa}\right]^{q(\tilde{\sigma}, \dot{\varepsilon}^{\text{in}})}}{(1-D)^{\kappa}} \tag{16}$$

$$\wedge \quad \tilde{\boldsymbol{\sigma}} = \frac{\boldsymbol{\sigma}}{1-D} \quad , \quad q(\tilde{\boldsymbol{\sigma}}, \dot{\boldsymbol{\varepsilon}}^{\text{in}}) = \left\langle 1 - \left(\frac{\dot{p}}{\dot{\varepsilon}_s^{\text{in}}} \psi(\boldsymbol{\sigma}, \dot{\boldsymbol{\varepsilon}}^{\text{in}}) \right)^{-\zeta} \right\rangle$$

with

$$\dot{p} = \sqrt{\frac{2}{3} \dot{\boldsymbol{\varepsilon}}^{\text{in}} : \dot{\boldsymbol{\varepsilon}}^{\text{in}}}$$

and

$$\psi(\boldsymbol{\sigma}, \dot{\boldsymbol{\varepsilon}}^{\text{in}}) = \left\langle \frac{\boldsymbol{\sigma}' : \dot{\boldsymbol{\varepsilon}}^{\text{in}}}{\|\boldsymbol{\sigma}'\| \, \|\dot{\boldsymbol{\varepsilon}}^{\text{in}}\|} \right\rangle^{\zeta^*} \qquad \wedge \|\mathbf{x}\| = \sqrt{\mathbf{x} : \mathbf{x}}$$

$\boldsymbol{\sigma}'$ is the stress deviator. $\dot{\varepsilon}_s^{\text{in}}$ corresponds to the stationary creep rate at uniaxial stress which equals the effective von Mises reference stress and a temperature equaling the current temperature[3]:

$$\dot{\varepsilon}_s^{\text{in}} = \hat{\dot{\varepsilon}}_s^{\text{in}}(\tilde{\sigma}_V, T)$$

In case of stationary creep ($\dot{p} = \dot{\varepsilon}_s^{\text{in}}$ and $\psi(\boldsymbol{\sigma}, \dot{\boldsymbol{\varepsilon}}^{\text{in}}) = 1$), the model goes over to that by Rabotnov for pure creep damage [16] generalised by Leckie and Hayhurst [23] for multiaxial loadings:

$$\dot{D} = \left\langle \frac{\chi(\tilde{\sigma})}{A^*} \right\rangle^r (1-D)^{-\kappa} \, \mathrm{d}t \tag{17}$$

[3] At low temperatures, where stationary creep does not occur, $\dot{\varepsilon}_s^{\text{in}}$ is substituted by an appropriate function on the effective von Mises stress and the temperature.

There, the internal state is a function of the external load. $\langle\,\rangle$ are the McCauly brackets introduced in Section 5.4.2. A^*, r *and* κ are material- and temperature-dependent parameters. $\chi(\boldsymbol{\sigma})$ is an equivalent uniaxial stress for the multiaxial stress state $\boldsymbol{\sigma}$ which is determined in accordance with the three-invariant criterion:

$$\chi(\boldsymbol{\sigma}) = a_1\,\mathcal{J}_0(\boldsymbol{\sigma}) + a_2\,\mathcal{J}_1(\boldsymbol{\sigma}) + (1 - a_1 - a_2)\,\mathcal{J}_2(\boldsymbol{\sigma}) \tag{18}$$

with

$$\mathcal{J}_0(\boldsymbol{\sigma}) = \max\,(\sigma_I, \sigma_{II}, \sigma_{III})$$
$$\mathcal{J}_1(\boldsymbol{\sigma}) = \sigma_I + \sigma_{II} + \sigma_{III}$$
$$\mathcal{J}_2(\boldsymbol{\sigma}) = \sqrt{\frac{3}{2}\,\boldsymbol{\sigma}' : \boldsymbol{\sigma}'}$$

σ_I, σ_{II} and σ_{III} denote the principal stresses. a_1 and a_2 are material- and temperature-dependent parameters. They can be determined by fitting the Rabotnov model (Eq. (6)) to isochronous surface – the set of stress states with the same creep rupture time – determined experimentally.

Besides the parameters of the Rabotnov model (A^*, r and κ), the ISRM-model contains ζ and ζ^* as further material- and temperature-dependent parameters. Here, ζ^* must be determined for the application of the model to multiaxial non-proportional loadings only. Furthermore, it is assumed that the damage rate disappears in the case of inelastic unloading (negative inelastic power). The function ψ allows to take into account this fact and additionally the distinction between proportional and non-proportional loadings with their effects on the damage rate.

To calculate the lifetime with the ISRM-model, the stress and strain evolutions over the whole lifetime are required. In a test, where deformations are homogeneous, they can be measured. But generally, they must be evaluated by a deformation model which has to be coupled with the ISRM-model to take into account the influence of damage on the deformation behaviour. The coupling can be realized in accordance with the strain equivalence principle [11].

5.5.3.3 Application of the ISRM-Model for Lifetime Prediction

The ISRM-model was applied for the description of the damage behaviour of the combustion chamber material NiCr22Co12Mo9 (trade name Nicrofer 5520 Co) under uniaxial isothermal cyclic loading. The studies focused on cyclic total strain- and stress-controlled isothermal experiments which were performed with this material at temperatures of 850 °C and 1000 °C under the partial project C5. The experiments were aimed at studying the influences of the test frequency and the load amplitude on the lifetime. The creep data obtained under the project C4 were also applied for parameter determination. In addition some non-isothermal cyclic experiments performed under the project C7 with the same materials were considered for further verifications.

5.5.3.3.1 Determination of the Parameters

When using the ISRM-model, it is assumed that the damage is not changed under uniaxial compression. Hence, the ISRM-model yields the following damage rate under uniaxial load:

$$\dot{D} = \left\langle \frac{\tilde{\sigma}}{A^*} \right\rangle^r \left\langle \frac{\dot{\varepsilon}^{\text{in}}}{\dot{\varepsilon}_s^{\text{in}}} \right\rangle \frac{\left[1 - (1 - D)^{1+\kappa}\right]^{q(\tilde{\sigma}, \dot{\varepsilon}^{\text{in}})}}{(1 - D)^\kappa} \tag{19}$$

$$\wedge \quad \tilde{\sigma} = \frac{\sigma}{1 - D} \quad , \quad q(\tilde{\sigma}, \dot{\varepsilon}^{\text{in}}) = \left\langle 1 - \left(\frac{\dot{\varepsilon}^{\text{in}}}{\dot{\varepsilon}_s^{\text{in}}} \right)^{-\zeta} \right\rangle$$

In case of stationary creep under tensile loads, this model is transferred to that of Rabotnov:

$$\dot{D} = \left(\frac{\sigma}{A^*} \right)^r (1 - D)^{-(\kappa + r)}$$

The parameters of the Rabotnov model (A^*, r and κ) were determined by applying the model to isothermal creep tests. Fitting of the lifetimes calculated with the model to the experimentally obtained lifetimes yielded the values of r and $\hat{A}$ ($\hat{A} = A^*(1 + \kappa + r)^{-1/r}$). When calculating the lifetimes, the increase of the actual stress was taken into account by the creep tests performed under constant load [4]. As the data of these experiments did not allow sufficiently precise damage measurements for the determination of $\kappa + r$ and, hence, A^* from $\hat{A}$, a pragmatical value was taken from literature [11]. The thus determined parameters of the creep term at the temperatures of 600 °C, 850 °C, and 1050 °C are listed in Tab. 5.5.2. For the tem-

Table 5.5.2: Parameters of the ISRM-model determined for the material NiCr22Co12Mo9.

T [°C]	B [$\text{MPa}^{-n} \cdot \text{s}^{-1}$]	m	A^* [$\text{MPa} \cdot \text{s}^{1/r}$]	r	κ	ζ
600	$9.32 \cdot 10^{-22}$	6.04	4248	7.38	7.62	–
850	$1.06 \cdot 10^{-18}$	6	1293	6.5	8.5	0.38
1000	$3.16 \cdot 10^{-17}$	6.09	897	5.73	9.27	0.64
1050	$5.04 \cdot 10^{-16}$	6.05	812	5.5	9.5	–

perature of 1000 °C, a similar direct determination of the parameters could not be carried out, as no creep tests until rupture had been run at this temperature. The values of the parameters at this temperature were obtained by interpolation between the values of 850 °C and 1050 °C (see Tab. 5.5.2). During interpolation, r and $\log \hat{A}$ were assumed to be linearly dependent on $1/T$ in accordance with the Larson-Miller parameter.

The parameter ζ remained to be determined for the application of the ISRM-model (Eq. (11)) to the material NiCr22Co12Mo9 at these temperatures. This parameter was obtained by fitting the number of cycles to failure computed with the ISRM-model to the experimental values determined by cyclic experiments. As the parameter ζ influences the dependence of the damage rate and, hence, of the lifetime on the inelastic strain rate, the strain- and stress-controlled cyclic experiments performed at various frequencies were used for the fitting procedure. To compute the number of cycles to failure, the ISRM-model was coupled with the viscoplastic deformation model by Chaboche according to the principle of strain equivalence (see Annex B). Furthermore, the stationary strain rate in the damage model was selected to be dependent on the actual effective stress according to Norton:

$$\dot{\varepsilon}_s^{\text{in}} = B\,\tilde{\sigma}^m$$

B and m are material- and temperature-dependent parameters which are determined on the basis of the creep data [24]. Their values are listed in Tab. 5.5.2 together with those of the other parameters. It is obvious from Fig. 5.5.8 that the dependence of the number of cycles to failure on the frequency in stress-controlled experiments is well described by the ISRM-model.

5.5.3.3.2 Verification of Fitted Model

For verification, the fitted ISRM-model coupled with the viscoplastic deformation model by Chaboche (see Annex B) was applied. It should be noticed that the results obtained have to be interpreted as the product of both models.

Application to Isothermal Loading

The fitted ISRM-model coupled with the viscoplastic deformation model by Chaboche (see Annex B) was applied to the remaining cyclic experiments performed at the frequency of 5 Hz. Here, good predictions were made for the experimentally determined Wöhler curves above all in the range of low cycle fatigue (LCF) (see Fig. 5.5.9). In the range of high cycle fatigue (HCF), an underestimation occurred and, hence, a reliable prediction of lifetime was made. When applied to strain-controlled experiments, good results were reached as well (see Fig. 5.5.10).

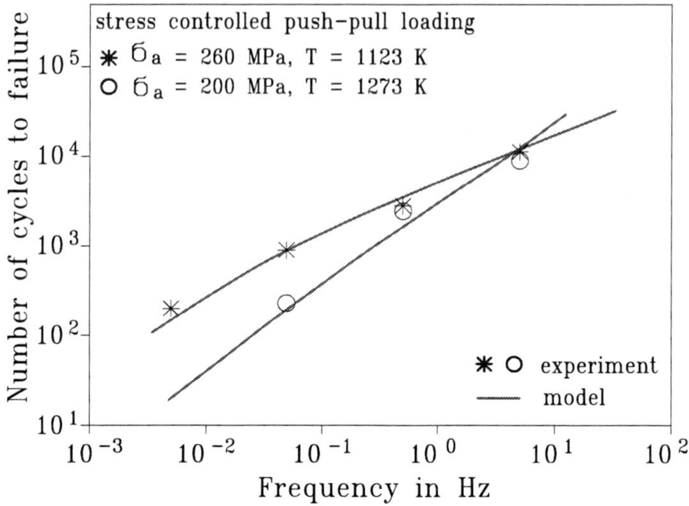

Figure 5.5.8: Frequency dependence of the number of cycles to failure: Comparison between the ISRM model and material response.

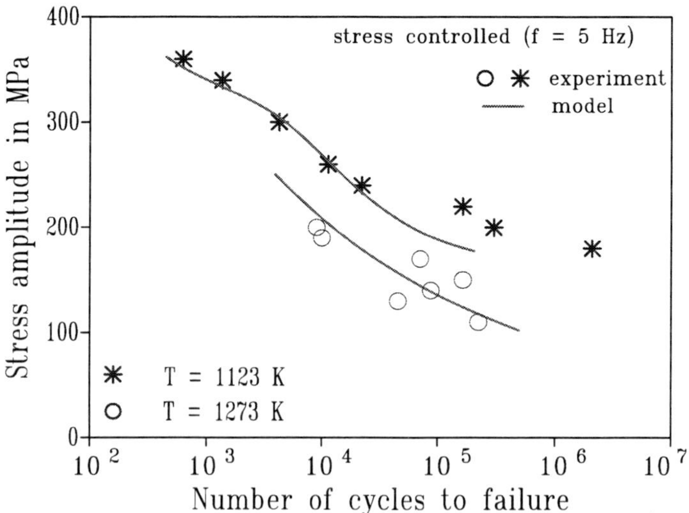

Figure 5.5.9: Comparison of the Wöhler curves computed with the ISRM-model and the experiment.

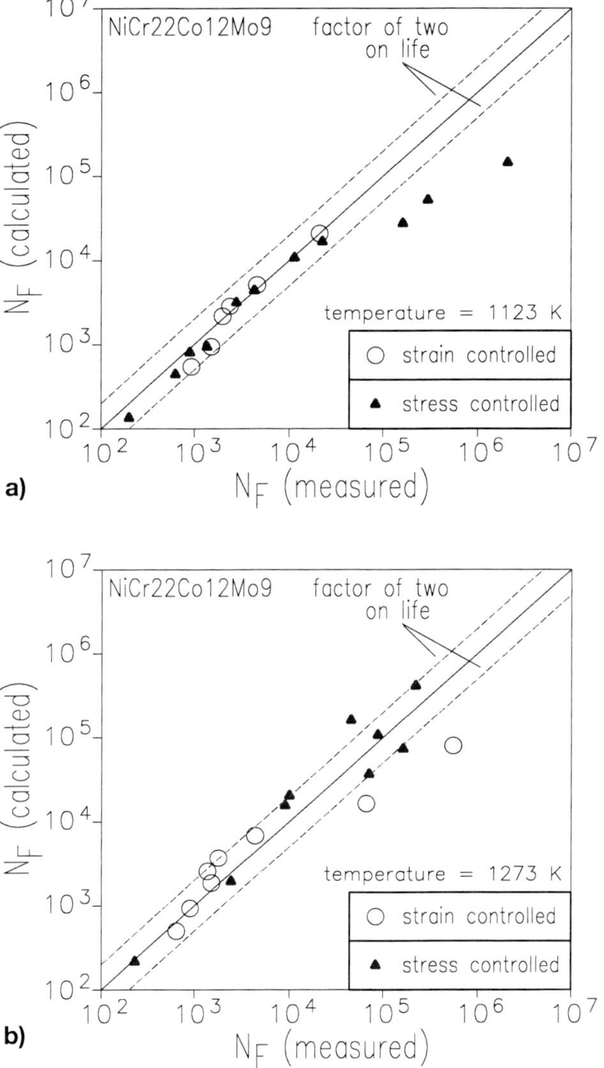

Figure 5.5.10: Comparison of the numbers of cycles to failure calculated with the ISRM-model and determined from the experiment.

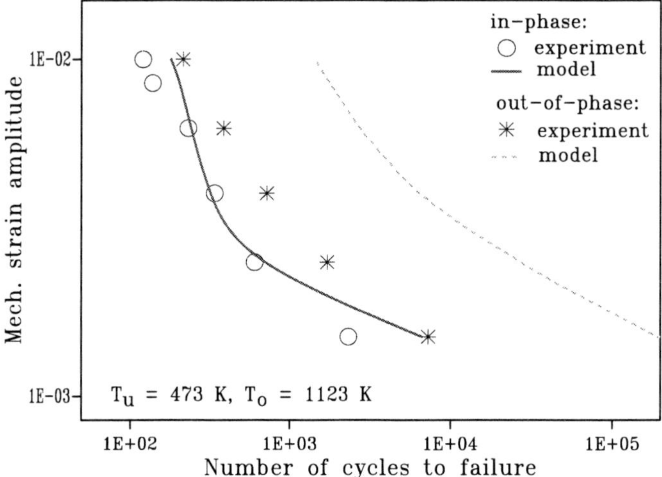

Figure 5.5.11: Comparison of the numbers of cycles to failure calculated with the coupled Chaboche/ISRM-model and determined from the non-isothermal experiments.

Application to Non-isothermal Loading

The coupled Chaboche/ISRM-model was applied to the thermomechanical tests mentioned in Section 5.5.2.2.2. In-phase and out-of-phase tests were considered which were performed with the same upper and lower temperatures of 200 °C and 850 °C, respectively, and different mechanical strain amplitudes ranging between 0.15 % and 1 %. As it can be seen from Fig. 5.5.11 the lifetimes of the in-phase tests are well predicted by the model. In contrast, poor predictions are obtained for the out-of-phase tests. The main reason for this is that the model parameters were determined at temperatures (>600 °C) higher than the lowest temperature of the tests. Consequently, the damaging tensile stresses appearing at low temperatures are underestimated and the lifetimes are overestimated.

5.5.4 Conclusions

It is demonstrated by the only partially satisfying results obtained for the verification of the deformation model of Chaboche for non-isothermal loads that the development of deformation models is by no means completed. It still remains to be settled, whether additional phenomena have to be taken into account for the description of such loads. Further detailed investigations are required.

The results obtained with the developed damage model have demonstrated the efficiency of the latter when applied to simple isothermal loads. Verifications for thermocyclic loadings yielded partially fairly well results. The poor results were mainly obtained in cases in which the models were extrapolated to the damage-relevant low-temperature range. Therefore, the models should be handled with care as far as extrapolation is necessary.

Acknowledgements

The contributions of Prof. Dr. B. Schinke, Dr. J. Schwertel, and Dr. A. Haupt within their former activities in the Collaborative Research Centre 167 are acknowledged. The theoretical calculations have been carried out with the help of the staff and the equipment of the Forschungszentrum Karlsruhe.

References

1. Haupt, P., Korzen, M., and Bumb, H. (1985): Time-independent constitutive theories for cyclic plasticity, Abschlußbericht BMFT-Vorhaben 1500 516, TU Darmstadt.
2. Lindholm, U. S., Chan, K. S., Bodner, S. R., Weber, R. M., Walker, K. P., and Cassenti, B. N. (1984): Constitutive modelling for isotropic materials, NASA Contract Report 174718, NASA.
3. Schwertel, J. (1992): Modellierung des einachsigen mechanischen Verhaltens von Werkstoffen durch viskoplastische Modelle, Dissertation, Universität Karlsruhe.
4. Aktaa, J. (1994): Kontinuumsmechanische Modellierung der zeitabhängigen Schädigung bei hohen Temperaturen, Dissertation, Universität Karlsruhe.
5. Chaboche, J. L. (1989): Constitutive equations for cyclic viscoplasticity, *Int. J. of Plasticity*, **5**, 247–302.
6. Chaboche, J. L. and Cailletaud, G. (1986): On the calculation of structures in cyclic plasticity or viscoplasticity, *Computers and Structures*, **23**, no. 1, 23–31.
7. James, F. and Roos, M. (1987): Function minimisation and error analysis, CERN Computer Centre, Program Library, Geneva/Switzerland.
8. Schwertel, J. and Schinke, B. (1996): Automated evaluation of material parameters of viscoplastic constitutive equations, *J. Engng. Mat. Tech.*, **118**, 273–280.
9. Kachanov, L. M. (1958): Time of the rupture process under creep conditions, TVZ Akad. Nauk. S.S.R. Otd. Tech. Nauk., no. 8.

10. Lemaitre, J. and Chaboche, J. L. (1978): Aspects phénoménologiques de la rupture par endommagement, *J. Méc. Appl.*, no. 2(3), 317–365.
11. Lemaitre, J. and Chaboche, J. L. (1990): Mechanics of Solid Materials, Cambridge University Press, Cambridge.
12. Aktaa, J., Munz, D., and Schinke, B. (1994): Phenomenological modelling of time dependent damage taking into account the deformation prehistory, *Z. angew. Math. Mech. (ZAMM)*, no. 74(4), T171–T173.
13. Aktaa, J. and Schinke, B. (1996): The influence of the hardening state on time dependent damage and its consideration in a unified damage model, *Fatigue Engng. Mater. Struct.*, **19**, 1143–1151.
14. Murakami, S. (1987): Anisotropic aspects of material damage and application of continuum damage mechanics, in: Continuum Damage Mechanics, Theory and Application, (D. Krajcinovic and J. Lemaitre, eds.), Springer.
15. Lemaitre, J. (1984): How to use damage mechanics, *Nuclear Engineering and Design*, **80**, 233–245.
16. Rabotnov, N. Y. (1968): Creep rupture, Proc. of the 12th Int. Congress of Applied Mechanics, (Stanford).
17. Chaboche, J. L. (1977): Viscoplastic constitutive equations for the description of cyclic and anisotropic behavior of metals, *Bulletin de l'Academie Polonaise des Sciences, serie des Sciences Techniques*, **XXV**, no. 1, 33–39.
18. Lemaitre, J. (1987): Damage constitutive equations, in Continuum Damage Mechanics, Theory and Application (D. Krajcinovic and J. Lemaitre, eds.), Springer.
19. Aktaa, J. and Schinke, B. (1992): A model for damage and lifetime prediction taking into account the backstress, Proc. of the 3rd Int. Conf. on Low-Cycle Fatigue and Elasto-Plastic Behaviour of Materials, Berlin.
20. Aktaa, J. and Schinke, B. (1997): Unified modelling of time dependent damage taking into account an explicit dependency on backstress, *Int. J. of Fatigue*, **19**, 195–200.
21. Franklin, C. J. (1978): Cyclic creep and fatigue life time prediction, in: High Temperature Alloys for Gas Turbines, (D. Coutsouradis et al., eds.), Appl. Sci Publ., Parking, UK.
22. Danzer, R. (1988): Lebensdauerprognose hochfester metallischer Werkstoffe im Bereich hoher Temperaturen, Gebrüder Borntraeger, Berlin, Stuttgart.
23. Leckie, F. A. and Hayhurst, D. R. (1974): Creep rupture in structure, in: Proc. R. Soc. Lond., 240–323.
24. Merckling, G. (1989): Kriech- und Ermüdungsverhalten ausgewählter metallischer Werkstoffe bei hohen Temperaturen, Dissertation, Universität Karlsruhe (TH).

Annex A: Parameters of Chaboche's model

CoCr22Ni22W14				NiCr22Co12Mo9*			
	20 °C	850 °C	1000 °C		600 °C	850 °C	1000 °C
k	177	1	1	k	83	11.5	0.35
n	1.105	19	3.5	n	46.7	15.8	4.82
Z	110	210	246.6	Z	100	199	347
H_1	750 000	163 966	379 865	H_1	162 000	300 000	82 600
D_1	2030	2750	3487	D_1	1330	3230	1490
R_1	0	0.000577	1	R_1	5.32E-6	0.445	4.87E-4
m_1	–	1.10015	0.9	m_1	1.51	–0.7	0.08
c	1	1	1	c	1	7.79	1.04
R_∞	–34	51	46	R_∞	187	20.4	15
H_2	4000	4580	5976	H_2	4720	1840	0.01
D_2	4.39	20	62.13	D_2	50.9	2.29	0.01
R_2	0	0.01577	7.7E-6	R_2	0	1.65E-3	0
m_2	–	0.5701	0.99	m_2	–	0.3	–
E	240 000	160 000	140 000	E	166 000	146 000	152 000

* Data according to Schwertel [3].

Annex B: Chaboche's model coupled with damage

The coupling of the viscoplastic deformation model of Chaboche with the ISRM model for damage takes place according to the hypothesis of strain equivalence by using the effective stress $\tilde{\boldsymbol{\sigma}}$ instead of the stress $\boldsymbol{\sigma}$ in the deformation model. For uniaxial isothermal loadings, the following coupled model is obtained:

$$\dot{\varepsilon} = \dot{\varepsilon}^{\mathrm{el}} + \dot{\varepsilon}^{\mathrm{in}} \qquad \wedge \quad \varepsilon^{\mathrm{el}} = \frac{\tilde{\sigma}}{E}$$

$$\dot{\varepsilon}^{\mathrm{in}} = \left\langle \frac{|\tilde{\sigma} - \Omega| - R_\infty - k}{Z} \right\rangle^n sgn(\tilde{\sigma} - \Omega) \qquad \wedge \quad \Omega = \sum_{i=1}^{I_R} \Omega_i$$

$$\dot{\Omega}_i = H_i \dot{\varepsilon}^{\mathrm{in}} - D_i \Omega_i |\dot{\varepsilon}^{\mathrm{in}}| - R_i |\Omega_i|^{m_i - 1} \Omega_i$$

$$\dot{D} = \left\langle \frac{\tilde{\sigma}}{A^*} \right\rangle^r \left\langle \frac{\dot{\varepsilon}^{\mathrm{in}}}{\dot{\varepsilon}^{\mathrm{in}}_s} \right\rangle \frac{\left[1 - (1 - D)^{1+\kappa}\right]^{q(\tilde{\sigma}, \dot{\varepsilon}^{\mathrm{in}})}}{(1 - D)^\kappa}$$

$$\wedge \quad \tilde{\sigma} = \frac{\sigma}{1 - D} \quad , \quad q(\tilde{\sigma}, \dot{\varepsilon}^{\mathrm{in}}) = \left\langle 1 - \left(\frac{\dot{\varepsilon}^{\mathrm{in}}}{\dot{\varepsilon}^{\mathrm{in}}_s} \right)^{-\zeta} \right\rangle$$

I_R is the number of variables for kinematic hardening contained in the model. E, k, R_∞, Z, H_i, D_i, R_i, and m_i are the parameters of the deformation model. Their values for the material NiCr22Co12Mo9 were taken from [3] (see also Annex A).

6 Thermal Barrier Coatings

High-Temperature Behaviour of Thermal Barrier Coatings

Petra A. Langjahr*, Rainer Oberacker*, and Michael J. Hoffmann*

This chapter focuses on the high-temperature behaviour of combustion chamber materials, especially on systems with thermal barrier coatings (TBCs). The benefit of coating combustion chamber components with TBCs consists either in an increased efficiency (reduced fuel consumption due to an increased inlet temperature) and/or in an increased lifetime of the metal alloy components. TBCs are subjected to thermal and mechanical stresses during service that may alter the materials properties and therefore limit the lifetime. During the time period of the Collaborative Research Centre 167, various concepts have been successfully investigated in order to understand and to improve the reliability and lifetime of TBCs and TBC/bond coat/metal alloy composites and also to define lifetime prediction models. The main goals of the investigations performed especially in the last period (1996–1998) included 1.) the assessment of new or modified TBC systems from different industrial partners (mainly project C13), 2.) the long-term behaviour of coated combustion chamber components (project C4), 3.) deformation analysis and failure prediction of typical coated and uncoated combustion chamber alloys (mainly project C7) and 4.) failure analysis and modelling of the deformation behaviour under typical service conditions (mainly project C9).

Within the work of Alaya [1], it could be demonstrated, that the resistance to failure during thermal cycling of plasma-sprayed TBC compounds can be improved by several methods, e. g. 1.) a reduction of the elastic modulus by an increased porosity or a controlled segmentation, 2.) a reduction of the thermal mismatch by choosing a different stabilizer or 3.) by using graded TBCs. In addition, an advanced adhesion of the TBCs could be realised by increasing the average roughness of the interface. Also, the stress state of the TBC could be reduced by heating the substrate during plasma-spraying. Besides the optimisation of the behaviour during thermal cycling, the long-term degradation of TBCs at high temperatures was studied. This investiga-

* Institut für Keramik im Maschinenbau, Universität Karlsruhe, Kaiserstr. 12, 76128 Karlsruhe, Germany

tions were carried on by Langjahr et al. (Section 6.1), who studied the long-term behaviour of CeO_2- and Y_2O_3-stabilized ZrO_2-TBCs at typical service temperatures. Tentative TTT (time-temperature-transformation) diagrams could be derived according to the amount of phase transformation. An influence of phase transformations and sintering effects on the mechanical properties could be clearly demonstrated. TBCs containing a stabilizer content of 19.5 mol-% CeO_2 and 1.5 mol-% Y_2O_3 showed a smaller degree of phase transformation and therefore may have a higher potential for long-term applications than Y_2O_3-stabilized TBCs with a stabilizer content as commercially applied. However, long-term degradation remains a serious problem with all plasma sprayed TBCs, which were developed so far.

These experimental investigations on thermal cycling behaviour and long-term degradation were supported by numerical modelling, carried out in project C9. Finite element calculations were performed to derive the temperature and stress fields of the thermal cycling experiments [2]. A microstructure oriented model was developed and applied to calculate the stresses at the interface between bond coat and TBC caused by oxidation [3]. Contributions of project B8 include the measurement of the emission behaviour of CeO_2- and Y_2O_3-stabilized TBCs [4].

The paper by Schmidt et al. (Section 6.2) deals with 1.) tensile tests and creep tests of combustion chamber alloys ($NiCr_{22}Co_{12}Mo_9$ and $CoCr_{22}Ni_{22}W_{14}$) to determine the microstructural features governing the macroscopic behaviour and to understand the influence of manufacturing processes and 2.) the response of plasma-sprayed ZrO_2-based TBCs to plastic deformation. The response consists in segmentation, spallation, or in a combined mode. A failure model could be successfully developed showing that the probability of either one is governed by the ratio of maximum shear stress (spallation) to maximum normal stress (segmentation), which itself is mainly influenced by the TBC thickness.

Besides plasma sprayed coatings, EB-PVD coatings were subject of the investigations. Failure mechanisms of TBCs manufactured by EB-PVD were studied in detail by Rostek et al. [5].

In summary, the investigations performed during the time period of the Collaborative Research Centre have led to significant improvements of the performance and lifetime of TBCs as well as to a better understanding of failure mechanisms. The close co-operation with gas turbine manufacturers enabled a direct technological transfer. Although the co-operative research still has to go on, the investigations of the Collaborative Research Centre allow to manufacture more powerful and reliable gas turbines during the first century of the new millennium.

References

1. M. Alaya (1997): Bewertung und Optimierung von Konzepten zur Verbesserung des Einsatzverhaltens von ZrO_2-Wärmedämmschichtsystemen. Dissertation Universität Karlsruhe, IKM-Bericht 019.
2. F. Thome, M. Alaya, E. Diegele, R. Oberacker (1989/1996): Berechnung von Temperatur- und Spannungsverteilung thermozyklisch beanspruchter Wärmedämmschichtsysteme. In: J. Kriegesmann (Ed.): Technische keramische Werkstoffe, Deutscher Wirtschaftsdienst, Köln, Chapter 5.2.9.2.1, 1–15.
3. K. Sfar (1997): FE-Berechnung von Eigenspannungsverteilungen in 8 % Y_2O_3-ZrO_2-WDS-Systemen unter Berücksichtigung der Oxidation der Haftvermittlerschicht. Studienarbeit Universität Karlsruhe, IZSM (to be published).
4. M. Alaya, R. Oberacker, M. J. Hoffmann, W. Krebs, R. Koch, S. Wittig (1997): Thermophysikalische Eigenschaften von CeO_2- und Y_2O_3-stabilisierten Wärmedämmschichtsystemen und ihre Auswirkung auf das thermozyklische Verhalten und den Strahlungswärmeübergang. In: H. W. Grünling (Ed.): Werkstoffe für die Energietechnik. DGM Informationsgesellschaft, Frankfurt, 243–248.
5. K. E. Rostek, D. Löhe, O. Vöhringer (1999): Oxidation und Versagen von EB-PVD-ZrO_2-Wärmedämmschichtsystemen für Gasturbinen unter Hochtemperaturbeanspruchungen. In: A. Kranzmann, U. Gramberg (Ed.): Werkstoffe für die Energietechnik/Werkstoffe und Korrosion. Wiley-VCH, Weinheim, 121–126.

6.1 Long-Term Behaviour and Application Limits of Plasma-Sprayed ZrO_2 Thermal Barrier Coatings

Petra A. Langjahr*, Rainer Oberacker*, and Michael J. Hoffmann*

Abstract

Changes in phase composition and mechanical properties of various zirconia-based plasma-sprayed thermal barrier coatings (TBCs) after long-term heat treatments at typical operation temperatures (1000–1400 °C) were investigated by X-ray diffraction (XRD) and mechanical testing. The experimental investigations clearly indicated significant changes with respect to the as-sprayed stage even after relatively short heat treatments.

TBCs containing 8 mol-% CeO_2 are not suitable for long-term applications due to a high degree of tetragonal-to-monoclinic phase transformation. According to the XRD-investigations, systems with a stabilizer content of 19.5 mol-% CeO_2/1.5 mol-% Y_2O_3 or 35 mol-% CeO_2 revealed a smaller amount of monoclinic phase after heat treatments compared to TBCs containing 4.5 mol-% Y_2O_3 and, therefore, they might have a higher potential for long-term applications. TBCs containing 35 mol-% CeO_2 seem to be less suitable than TBCs with 19.5 mol-% CeO_2/1.5 mol-% Y_2O_3 with respect to a higher increase of elastic modulus and the related decrease of strain tolerance.

6.1.1 Introduction

ZrO_2-based plasma-sprayed coatings are currently applied as thermal barrier coatings (TBCs) in hot gas sections of gas turbines [1]. Common TBCs are Y_2O_3-stabilized. Only few studies have been performed with CeO_2 as stabilizer [2–6] although CeO_2-stabilized ZrO_2 may own certain advantages to

* Institut für Keramik im Maschinenbau, Universität Karlsruhe, Kaiserstr. 12, 76128 Karlsruhe, Germany

Y_2O_3-stabilized ZrO_2. For example, it has been shown that Ce-TZP ceramics can attain higher fracture toughness than Y-TZP [7, 8]. Furthermore, TBCs with a CeO_2 content above ~10 mol-% show a lower thermal conductivity and a lower thermal expansion mismatch on common metal superalloys than commercial Y_2O_3-stabilized TBCs [3, 4]. It was also demonstrated that CeO_2-stabilized TBCs may own a higher lifetime in thermal cycling tests [4, 5].

During service, TBCs are subjected to thermal and mechanical stresses that may alter the microstructure and therefore limit the lifetime. Previous studies on damage (e.g. [2, 6, 9, 10]) do not include CeO_2-stabilized TBCs or focus on changes after short-term investigations such as thermal cycling. Therefore, the present investigation mainly deals with the question of long-term behaviour, especially after heat treatments at typical operation temperatures. In ZrO_2-based systems, two types of structural changes that are not independent of each other may occur during operation at high temperatures: 1.) sintering and grain growth effects and 2.) phase transformations. The structural changes influence the mechanical properties of the TBCs. For example, even relatively short heat treatments at typical operation temperatures can lead to the formation of sintering necks between the single lamellae of porous TBCs and therefore cause an increase of the elastic modulus [5]. However, a higher elastic modulus causes a reduced strain tolerance of the TBCs.

Previous investigations on Y_2O_3-stabilized ZrO_2 have shown that mechanical properties and lifetime are optimised when the maximum amount of the non-equilibrium tetragonal t'-phase can be obtained by the plasma-spraying process, that is, for ~6–8 wt.-% Y_2O_3 [1, 11]. The metastable tetragonal t'-phase that contains a higher stabilizer content than the tetragonal equilibrium phase t occurs in Y_2O_3-, as well as in CeO_2-stabilized plasma-sprayed coatings (e.g. for ~20 mol-% CeO_2 [2]). According to Yashima et al. [12], within the CeO_2-ZrO_2 system, a second metastable phase exists in addition to the t'-phase: the cubic c'-phase. The c'-phase has an oxygen deficit due to partial reduction of Ce^{4+} to Ce^{3+}. A heat treatment at a few 100 °C can lead again to the cation-diffusionless c' $\rightarrow$ t' transformation (Fig. 6.1.1). The t'-phase does not undergo the detrimental tetragonal-to-monoclinic (t $\rightarrow$ m) transformation that is related with large stresses. Therefore, the t'-phase is stable during thermal cycling above and below the t $\rightarrow$ m transformation temperature [2]. However, during heat treatment at typical service temperatures, a cation-diffusion controlled transformation of the t'-phase to the tetragonal and cubic equilibrium phases (t' $\rightarrow$ t + c) may take place (Fig. 6.1.1). In contrast to the t'-phase, the t-phase may undergo the t $\rightarrow$ m phase transformation that limits the lifetime of the TBC.

Within the present study, TBCs were heat treated in the range of typical service temperatures to evaluate the long-term behaviour and application limits due to structural changes. X-ray diffraction (XRD) and mechanical testing was performed to investigate changes in phase composition and mechanical properties for TBCs with different stabilizers: 1.) 4.5 mol-% Y_2O_3 (a com-

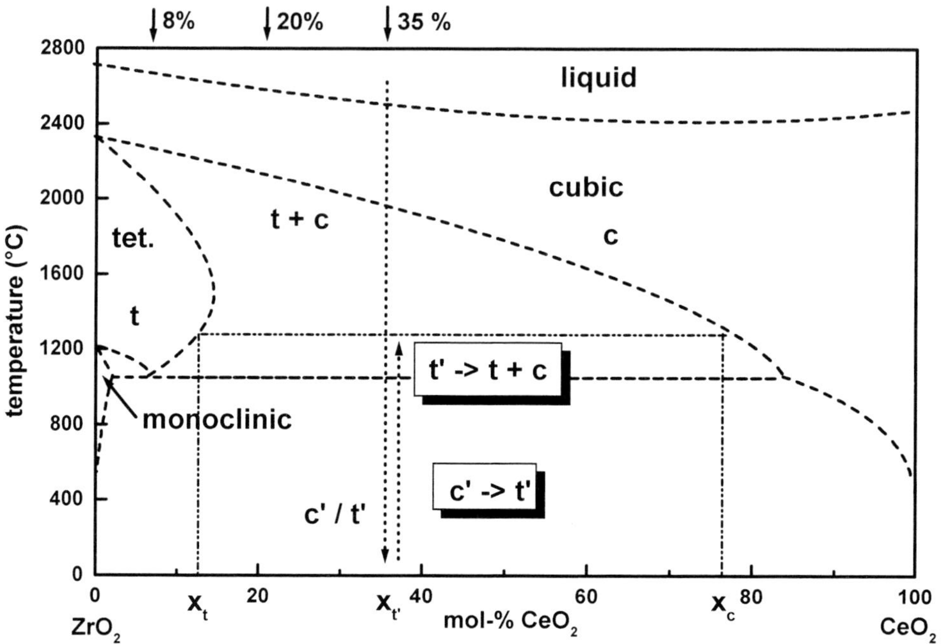

Figure 6.1.1: Phase diagram of the system ZrO_2-CeO_2 according to Yashima et al. [12]. After quenching due to the plasma-spraying process, metastable c'- and/or t'-phases are present with a CeO_2 content of $x_{t'}$. During heating, the c'-phase transforms to the t'-phase and t' transforms diffusion-controlled to the tetragonal and cubic equilibrium phases t and c with CeO_2 contents of x_t and x_c. The compositions investigated in the present study are marked.

position similar to the one used in commercial TBCs with a high content of t'-phase) and CeO_2-stabilized TBCs with 2.) 8 mol-% CeO_2, 3.) 19.5 mol-% CeO_2/1.5 mol-% Y_2O_3, and 4.) 35 mol-% CeO_2 (Fig. 6.1.1). For the investigation of the long-term behaviour, TBCs were separated from their substrates for two reasons: 1.) the exposition temperature of the TBCs is well above the application temperature of the metal-alloy substrates and 2.) the structural changes of the TBCs can be investigated independently of any interaction with a specific bond coat or substrate material, like interdiffusion or oxidation [13].

Complementary to the study of the long-term behaviour after heat treatments, the influence of thermal cycling on complete TBC systems with duplex structure is compared for materials with different stabilizer contents.

6.1.2 Experimental Procedure

Plasma-sprayed TBC systems with a typical duplex structure consisting of stabilized ZrO_2 (~0.6 mm), bond coat (~0.15 mm) and substrate were investigated. In order to study the influence of thermal treatments, TBCs were cut into pieces of ~50 mm x 10 mm and afterwards separated from their substrates by dissolving the bond coat in concentrated hydrochloric acid. The deposition time in hydrochloric acid was 24 h. Subsequently, the TBCs were thoroughly cleaned by rinsing and infiltration of water. The TBCs were afterwards heat treated between 1 h and 1000 h at temperatures between 1000 °C and 1400 °C.

Phase transformations were studied by XRD (Siemens D500, equipped with CuK_α X-ray source and graphite secondary monochromator). The quantification of monoclinic, tetragonal and cubic phases was performed by determining integral intensities I of the following X-ray reflections by profile refinement with a computer program (XPLOT [14]): 1.) monoclinic: $(111)_m$ and $(11\text{-}1)_m$, 2.) tetragonal: $(111)_t$, $(004)_t$ and $(400)_t$, and 3.) cubic: $(111)_c$ and $(004)_c$. The volume fraction of the monoclinic phase was estimated in a reasonable good approximation from the ratio of the {111} reflections: $I_m = I_m(111) + I_m(11\text{-}1)$ and $I_{c+t} = I_t(111) + I_c(111)$. Correspondingly, the volume fraction of the tetragonal and cubic phases were determined from the intensities of the {004} reflections: $I_t = I_t(004) + I_t(400)$ and $I_c = I_c(004)$. Deviations from linearity between the intensity ratios and the volume fractions as they were considered by various authors [15, 16] were neglected as they are not significant (e. g. $< 7\,\%$ in a tetragonal-monoclinic mixture of ZrO_2 according to Toraya et al. [15]) and especially as the different materials investigated would require the measurement of different calibration curves, which was beyond the scope of the present study.

The sintering behaviour of the TBCs was characterized by a geometrical determination of the shrinkage. The elastic modulus was measured by the deflection in three-point bending tests within the elastic range (span distance: 40 mm, load: 0,5 N). The bending strength was investigated in four-point bending tests with an outer and inner span of 40 mm and 20 mm, respectively.

In addition, the lifetime of TBC-bond coat-substrate-systems during thermal cycling was compared in burner-rig tests. After heating with an acetylene-oxygen burner to the maximum temperature, the TBCs were rapidly cooled to 60 °C by compressed air. The following two conditions were chosen: 1.) maximum temperature on the surface 1250 °C/maximum temperature on the backside of the substrate 800 °C and 2.) 1350 °C/1000 °C, respectively. The tests were conducted for 1000 cycles if no failure occurred.

6.1.3 Experimental Results

6.1.3.1 X-Ray Diffraction Analysis

The course of the diffusion-controlled phase transformation t′ → t + c is demonstrated at the example of a TBC with 35 mol-% CeO_2: Figure 6.1.2 shows the XRD spectra in the 2Θ-range between 28 and 36° including the {111} and {002} reflections. The as-sprayed coating consists of the metastable cubic phase c′. After heat treatment at 1000 °C for 10 h, the reflections of the c′-phase disappeared, whereas reflections of the t′-phase occurred. This indicates that the c′ → t′ phase transformation takes place during heating-up. Already after a heat treatment of 1 h at 1250 °C, a significant part of the metastable t′-phase transformed to the cubic and tetragonal equilibrium phases c and t. The ratios of t′ : t : c, determined from the intensity ratios are $I_{t'} \approx 60\,\%$: $I_t \approx 30\,\%$: $I_c \approx 10\,\%$. After 1000 h at 1250 °C, the phase transformation is complete and the t′-phase fully disappeared. According to the XRD reflections, the relative fractions of the tetragonal and cubic equilibrium phases are $I_t \approx 56\,\%$: $I_c \approx 42\,\%$.

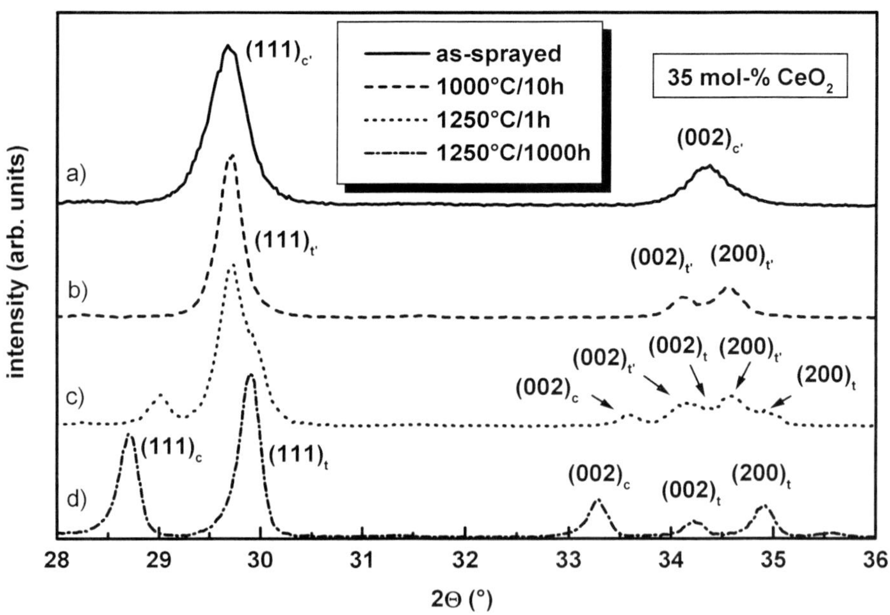

Figure 6.1.2: XRD spectra of coatings a) as-sprayed, b) after heat treatment at 1000 °C for 10 h, c) after heat treatment at 1250 °C for 1 h and d) after heat treatment at 1250 °C for 1000 h.

The phase compositions of this system and the other ones in the as-sprayed stage and after heat treatments for 1 h, 10 h, 100 h and 1000 h at 1250 °C are compared in form of area diagrams (Fig. 6.1.3). The amount of the detrimental monoclinic phase is highlighted in grey. TBCs containing 8 mol-% CeO_2 (Fig. 6.1.3b) are almost fully converted to the monoclinic phase after heat treating for 1000 h. The high fraction of transformation occurs due to the high amount of transformable tetragonal phase t. Due to the low CeO_2 content of 8 %, the non-transformable t′-phase does not exist in the as-sprayed stage. The Y_2O_3-stabilized system (Fig. 6.1.3a) shows a strong increase of the amount of monoclinic phase to about 30 % after a heat treatment for 1000 h. For shorter heat treatments (≤100 h), only a few percent of monoclinic phase is present. The CeO_2-stabilized systems with 19.5 mol-% CeO_2/1.5 mol-% Y_2O_3 (Fig. 6.1.3c) and 35 mol-% CeO_2 (Fig. 6.1.3d) show only a few percent of monoclinic phase, even after heat treatment for 1000 h, in contrast to the Y_2O_3-stabilized system. TBCs with 19.5 mol-% CeO_2/1.5. mol-% Y_2O_3 contain less amounts of cubic phase after 1000 h than TBCs with 35 mol-% CeO_2, as expected from the phase diagram [12] (Fig. 6.1.1).

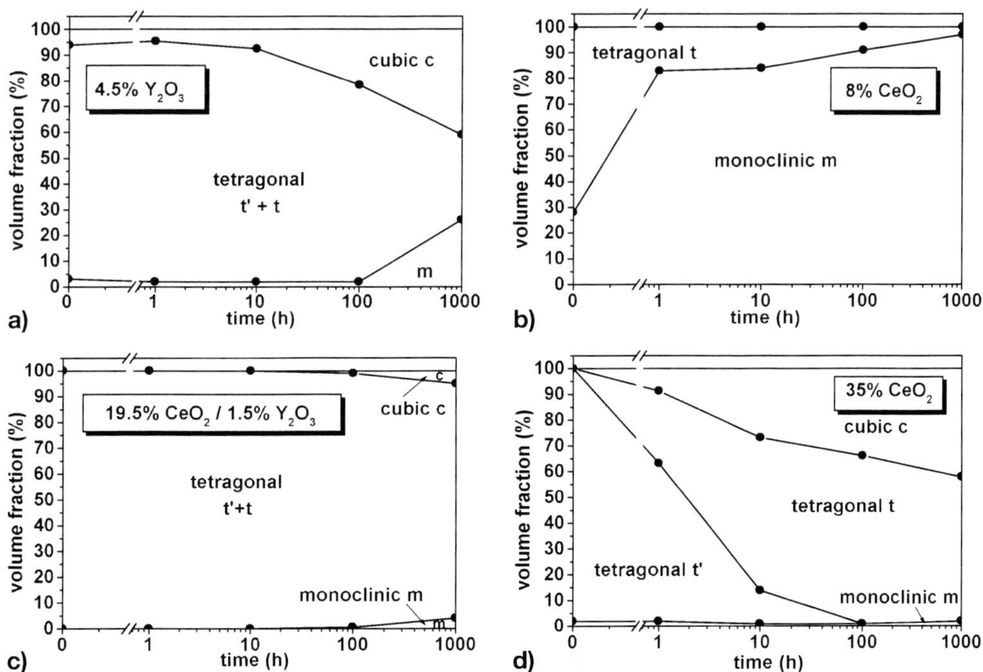

Figure 6.1.3: Area diagrams of the relative volume fractions of monoclinic, cubic, and tetragonal phases in the TBCs as-sprayed and after heat treatments of 1 h, 10 h, 100 h and 1000 h at 1250 °C.

6.1.3.2 Mechanical Properties

Figure 6.1.4 shows the change of the elastic modulus after heat treatments for various times at different temperatures between 1000 °C and 1400 °C at the example of TBCs containing 19.5 mol-% CeO_2/1.5 mol-% Y_2O_3. The TBC reveals a continuous increase of the elastic modulus compared to the as-sprayed stage with increasing heat treatment temperature and increasing heat treatment times. An increase of the elastic modulus, however, is related with a decrease in strain tolerance.

Figure 6.1.5 shows a comparison of the change in elastic modulus for all systems after heat treatments for different times (1 h, 10 h, 100 h, and 1000 h) at 1250 °C. Two tendencies become obvious: 1.) All systems show an increase of the elastic modulus even after short heat treatments (1 h) compared to the as-sprayed condition. In case of the systems containing 19.5 mol-% CeO_2/1.5 mol-% Y_2O_3 and 35 mol-% CeO_2, the increase exists up to the maximum exposition time of 1000 h, whereas in case of the system containing 4.5 mol-% Y_2O_3 only up to ~100 h. The system containing 35 mol-% CeO_2 shows the strongest increase of the elastic modulus. 2.) A subsequent decrease of the elastic modulus is observed for the system containing ~4.5 mol-% Y_2O_3 after 1000 h as well as in case of the system with 8 % CeO_2 after 10 h. This decrease can be attributed to the formation of the monoclinic phase (Fig. 6.1.3).

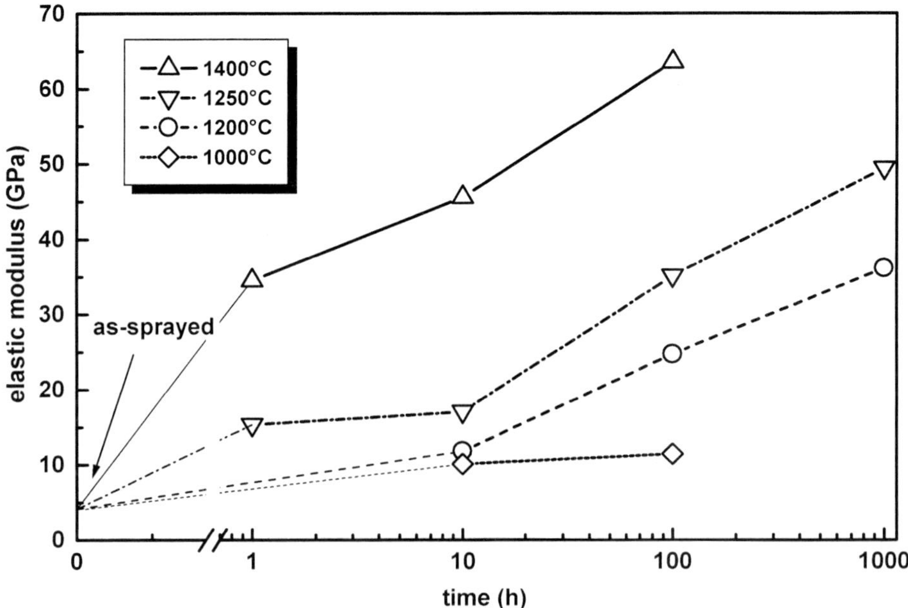

Figure 6.1.4: Elastic modulus of a TBC containing ~19.5 mol-% CeO_2/1.5 mol-% Y_2O_3 as-sprayed and after heat-treatments at temperatures between 1000 °C and 1400 °C.

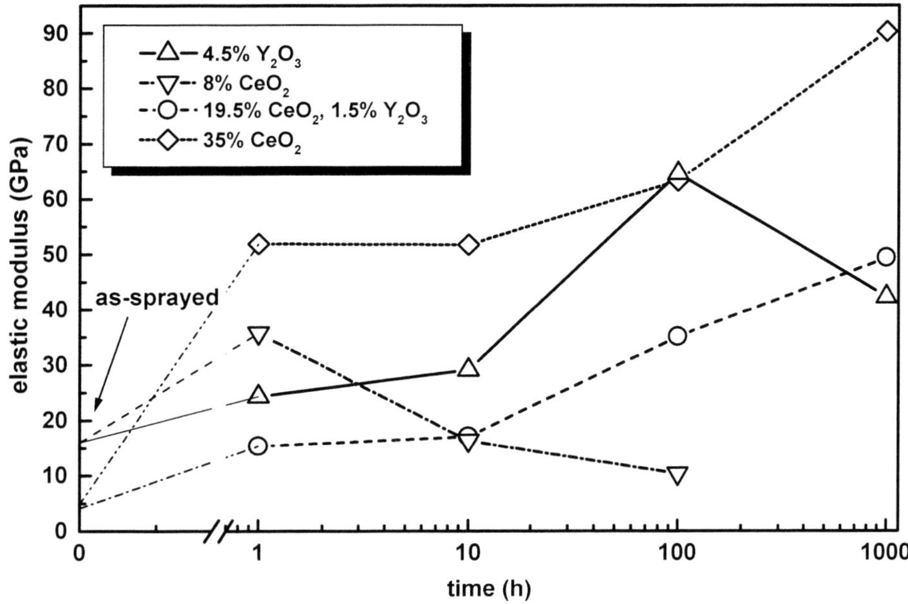

Figure 6.1.5: Elastic modulus of different TBCs as-sprayed and after heat treatments for various times at 1250 °C.

Figure 6.1.6 shows the change in four-point bending strength after heat treatments from 1 h to 1000 h at various temperatures compared to as-sprayed samples for TBCs containing 19.5 mol-% CeO_2/1.5 mol-% Y_2O_3. The results clearly indicate that the four-point bending strength increases with temperature as well as with heating time, similar to the elastic modulus.

Changes of the four-point bending strength for all investigated systems after heat treatments at 1250 °C are shown in Fig. 6.1.7. The course of the four-point bending strength resembles the one of the elastic modulus, showing two tendencies: 1.) a general increase of the four-point bending strength with increasing heating time and 2.) a decrease of the four-point bending strength for cases with a significant formation of the monoclinic phase (Fig. 6.1.3a and 6.1.3b). A subsequent decrease of the four-point bending strength is also observed in the system containing 35 % CeO_2 for heat treatments between 1400 °C/10 h and 1400 °C/100 h. The reason is not fully understood, possibly due to a decreasing ratio of cubic to tetragonal phase observed in the latter case (not shown).

The observed increase of the elastic modulus and the four-point bending strength can be correlated with shrinkage and sintering effects. In Fig. 6.1.8, the relative decrease in length of the higher stabilized CeO_2-systems is compared for two different heating temperatures (1250 °C and 1400 °C). It turns out that the system containing 35 mol-% CeO_2 shows a higher degree of shrinkage than the system containing 19.5 mol-% CeO_2/1.5 mol-% Y_2O_3.

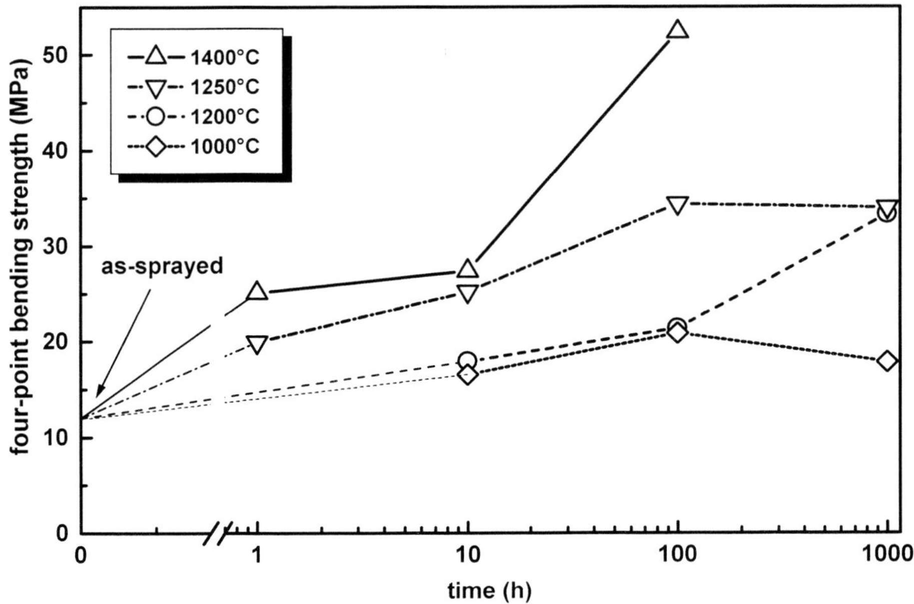

Figure 6.1.6: Four-point bending strength of a TBC containing ~19.5 mol-% CeO_2/1.5 mol-% Y_2O_3 as-sprayed and after heat treatments between 1000 °C and 1400 °C.

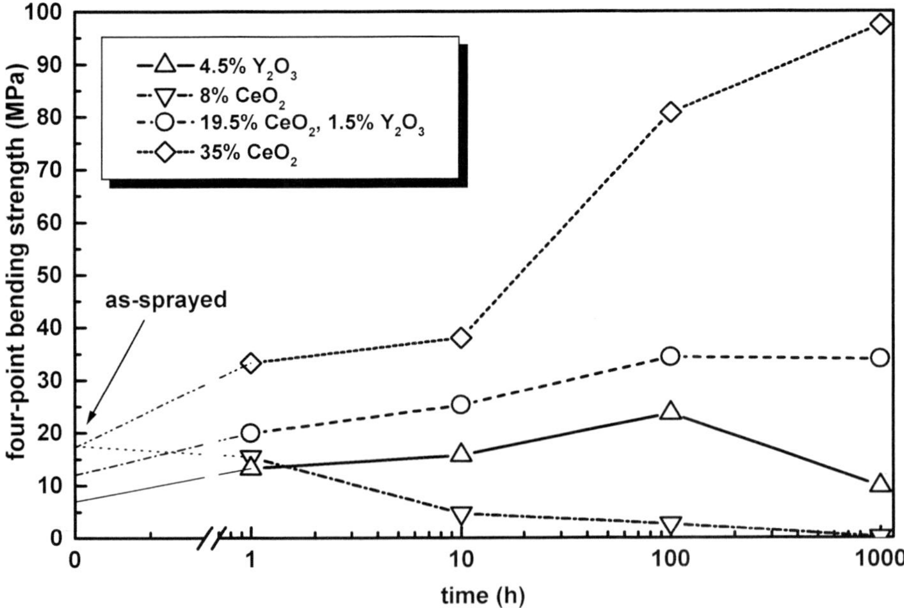

Figure 6.1.7: Four-point bending strength of different TBCs as-sprayed and after heat-treatments for various times at 1250 °C.

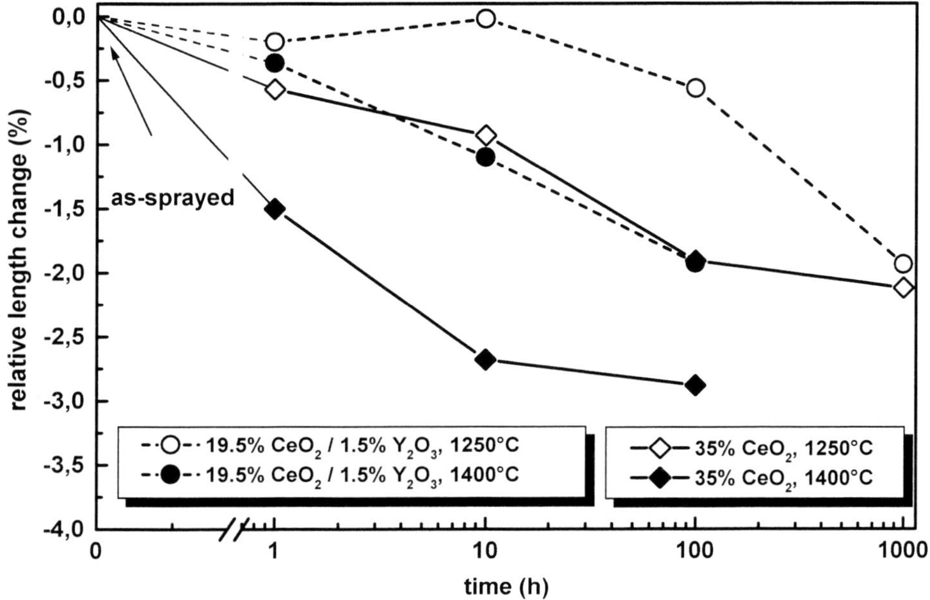

Figure 6.1.8: Shrinkage of TBCs containing ~19.5 mol-% CeO_2/1.5 mol-% Y_2O_3 and ~35 mol-% CeO_2 after heat treatments at 1250 °C and 1400 °C compared to the as-sprayed stage.

Hence, the 35 mol-% CeO_2-TBCs own a higher bending strength as well as a higher elastic modulus (Fig. 6.1.5).

6.1.3.3 Thermal Cycling

Figure 6.1.9 shows the number of cycles performed in burner-rig tests for the conditions 1250 °C/800 °C and 1350 °C/1000 °C. The TBC containing 8 mol-% CeO_2 already failed after 500 cycles under the condition of a maximum surface temperature on the TBC of 1250 °C and 800 °C on the backside of the substrate. The burner-rig test clearly demonstrates that the system containing 8 mol-% CeO_2 is not suitable for applications. XRD measurements before the test and after 500 cycles at 1250 °C/800 °C indicate an increase of the intensity fraction of the $\{111\}_m$ reflections from ~29 % (as-sprayed) to ~82 % (after failure). The increase in monoclinic phase reveals that thermal cycling promotes the t $\rightarrow$ m phase transition. Therefore, it is assumed that the early failure is caused by stresses arising from the phase transformation.

The TBCs containing 1.) 4.5 mol-% Y_2O_3, 2.) 19.5 mol-% CeO_2/1.5 mol-% Y_2O_3, and 3.) 35 mol-% CeO_2 did not fail in burner-rig tests within 1000 cycles under the condition 1250 °C/800 °C. At an increased temperature

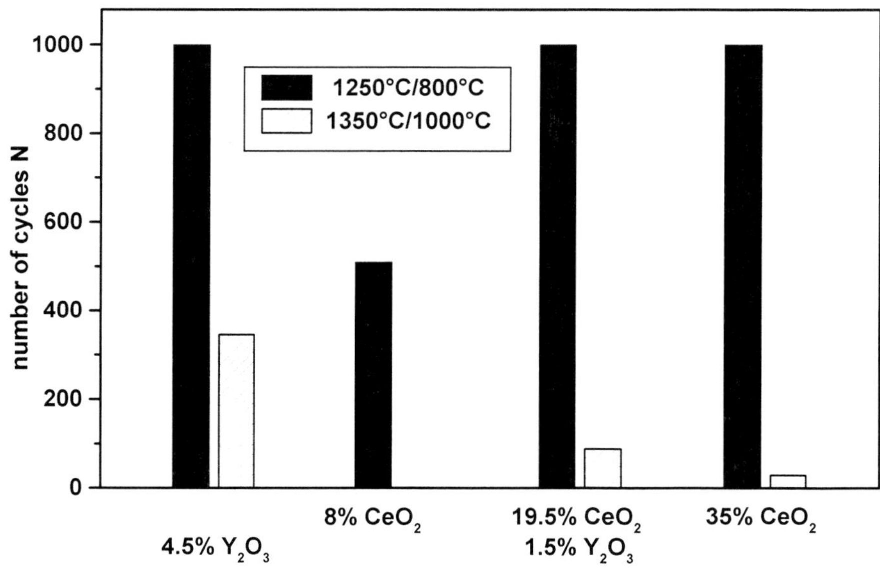

Figure 6.1.9: Number of thermal cycles performed in burner rig-tests for the two conditions 1250 °C (TBC-surface)/800 ° (substrate) and 1350 °C/1000 °C.

(1350 °C/1000 °C), the systems failed after 1.) 350, 2.) 90, and 3.) 30 cycles. The system with 4.5 mol-% Y_2O_3 owned the highest lifetime. However, a direct comparison with the other systems is critical, as the latter ones own a higher thermal conductivity $\lambda(\lambda_{35\,\%\ CeO_2} < \lambda_{\sim 20\,\%\ CeO_2} < \lambda_{4.5\,\%\ Y_2O_3})$ at room temperature [3] and presumably also at higher temperatures. Because of the similar thickness of all TBCs, the thermomechanical stresses are expected to be the highest for the systems with the lowest thermal conductivity.

Figure 6.1.10 shows a typical SEM micrograph of a TBC containing 35 % CeO_2 after failure. One can identify cracks normal and parallel to the interface. Cracks parallel to the interface lead to spallation of the TBC. Obviously, the cracks run along the boundary between two plasma-sprayed layers, probably due to a low cohesive strength between the layers.

Measurements by XRD demonstrated, that the TBC systems containing 4.5 mol-% Y_2O_3 and 19.5 mol-% CeO_2/1.5 mol-% Y_2O_3 did not exhibit an increase of the monoclinic phase content after thermal cycling, indicating the stability of the t′-phase during the test. Obviously, the diffusion rates are too low for the occurrence of the diffusion-controlled transformation t′ → t + c during thermal cycling. As expected, a part of the TBC containing 35 % CeO_2 is converted from the metastable cubic c′-phase ($I_{c'} \approx 99\,\%$) to the tetragonal t′-phase ($I_{t'} \approx 23\,\%$, $I_{c'} \approx 71\,\%$) after thermal cycling at 1250 °C/800 °C (Fig. 6.1.1). In addition, the system shows an increase of the intensity fraction of the monoclinic phase from ~1 % to ~6 % after thermal cycling at these conditions.

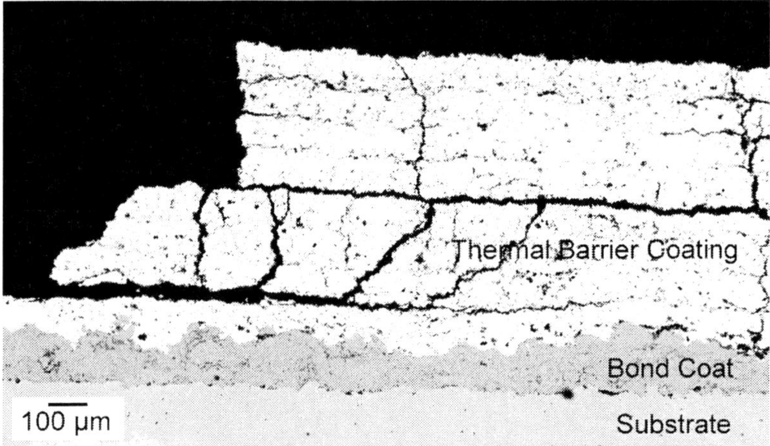

Figure 6.1.10: SEM micrograph of a failured TBC (~35 % CeO_2) after 30 thermal cycles at 1350 °C/1000 °C.

6.1.4 Discussion

The long-term investigations demonstrated clearly that significant changes of the phase composition and mechanical properties of the TBCs occur already after extremely short heat treatments at typical service temperatures in gas-turbines. Figure 6.1.11 represents a tentative time temperature transformation diagram for TBCs containing 35 mol-% CeO_2 derived from XRD measurements as shown in Fig. 6.1.2. Already after extremely short heat treatments, the amount of non-transformable t'-phase is significantly reduced e. g. from ~98 % to ~60 % after 1 h at 1250 °C or to ~74 % even after 1000 h at a relatively low temperature of 1000 °C.

The phase transformation t' → t + c is one of the limiting factors for the lifetime of the TBCs as the stable tetragonal phase t may undergo a transformation to the monoclinic phase m during cooling and lead to a strong decrease of the bending strength and correspondingly promote rapid failure. It could be demonstrated in agreement with previous studies [2] that TBCs with a relatively low CeO_2 content, e. g. 8 mol-% are not suitable at all for applications due to t → m transformations. Those TBCs do not contain the non-transformable t'-phase but the transformable t phase (Fig. 6.1.1). The result from XRD that the higher stabilized CeO_2-TBCs showed a smaller degree of t → m transformation compared to the 4.5 mol-% Y_2O_3-stabilized system suggests that the first ones may own a higher potential for long-term applications and therefore be of important technological interest. As the amount of t → m transformation observed after the maximum heating times of ~1000 h is small in the higher stabilized CeO_2-TBCs, further

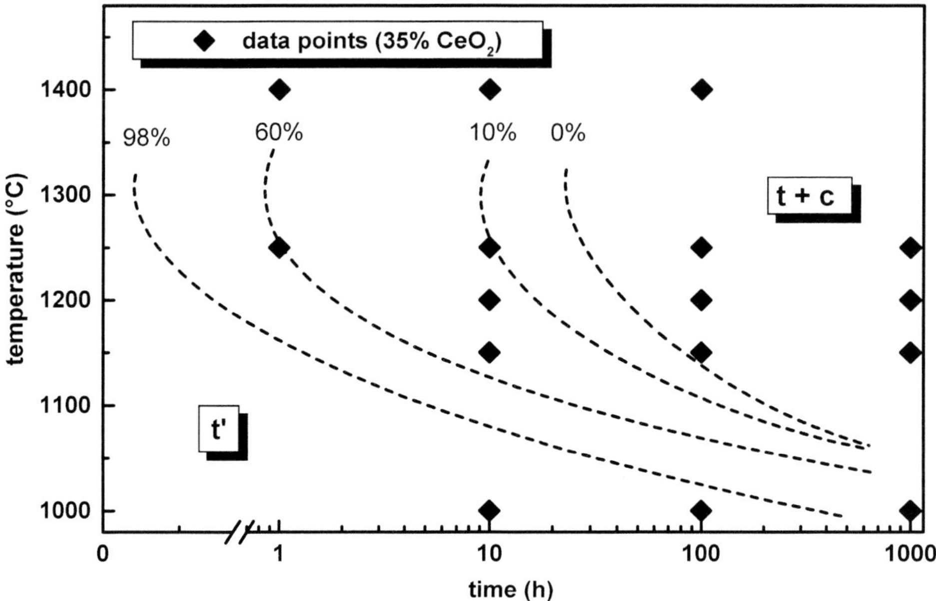

Figure 6.1.11: Tentative time temperature transformation (TTT) diagram of the phase transformation t' → t + c within the system containing ~35 % CeO_2. The amount of t'-phase is shown.

experimental investigations are necessary to evaluate the long-term behaviour. It is not possible yet to extrapolate to typical operation times in the range of ~40 000 h. In the case of the 4.5 mol-% Y_2O_3-stabilized system, according to the high amount of monoclinic phase e.g. after 1000 h at 1250 °C, the maximum operation conditions should not much exceed this time/temperature values. A tendency to failure becomes obvious by the extremely low bending strength after 1000 h at 1250 °C (Fig. 6.1.7). TBCs with a higher stabilizer content of 35 mol-% CeO_2 reveal a lower thermal conductivity compared to TBCs with lower CeO_2 contents [3] as for example the system with 19.5 mol-% CeO_2/1.5 mol-% Y_2O_3, however, they show a higher degree of shrinkage and a higher elastic modulus after heat treatments and therefore a lower strain tolerance (Fig. 6.1.5). In addition, the 35 mol-% CeO_2-stabilized TBCs showed the higher fraction of monoclinic phase after thermal cycling.

Regarding thermal cycling, the 4.5 mol-% Y_2O_3-stablized TBCs showed the highest lifetime. However, further experimental investigations taking into account the same ratio of thickness and thermal conductivity for all materials would be necessary to clarify the influence of different thermal conductivities on the observed life-time. As shown in Section 6.1.3.3, failure in thermal cycling might not only be promoted by a high degree of t → m transformation, as in the case of the system containing 8 mol-% CeO_2, but also due to a low cohesive strength between the plasma-sprayed layers (Fig. 6.1.10): The other investigated systems did not exhibit significant increases of the monoclinic phase after failure.

The experimental investigations of phase changes in long-term heat treated TBCs have been often neglected in previous studies that focused only on changes during thermal cycling [2]. In practical use, e. g. in gas turbines, both factors are important: 1.) a high resistivity to thermal cycling and 2.) a high lifetime during extended operation times at high temperatures.

The experimental results of phase transformations and the related changes in mechanical properties could be obtained independent of any interaction with a substrate or bond coat material, as the TBCs were separated from their substrates. The next more complicated step must include the influence of interactions at TBC-bond coat/substrate couples. However, for this purpose TBCs have to be heated from the top side only and cooled from the backside as the operation temperatures are well above the application temperatures of the metal alloy substrates. An advantage of this measurement would be that the shrinkage of the TBCs due to sintering effects is restricted by the substrate. This should have an influence on the mechanical properties.

6.1.5 Conclusion

Phase composition and mechanical properties of ZrO_2-based plasma-sprayed TBCs are significantly altered even after relatively short heat treatments at typical service temperatures. TBCs with 8 mol-% CeO_2 show a high degree of transformation of the tetragonal phase t to the monoclinic phase m and therefore are not suitable for long-term applications. TBCs consisting almost entirely of the non-transformable t'-phase and/or the c'-phase showed a higher stability during thermal cycling. However, heat treatments at typical service temperatures lead within relatively short times to a diffusion-controlled phase transformation of the t'-phase to the tetragonal and cubic equilibrium phases. For the systems investigated, the amount of t $\rightarrow$ m transformation of the tetragonal phase was smaller in the case of the higher stabilized CeO_2-TBCs (19.5 mol-% CeO_2/1.5 mol-% Y_2O_3 or 35 mol-% CeO_2) than in the 4.5 mol-% Y_2O_3-stabilized TBCs suggesting a higher potential for long-term-applications of the CeO_2-stabilized systems.

Acknowledgements

The authors would like to thank E. Geiger and V. Pinjo for their assistance during a part of the experimental investigations and Dr. M. Alaya for the initial work on this topic. We also thank Dr. M. Sommer and Dr. A. Kranzmann from the company ABB for providing plasma-sprayed specimen. Financial support from the Collaborative Research Centre 167 is gratefully acknowledged.

References

1. Heimann, R.B. (1996): Plasma-Spray Coating: Principles and Applications, VCH (Weinheim, New York), 209–224.
2. Harmsworth, P.D., Stevens, R. (1991): Microstructure and phase compositions of ZrO_2-CeO_2 thermal barrier coatings, *J. Mater. Sci.* **26**, 3991–3995.
3. Sodeoka, S., Suzuki, M., Ueno, K., Sakuramoto, H., Shibata, T., and Ando, M. (1997): Thermal and Mechanical Properties of ZrO_2-CeO_2 Plasma-Sprayed Coatings, *Journal of Thermal Spray Technology* **6**, 361–367.
4. Alaya, M., Oberacker, R., Hoffmann, M.J., Krebs, W., Koch, R., Wittig, S. (1997): Thermophysikalische Eigenschaften von CeO_2- und Y_2O_3-stabilisierten ZrO_2-Wärmedämmschichten und ihre Auswirkung auf das thermozyklische Verhalten und den Strahlungswärmeübergang, in: Tagungsband Werkstoffwoche '96, Symposium 3, Werkstoffe für die Energietechnik, edited by H.W. Grünling (DGM Informationsgesellschaft mbH, Frankfurt), 243–248.
5. Alaya, M. (1997): Bewertung und Optimierung von Konzepten zur Verbesserung des Einsatzverhaltens von ZrO_2-Wärmedämmschichten, PhD thesis, Universität Karlsruhe.
6. Lee, E.Y., Biedermann, R.R., and Sisson, R.D., Jr. (1989): Effect of Thermal Cycling on the Microstructure of Ceria-stabilized Zirconia Thermal Barrier Coatings, *Microstructural Science* **17**, 505–513.
7. Coyle, T., Coblenz, W., and Bender, B.A. (1983): Toughness, Strength and Microstructures of Sintered CeO_2-doped ZrO_2 Alloys, *Amer. Ceram. Soc. Bull.* **62**, 966.
8. Tsukuma, K., Shimada, M. (1985): Strength, Fracture Toughness and Vickers Hardness of CeO_2-Stabilized Tetragonal ZrO_2 Polycrystals (Ce-TZP), *J. Mater. Sci.* **20**, 1178–1184.
9. DeMasi, J.T., Oritz, M., and Sheffler, K.D.(1989): Thermal Barrier Coating Life Prediction Model Development, Phase I Final Report, Contract NAS3-23944, NASA CR-182230.
10. Crostack, H.-A., Beller, U., Steffens, H.-D., and Gramlich, M. (1995): Herstellung von dicken Wärmedämmschichten und das Schädigungsverhalten bei thermischer und mechanischer Belastung, *Fortschrittsberichte der Deutschen Keramischen Gesellschaft* **10**, 11–21.
11. Suhr, D.S., Mitchell, T.E., and Keller, R.J. (1984): Microstructure and Durability of Zirconia Thermal Barrier Coatings, *Advances in Ceramics*, 503–516.

12. Yashima, M., Morimoto, M., Ishizawa, N., and Yoshimura, M. (1993): Cubic-Tetragonal Phase Transformations in the ZrO_2-CeO_2 System, in: Science and Technology of Zirconia V, edited by S. P. S. Badwal, M. J. Bannister, and R. H. Hannink, (Technomic Publishing Co., Inc., Lancaster), 108–116.
13. Lee, E. Y., Biedermann, R. R. and Sisson, R. D., Jr. (1989): Diffusional Interactions and Reactions between a Partially Stabilized Zirconia Thermal Barrier Coating and the NiCrAlY Bond Coat, *Materials Science and Engineering*, **A 121**, 467–473.
14. Ritter, H., Ihringer, J., and Maichle, J. (1995): XPLOT, Refinement Program for X-ray data, Institut für Kristallographie, Universität Tübingen.
15. Toraya, H., Yoshimura, M., and Somiya, S. (1984): Calibration Curve for Quantitative Analysis of the Monoclinic-Tetragonal ZrO_2 System by X-ray diffraction, *J. Am. Ceram. Soc.* **67**, C119–C121.
16. Schmid, H. K. (1987): Quantitative Analysis of Polymorphic Mixes of Zirconia by X-ray diffraction, *J. Am. Ceram. Soc.* **70**, 367–376.

6.2 The Creep Damage Behaviour of a Plasma-Sprayed Thermal Barrier Coating System for Combustion Chambers

Uli T. Schmidt, Otmar Vöhringer, Detlef Löhe,
and Eckard Macherauch*

Abstract

Creep tests were carried out on the thermal barrier coating system (TBC) NiCr22Co12Mo9-NiCoCrAlY-ZrO_2/7 % Y_2O_3. During creep under tensile loading the nickel base substrate imposes deformation on the deposited TBC. Strain accommodation of the TBC is attained by means of crack initiation, crack opening, crack propagation, or sliding of adjacent crack faces. The TBC can respond to creep deformation by segmentation (crack propagation perpendicular to the surface) or/and spallation (crack propagation parallel to the substrate-coating interface). The probability of either one is governed by the ratio of maximum shear stress (spallation) to maximum normal stress (segmentation) which itself is mainly influenced by the thermal barrier coating thickness. It can be stated that spallation failure probability increases with increasing creep rate, creep deformation, and layer thickness. The presence of pores between single spraying layers also strongly augments the likelihood of spallation. No significant influence of temperature on spallation failure probability can be found in the range from 850 °C to 1050 °C. A relationship between microstructural changes in the TBC and the emission of acoustic signals recorded during creep is presented.

* Institut für Werkstoffkunde I, Universität Karlsruhe, Kaiserstr. 12, 76128 Karlsruhe, Germany

6.2.1 Introduction

Plasma-sprayed thermal barrier coating systems composed of MCrAlY (M for Ni, Co, Fe) bond coat and a ZrO_2/Y_2O_3 thermal barrier coating (TBC) are now a state-of-art feature in modern combustion chambers for turbines in both, aviation and power generation. Although much empirical knowledge is available about the optimisation of the spray process in respect to the behaviour under thermal cycling or thermo-shock, little is known about the failure mechanisms which determine the degradation process of these coating systems. However, an understanding of these failure mechanisms is necessary for the modelling of coating degradation and lifetime prediction of coated engine parts. Therefore, elementary creep tests were carried out in order to separate basic parameters which influence the damage process and to determine their degree of influence.

6.2.2 Experimental Details

Failure behaviour of the ceramic TBC (ZrO_2, partially stabilized with 7 weight-% Y_2O_3), was studied in creep tests at 850 °C, 950 °C, and 1050 °C in air at constant tensile load resulting in creep rates of approximately $\dot{\varepsilon} = 5 \cdot 10^{-5}\ s^{-1}$ (fast) and $\dot{\varepsilon} = 5 \cdot 10^{-7}\ s^{-1}$ (slow). After creep strains to $\varepsilon_p = 1\,\%$, 3 %, and 5.5 %, respectively, using traverse microsections of the gauge length, the crack densities were determined by means of an intercepted-segment method on light microscope and scanning electron microscope micrographs. The TBC parameter variegated were ceramic thermal barrier layer thickness (300 μm, 500 μm, 1500 μm) and spraying process parameters. Furthermore, accompanying investigations using an acoustic emission technique were performed to obtain more information of the damage behaviour of the TBC.

6.2.3 Results and Discussion

During creep loading metallic substrates impose deformation on deposited ceramic thermal barrier coatings. Strain accommodation of the TBC is not attained by plastic deformation, but by means of crack initiation, crack opening, crack propagation, or sliding of adjacent crack faces. Thermal barrier coatings can respond to creep deformation by segmentation or spallation, the latter being referred to as failure [1, 2].

In Figs. 6.2.1 to 6.2.3 it is shown that segmentation and spallation can either occur separately or in a combined mode. Provided a strong bond between bond coat and TBC and between single layers of the TBC the coating system responds to creep strain in a mode depicted in Fig. 6.2.1. Here, an increase in creep strain (Fig. 6.2.1a–c) leads to the formation and growth of cracks perpendicular to both, the coating plane and the strain direction.

The cracks perpendicular to the surface visible in Fig. 6.2.2 are caused by the spraying process and the subsequent cooling and temperature equalization rather than by creep deformation of the substrate. During spraying, usually performed in several layers, a temperature gradient develops which leads to a higher thermal contraction in the outermost layers and thus to a thermally induced crack initiation and crack growth. Despite of these cracks

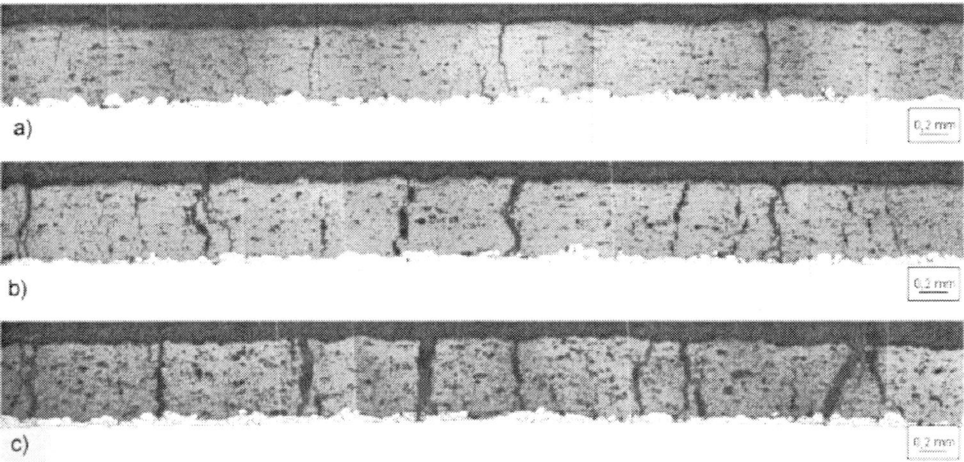

Figure 6.2.1: Pure segmentation failure mode of plasma-sprayed TBC (500 μm thickness, $T = 1050\,°C$, $\sigma_n = 35$ N/mm^2). a) $\varepsilon = 1\,\%$, b) $\varepsilon = 3\,\%$, and c) $\varepsilon = 5.5\,\%$.

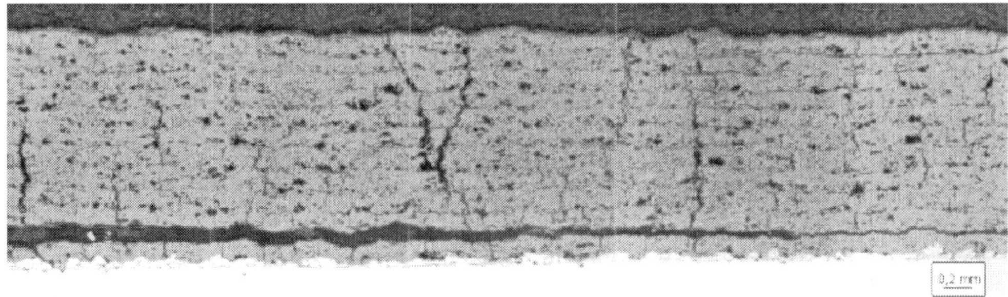

Figure 6.2.2: Pure spallation failure mode of plasma-sprayed TBC (1500 μm thickness, $T = 850\,°$, $\sigma_n = 100$ N/mm^2, $\varepsilon = 5.5\,\%$).

the coating system in Fig. 6.2.2 responds to creep deformation of the substrate by initiation and propagation of a single crack along a row of pores between the first and second layer of the TBC leading to a complete loss of the TBC above it.

A combination of these two extreme modes of failure is shown in Fig. 6.2.3. In this combined mode the cracks already existing after the spraying process propagate from the surface towards the bond coat/TBC interface. The distance between these cracks is not a function of creep conditions but of spraying conditions. Thus, as an adaptation to creep temperature, creep rate, and creep strain, additional cracks originate at the bond coat/TBC interface and propagate to the surface. Final failure, i. e. spallation, occurs pro-

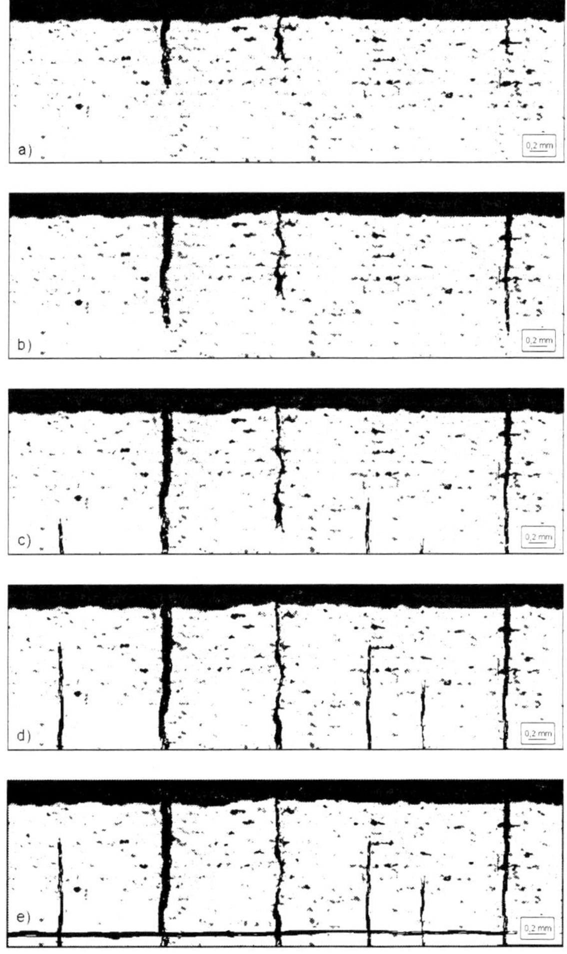

Figure 6.2.3: Schematic depiction of combined segmentation/spallation failure mode of plasma-sprayed TBC.

vided the stress relief achieved by perpendicular cracks does not insure a subcritical shear stress distribution in the first/second layer interface.

It was found that increasing TBC thickness, creep strain, and creep rate lead to a higher probability of spallation. Quantitative statements cannot be made due to the scatter in microstructure of the specimens. No influence of temperature on spallation failure probability was observed.

Quantitative information about the development of perpendicular cracks can be obtained from Fig. 6.2.4 for a 500 μm TBC. Crack density values were calculated by normalizing the number of intersection points of cracks and an intersection line to the length of this line. With increasing creep strain the density of cracks visible at 50-fold magnification, in the following referred to as macro-cracks, increases tending to a saturation value. Only in very few cases did spallation occur in TBC of 500 μm thickness after 5.5 % creep strain.

The failure behaviour of a TBC of 1500 μm thickness is shown in Fig. 6.2.5. Distinction is made between big macro-cracks clearly separating two neighbouring segments and small macro-cracks within a single segment and also between the locations near the surface (top), mid-coating thickness (centre), and near the bond coat/TBC interface (bottom).

At the top position as the density of big macro-cracks (Fig. 6.2.5b) increases with the same rate as the density of small macro-cracks (Fig. 6.2.5a) decreases, it can be stated that no crack initiation takes place, which causes a constant total macro-crack density (Fig. 6.2.5c) with increas-

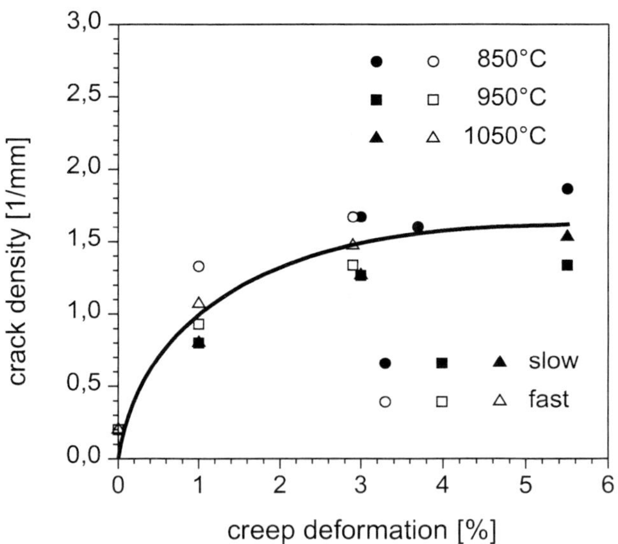

Figure 6.2.4: Macro-crack density as a function of creep strain for different creep temperatures and creep rates (500 μm thickness, slow $\triangleq \dot{\varepsilon} = 5 \cdot 10^{-7}\ \mathrm{s}^{-1}$, fast $\triangleq \dot{\varepsilon} = 5 \cdot 10^{-5}\ \mathrm{s}^{-1}$).

ing creep strain. Near the bond coat/TBC interface strain accommodation can only be obtained by formation of additional cracks, as can be seen in a steep increase in small macro-crack density (Fig. 6.2.5g) with increasing creep strain. This formation of additional small macro-cracks also causes an increase in total macro-crack density (Fig. 6.2.5i), whereas the density of big macro-cracks (Fig. 6.2.5h) shows the same dependence on creep strain as in the top section.

In the centre section the density of small macro-cracks (Fig. 6.2.5d) remains nearly constant. Therefore, new small macro-cracks must form mainly as small macro-cracks from the bottom section propagating upwards

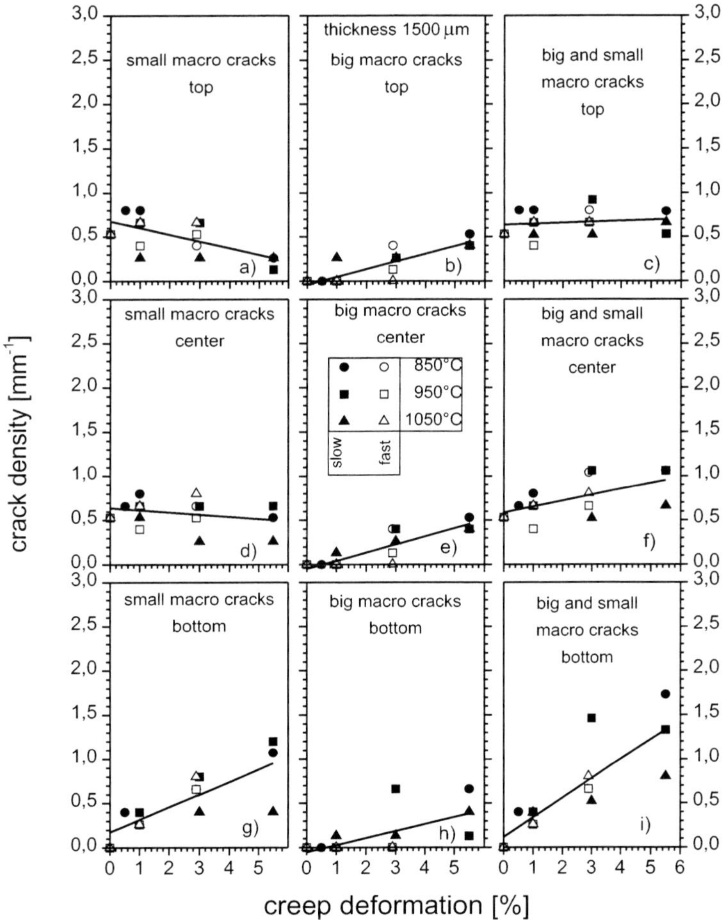

Figure 6.2.5: Small (a, d, g) and big (b, e, h) macro-crack density at the top (a–c), centre (d–f) and bottom (g–j) section as a function of creep strain for different creep temperatures and creep rates (1500 μm thickness, slow $\triangleq \dot{\varepsilon} = 5 \cdot 10^{-7}$ s^{-1}, fast $\triangleq \dot{\varepsilon} = 5 \cdot 10^{-5}$ s^{-1}).

to the centre to compensate the small macro-cracks turning into big macro-cracks (Fig. 6.2.5e).

Generally speaking it is observed that strain accommodation is accomplished by means of macro-crack initiation and macro-crack growth as well as macro-crack opening which causes stress relief in the sections between the macro-cracks. As a consequence, micro-cracks visible at 500-fold magnification within these sections are stress free. Figure 6.2.6 shows the density of micro-cracks parallel and perpendicular to the surface as a function of creep strain. To separate the influence of the heating process and temperature equalization time prior to the application of the creep load from the influence of creep strain, crack densities are normalized by subtracting the micro-crack density after 1 % creep strain. Within a considerable scatter band the general tendency can be found that the micro-crack density both, in parallel and perpendicular direction decreases with increasing creep strain.

These findings which are obviously caused by sintering processes can also be observed within thermal barrier coatings in the as received condition detached by etching in hydrochloric acid as shown in Fig. 6.2.7. The micro-crack density decreases with increasing annealing time and temperature even at a temperature as low as 850 °C.

The afore mentioned microstructural processes such as crack initiation, crack propagation, and also stick-slip movements of spalled segments relative to the substrate are all discontinuous processes at which elastic energy is set free at discrete times.

Thus an acoustic emission system was set up to detect, record, and analyse the emission of the elastic energy. In order to locate the sources of acous-

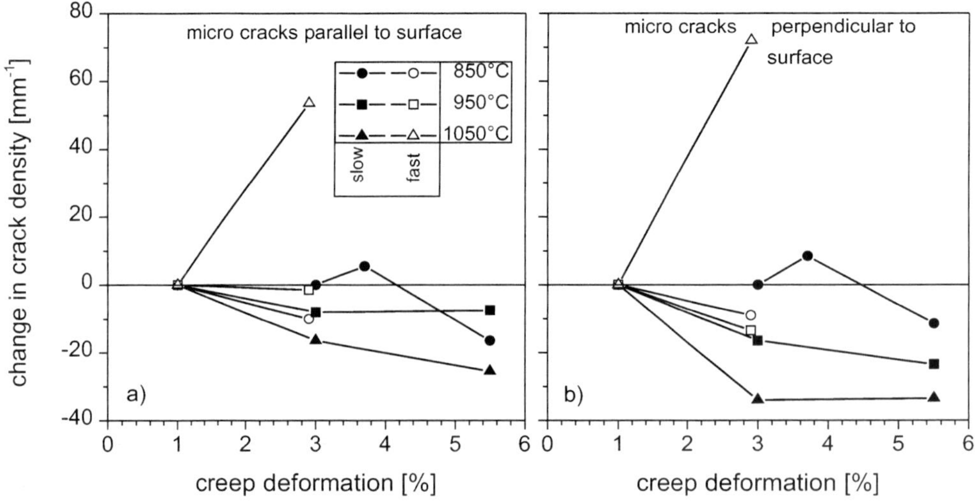

Figure 6.2.6: Change of micro-crack density as a function of creep strain for different creep temperatures and creep rates (500 μm thickness, slow $\triangleq \dot{\varepsilon} = 5 \cdot 10^{-7}$ s^{-1}, fast $\triangleq \dot{\varepsilon} = 5 \cdot 10^{-5}$ s^{-1}).

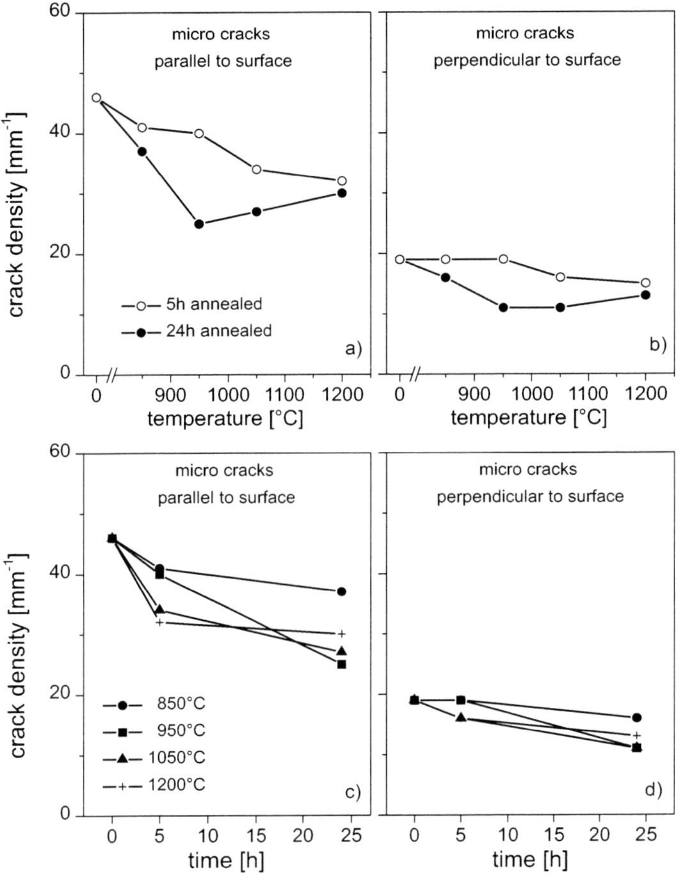

Figure 6.2.7: Micro-crack density of detached TBC with 500 μm thickness for different annealing times as a function of annealing temperature.

tic emission a linear chain arrangement of two piezoelectric transducers was coupled to the water cooled heads of the creep specimens using roller bearing grease as couplant. After the creep tests the recorded signals were analysed using logical filters such as location filters to separate acoustic signals which originated from the thermal barrier coating from acoustic or electromagnetic noise. Several characteristic parameters were used to analyse the acoustic emission data. A visualization of these parameters is given in Fig. 6.2.8.

A selection of cumulative frequencies of the parameter rise time, i. e. the time elapsed from the first threshold crossing to the arrival of the maximum peak, is shown in Fig. 6.2.9. Cumulative frequencies of the total signals from beginning of the creep test i. e. after complete application of creep load to 1 % (solid line), 3 % (dashed line) and 5.5 % (dotted line) are pres-

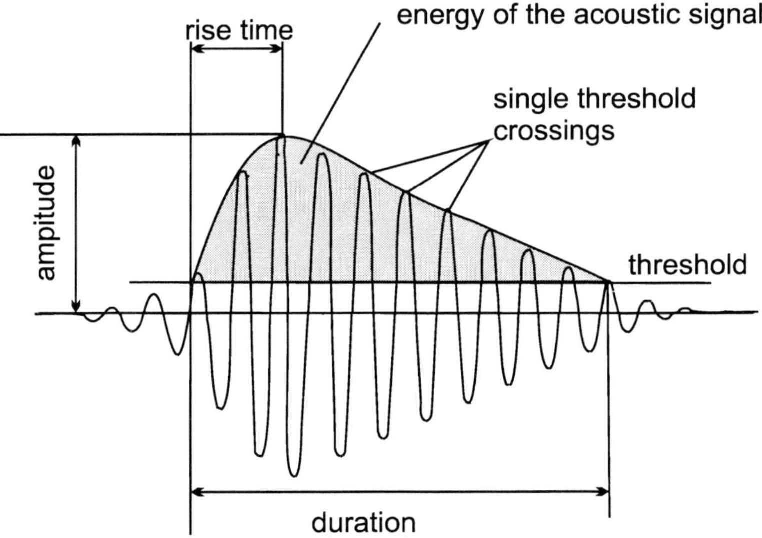

Figure 6.2.8: Visualization of characteristic parameters to describe acoustic emission data.

ented for different temperatures and TBC thickness. It can be stated that an increase in creep strain causes an increase in rise time as the cumulative frequency shifts to higher rise time values. This increase in rise time is most significant for creep tests at 850 °C or 1050 °C and thin (300 µm) or thick (1500 µm) thermal barrier coatings.

The changes in acoustic emission characteristics are explained by the changes in microstructural damage behaviour during creep. Both, segmentation close to the point of saturation and spallation cause a decrease in crack initiation and crack propagation activity. These acoustic events are related to single distinguished signals graphically described as a click. As this acoustic activity fades other events come to the fore such as a stick-slip movement of adjacent crack faces parallel to the surface. This grating stick-slip movement emits cascades of single bursts that cannot be distinguished. Thus the time between first threshold crossing and arrival of the highest amplitude oscillation of a subsequent burst can be significantly high, which leads to a shift of the rise time distribution to higher values.

The increase of the frequency of the low rise time signals in the creep test of the 1500 µm TBC systems at 1050 °C (Fig. 6.2.9i) is related to cracking of the oxide scale growing underneath the spalled TBC. The growth rate of the oxide scale is highest because under these conditions (high temperature/thick TBC) premature spalling causes a loss of oxidation protection by the TBC and the oxidation attack is strongest.

Creep strain of the substrate imposes stresses on applied thermal barrier coatings. However, these stresses are relatively low on a macroscopic scale

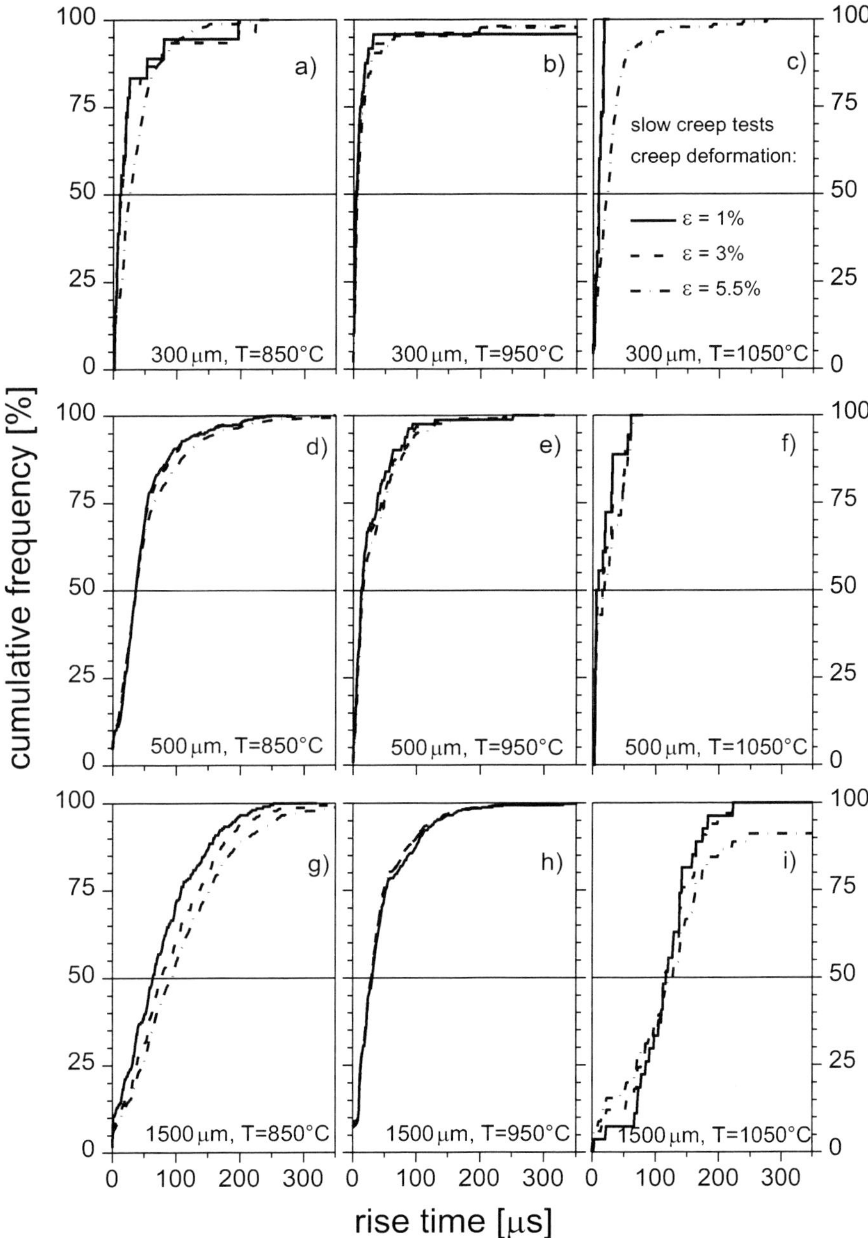

Figure 6.2.9: Cumulative frequencies of rise time distributions as a function of creep strain for different creep temperatures and TBC thickness ($\dot{\varepsilon} = 5 \cdot 10^{-7}$ s^{-1}).

compared to the creep stresses in the substrate because of the high density of pores, micro and macro-cracks in plasma-sprayed coatings and, therefore, do not influence the stress distribution in the substrate.

With this presumption the boundary conditions at the substrate coating interface can be formulated that the displacement of each point of the coating is equal to the creep displacement of the substrate. With increasing creep strain of the substrate a distribution of shear stresses and a distribution of direct stresses develop. These stresses are proportional to creep strain. The maximum shear stress is found at the end of the coated section or when segmentation already took place at the edge of segments, maximum direct stress is found in the symmetry plane of the coated section or the formed segments.

Comparison of the ratio of maximum shear stress to maximum direct stress and the ratio of the critical shear stress for mode II cracking, i. e. spallation, to the critical direct stress for mode I cracking, i. e. segmentation, has to be made in order to gain information which failure mode is more likely to occur.

A first approach to calculate the shear stress to direct stress ratio was done on the assumption that direct stress is constant in a cross section perpendicular to the strain direction. An analysis of the equilibrium of forces of a cut free element of the TBC leads to a differential equation already known in the field of fibre reinforced materials [3]:

$$\frac{\partial^2 \sigma}{\partial x^2} = \frac{2\,G}{t^2}\left(\frac{\sigma}{E} - \varepsilon_0\right) \tag{1}$$

or

$$\frac{\partial^2 \tau}{\partial x^2} = \frac{2\,G}{t^2}\,\frac{\tau}{E} \tag{2}$$

where x = distance from the symmetry plane of the coated section or segment, t = thickness of the TBC, ε_0 = creep strain of the substrate, E = Young's modulus of the TBC and G = shear modulus of the TBC. Equations (1) and (2) can be solved by using

$$\sigma(x) = \sigma_0 \cosh(ax) + c \tag{3}$$

and

$$\tau(x) = \sigma_0\, t\, a \sinh(ax) \tag{4}$$

with σ_0, a, and c as integration constants.

Thus the maximum stress ratio becomes

$$\frac{\tau_{max}}{\sigma_{max}} = \frac{\tau\left(x = \frac{w}{2}\right)}{\sigma\,(x = 0)} = \frac{t\,a \sinh \frac{aw}{2}}{1 + \frac{c}{\sigma_0}} \tag{5}$$

with w = width of the coated section or segment.

The boundary condition

$$\sigma\left(x = \frac{w}{2}\right) = 0 \tag{6}$$

leads to

$$\frac{c}{\sigma_0} = -\cosh\left(\frac{aw}{2}\right) \tag{7}$$

Substituting Eq. (7) in Eq. (5):

$$\frac{\tau_{\max}}{\sigma_{\max}} = \frac{t\ a \sinh \dfrac{aw}{2}}{1 - \cosh \dfrac{aw}{2}} \tag{8}$$

From Eq. (7) it can be seen that the maximum shear stress to maximum direct stress increases with increasing TBC thickness. Further statements about the maximum stress ratio are difficult because the integration constant *a* shows a high order dependence on both, thickness and width.

The strong dependence of the maximum stress ratio on TBC thickness found in the analytical approach is confirmed by the experimental findings. An increase in TBC thickness strongly augments the fault probability of spallation. Another experimental finding is that spallation occurs along a row of pores between the first and second spraying layer. This can be discussed as a locally very low critical shear stress to critical direct stress ratio. The influence of creep rate on spallation failure probability has to be explained by the ability of the bond coat to relax the shear stress peaks due to its very low creep strength. This effect is not included in the basic stress model presented. The missing temperature influence is either caused by a constant critical shear stress to critical direct stress ratio or is the result of the interplay of different mechanisms such as change in critical stress ratio, maximum stress ratio, oxidation, and creep strength of the bond coat.

As long as segmentation is the governing failure mode an increase in creep strain causes an increase in direct stress above the critical value relieved by cracking. With decreasing segment width the bond coat gains more and more the ability to relax the stresses in the TBC so that a saturation value as shown in Fig. 6.2.4 is aspired.

The investigation shows that stress relief is basically achieved by means of macro-cracking and that the influence of pores and micro-cracks has to be discussed as an influence on macroscopic Young's modulus and shear modulus as well as an influence on critical direct stress and critical shear stress. The latter is quite obvious when looking at the rows of pores between single spraying layers as a major location of severe spalling.

As stress is relieved by macro-cracks the observed sintering processes even in specimens under creep load can be understood. This finding is supported by investigations of Thurn [4, 5] who detected a shrinkage of detached plasma-sprayed TBC in the first heating cycle in a differential dilatometer.

Acknowledgements

The support of the investigations by the DFG is gratefully acknowledged.

References

1. Schmidt, U. (1998): Das Kriechverformungs-, Zeitstand- und Versagensverhalten plasmagespritzter Wärmedämmschichtverbunde, Dissertation Universität Karlsruhe (TH).
2. Schmidt, U. T., Vöhringer, O., Löhe, D. (1999): The Creep Damage Behaviour of the Plasma-sprayed Thermal Barrier Coating System NiCr22Co12Mo9-NiCoCrAlY-ZrO_2/7 % Y_2O_3, Trans. *ASME J. of Engineering for Gas Turbines and Power* **121**, 678–682.
3. Holister, G. S., and Thomas, C. (1966): Fibre Reinforced Materials, Elsevier Publishing Co. LTD. Amsterdam, p. 15.
4. Thurn, G., Schneider, G. A., and Aldinger, F. (1997): High-Temperature Deformation of Plasma-Sprayed ZrO_2 Thermal Barrier Coatings, *Materials Science and Engineering* **A233**, 176–182.
5. Thurn, G. (1997): Hochtemperatureigenschaften und Schädigungsverhalten plasmagespritzter ZrO_2-Wärmebarrieren, Dissertation Universität Stuttgart.

7 Projects, Organization, Structure, Members and Participants of the Collaborative Research Centre 167

7.1 Research Projects

Projectfield A: Brennstoffaufbereitung, Verbrennung und Schadstoffbildung

Project No.	Title	Project Leader	Funding period
A1	Aerodynamische Untersuchungen an kalten Strömungsmodellen zur Optimierung von Flammenstabilisierungsmechanismen	B. Lenze	1984–1987
A1/A2	Wechselseitige Beeinflussung von Strömungs-, Mischungs- und Reaktionsfeld bei staukörper- und drallstabilisierten Flammen im Bereich des Anhebens und Verlöschens	B. Lenze	1987–1989
A1/2	Wechselseitige Beeinflussung des Strömungs-, Mischungs- und Reaktionsfeldes bei staukörperstabilisierten Flammen: Übergang zu eingeschlossenen und Typ II Flammen	B. Lenze	1990–1992
A2	Einfluss von Reaktion auf aerodynamische Stabilisierungsmechanismen	B. Lenze	1984–1987
A3	Flüssigbrennstoffzerstäubung mit Einstoffdüsen im Hinblick auf deren Einsatz für industrielle Flammen	W. Leuckel	1990–1992
A4	Brennstoffzerstäubung in Düsen mit geringen Abmessungen und Brennstoffdurchsätzen – Optimierung der Parameter bei luftgestützter Zerstäubung unter Laständerung	K. Sakbani S. Kim B. Noll	1984–1989
A4	Brennstoffaufbereitung durch Zerstäubung und Verdunstung schubspannungsgetriebener Flüssigkeitsfilme bei erhöhten Drücken und Temperaturen	B. Noll S. Kim J. Himmelsbach	1990–1995
A5	Grenzen der Abmessungen von Ring- und Rohrbrennkammer kleiner Baugröße – Abstimmung von Zerstäubung und Strömungsparametern bei hohen Drücken und Temperaturen: Sprühstrahlausbreitung	S. Wittig A. Schulz K. Sakbani M. Scheurlen J. Himmelsbach	1984–1995
A6	Untersuchungen zur Rußbildung und zum Rußabbrand unter den Bedingungen industrieller Kohlenwasserstoff-Flammen	W. Leuckel	1984–1987
A6	Untersuchungen zum Rußwachstum und zum Rußabbrand unter den Bedingungen industrieller Kohlenwasserstoff-Flammen	W. Leuckel	1987–1989

Project No.	Title	Project Leader	Funding period
A6	Untersuchungen zum Rußwachstum unter den Bedingungen industrieller Diffusionsflammen	W. Leuckel	1990–1992
A6	Untersuchungen zum Rußwachstum und Rußabbrand unter den Bedingungen industrieller Diffusionsflammen	W. Leuckel	1993–1995
A6	Untersuchungen zum Rußwachstum und Rußabbrand unter den Bedingungen industrieller Diffusionsflammen Kennwort: Rußbildung und -oxidation	W. Leuckel	1996–1997
A7	Einfluss der Tropfengrößenverteilung von Brennstoffdüsen auf die Brennkammer-Abgasemissionen: Überprüfung und Anpassung charakteristischer Zeitmodelle	S. Wittig R. Zahoransky B. Noll	1984–1988
A8	Stickoxidbildung aus brennstoffgebundenem Stickstoff in industriellen Verbrennungssystemen	W. Leuckel	1984–1987
A8	Stickoxidbildung aus brennstoffgebundenem Stickstoff in industriellen Diffusionsflammen unterschiedlicher Strömungs- und Reaktionsstruktur	W. Leuckel	1987–1989
A8	Experimentelle Bestimmung und reaktionskinetische Modellierung der NO-Bildung aus brennstoffgebundenem Stickstoff in industriellen Diffusionsflammen	W. Leuckel	1990–1995
A8	Modellierung der NO_x-Bildung in technischen Flammen unter dem Einfluss turbulenter Schwankungen von Temperatur und Stöchiometrie Kennwort: Stickstoffoxidbildung	W. Leuckel B. Lenze	1996–1998
A9	Rechenmodell für aerodynamisch stabilisierte Flammen	H. Eickhoff W. Leuckel	1984–1995
A9	Modellierung und experimentelle Validierung von eingeschlossenen Vormisch- und Diffusionsdrallflammen Kennwort: Reaktionsmodellierung	H. Eickhoff B. Lenze W. Leuckel	1996–1998
A10	Einfluss der Tangentialgeschwindigkeit auf den turbulenten Impuls- und Stoffaustausch in eingeschlossenen Drallströmungen, Messungen turbulenter Austauschgrößen und Weiterentwicklung des physikalischen Turbulenzmodells	B. Lenze	1987–1989

Project No.	Title	Project Leader	Funding period
A10	Feldmessungen der turbulenten Austauschgrößen und der mittleren Geschwindigkeiten in Modellen von Drallbrennkammern zur Validierung von Turbulenzmodellen	B. Lenze	1990–1992
A10	Modifizierung von Turbulenzmodellen zur Beschreibung von eingeschlossenen Drallströmungen auf der Basis von Feldmessungen der turbulenten Austauschgrößen	B. Lenze	1993–1995
A10	Einfluss der drallgesteuerten Intermittenz auf den turbulenten Impulsaustausch in Drehströmungen	B. Lenze	1996–1997
A11	Untersuchungen zur Stabilität und zum Ausbrandverhalten von stauscheiben- und drallstabilisierten Ölflammen	W. Leuckel B. Lenze	1993–1995
A11	Untersuchungen zur Stabilität und zum Ausbrandverhalten von drallstabilisierten Flüssigbrennstoffflammen Kennwort: Flüssigrückstandsverbrennung	W. Leuckel	1996–1998
A12	Brennstoffverdunstung bei überkritischem Druck	S. Wittig J. Himmelsbach	1993–1995
A13	Abstimmung von Zerstäubung und Strömungsparametern bei hohen Drücken und Temperaturen: Tropfenbildung und Sprühstrahlausbreitung Kennwort: Sprühstrahlausbreitung	H.-J. Bauer S. Kim	1996–1998
A14	Verdunstung von Mehrkomponententropfen bei überkritischem Druck Kennwort: Tropfenverdunstung	K. Dullenkopf S. Wittig	1996–1998

Projectfield B: Strömung und Wärmeübergang

Project No.	Title	Project Leader	Funding period
B1	Zwei- und dreidimensionale Strömungsvorgänge: Entwicklung des Strömungszustandes und der Turbulenz in und am Austritt von Brennkammern bei Variation der Betriebszustände	S. Wittig B. Noll	1984–1989
B1	Dreidimensionale Strömungsvorgänge in Brennkammern mit komplexer Geometrie	S. Wittig B. Noll M. Scheurlen	1990–1993

Project No.	Title	Project Leader	Funding period
B2	Neue Konzepte zur Brennkammer-Flammrohrkühlung: Kombinationsverfahren	S. Wittig S. Kim A. Schulz	1984–1995
B3	Verlustarme stabilisierende Diffusoren im Übergangsteil und in der Primärzone hochbelasteter Brennkammern	S. Kim B. Noll	1984–1992
B3	Einfluss der Zuströmung und der Wandkontur auf die Strömung im Vordiffusor und im Übergangsbereich hochbelasteter Brennkammern	S. Kim	1993–1995
B3	Strömungsgerechte Gestaltung von Brennkammerdiffusoren Kennwort: Brennkammerdiffusoren	S. Kim A. Schulz	1996–1997
B4	Optische Messung von Geschwindigkeits- und Dichtefeldern in Brennräumen	F. Mesch	1984–1987
B4	Optische Messung von Dichtefeldern in Brennräumen	F. Mesch H. Braun	1987–1989
B4	Optische Messung von Dichte- und Temperaturfeldern mit tomografischen Methoden	F. Mesch H. Braun	1990–1992
B4	Optische und akustische Messung von Dichte- und Temperaturfeldern mit tomografischen Methoden	F. Mesch	1993–1995
B4	Optische Messverfahren in Flammen mit tomografischen Methoden Kennwort: Tomografie	F. Mesch	1996
B5/B7	Untersuchung des mehrdimensionalen Wärmestrahlungstransports in absorbierenden und streuenden Medien	B. Noll R. Koch S. Wittig	1990–1995
B6	Filmbildung und -verdampfung auf keramischen Oberflächen	A. Schulz	1990–1995
B8	Berechnung des Wärmeübergangs an und durch die Brennkammerwände von Brennkammern komplexer Geometrie Kennwort: Wärmeübergang	R. Koch S. Wittig	1996–1998

Projectfield C: Hochtemperatur-Werkstoffverhalten und Schutzschichten

Project No.	Title	Project Leader	Funding period
C1	Heißkorrosionsbeständige Wärmedämmschichten für hochbelastete Brennräume bei stationärer Gleichdruckverbrennung	E. Fitzer G. Emig	1984–1989
C2	Analytische Erfassung des Schadensablaufs in thermozyklisch beanspruchten Brennkammerwerkstoffen	D. Munz	1984–1987
C2/C3	Inelastische Analyse zur Beurteilung des Versagens von thermozyklisch beanspruchten Brennkammerstrukturen	D. Munz H. Stamm	1987–1989
C2/C3	Inelastische Analyse und Schädigungsrechnung zur Beurteilung des Versagens thermozyklisch beanspruchter Brennkammerwerkstoffe	D. Munz B. Schinke	1990–1995
C4	Festigkeitsverhalten unter statischer und quasistatischer Beanspruchung	E. Macherauch O. Vöhringer D. Löhe	1984–1987
C4	Festigkeitsverhalten von Brennkammerwerkstoffen unter statischer und quasistatischer Beanspruchung	O. Vöhringer D. Löhe	1987–1989
C4	Statisches und quasistatisches Verformungsverhalten von Hochtemperaturwerkstoffen im Anlieferungs- und technologisch beeinflussten Zustand	O. Vöhringer E. Macherauch	1990–1992
C4	Festigkeitsverhalten von Brennkammerwerkstoffen unter statischer und quasistatischer Beanspruchung	O. Vöhringer E. Macherauch	1993–1995
C4	Langzeitverhalten beschichteter Brennkammerelemente Kennwort: Langzeitverhalten	O. Vöhringer D. Löhe	1996–1998
C5	Schwingfestigkeitsverhalten unter wechselnder mechanischer und thermischer Beanspruchung	E. Macherauch O. Vöhringer D. Eifler	1984–1987
C5/C11	Schwingfestigkeitsverhalten von Brennkammerwerkstoffen unter isothermer Wechselbeanspruchung	D. Eifler E. Macherauch B. Scholtes	1987–1995
C6	Werkstoffkundliche Untersuchungen an keramikbeschichteten Brennkammern	F. Thümmler H. Cohrt	1984–1989
C6	Werkstoffkundliche Untersuchungen an keramischen Wärmedämmschichtsystemen	F. Thümmler	1990–1992
C7	Festigkeitsverhalten von Brennkammerwerkstoffen unter wechselnder thermischer Beanspruchung	D. Löhe D. Eifler K.-H. Lang	1987–1995

Project No.	Title	Project Leader	Funding period
C7	Verformungs- und Versagensverhalten von Brennkammerwerkstoffen unter komplexer thermisch-mechanischer Beanspruchung Kennwort: Thermisch-mechanische Beanspruchung	K.-H. Lang D. Löhe	1996–1998
C8	Theoretische Beschreibung von Bauteilen mit Rissen bei viskoplastischem Materialverhalten	D. Munz H. Stamm	1987–1991
C9	Versagensanalyse thermozyklisch beanspruchter Brennkammerstrukturen Kennwort: Versagensanalyse	D. Munz B. Schinke J. Aktaa	1993–1998
C10	Mikrostrukturelle und analytische Charakterisierung von Brennkammer-werkstoffen nach mechanischer und thermischer Beanspruchung	H. Oettel U. Martin	1991–1995
C12	Prüfung und Charakterisierung keramischer Wärmedämmschichtsysteme	G. Grathwohl	1993–1995
C13	Prüfung und Charakterisierung keramischer Wärmedämmschichtsysteme Kennwort: Wärmedämmschichtsystem	R. Oberacker	1996–1998

Projectfield Z: Zentrale Aufgaben

Project No.	Title	Project Leader	Funding period
Z1	Verwaltung des Sonderforschungsbereichs	S. Wittig O. Vöhringer	1984–1998

7.2 Scientific Committee

Members

Dr.-Ing. Jarir Aktaa
Dr.-Ing. Hans-Jörg Bauer
Dr.-Ing. Hans Braun
Dr.-Ing. Henri Cohrt
Dr.-Ing. Klaus Dullenkopf
Prof. Dr.-Ing. habil. Heinrich Eickhoff
Prof. Dr.-Ing. Dietmar Eifler
o. Prof. Dr.-Ing. Gerhard Emig
o. Prof. Dr. techn. Erich Fitzer
Prof. Dr.-Ing. Georg Grathwohl
Dr.-Ing. Johann Himmelsbach
Prof. Dr. rer. nat. Michael J. Hoffmann
Dr.-Ing. Soksik Kim
Dr.-Ing. Reinhold Kneer
Dr.-Ing. Rainer Koch
Dr.-Ing. Karl-Heinz Lang
Prof. Dr.-Ing. habil. Bernhard Lenze
o. Prof. Dr.-Ing. Wolfgang Leuckel
o. Prof. Dr.-Ing. Detlef Löhe
o. Prof. Dr. rer. nat. Dr.-Ing. E. h. mult. Eckard Macherauch
Dr.-Ing. Jürgen Meisl
o. Prof. Dr.-Ing. Franz Mesch
o. Prof. Dr. rer. nat. Dietrich Munz
Dr.-Ing. Berthold Noll
Dr.-Ing. Michael Scheurlen
Dr.-Ing. Bernd Schinke
Prof. Dr.-Ing. Berthold Scholtes
Dr.-Ing. Achmed Schulz
Dr. rer. nat. Herman Stamm
o. Prof. Dr.-Ing. Fritz Thümmler
Prof. Dr. rer. nat. Otmar Vöhringer
o. Prof. Dr.-Ing. Dr.-Ing. E. h. Dr. h. c. mult. Sigmar Wittig

Associated Members

Prof. Dr. Klaus J. Hüttinger
Dr. rer. nat. Ulrich Martin
Prof. Dr.-Ing. Heinrich Oettel

Chairman

o. Prof. Dr.-Ing. Dr.-Ing. E. h. Dr. h. c. mult.
Sigmar Wittig — 2. HJ 1984–2. HJ 1994
Prof. Dr. rer.nat. Otmar Vöhringer — 1. HJ 1995–2. HJ 1998

Vice-chairman

o. Prof. Dr.-Ing. Wolfgang Leuckel — 2. HJ 1984–2. HJ 1998

Executive director

Dr.-Ing. Soksik Kim — 2. HJ 1984–2. HJ 1998

7.3 Visiting Researchers

Prof. Dr. Mohsen Abdel Aal Al Azhar University, Cairo, Egypt	25.08.1997–28.02.1998 01.03.1998–30.06.1998
Prof. Dr. Dr. habil. Juri Boiko Kharkov University, Ukraine	01.11.1997–31.12.1997
Dipl.-Ing. Do Young Byun Korean Advances Institute of Science and Technology, Taejon, Korea	01.09.1998–31.10.1998
Dr. W. John Chew Rolls Royce olc, Derby, England	20.08.1990–07.09.1990
Prof. Dr. Michael E. Crawford University of Texas, Austin, U.S.A.	05.06.1990–09.06.1990 11.06.1991–11.07.1991 07.06.1992–21.06.1992
Prof. Dr. Valeri Kharitonov Ufa State Aviation Technical University, Ufa, Russia	27.11.1996–23.12.1996 21.04.1998–28.05.1998
Laurent Desbat Université Joseph-Fourier, Grenoble, Switzerland	07.07.1992–09.07.1992
Prof. Mohamed Mohmoud Elkotb Cairo University, Cairo, Egypt	15.08.1985–14.10.1985
Prof. Dr.-Ing. M. Hackeschmidt University of Dresden, Germany	12.07.1990–20.07.1990
Dr. Abdelkrim Haddad Guelma University, Guelma, Algeria	25.06.1998–15.07.1998
Prof. Dr. Vladimir Karpow Semenov Institute of Chemical Physics, Moscow, Russia	01.04.1995–31.05.1995
Prof. Dr. Joon Sik Lee Seoul National University, Seoul, Korea	03.01.1994–28.02.1994

Prof. Dr. P. Lukas Czechoslovak Academy of Sciences, Brno, Czech Republic	01.04.1993–30.06.1993
Prof. Dr. Robert E. Mayle Rensselaer Polytechnic Institute, Troy, U.S.A.	14.06.1990–14.07.1990 11.06.1991–28.06.1991 05.06.1992–03.07.1992 02.08.1993–20.08.1993 10.01.1994–04.03.1994 01.04.1995–30.04.1995 21.08.1995–15.09.1995 17.06.1996–17.07.1996 18.05.1998–18.07.1998
Prof. Dr. Paul G. McCormick University of Western Australia, Nedlands, Perth, Australia	01.04.1991–31.05.1991
Prof. Dr. Arthur McEvily, Jr. University of Connecticut, Storrs, U.S.A.	11.01.1988–20.04.1988
Prof. Dr. Darryl Metzger Arizona State University, Tempe, U.S.A.	01.06.1990–30.06.1990 17.05.1991–31.05.1991 10.11.1991–16.11.1991 15.05.1992–31.05.1992
Prof. Dr. habil. Klaus P. Meyer Akademie der Wissenschaften, Berlin, Germany	11.06.1990–23.06.1990
Prof. Dr.-Ing. Nady N. Mikhael Suez Canal University, Port Said, Egypt	01.08.1991–30.09.1991
Prof. Dr. Knox Millsaps, Jr. Naval Postgraduate School, Monterey, CA, U.S.A.	15.06.1992–12.08.1992 07.12.1998–03.01.1999
Prof. Dr. Werner Neidel Technische Universität Magdeburg, Germany	26.05.1990–30.05.1990
Dr. Jerzy Pacyna St. Stszic Academy of Mining and Metallurgy, Cracow, Poland	23.10.1990–25.10.1990
Prof. Dr. Jim E. Peters University of Illinois, Urbana, U.S.A.	18.06.1990–20.06.1990

Prof. Dr. Gennady A. Philippov Institute for the All-Russia Nuclear Power Engineering Research and Development, Moscow, Russia	14.10.1996–18.10.1998
Prof. Dr. Anatoli M. Russak Ufa State Aviation Technical University, Ufa, Russia	10.09.1995–10.10.1995
Dr. Michael N. Rytschagow Moscow University, Moscow, Russia	21.01.1992–21.02.1992
Prof. Dr. Jeong-In Ryu Chungnam National University, Daejeon, Korea	14.12.1990–02.02.1991
Dr. Premananda U. S. Shet Indian Institute of Technology, Madras, India	26.06.1990–01.10.1990
Dr. Karen Thole Carnegie Mellon University, Pittsburgh, Pennsylvania, U.S.A.	01.01.1993–31.08.1994 17.07.1995–11.08.1995
Prof. Dr. Genki Yagawa University of Tokyo, Japan	20.07.1992–28.08.1992
Prof. Dr.-Ing. Maozheng Yu Xian Jiaotong University, Xian, China	18.09.1990–17.11.1990
Dr.-Ing. Han Jong Yu Fa. Rockwell, Los Angeles, U.S.A.	21.05.1995–20.07.1995
Prof. Dr. Vladimir Zimont Moscow Institute of Physics and Technology, Moscow, Russia	01.04.1995–31.05.1995

7.4 Financial Support by Means of the Deutsche Forschungsgemeinschaft

The Collaborative Research Centre 167 was supported by grants of the Deutsche Forschungsgemeinschaft totalling DM 37.823.800,–.

1984	1.120.600,–
1985	2.553.000,–
1986	2.215.300,–
1987	2.965.000,–
1988	2.762.500,–
1989	2.361.200,–
1990	2.973.400,–
1991	2.893.200,–
1992	2.642.200,–
1993	3.575.100,–
1994	2.963.100,–
1995	2.879.700,–
1996	2.424.800,–
1997	1.958.300,–
1998	1.536.400,–